W9-CRZ-215

Physics of the Earth's Space Environment

Gerd W. Prölss

Physics
of the Earth's Space Environment

An Introduction

With 263 Figures
Including 4 Color Figures

 Springer

Professor Dr. Gerd W. Prölss
Universität Bonn
Institut für Astrophysik und Extraterrestrische Forschung
Auf dem Hügel 71
53121 Bonn, Germany

Translated by: Dr. Michael Keith Bird
Universität Bonn
Institut für Radioastronomie
Auf dem Hügel 71
53121 Bonn, Germany

Cover picture: Dayglow and aurora imaged from the DE1 satellite at an altitude of about 20,000 km. Further information may be found in Sections 3.3.8 and 7.4 (L.A. Frank, University of Iowa).

Library of Congress Control Number: 2004102968

ISBN 3-540-21426-7 Springer Berlin Heidelberg New York

This work is subject to copyright. All rights are reserved, whether the whole or part of the material is concerned, specifically the rights of translation, reprinting, reuse of illustrations, recitation, broadcasting, reproduction on microfilm or in any other way, and storage in data banks. Duplication of this publication or parts thereof is permitted only under the provisions of the German Copyright Law of September 9, 1965, in its current version, and permission for use must always be obtained from Springer. Violations are liable to prosecution under the German Copyright Law.

Springer is a part of Springer Science+Business Media
springeronline.com

© Springer-Verlag Berlin Heidelberg 2004
Printed in Germany

The use of general descriptive names, registered names, trademarks, etc. in this publication does not imply, even in the absence of a specific statement, that such names are exempt from the relevant protective laws and regulations and therefore free for general use.

Print data prepared by LE-TeX Jelonek, Schmidt & Vöckler GbR, Leipzig
Cover design: Erich Kirchner, Heidelberg

Printed on acid-free paper   57/3141/ts   5 4 3 2 1 0

Preface

This book was written for readers interested in learning about the disciplines, methods and results of space research, perhaps because they happened upon the field during the course of their higher education or professional career, or perhaps because they simply feel an urge to know more about the space environment of the Earth. The present monograph is based on lectures covering the same topic, which have been held regularly over the past years at the University of Bonn. Like the lecture series, the book is directed at a relatively broad group of students and interested laypersons, the only prerequisite being knowledge of fundamental physics and mathematics, as usually acquired from introductory college courses in science or engineering curricula. More specific knowledge is derived in association with each phenomenon considered. These derivations are kept as simple as possible, adhering to the principle that, when conflicts arise, physical insight is preferable to mathematical precision. As a rule, I strived to avoid the trite phrase 'It may be easily shown that …' and tried to present all derivations in readily verifiable steps, even if this may seem somewhat tedious to the more advanced readers. Also serving clarity and insight are the many illustrations, which do indeed often say more than 'a thousand words'.

Our knowledge of the Earth's space environment has grown exponentially during the last few decades and an attempt to cover all aspects of the field would extend way beyond the scope of an introductory text. Acknowledging this fact, the book does contain some unavoidable gaps and even topics of special interest to the author have been omitted for lack of space. In particular, measurement techniques, although constituting a cornerstone of space research (and of physics in general), could only be described in passing. We content ourselves here with presenting the experimental results and then trying to explain the underlying physics on the basis of simple reasoning and argumentation. It is fair to say that this introduction to the field will have fulfilled its purpose if its readers are inspired to investigate a topic in more detail on their own, referring to the pertinent literature.

It is a pleasure to thank all those who directly or indirectly participated in the preparation and production of this book. I would like to thank my mentor W. Priester and my colleagues M. Roemer, H.J. Fahr and H. Volland for their support and the pleasant work environment in our institute. Parts of this book were reworked during a lecture series presented at the University of Innsbruck and during a research fellowship at the University of Nagoya. I extend my sincere gratitude here to colleagues M. Kuhn and Y. Kamide for the invitations. Valuable suggestions and various forms of assistance were provided by colleagues and coworkers S.J. Bauer, M.K. Bird, H. Fichtner, G. Lay, C.A. Loewe, K. Schrüfer, R. Treumann and S. Werner. I would like to make special mention of contributions from S. Noël, M. Kilbinger, B. Kuhlen, J. Pielorz and B. Winkel, who competently and patiently edited the manuscript into its final form. Without their help this book would surely have been a never ending story.

Bonn, December 2003   *Gerd W. Prölss*

Contents

List of Frequently Used Symbols

A	area
α	generic angle; pitch angle
$\vec{\mathcal{B}}$	magnetic flux density, here denoted as magnetic field
$\mathcal{B}_{00}$	Earth's equatorial surface magnetic field intensity
$\vec{c}$	random velocity (thermal velocity, peculiar velocity)
c_0	speed of light
c_p, c_V	specific heat capacity at constant pressure, volume
χ	zenith angle; spiral angle
d	thickness; transport term in the equations of balance
D	diffusion coefficient; declination
e	electron
e	elementary charge; base of the natural logarithm
E	energy
$\vec{\mathcal{E}}$	electric field
ε_0	permittivity of free space
ε_r	relative permittivity (dielectric constant)
f	degree of freedom; frequency
f	distribution function
$\vec{F}$	force
$\vec{F}^*$	force per unit volume
$\vec{g}$	gravitational acceleration ($\vec{g}_E, \vec{g}_S$: terrestrial, solar acceleration)
g	velocity distribution function
G	gravitational constant
γ	adiabatic exponent
γ^*	polytropic index
h	height (altitude)
h_P	Planck constant
$h(c)$	speed (velocity magnitude) distribution function
H	scale height; horizontal component of the Earth's magnetic field
$\vec{\mathcal{H}}$	magnetic field
I	momentum; inclination
$\vec{\mathcal{I}}$	current
$\vec{\mathcal{I}}^*$	surface current density

$\vec{j}$ current density

J_X ionization rate coefficient for the species X

k Boltzmann constant

$k_{s,t}$ reaction constants

K generic constant; eddy diffusion coefficient

κ heat conductivity

l length; loss rate per unit volume

$l_{1,2}$ mean free path

λ wavelength; geographic, heliographic longitude

L shell parameter

$\mathcal{L}$ induction constant

$\ln \Lambda$ Coulomb logarithm

m particle mass

m_u atomic mass unit

M generic mass (M_E, M_S: mass of Earth, Sun); Mach number

$\mathcal{M}$ mass number (atomic, molecular)

$\vec{\mathcal{M}}$ magnetic dipole moment ($\vec{\mathcal{M}}_E$: Earth; $\vec{\mathcal{M}}_g$: gyromoment)

μ_0 permeability of free space

n particle number density

n_{ref} reference density; index of refraction

$\hat{n}$ surface normal

N number of particles

$\mathcal{N}$ column density

$\nu_{1,2}$ collision frequency ($\nu_{1,2}^{Cb}$: Coulomb collision frequency)

$\nu_{1,2}^*$ momentum transfer collision frequency (frictional frequency)

ω angular velocity; rotation rate

ω_g Brunt-Väisälä frequency

$\omega_{\mathcal{B}}$ gyrofrequency (Larmor frequency)

ω_p plasma frequency

$\Omega_{E,S}$ angular rotation rate of Earth, Sun

p thermodynamic pressure

p_d dynamic pressure

$p_{\mathcal{B}}$ magnetic pressure

p   proton

P power

φ latitude

$\vec{\phi}$ flux of a scalar quantity

Φ magnetic flux

q charge per particle; production rate per unit volume

Q heat

$\mathcal{Q}$ electrical charge

r particle radius; radial distance

$r_{\mathcal{B}}$ gyroradius (Larmor radius)

$\vec{r}$ position vector

R_E, R_S	Earth's radius, Sun's radius
ρ	mass density
ρ_c	radius of curvature
s	distance; species index (e, i, n for electrons, ions and neutral gas particles)
$\sigma_{1,2}$	collision, interaction cross section
σ^A	absorption cross section
$\sigma_B, \sigma_H, \sigma_P$	Birkeland, Hall and Pedersen conductivities
t	time
T	temperature (T_∞: thermopause or exospheric temperature)
τ	time constant; period; optical thickness
$\vec{u}$	bulk velocity
U	internal energy
$\mathcal{U}$	voltage
$\vec{v}$	particle velocity
$\vec{v}_S, \vec{v}_A, \vec{v}_{MS}$	velocity of sound, Alfvén velocity, magnetosonic velocity
$\vec{v}_{ph}, \vec{v}_{gr}$	phase, group velocity
V	volume
w	probability
$\mathcal{W}$	work; index for eddy parameter
x, y, z	Cartesian coordinates; ($\hat{x}, \hat{y}, \hat{z}$ unit vectors); variables

1. Introduction

We begin by defining and constraining the topic of the book. This is followed by some comments on the scope of the subject matter and its organization. We conclude with a short introduction to the history of space research.

1.1 Definitions and Constraints

The broad topic 'Physics of the Earth's Space Environment', designated *space physics* in the following, is understood to mean the physics of particles and fields within the space regions of the solar system and its immediate vicinity. The 'physics' in this definition has the commonly accepted meaning, namely the study of natural phenomena that can be explored empirically, described mathematically and are subject to underlying universal laws. 'Particles' will be taken here to mean gas constituents such as atoms, molecules, ions and electrons, but not dust particles. 'Fields' are primarily of the electric and magnetic variety; gravitational fields are present, of course, but assumed to be given quantities. Concerning the 'space regions of the solar system', Fig. 1.1 provides an illustrative overview. Prominent among these regions are the outer gas envelopes of the Sun, planets, moons and comets. Particularly for the planets, it is essential that we distinguish between the *neutral upper atmosphere* and the *ionosphere*, the ionized component of the outer gas envelope. *Magnetospheres* represent other prominent entities of the solar system. As the name implies, these are the regions dominated by the magnetic fields of the planets. Outside of the magnetospheres, or outside the upper atmosphere in the absence of a planetary magnetic field, or directly adjacent to the surface of the solar system object in the absence of an atmosphere, is the region of *interplanetary space*. This is primarily occupied by a particle flow, the solar wind, which blows continually outward from the Sun. Another key role here is played by the interplanetary magnetic field. Similar to the way the pressure of the solar wind constrains the extent of planetary magnetic fields to their respective magnetospheres, the pressure of the *interstellar wind* constrains the gases and fields of interplanetary space to a finite volume called the *heliosphere*. The boundary of this region, the heliopause, is synonymous with the outer boundary of the 'solar system' in our above definition. It fol-

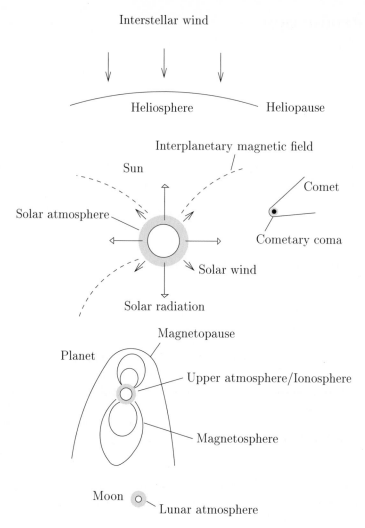

Fig. 1.1. Space regions of the solar system

lows that the 'immediate vicinity' is determined by the properties of the local interstellar medium.

The scientific disciplines of space physics, as we have chosen to define them, also imply some constraints. One of these is that we will not discuss topics related to the solid bodies of the solar system. For planetary bodies, moons, asteroids, comets, meteorites, planetary rings and interplanetary dust, this is the domain of planetology; the equivalent for the solar interior is generally handled by solar physics. Space research should also not be extended downward into the established domain of the 'weather watchers' – the thick lower atmospheres of planets and moons are traditionally relegated to the realm of meteorology. Finally, space physics, with its limitation to the solar system and its immediate vicinity, is well distinguished from the field of astronomy, for which the objects of interest are located far beyond this local foreground. The demarcation of these scientific disciplines thus follows the scheme

$$
\begin{matrix}
\text{Solar physics} \\
\text{Planetology} \\
\text{Meteorology}
\end{matrix}
\quad < \quad \text{Space physics} \quad < \quad \text{Astronomy}
$$

and, referred to our planet, this encompasses roughly the following range of distances

$$
100 \text{ km (Earth)} \quad \lesssim \quad \text{Space physics} \quad \lesssim \quad 1000 \text{ AU}
$$

where an astronomical unit (AU) corresponds to the mean Sun-Earth distance. For comparison: the mean heliocentric distance of the most distant planet Pluto amounts to roughly 40 AU (see Appendix A.3); the minimum distance to the heliopause is ca. 100 AU.

Separating the physics of the lower atmosphere (meteorology) from that of the outer gas envelope (space physics) may seem arbitrary and is in need of justification. Some of the arguments for incorporating outer gas envelopes into the realm of space physics are that they have a different chemical composition; are dominated by different transport processes; are often subject to evaporation effects; and are partially ionized and thus possess the physical properties of a plasma. The main reason for this separation, however, is that the energetics and dynamics of this region, in stark contrast to the lower atmosphere, are greatly influenced by the solar activity and the properties of the interplanetary medium.

Setting an outer boundary of jurisdiction in the direction of astronomical pursuits is also worthy of explanation. After all, what we now call space physics was once a most active branch of astronomy. Some reasons why this relatively young field of study should be recognized as an independent scientific discipline are that it covers a well delineated and quite extensive area of research; there is a sufficiently large number of scientists actively working in the field; these scientists have formed their own national and international organizations and hold their own conferences; and finally they publish their own scientific journals and have founded their own specialized research centers. Most of the above reasons, of course, also have a monetary component. There can be no doubt that substantial resources have been invested in this area over the past decades. To quote some nominal sums in this game, it may be noted that a quality rocket launch will cost about a half million US dollars or more; a research satellite requires an investment of at least a few tens of millions; and an interplanetary spacecraft will easily run upwards of a few hundred million US dollars.

A legitimate question is whether or not the above stated boundary between space physics and astronomy will expand outward with time. For example, will space research later claim hegemony over the regions beyond the local interstellar medium? The short answer is 'no'. After all, it is a quantum leap from the heliopause region ($\simeq$ 100 AU) to the nearest star ($\simeq$ 270 000 AU). Just to get a feeling for the distance to the heliopause, note that the fastest still active spacecraft, VOYAGER 1, will have been underway for almost 30 years when it reaches a distance of 100 AU in the year 2006. For all practical purposes, space physics is thus confined to that part of the universe accessible to *in situ* exploration.

A number of other definitions of space research may be found in the literature. Sometimes everything explored with the aid of space technology is included within this concept. In this case the orientation is aligned more with the research techniques rather than with the research objects. Such a definition may be useful when organizing all possible activities related to space technology, but it is not particularly well suited for summarizing a branch of research. This is demonstrated by the many independent disciplines of research, whether it be meteorology, oceanography, geodesy, materials science, biology or astronomy, just to name a few, that make use of satellite technology today. On the other hand, space research frequently utilizes ground-based measurements such as magnetic field recordings, radio soundings, photometry and radar measurements. Still another definition would understand space research to be plasma physics of the solar system, and this interpretation would indeed cover many of the phenomena treated in this book. It would leave out a number of very important topics, however, such as the neutral upper atmosphere, neutral interplanetary and interstellar gases, and high energy particles in the solar system. These are the primary reasons why we here prefer the more precise (albeit somewhat lengthy and less elegant) definition of space physics proposed at the beginning of this chapter.

1.2 Scope and Organization of the Material

As indicated in the title of the book, our discussion is confined to a description of the Earth's space environment. First, this is the region closest to us in every conceivable way; second, it is by far the most extensively explored and investigated; and third, trying to include the space environments of other planets would extend the scope of this book beyond that of the intended introduction. Even so, the material selected for discussion is quite voluminous and demands concentration on the essentials. These essentials, on the other hand, can be directly applied in many cases to the space environments of other planets (see, for example, some of the exercises in the problem sets at the end of each chapter). If one is familiar with the physics of the terrestrial ionosphere, for example, the structure of the Venusian ionosphere will be quickly understood.

The organization of the subject matter proceeds at first in its natural order, treating each region in turn as they are encountered with increasing distance from the Earth. The two subsequent chapters are thus devoted to a description of the neutral upper atmosphere and the absorption of solar radiation in this region. This is followed by descriptions of the ionosphere, the magnetosphere and the interplanetary medium in separate chapters. The concluding chapters, devoted to solar-terrestrial relations, treat the topics of transfer and dissipation of solar wind energy in the polar upper atmosphere and geospheric storms, respectively.

The description of the neutral upper atmosphere is given a relatively generous allocation within this organizational structure. One of the reasons is that this important topic is frequently neglected in introductory treatises. This, in spite of the fact that the vast majority of satellites, the fleet of space shuttles as well as the international space station ISS all operate within the Earth's upper atmosphere. Furthermore, many important concepts and relationships of gas physics are recapitulated in these first chapters. This is of benefit to the subsequent chapters, which can then be presented more concisely. This step by step procedure does mean that the book is probably less suitable as a reference volume. It also implies that a person interested in ionospheric physics, for example, would be advised to first read the chapter on the neutral upper atmosphere. Similarly, long period waves in magneto-plasmas will be understood better, if one first studies the section on acoustic waves.

As with all introductions, this one is also based on a great volume of primary and secondary references from the literature. It would be improper to cite all of these here individually. The source is stated, of course, whenever results from research are presented as illustrations or in tabular form. Otherwise we refer to the numerous textbooks, monographs and journals devoted to the field of space research as a whole or to one of the associated subfields. A short list of such references, subject certainly to personal preferences, is included at the end of each separate chapter. Note that some of the older treatises, although dated to some extent, are still very impressive from the standpoint of their physical insight and didactic skill.

1.3 Brief History of Space Research

Space research is a remarkably older science than one might at first imagine. Humankind has investigated some of the topics and fields for several hundred years and has spawned a number of brilliant ideas along the way. Prior to the advent of rockets and satellites, of course, one was limited to ground-based observations, i.e. one could only measure and interpret those signatures imposed on the Earth by a given phenomenon in space.

Some of the earliest activities that would be classified as space research today were undoubtedly connected with studies of the Earth's magnetic field. Some initial notions concerning its nature were obtained with terrella experiments. In these investigations the Earth was modeled as a small magnetized sphere and its external magnetic field was determined with the help of a compass needle (Gilbert, 1600; note that here and in the following the specified year only serves to order an event chronologically, not to identify a citation in a reference list). About one hundred years later (1722/23) the London watchmaker Graham discovered that the Earth's magnetic field was not really as constant as previously assumed, but rather displayed short-period fluctuations. His scientific apparatus consisted of a very fine, free-hanging compass needle that he observed with the help of a magnifying glass. Irregular disturbances of strong intensity were later designated as 'magnetic storms' (Humboldt, 1808) and are a subject of intensive research up to the present day. It was Gauss (1839), more than one hundred years after Graham, who first succeeded in showing that a small part of the magnetic field measured on the surface was not intrinsic to the Earth. He considered currents in the upper atmosphere to be the likely source of this extraterrestrial component. Another fifty years later Stewart (1883) formulated the hypothesis, which is still valid today, that the regular magnetic variations are caused by tidal winds in the dynamo region of the upper atmosphere. The first solar-terrestrial relations pertaining to magnetic activity were discovered by Sabine (1852) and Carrington (1859). Sabine showed that a relation exists between the intensity of magnetic disturbances and the solar sunspot cycle that had been discovered a few years earlier by Schwabe (1842). Carrington observed that solar flares can be followed by magnetic storms. This was also the time when the first 'space weather' effect was discovered: disturbances of telegraphic communications during geomagnetic storms were reported by Barlow (1849).

Parallel to the investigations of the Earth's magnetic field, there were numerous studies of another phenomenon that is surely among the most spectacular of all space research, the *aurora* (also called the *polar lights*). One of the first definitive works on this subject was written by de Mairan. His book, which first appeared in the year 1733, contained drawings of polar light apparitions, a list of known events, a discussion on the height of the auroral arcs and speculation about their origin. These speculations sound remarkably modern even today: de Mairan assumed that the Earth moved within the extended atmosphere of the Sun and that penetration of solar particles into the Earth's atmosphere was the cause of the observed polar lights. A short time later Hjorter and Celsius (1741) discovered that particularly intense magnetic field perturbations were observed during periods of enhanced auroral activity. They thus established a bridge between two areas of study that had previously been considered separate. The spectrum of the polar lights, specifically the prominent greenish yellow auroral line at 557.7 nm, was first recorded by Ångström (1866). By the end of the 19th century, Birkeland was even carrying out the first experimental simulations of aurorae. In his experiment, a terrella in an evacuated vessel was bombarded with cathode ray particles (just discovered at the time) and optical emission similar to the aurora was observed. At about the same time, Størmer, a colleague of Birkeland, began to calculate trajectories of electrically charged particles in the Earth's dipole field and thereby established the theoretical basis for describing radiation belts and other related phenomena.

A third branch of space research began in the year 1924 as two teams of scientists (Breit and Tuve in the USA and Appleton and Barnett in England) independently succeeded in proving the existence of the ionosphere. Although Marconi (1901) had demonstrated the feasibility of transatlantic radio communication much earlier, thereby instigating speculation about a conducting layer in the upper atmosphere, the first definitive experimental confirmation of this layer is attributed to the two research groups cited above. Appleton later received a Nobel prize for this and his subsequent work – one of two such awards bestowed on scientists engaged in space research (the other went to Alfvén). Following the discovery of the ionosphere, it was Chapman who took the lead in developing the theory of this phenomenon. The first correlations between ionospheric disturbances and geomagnetic activity, however, were observed by radio engineers (e.g. Espenschied and others, 1925). Alongside active radio sounding techniques, passive radio recordings were employed. As an example, Storey (1953) succeeded in detecting distant plasma populations while measuring *whistlers*, i.e. a special type of low-frequency radio wave that propagates in magnetized ionized gases. This extension of the ionosphere into the magnetospheric region of space is known today as the plasmasphere.

The exploration of the interplanetary medium is another discipline of space research that traces its beginnings to ground-based observations. Alfvén (1942) had shown that the Sun possesses an extremely hot and highly ionized outer atmosphere. The fact that this outer atmosphere is not a static gas envelope, but rather a plasma flow streaming outward from the Sun at all times, was first shown by Biermann (1951) on the basis of comet tail observations. This hypothesis stimulated Parker (1958) to propose his famous model of the solar wind and interplanetary magnetic field, the validity of which was confirmed impressively a number of years later.

In full appreciation of the many success stories associated with ground-based observations, however, the decisive impetus for space research came from a direct exploration of the Earth's space environment with the help of rockets and satellites. It could be said that the era of space flight began shortly after World War II, when the USA refurbished unused German V2 rockets and launched them into the upper atmosphere as research probes. Among these projects, as an example, was the first recorded spectrum of solar UV and X-ray radiation in the year 1949. Although lacking the appropriate rocket technology, systematic investigations of the neutral upper atmosphere were also initiated at this time.

A veritable explosion of activity followed the launch of the first artificial orbiting satellite, SPUTNIK 1, on 4 October 1957. Shocked by this unexpected Soviet success, the Americans promptly launched their own artificial satellite, EXPLORER 1, in January 1958, into Earth orbit and, two months later, announced the first important *in situ* discovery of space research, the detection of the Van Allen radiation belt. Meanwhile a large number of research satellites have been propelled into near-Earth space, many of which have provided important advances in our knowledge on the state of this fascinating region. The following presentation of space physics, in fact, would not be possible without having achieved the present day wealth of results accumulated by these missions.

References

Space Physics: General Literature

M.D. Papagiannis, *Space Physics and Space Astronomy*, Gordon and Breach Science Publishers, New York, 1972

S.-I. Akasofu and S. Chapman, *Solar-Terrestrial Physics*, Clarendon Press, Oxford, 1972

A. Egeland, Ø. Holter and A. Omholt (eds.), *Cosmical Geophysics*, Universitetsforlaget, Oslo, 1973

R.L. Carovillano and J.M. Forbes (eds.), *Solar-Terrestrial Physics*, Reidel Publishing Company, Dordrecht, 1983

G.K. Parks, *Physics of Space Plasmas*, Addison-Wesley Publishing Company, Redwood City, 1991

J.K. Hargreaves, *The Solar-Terrestrial Environment*, Cambridge University Press, Cambridge, 1992

M.G. Kivelson and C.T. Russell (eds.), *Introduction to Space Physics*, Cambridge University Press, Cambridge, 1995

W. Baumjohann and R.A. Treumann, *Basic Space Plasma Physics*, Imperial College Press, London, 1996

T.E. Cravens, *Physics of Solar System Plasmas*, Cambridge University Press, Cambridge, 1997

M.B. Kallenrode, *Space Physics*, Springer-Verlag, Berlin, 1998

T.I. Gombosi, *Physics of the Space Environment*, Cambridge University Press, Cambridge, 1998

S. Biswas, *Cosmic Perspectives in Space Physics*, Kluwer Academic Publishers, Dordrecht, 2000

Space Technology

M. Rycroft (ed.), *The Cambridge Encyclopedia of Space*, Cambridge University Press, Cambridge, 1990

O. Montenbruck and E. Gill, *Satellite Orbits*, Springer-Verlag, Berlin, 2000

Journals

Journal of Geophysical Research – Space Physics, American Geophysical Union, Washington/DC

Annales Geophysicae – Atmospheres, Hydrospheres and Space Sciences, European Geophysical Society, Katlenburg-Lindau

Journal of Atmospheric and Solar-Terrestrial Physics, Pergamon, Elsevier Science Ltd., Oxford

Advances in Space Research, Committee on Space Research (COSPAR), Pergamon, Elsevier Science Ltd., Oxford

Geophysical Research Letters, American Geophysical Union, Washington/DC

Reviews of Geophysics, American Geophysical Union, Washington/DC

Space Science Reviews, Kluwer Academic Publishers, Dordrecht

2. Neutral Upper Atmosphere

The description of an atmosphere makes use of a number of characteristic parameters that are introduced in this chapter. We describe the height variations of these parameters in the Earth's atmosphere and the subdivisions based on them. The main part of the chapter is devoted to an explanation of the observed density height profile. In this context it is necessary to first understand the density variation in a homogeneous atmosphere and then in a gravitationally settled atmosphere. Finally, we explain the calculation of the density profile for the outermost gas envelope of the Earth, a region characterized by evaporation effects.

2.1 State Parameters of Gases and their Gas Kinetic Interpretation

The general state of an atmosphere – like the state of any gas – is conveniently described by a few selected parameters. Among these so-called *state parameters* are, for example, the mass density, the flow velocity, the temperature and the pressure. Representative values of these parameters near the Earth's surface and in the upper atmosphere are collected in Table 2.1.

While these state parameters are sufficient for describing the macroscopic characteristics of an atmosphere, there are phenomena that can only be understood with the help of gas kinetics. Among these phenomena are diffusion, viscosity and heat conduction. Moreover, gas kinetics provides a physical interpretation of the state parameters as well as physically well-founded connections between them. The behavior of gas particles in the gas kinetic approach is determined by three types of variability: first by atomic parameters; second by dynamical variables; and third by external boundary conditions. An extremely simple description is usually adequate for the *atomic parameters*. In the simplest case, the gas particles are considered as elastic (billiard) balls with a certain radius and mass, the inner structure of which is characterized solely by the number of degrees of freedom. Values of atomic parameters for the gas species of interest here are summarized in Table 2.1. The meaning of the degrees of freedom will be further elucidated in our definition of the temperature.

Table 2.1. Atmospheric parameters. Note that units of the Système International d'Unités (SI units) are used here and in the following – with few exceptions.

1. ATOMIC PARAMETERS

Parameter	Symbol	Gas type						unit
		H	He	O	N_2	O_2	Ar	
Particle radius	r			$1 - 3 \cdot 10^{-10}$				m
Mass number	$\mathcal{M}$	1	4	16	28	32	40	
Particle mass	m			$= m_u \, \mathcal{M} = 1.66 \cdot 10^{-27} \mathcal{M}$ [1]				kg
Degrees of freedom	f	3	3	3	5	5	3	

[1] m_u = atomic mass unit

2. GAS KINETIC PARAMETERS

Parameter	Symbol	Height		Unit
		0 km (300 K; N_2)	300 km ($T_\infty = 1000$ K; O)	
Particle density	n	$2 \cdot 10^{25}$ [1]	10^{15}	$1/m^3$
Random velocity	$\vec{c}$	470 [2]	1100 [2]	m/s
Collision frequency	ν	$6 \cdot 10^9$	0.4	1/s
Mean free path	l	$8 \cdot 10^{-8}$	3000	m

[1] For normal conditions (0°C, 101 kPa) the *total* particle number density $n \simeq 2.69 \cdot 10^{25} \mathrm{m}^{-3}$ (*Loschmidt number*); [2] Mean absolute value

3. MACROSCOPIC STATE PARAMETERS

Parameter	Symbol	Gas kinetic meaning	Height		Unit
			0 km	300 km	
Chemical Composition		n_i/n	78% N_2 21% O_2 1% Ar	78% O 21% N_2 1% O_2	
Mass Density	ρ	$\sum_i m_i n_i$	1.3	$2 \cdot 10^{-11}$	kg/m^3
Flow velocity	$\vec{u}$	$\langle \vec{v} \rangle$	0 - 50	0 - 1000	m/s
Temperature	T	$(2/3k)m\,\overline{c^2}/2$	200 - 320	600 - 2500	K
Pressure	p	$n\,m\,\overline{c^2}/3$	10^5	10^{-5}	Pa

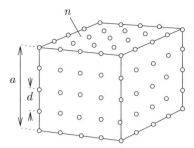

Fig. 2.1. Estimating the mean distance between particles

For cases when the statistical behavior of the gas particles is known, important *gas kinetic parameters* can be derived. Of special interest here are the particle number density, the characteristic particle velocities, the mean collision frequency and the mean free path. Representative values of these parameters near the ground and in the upper atmosphere are also given in Table 2.1.

2.1.1 Definition and Derivation of Gas Kinetic Parameters

Particle Number Density. The particle number density (or simply particle density) $n(\vec{r}, t)$ of a gas at the position $\vec{r}$ at the time t is defined as the limiting value of the number of particles ΔN divided by the volume element ΔV occupied by them at this position and time as the volume shrinks to a differentially small element dV

$$n(\vec{r}, t) = \lim_{\Delta V \to dV} \left(\frac{\Delta N}{\Delta V} \right)_{\vec{r}, t} \tag{2.1}$$

It is necessary here that the dimensions of the differential volume are small with respect to typical scales of density variations H_n, but at the same time big enough that dV still contains a large number of particles

$$H_n \gg \sqrt[3]{dV} \gg d$$

where d denotes the mean distance between gas particles. This distance can be estimated as follows. On the average, let the gas particles assume positions at equal distances from their neighboring particles and be located at the lattice points of a cubic crystal system; see Fig. 2.1. The number of particles along each edge is $(a/d) + 1$, so that the total number of particles in the cube must be $N = (a/d + 1)^3$. By definition, however, the number of particles in the cube is $N = n\, a^3$, so if $a \gg d$ we obtain the following estimate for the mean distance between particles

$$d \simeq 1/\sqrt[3]{n} \tag{2.2}$$

Sometimes more and sometimes fewer particles may be present in the given volume element dV. Requiring that these fluctuations do not exceed a certain limit, one may determine a minimum size for dV. It can be shown from statistical physics that the number of particles in the volume will fluctuate about the mean value N by an amount $1/\sqrt{N}$ (relative standard deviation). Demanding that this deviation be not greater than, say, 0.1%, requires that the volume contain at least 10^6 particles. For a relatively low density of $n = 10^4$ m^{-3}, the edge of the cubic volume dV would thus need to have a length of ca. 5 m. This dimension is substantially smaller than the variational scale lengths typically observed in such low density regions ($H_n > 1000$ km) and at the same time much larger than the mean distance between gas particles ($d \simeq 0.05$ m).

In addition to the spatial mean, one could also compute a temporal mean. In this case dV could indeed be chosen very small, but at the cost of the time resolution, which should remain compatible with the typical time scales of density variations.

Flow and Random velocity. When describing an ensemble of gas particles, it is important to distinguish between the following three types of velocities

- the actual gas particle velocity $\vec{v}$
- the flow (bulk, wind) velocity $\vec{u}$
- and the random (thermal) velocity of a gas particle $\vec{c}$

The flow velocity describes the motion of the entire ensemble of gas particles and is defined as follows

$$\vec{u}(\vec{r}, t) = \langle \vec{v} \rangle_{\vec{r}, t} = \lim_{\Delta V \to dV} \left(\frac{1}{n\,\Delta V} \sum_{i=1}^{n\Delta V} \vec{v}_i \right)_{\vec{r}, t} \tag{2.3}$$

As usual, the angular brackets denote the mean value of a quantity, and dV is the differentially small volume element at position $\vec{r}$ and time t over which the mean value is calculated. Vector addition is performed, of course, in the indicated sum and $n\Delta V$ is the number of particles in the given volume element.

Concerning the size of dV, the same arguments apply as those discussed in connection with the definition for the density. One complicating detail is that particles with velocities far above the mean value will also pass through the volume dV. Whereas the mean density is changed here only by an amount proportional to the ratio $1/\Delta N$, the effect on the mean velocity is much greater. This again raises the question whether a temporal mean is desirable in addition to the spatial mean.

The random velocity describes the translational motion of the gas particle relative to the flow velocity

$$\vec{c} = \vec{v} - \vec{u} \tag{2.4}$$

a. Real scenario

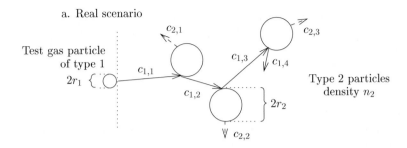

b. Statistically equivalent scenario

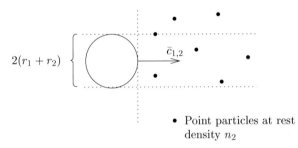

- Point particles at rest
  density n_2

Fig. 2.2. Calculating the collision frequency

The random velocity plays a key role in the definition of gas kinetic and macroscopic parameters. Obviously, its name is derived from the fact that it describes the deviation from the 'ordered' mean (flow) velocity. Since the random velocity determines the temperature of a gas, it is also called the thermal velocity. It follows immediately from the above definition of $\vec{c}$ that

$$\langle \vec{c} \rangle = \langle \vec{v} - \vec{u} \rangle = \langle \vec{v} \rangle - \vec{u} = 0$$

Collision Frequency and Mean Free Path. The collision frequency is understood to be the mean number of collisions experienced by a particle in a gas per unit time. As illustrated in Fig. 2.2a, a test particle of gas type 1 moves with its random velocity $\vec{c}_{1,i}$ in a gas of type 2 at rest. Particles of type 2 move at their own various random velocities $\vec{c}_{2,i}$. Collisions occur often, producing changes in the magnitude as well as the direction of the velocity of both collision partners. In order to facilitate a calculation of the mean number of collisions per unit time, the situation sketched in Fig. 2.2a is reduced to a simpler, but statistically equivalent, scenario. As shown in Fig. 2.2b, the interaction between the test particle of radius r_1 and a type 2 gas particle of radius r_2 is replaced by a test particle of radius $r_1 + r_2$ with point particles of type 2. This is allowed, because collisions occur in both cases if the distance between the particle centers becomes equal to $r_1 + r_2$. Furthermore, the interaction between the test particle with random velocity $\vec{c}_{1,i}$ and a type 2 gas particle with random velocity $\vec{c}_{2,j}$ may be replaced by an interaction between a test particle with the mean relative velocity $\bar{c}_{1,2} = \langle |\vec{c}_1 - \vec{c}_2| \rangle$ and a stationary point particle. This is allowed, because the collision frequency depends only on the relative velocity of the collision partners. Finally, the zig-zag course of the test particle can be replaced by a straight trajectory, unaffected by collisions. This is allowed, assuming the gas is homogeneous over the spatial scales considered, because then the number of collisions is independent of the direction taken by the test particle.

In this simplified picture, the modified test particle cuts out a cylinder of base area $\pi \, (r_1 + r_2)^2$ as it ploughs through the type 2 gas. It flies a distance $\bar{c}_{1,2} \, \Delta t$ in the time Δt, colliding with all point particles located within the cylindrical volume. The number of collisions per time interval Δt, i.e. the collision frequency, is thus found to be

$$\nu_{1,2} = \pi \, (r_1 + r_2)^2 \, \bar{c}_{1,2} \, \Delta t \, n_2 \, / \, \Delta t = \sigma_{1,2} \, n_2 \, \bar{c}_{1,2} \qquad (2.5)$$

where we have introduced the collision cross section $\sigma_{1,2}$. This quantity is explicitly dependent on the particle types 1 and 2, but, at least for the idealized collision model used here (billiard ball collisions!), is assumed independent of the velocity and energy of the interacting particles

$$\sigma_{1,2} = \pi \, (r_1 + r_2)^2 \qquad (2.6)$$

In order to derive an explicit formula for the collision frequency, we need an expression for the mean relative velocity $\bar{c}_{1,2}$. This quantity, in turn, depends on the particular velocity distribution of the test and background gas particles. Because realistic velocity distributions are not discussed until Section 2.4.3, we limit ourselves here to an extremely simple model that yields sufficiently accurate results in many situations. This model, denoted the *reduced velocity distribution*, is based on the following assumptions

- all gas particles i move with their mean random velocity $\bar{c} = \langle |\vec{c}_i| \rangle$
- their direction is chosen such that 1/6 of the particles flies in each of the 6 principal directions $\pm x$, $\pm y$ and $\pm z$ of a Cartesian coordinate system

With the help of Fig. 2.3, the mean relative velocity calculated from the reduced velocity distribution is found to be

$$\bar{c}_{1,2} = \langle |\vec{c}_1 - \vec{c}_2| \rangle \simeq \frac{1}{6}|\bar{c}_1 - \bar{c}_2| + \frac{1}{6}(\bar{c}_1 + \bar{c}_2) + \frac{4}{6}\sqrt{(\bar{c}_1)^2 + (\bar{c}_2)^2}$$

Note here the way the mean value is calculated: the various relative velocity magnitudes are first multiplied with their fractional occurrence probability and then added together. This method also works in more complicated situations.

The above expression may be reduced to the following approximate, but much simpler, form

$$\bar{c}_{1,2} = \bar{c}_1 \sqrt{1 + (\bar{c}_2/\bar{c}_1)^2} \qquad (2.7)$$

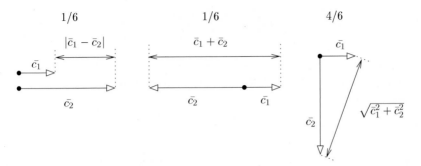

Fig. 2.3. Estimating the mean relative velocity

Consider, for example, the case $\bar{c}_1 \geq \bar{c}_2$. Under this condition our more complicated expression for $\bar{c}_{1,2}$ reduces to

$$\bar{c}_{1,2} = \bar{c}_1 \left(1/3 + 2/3\sqrt{1 + (\bar{c}_2/\bar{c}_1)^2}\right) \tag{2.8}$$

Using various values for $\bar{c}_2/\bar{c}_1$, it can be readily verified that the ratio of the factors of $\bar{c}_1$ in Eq. (2.8) and Eq. (2.7)

$$\frac{1/3 + 2/3\sqrt{1 + (\bar{c}_2/\bar{c}_1)^2}}{\sqrt{1 + (\bar{c}_2/\bar{c}_1)^2}}$$

attains values between 0.9 and 1.0 and can thus be set equal to 1 under the approximation considered here. This is also true for the case $\bar{c}_2 > \bar{c}_1$. Furthermore, it may be shown that the simplified expression for the relative velocity is exactly valid for the case of two gases, if each of these is in thermodynamic equilibrium. As such, Eq. (2.7) is used in a broad range of applications.

Jumping ahead for a moment to Eq. (2.94) of Section 2.4.3, we retrieve the following relation between the temperature and the mean random velocity for a gas in thermodynamic equilibrium

$$\bar{c} = \sqrt{8\,k\,T/\pi\,m}$$

where, as usual, k denotes the Boltzmann constant, T the gas temperature, and m the mass of the gas particles. For two coexisting gases with temperatures T_1, T_2 and particle masses m_1, m_2, the mean relative velocity may therefore be written as follows

$$\bar{c}_{1,2} = \sqrt{\frac{8\,k}{\pi}\left(\frac{T_1}{m_1} + \frac{T_2}{m_2}\right)} = \sqrt{\frac{8\,k\,T_{1,2}}{\pi\,m_{1,2}}} \tag{2.9}$$

where we have introduced the so-called *reduced temperature* $T_{1,2}$ and the *reduced mass* $m_{1,2}$ of the collision partners

$$T_{1,2} = \frac{m_2\ T_1 + m_1\ T_2}{m_1 + m_2} \tag{2.10}$$

$$m_{1,2} = m_1\ m_2/(m_1 + m_2) \tag{2.11}$$

Inserting Eq. (2.9) into Eq. (2.7), we obtain the following expression for the collision frequency

$$\nu_{1,2} = \sigma_{1,2}\ n_2\ \sqrt{\frac{8\ k\ T_{1,2}}{\pi\ m_{1,2}}} \tag{2.12}$$

In order to obtain the *mean free path* of a test particle between two collisions, we divide the distance traveled during a given time interval $(= \bar{c}_1 \Delta t)$ by the number of collisions occurring within that interval $(= \nu_{1,2}\Delta t)$

$$\begin{aligned}
l_{1,2} &= \frac{\bar{c}_1}{\nu_{1,2}} = \frac{1}{\sigma_{1,2}\ n_2\ \sqrt{1 + (\bar{c}_2/\bar{c}_1)^2}} \\
&= \frac{1}{\sigma_{1,2}\ n_2\ \sqrt{1 + (m_1\ T_2)/(m_2\ T_1)}}
\end{aligned} \tag{2.13}$$

For the special case when the collision partners are of one and the same type (radius r, mass m, density n, temperature T), we obtain the following simple expressions

$$\sigma_{1,1} = 4\ \pi\ r^2, \quad \nu_{1,1} = 4\ \sigma_{1,1}\ n\ \sqrt{k\ T/\pi\ m} \quad \text{and} \quad l_{1,1} = 1/(\sigma_{1,1}\ n\ \sqrt{2}) \tag{2.14}$$

2.1.2 Macroscopic State Parameters

Flux of a Scalar Quantity. A concept incorporating density and velocity, important for the gas kinetic as well as the macroscopic formalism, is the *flux* or *current density* of a scalar quantity. This is understood to be the net amount of a scalar quantity transported per unit area and per unit time through a reference surface oriented perpendicular to the flow. The scalar quantity may be, for example, number of particles, mass, heat, charge, or also the component of a vector quantity. The flux itself is obviously a vector quantity, the direction of which is determined by the transport velocity, and is denoted here by the symbol $\vec{\phi}$. A subscript, as customary, denotes the given component of this vector, ϕ_s. The type of flux can be defined by a superscript that denotes either the transported scalar quantity or the process responsible for the transport, i.e. $\vec{\phi}^E$ for the energy flux or $\vec{\phi}^D$ for the particle flux generated by diffusion. If the flux involves the transport of a single component t of a vector quantity, this is given in an additional subscript $\phi^j_{s,t}$. Should we be dealing with the transport of a complete vector quantity, rather than the transport of one component of this vector quantity, then the flux becomes a tensor quantity, $[\phi]$.

To illustrate the above definition, we calculate the flux of a particle current. Let the particle density be n and the flow velocity be u_x; see Fig. 2.4.

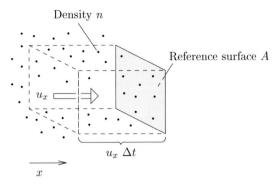

Fig. 2.4. Deriving the particle flux

Of course, only particles closer than the distance $u_x \Delta t$ to the surface A can reach the surface within the time Δt. Accordingly, only those particles within the volume element $u_x \Delta t\, A$ will pass through the surface during the time Δt. The particle flux is thus given by

$$\phi_x = (u_x\ \Delta t\ A)\ n/A\ \Delta t = n\ u_x$$

or when extended to three dimensions

$$\vec{\phi} = n\ \vec{u} \tag{2.15}$$

The momentum flux associated with this particle flux is

$$\phi^{I(u)}_{x,x} = I_x(u)\ n\ u_x = m\ n\ u_x^2 \tag{2.16}$$

where $I_x(u) = m\ u_x$ is the x-component of the kinetic momentum carried by each particle with velocity u_x. While the double subscripting of $\phi^I_{x,x}$ is unnecessary in the situation considered here, there are cases in which the transport of the y- or z-component, rather then that of the x-component of the kinetic momentum, is considered. The annotation $\phi^I_{x,y}$ or $\phi^I_{x,z}$ would be appropriate in these cases.

As a further demonstration of the flux concept we determine now the change in density produced by an inhomogeneous particle flux. We consider a volume with base area A and height Δz; see Fig. 2.5. A total of ΔN^+ particles flow into the volume and ΔN^- particles flow out during the time Δt. We may thus write

$$\Delta N^+ = -\phi_z(z + \Delta z/2)\ A\ \Delta t$$
$$\simeq -(\phi_z(z) + \frac{\partial \phi_z}{\partial z}\frac{\Delta z}{2} + \cdots)\ A\ \Delta t$$

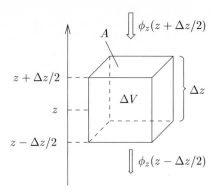

Fig. 2.5. Deriving the rate of change in the density produced by an inhomogeneous particle flux

and analogously

$$\Delta N^- \simeq -\left(\phi_z(z) - \frac{\partial \phi_z}{\partial z} \frac{\Delta z}{2} + \cdots\right) A \, \Delta t$$

where the minus sign on ϕ_z accounts for the flux direction (negative z-direction) and $\phi_z(z \pm \Delta z/2)$ has been approximated by a Taylor series, truncated after the second term (see Eq. (A.13) in Appendix A.1). The net gain (or loss) in the number of particles in the volume is thus

$$\Delta N = \Delta N^+ - \Delta N^- = -\frac{\partial \phi_z}{\partial z} \Delta V \, \Delta t$$

Referring the net gain (or loss) to the size of the given volume element ΔV and to the length of the given time interval Δt, the rate of change of density in the volume produced by the particle transport may be written as

$$\left(\frac{\partial n}{\partial t}\right)_{\phi_z} = \lim_{\Delta V \to dV, \, \Delta t \to dt} \left(\frac{\Delta N/\Delta V}{\Delta t}\right)_{\phi_z} = -\frac{\partial \phi_z}{\partial z} = d_z \qquad (2.17)$$

or, upon extending to three dimensions

$$\left(\frac{\partial n}{\partial t}\right)_{\phi} = -\left(\frac{\partial \phi_x}{\partial x} + \frac{\partial \phi_y}{\partial y} + \frac{\partial \phi_z}{\partial z}\right) = -\mathrm{div}\vec{\phi} = -\nabla\vec{\phi} = d \qquad (2.18)$$

As usual, div $\vec{\phi}$ denotes the divergence of the particle flux $\vec{\phi}$ and ∇ is the nabla (or del) operator; see Eq. (A.20) and (A.22) in Appendix A.1. Note that the rate of density change is not dependent on only the magnitude of the particle flux. In the case of a strong, spatially constant particle flux, many particles will flow into the given volume element, but just as many will flow out, essentially yielding a null result. Equally important here is the spatial rate of change, or divergence, of the particle flux. Taking the first letter of the word *divergence*, we use the symbol d to mean this transport-driven rate of density change. Note also that, for positive divergence of the particle flux (i.e. an increase of the flux inside the given volume element), the transport term d corresponds to a density loss. Conversely, for negative divergence (a decrease in flux in the given volume), d acts as a density source term. Obviously, more particles flow out than into the volume in the first case, and vice versa in the second case. With $\vec{\phi} = n\vec{u}$ and dropping the explicit indexing, we can write the transport-driven rate of density change as follows

$$\frac{\partial n}{\partial t} = -\frac{\partial(nu_z)}{\partial z} \tag{2.19}$$

or generally

$$\frac{\partial n}{\partial t} = -\text{div}(n\vec{u}) = -\nabla(n\vec{u}) \tag{2.20}$$

This relation is called the *continuity equation* and we will be making frequent use of it.

Pressure. In order to interpret the thermodynamic pressure of a gas in a manner consistent with experience and theory, we consider the momentum transport delivered by the thermal motion of the individual gas particles. Using the pressure on a wall as an example, the pressure force is attributed to the momentum transfer delivered by the reflected gas particles. Here, a more general definition of the state parameter pressure is presented. It is shown that this definition is consistent with our traditional notion of pressure on a wall.

The *thermodynamic (internal, static, scalar) pressure* is understood to be the mean net transport of kinetic momentum through a surface per unit area and per unit time (=momentum flux), produced by the thermal motion of the gas particles, whereby only the kinetic momentum parallel to the surface normal is considered and the averaging is performed over all directions. This definition can be formally expressed in a Cartesian coordinate system in the following way

$$p = \frac{1}{3}(\phi_{x,x}^{I(c)} + \phi_{y,y}^{I(c)} + \phi_{z,z}^{I(c)}) \tag{2.21}$$

where $\phi_{i,i}^{I(c)}$ denotes the (net) momentum flux in the direction i. Only those particle momentum components pointing in this same direction and associated with the random velocity c are considered, and the average is taken over the three principal directions of the Cartesian coordinate system.

In order to calculate p explicitly, the velocity distribution of the gas particles must be known. For the moment, we refer back to the reduced velocity distribution introduced in Section 2.1.1, where all particles have the same speed $\bar{c}$ and $1/6$ of the particles moves in each of the 6 principal directions $\pm x$, $\pm y$ and $\pm z$. It can be shown that one obtains the same result with this simple model as with the more realistic Maxwell velocity distribution

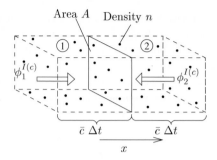

Fig. 2.6. Calculating the thermodynamic pressure

discussed later. Figure 2.6 illustrates the situation considered for calculating $\phi_{x,x}^{I(c)}$. As shown, only those particles located within the volumes 1 and 2 with velocities parallel to the x-direction can contribute to the momentum transport through the surface A during the time interval Δt. The momentum transport from volume 1 to volume 2 per unit area and unit time is thus given by

$$\phi_1^{I(c)} = \frac{1}{6}(\bar{c}\,\Delta t\,A)\,n\,(\bar{c}\,m)/A\,\Delta t = \frac{1}{6}\,n\,m\,\bar{c}^2$$

By analogy, $\phi_2^{I(c)}$ may be written

$$\phi_2^{I(c)} = -\frac{1}{6}\,n\,m\,\bar{c}^2$$

where the minus sign indicates the direction of the transported momentum, not the direction of the momentum flux. The net momentum flux in the x-direction is thus

$$\phi_{x,x}^{I(c)} = \phi_1^{I(c)} - \phi_2^{I(c)} = \frac{1}{3}\,n\,m\,\bar{c}^2 = \frac{1}{3}\,n\,m\,\overline{c^2} \qquad (2.22)$$

where the last equality is only valid for the reduced velocity distribution considered here. It is noted that the effects of the two momentum fluxes add together. Kinetic momentum in the x-direction is conveyed into volume 2 via the term $\phi_1^{I(c)}$. Simultaneously, kinetic momentum in the negative x-direction is extracted from volume 2 via the term $\phi_2^{I(c)}$, corresponding to a gain in positively directed kinetic momentum.

The identities $\phi_{x,x}^{I(c)} = \phi_{y,y}^{I(c)} = \phi_{z,z}^{I(c)}$ hold for the reduced velocity distribution considered here, but also for the Maxwell velocity distribution (isotropic momentum flux, independent of direction). In both cases, the thermodynamic pressure is therefore found to be $p = \phi_{x,x}^{I(c)} = \phi_{y,y}^{I(c)} = \phi_{z,z}^{I(c)}$ or, with Eq. (2.22)

$$p = \frac{1}{3}\,n\,m\,\overline{c^2} \qquad (2.23)$$

This particular form for the pressure, first derived by Daniel Bernoulli, also represents a measure of the energy density stored in the thermal translational motion of the gas particles.

The above definition of internal pressure is consistent with the usual interpretation of the pressure on a wall, as shown by the following arguments. Assuming the surface A shown in Fig. 2.6 represents a solid wall, then the incident particles will each deliver an amount of momentum $2m\bar{c}$ (direction reversal!) upon reflection. The associated pressure on the wall is thus

$$p_{\text{wall}} \left(= \frac{\text{force}}{\text{surface area}}\right) = \frac{\Delta I/\Delta t}{A} = \frac{1}{6} \frac{(\bar{c}\,\Delta t\,A)\,n\,(2\,m\,\bar{c})}{A\,\Delta t}$$
$$= \frac{1}{3}\,n\,m\,\overline{c^2} = p$$

Momentum in a moving gas will not be transported by just the random velocity, but also by the flow velocity. For a surface oriented perpendicular to the flow direction, the mean net transport of kinetic momentum per unit area and unit time associated with the flow velocity is

$$p_d = (m\,\vec{u})\,n\,\vec{u} = m\,n\,u^2 \tag{2.24}$$

where p_d denotes the *dynamic pressure* of the gas flow. Alternatively, and conforming formally more with the definition of the thermodynamic pressure, the dynamic pressure can be written as follows

$$p_d = \phi_{x,x}^{I(u)} + \phi_{y,y}^{I(u)} + \phi_{z,z}^{I(u)}$$

see Eq. (2.16). Because the momentum flux points exclusively along the flow direction in this case, the net value and directional averaging calculations are unnecessary.

When determining the thermodynamic pressure in a gas flow, it should be noted that the momentum flux associated with the dynamic pressure must not be included in the calculation. In this case, one should instead consider the momentum flux through a reference surface moving with the gas flow.

Temperature. Phenomena connected with temperature or heat are accessible from common experience, if only indirectly from the intensity of the heating or cooling effects. Physically, heat is nothing more than the kinetic energy of the random (i.e. *thermal*) motion of the particles, transferable via interactions. Random motion implies deviations from the mean velocity $\vec{u}$, which can occur in various forms. For gases this always includes the translational motion of the particles characterized by the velocity $\vec{c}$. Should the gas be composed of molecules, the repertoire of random motions is enhanced by molecular rotations and vibrations. To account for these differences, the *degree of freedom f* gives the number of mutually independent types of motion available for storage of heat. All gas particles possess at least the three degrees of freedom associated with their translational motion: the velocity

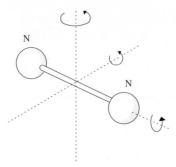

Fig. 2.7. Rotational degrees of freedom for a diatomic dumbbell type molecule such as N_2 and O_2

components in the x-, y- and z-directions, which are, of course, independent of each other. If the gas is composed of dumbbell-type molecules (see Fig. 2.7), two additional degrees of freedom are available. In this case, rotations about the axis along the line connecting the two atoms do not contribute to the thermal energy because of the vanishingly small moment of inertia. Molecular nitrogen and molecular oxygen are examples of molecules possessing five degrees of freedom (see Table 2.1). Only under special circumstances are the atoms of these molecules excited to vibrate along their connecting dumbbell axis, increasing the number of degrees of freedom to six (e.g. in vibrationally excited nitrogen).

There exist a number of rather abstract approaches for the definition of temperature. Fortunately, all lead to a very graphic interpretation of this concept for gases. The temperature T in this case is understood as a measure of the mean amount of heat stored per degree of freedom

$$T = \left(\frac{2}{k}\right) \overline{U}_f \tag{2.25}$$

where $\overline{U}_f$ denotes the average amount of heat (or *internal energy*) per degree of freedom, and k is the Boltzmann constant. When the heat is equally distributed over all degrees of freedom (= equipartition), as is the case for thermodynamic equilibrium, or when all of the energy is stored in the translational motion of the particles – as with atomic gas particles or electrons – then we have

$$T = \left(\frac{2}{3k}\right) \left(\frac{1}{2} m \,\overline{c^2}\right) \tag{2.26}$$

This relation, which will usually suffice for our purposes, shows that the temperature can also be understood as a measure of the mean energy stored in the thermal translational motion of a gas particle.

If the heat energy is not uniformly distributed over all degrees of freedom, a specific temperature can be introduced for each degree of freedom

$$T_f = \left(\frac{2}{k}\right) U_f \tag{2.27}$$

We make use of this possibility in our description of the solar wind (Chapter 6).

Comparing the gas kinetic interpretations of pressure and temperature as given by the Eqs. (2.23) and (2.26), one directly obtains the *equation of state for an ideal gas*, also called the *ideal gas law* or *universal gas law*

$$p = n\, k\, T \tag{2.28}$$

This relation between the pressure, density and temperature of a gas will play an important role in the subsequent discussions.

Specific Heat Capacity. Since the temperature is a measure of the heat content of a body, heat must be supplied in order to raise its temperature. For a gas enclosed in a fixed volume, Eq. (2.25) may be used to derive a relation between the heat ΔQ needed to raise the temperature by an amount ΔT

$$\Delta T = \frac{2}{k}\, \Delta \overline{U}_f = \frac{2}{k}\, \frac{\Delta Q}{N\, f} = \frac{2}{k}\, \frac{\Delta Q}{(M/m)\, f}$$

from which

$$\Delta Q (= \Delta U) = \frac{M}{m}\, f\, \frac{k}{2}\, \Delta T$$

Here, M is the total mass of the gas volume, N is the total number of gas particles, and f is the number of degrees of freedom available for heat storage. The proportionality factor $Mfk/2m$ is denoted the heat capacity of the gas at constant volume. This can be *referred to a unit mass* of the gas under consideration (i.e. 1 kg in SI units)

$$\Delta Q' (= \Delta U') = \Delta Q/M = c_V\, \Delta T \tag{2.29}$$

from which we obtain the definition of the *specific heat capacity at constant volume*

$$c_V = \frac{k\, f}{2\, m} \tag{2.30}$$

At constant pressure, but variable volume, we need more energy to raise the temperature of the gas. This is because, in addition to the increase in internal energy, work must be done to expand the volume. This work of expansion can be determined with the help of Fig. 2.8. Consider a gas in a cylinder that is enclosed at the upper end by a movable piston. The constant pressure force F_p acts on the piston from above. Supplying heat to the gas from outside, not only the temperature, but also the pressure will increase, and the piston will rise a distance Δz. The gas has performed work of expansion in the amount ΔW given by

$$\Delta W = -F_p\, \Delta z = -p\, \Delta V$$

where the minus sign indicates that this energy is lost to the gas.

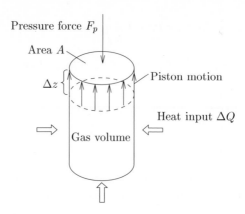

Fig. 2.8. Determining the work of expansion

Recalling the universal gas law, we have $p = n\,k\,T = N\,k\,T/V = \text{constant}$, i.e. $V/T = \text{constant}$ and therefore $\Delta V = V\,\Delta T/T$. Substituting this into the above equation, we obtain

$$\Delta W = -\frac{p\,V}{T}\,\Delta T = -k\,N\,\Delta T$$

which, when *referred to a unit mass*, yields

$$\Delta W' = -\frac{k}{m}\,\Delta T \tag{2.31}$$

In order to raise the temperature by ΔT, one thus needs the amount of heat per unit mass

$$\Delta Q' = \Delta U' - \Delta W' = c_V\,\Delta T + \frac{k}{m}\,\Delta T = c_p\,\Delta T \tag{2.32}$$

where the *specific heat capacity at constant pressure* is defined by

$$c_p = \frac{k}{m}\,(f/2 + 1) \tag{2.33}$$

Eq. (2.32) evidently corresponds to the first law of thermodynamics.

A frequently used quantity is the ratio of specific heat at constant pressure to the specific heat at constant volume, also called the *adiabatic exponent*, given by

$$\gamma = \frac{c_p}{c_V} = \frac{f+2}{f} \tag{2.34}$$

Adiabatic Changes of State. In many situations it is reasonable to assume that the change of state of a gas proceeds *adiabatically*. One thus assumes that no heat exchange takes place between an expanding or contracting gas volume and its environment ($\Delta Q = 0$). In this case the work performed by a gas volume against external pressure during expansion is done at the expense of its own internal energy. This demands that

$$d\mathcal{W} = -p \, dV = dU = N \, f \left(\frac{k}{2} \, dT \right)$$

or, with $N = n V$ and $p = n k T$

$$\frac{dT}{T} = -\frac{2}{f} \frac{dV}{V}$$

Integration of this equation yields the familiar *adiabatic law* which relates changes in temperature with those in volume

$$T = T_0 \left(\frac{V}{V_0} \right)^{-2/f} \quad \text{or} \quad T \, V^{2/f} = \text{const.} \tag{2.35}$$

Applying the universal gas law to a constant particle number N, an alternative form for this relation may be written as

$$n = \text{const.} \, p^{1/\gamma} \quad \text{or} \quad p \, \rho^{-\gamma} = \text{const.} \tag{2.36}$$

We often make use of these various forms of the adiabatic law. Moreover, as shown in Appendix A.7, the adiabatic law represents a particularly simple form of the energy balance equation.

2.2 Height Profiles of the State Parameters and Atmospheric Subdivisions

Because the atmosphere is a relatively complex phenomenon, it makes good sense to partition it into more easily comprehendible subdivisions. The state parameters of the gases and the physically important processes serve here as convenient guides for defining such subregions. Different classifications and nomenclature are introduced according to the particular parameter or process chosen.

The most common classification is based on the vertical temperature profile. As shown in Fig. 2.9, this is characterized by three maxima and two minima, together with the connecting layers of increasing or decreasing temperature in between. The first maximum arises from heating of the lowest air layers by the Earth's surface. This, in turn, receives most of its heat from the direct absorption of solar radiation. Additional heat comes from reabsorption of its own infrared radiation reflected primarily from atmospheric water vapor (greenhouse effect). Considered together, these effects produce a mean surface temperature of 288 K.

Radiative cooling causes the atmospheric temperature to decrease with increasing distance from the warm surface of the Earth. This region of falling

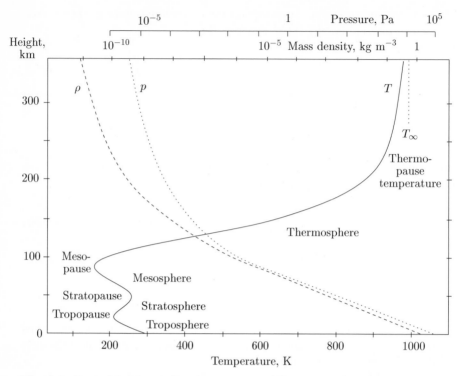

Fig. 2.9. Typical height profile of temperature (T), pressure (p) and mass density (ρ) in the Earth's atmosphere. The breakdown into atmospheric regions is based on the temperature profile.

temperature, designated the *troposphere*, extends up to about 10 km, where a broad minimum in the temperature is reached, the so-called *tropopause*. Continuing upward, the temperature rises again due to absorption of ultraviolet solar radiation at wavelengths above 242 nm by the trace gas ozone. This region of increasing temperature is designated the *stratosphere* and its upper boundary at roughly 50 km is called the *stratopause*. The temperature reaches the same values here as observed on the Earth's surface.

Absorption of radiation is less important above the stratopause and radiative cooling, particular that due to the trace gas carbon dioxide, becomes quite effective. The temperature thus falls again, reaching its absolute minimum at a height of 80–90 km. This region of decreasing temperature and its associated upper boundary are denoted the *mesosphere* and *mesopause*, respectively. The mean temperature minimum here is 160 K, but temperatures of less than 120 K have been measured under extreme conditions.

The temperature increases dramatically above this minimum, making the previously described temperature variations appear quite modest by comparison. This region is appropriately designated the *thermosphere*. The temper-

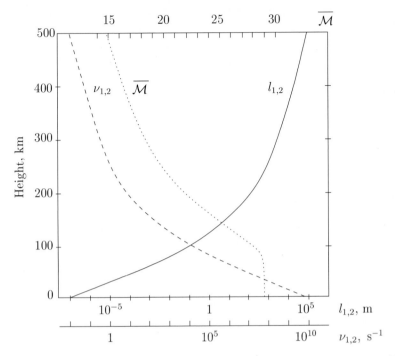

Fig. 2.10. Height profiles of the collision frequency ($\nu_{1,2}$), the mean free path ($l_{1,2}$) and the mean mass number ($\overline{\mathcal{M}}$) in the terrestrial atmosphere

ature increase is caused by the absorption of solar ultraviolet radiation at wavelengths below 242 nm, in combination with the absence of effective heat loss processes.

Finally, at heights above about 200 km, the temperature asymptotically approaches a limiting value referred to as the *thermopause temperature* or *exospheric temperature*, usually designated as T_∞. The thermopause temperature is typically 1000 K, but can vary from 600 to 2500 K.

A somewhat cruder classification based on the temperature profile differentiates only between the *lower atmosphere* (= troposphere), the *middle atmosphere* (= stratosphere and mesosphere), and the *upper atmosphere* (= thermosphere). We will be concerned exclusively with the physics of the upper atmosphere in the following.

Also shown in Fig. 2.9 are the vertical profiles of pressure and mass density. As expected, these two state parameters decrease with height, the rate of decrease clearly being determined by the prevailing temperature. Easy to remember is that the pressure and mass density at a height of 100 km are about one millionth of their values at the Earth's surface.

The height profiles of collision frequency and mean free path are shown in Fig. 2.10. It is interesting to compare the values at 500 km with those on the

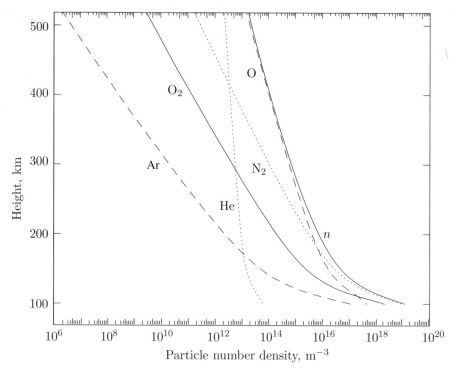

Fig. 2.11. Height profiles of various gas constituents in the heterosphere between 100 and 500 km. The associated thermopause temperature is 1000 K. n is the total particle number density. Numerical values are compiled in Appendix A.4.

ground. While the collision frequency falls from nearly 10^{10} s^{-1} to 10^{-2} s^{-1}, the mean free path increases from 0.1 µm to 100 km.

An alternative classification of atmospheric regions is based on the variation of the mean mass number (relative atomic mass) $\overline{\mathcal{M}}$. Figure 2.10 shows that this quantity is constant up to about 100 km and then decreases above this height. The atmosphere is evidently well mixed below 100 km and all constituents maintain their surface fractional abundance. This region is designated the *homosphere*, and its upper boundary is the *homopause*. Gravitational settling, however, for which heavy gases fall off rapidly and light gases gradually with height, becomes important above 100 km. This is illustrated in Figs. 2.11 and 2.12 for two different height regions and numerically documented by the model atmosphere presented in Appendix A.4.

As seen, molecular nitrogen dominates below 180 km, atomic oxygen becomes the prevailing species at altitudes between 180 and 700 km, helium takes over in the range from 700 to 1700 km, and atomic hydrogen at higher altitudes. Appropriate to the changing mixing ratios, the altitude range from 100 to 1700 km is designated as the *heterosphere*. The region above this is the *hydrogensphere* or *geocorona*. The above cited boundary heights are valid

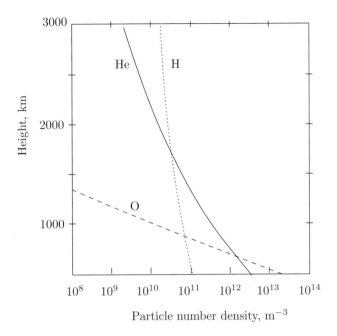

Fig. 2.12. Height profiles of gas constituents in the heterosphere and hydrogen geocorona between 500 and 3000 km. The thermopause temperature $T_\infty \simeq 1000$ K.

for a thermopause temperature of roughly 1000 K and can shift upwards or downwards significantly at other temperatures.

In addition to the temperature and neutral composition, vertical transport processes, evaporation phenomena, and ionic composition all offer possible criteria for classification of the atmosphere. These are summarized in Fig. 2.13 and will be described later in more detail.

2.3 Barospheric Density Distribution

The *barosphere* is that part of the atmosphere gravitationally bound to the Earth. In order to determine the density profile in this region, we consider an equilibrium relation between the pressure gradient and gravitational forces, the aerostatic equation. Here, the pressure gradient force is interpreted as the divergence of a momentum flux. The solution to the aerostatic equation is the barometric law, which can be applied immediately only to the case of a homogeneous gas mixture. When determining the density profile in a gravitationally separated heterosphere, we must consider the force balance relation for a single gas within a gas mixture. It turns out that frictional forces between the various gas constituents vanish in the static state, so that the aerostatic equation and barometric law are also valid for an individual

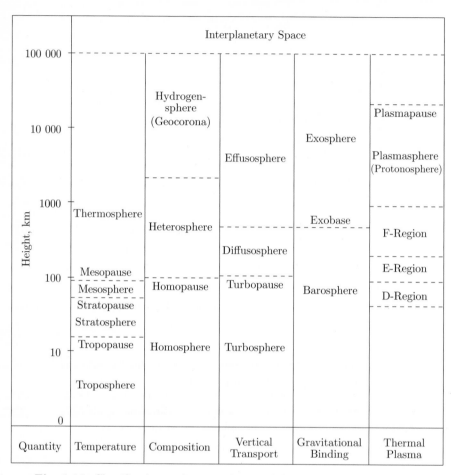

Fig. 2.13. Classification and nomenclature of the terrestrial atmosphere

gas in the heterosphere. Alternatively, the barospheric density distribution can be understood as the result of a dynamic equilibrium between upward and downward fluxes. From a gas kinetic viewpoint, the upward flux can be interpreted as a molecular diffusion flux and the barometric law as a diffusive equilibrium relation. Of special interest here is the transition from the homosphere to the heterosphere. Molecular transport evidently attains the same effectiveness as eddy diffusion at the homopause altitude. Finally, the (often dominant) presence of atomic oxygen and atomic hydrogen in the upper atmosphere, both of which are not natural constituents of the lower atmosphere, is explained.

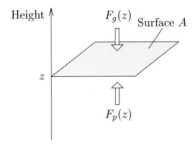

Fig. 2.14. Deriving the aerostatic equation

2.3.1 Aerostatic Equation

To assist our derivation of the aerostatic equation, we consider a massless membrane of area A that is subjected to the pressure from the gas below on its bottom side and to the weight of the gas column above on its upper side; see Fig. 2.14. We consider the static case, for which the forces acting on the membrane are in equilibrium, i.e. the pressure and gravity forces are equal and oppositely directed. For the pressure force at the height z we have

$$F_p(z) = A\, p(z)$$

and the weight of the gas column above z is calculated from

$$F_g(z) = A \int_z^\infty \rho(z')\, g(z')\, \mathrm{d}z'$$

where ρ is the mass density and g the acceleration of Earth's gravity. For static equilibrium, when $F_p = F_g$, we obtain

$$p(z) = \int_z^\infty \rho(z')\, g(z')\, \mathrm{d}z' \tag{2.37}$$

This equation is much better known in its differential form. Differentiating both sides with respect to z (see also Eq. (A.1)), we obtain the very important relation

$$\frac{\mathrm{d}p(z)}{\mathrm{d}z} = -\rho(z)\, g(z) \tag{2.38}$$

which is known as the *hydrostatic* or *aerostatic equation* (Laplace, 1805). Since we are working here exclusively with gases, we prefer the latter designation. The aerostatic equation describes the change of pressure with height as a function of the mass density and Earth's gravitational acceleration. Provided the relation between p and ρ is known, it allows a calculation of the height dependence of these two quantities.

The aerostatic equation has a wide range of applications, valid for a single gas atmosphere as well as an atmosphere composed of various gases, regardless of whether the atmosphere is well mixed or gravitationally separated. As

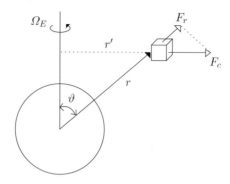

Fig. 2.15. Deriving the centrifugal force F_c acting on a gas volume that rotates with the Earth

such, the equation may be applied to the interiors of the Sun and gaseous planets, as well as to the solar atmosphere and to planetary homospheres and heterospheres. It is even valid when the mass density is nearly constant, e.g. for oceans and solid celestial bodies.

Two assumptions were made when deriving this relation that should be verified before going further. The first point is that only the weight, but not the thermodynamic pressure of the gas, was considered to act on the upper side of the membrane. A 'thought experiment' can justify this assumption. If we could push a button and turn the gravitational attraction off, the gas above the membrane would quickly vanish – after all, the entire universe is available for its expansion. As the density gets smaller, however, the thermodynamic pressure would also vanish. It follows that the pressure on the upper side of the membrane is maintained solely by the gravitational acceleration: the gas particles 'fall' onto this boundary surface, producing momentum transfer and pressure. The gas below the membrane, however, is confined to a finite volume and acts on the boundary surface only by means of its thermodynamic pressure.

The other assumption is related to the neglect of the centrifugal force acting on the gas volume in the rotating frame of the Earth. According to Fig. 2.15, the radial component of this force (only this component can influence the vertical distribution of the gas) may be written in the following form

$$F_r = F_c \; \sin \vartheta = M \; \Omega_E{}^2 \; r' \sin \vartheta = M \; \Omega_E{}^2 \; r \sin^2 \vartheta \qquad (2.39)$$

where r is the geocentric distance, ϑ the colatitude, M the mass of the gas volume, and Ω_E the Earth's angular rotation velocity. This centrifugal force component opposes the gravitational attraction of Earth and can be easily incorporated into an *effective* Earth's acceleration as

$$g(r, \vartheta) = g_m(r, \vartheta) - \Omega_E{}^2 \; r \; \sin^2 \vartheta \qquad (2.40)$$

Table 2.2. Terms contributing to the effective gravitational acceleration at the equator and the escape velocity as a function of height. The given height refers to distance above the mean radius of the Earth, rather than the equatorial radius.

Height	$g \simeq GM_E/r^2$	$C/2r^4$	$\Omega_E{}^2 r$	$v_{es} \simeq \sqrt{2\,GM_E/r}$
[km]	[m/s^2]	[m/s^2]	[m/s^2]	[km/s]
0	9.82	0.02	0.03	11.2
100	9.52	0.02	0.03	11.1
500	8.44	0.01	0.04	10.8
1 000	7.34	0.01	0.04	10.4
10 000	1.49	$< 10^{-3}$	0.09	7.0

Here the gravitational acceleration of Earth is given by

$$g_m(r, \vartheta) = \frac{G\,M_E}{r^2} - \frac{C}{r^4}\left(\frac{3}{2}\cos^2\vartheta - \frac{1}{2}\right) + \cdots \qquad (2.41)$$

where G is the gravitational constant, M_E the mass of Earth, and C is another constant. Numerical values of these constants may be found in Appendix A.2. The first term in the expression for g_m describes the acceleration exerted on a body from a perfectly round Earth. The second term accounts for the fact that the Earth is not really round, but rather, to a first approximation, an oblate spheroid with its polar radius about 21 km shorter than its equatorial radius, see again Appendix A.2. Additional terms describe the Earth's shape and its gravitational acceleration with ever increasing accuracy.

In order to assess the significance of the various terms appearing explicitly in the Eqs. (2.40) and (2.41), numerical values of these terms at the equator ($\vartheta = 90°$) are summarized in Table 2.2 for various heights. The higher orders of the gravitational potential and the centrifugal acceleration are clearly only small corrective factors. They can be safely neglected in the following discussions, which deal with only basic features. Further calculations are therefore based on the approximation

$$g \simeq \frac{G\,M_E}{r^2} = \frac{g_0}{(1 + h/R_E)^2} \qquad (2.42)$$

where g_0 denotes the *mean* acceleration of Earth at sea level ($\simeq 9.81$ m/s^2), h is the height ($r = R_E + h$) and R_E is the mean radius of the Earth ($\simeq$ 6371 km).

2.3.2 Pressure Gradient Force

Since the gravitational force per unit of gas volume is given on the right hand side of the aerostatic equation, it is clear that the force on the left hand side

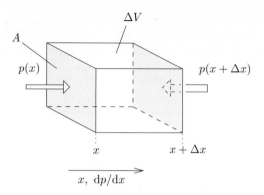

Fig. 2.16. Deriving the pressure gradient force

must also be a force per unit gas volume. The form of this *pressure gradient force* may be understood as follows. Consider the gas volume sketched in Fig. 2.16 bounded on the left and right sides by massless membranes with surface areas A. The box volume is placed in a gas with a pressure gradient. The force acting on this volume in the x-direction is given by

$$F_x = [p(x) - p(x + \Delta x)] \, A$$

Expanding $p(x + \Delta x)$ in a Taylor series and truncating after the second term, one obtains

$$F_x = \left[p(x) - \left(p(x) + \frac{\partial p}{\partial x} \Delta x + \cdots \right) \right] A \simeq -\frac{\partial p}{\partial x} \Delta V$$

It follows that the force per unit gas volume is

$$F_x^* = F_x / \Delta V = -\frac{\partial p}{\partial x} \tag{2.43}$$

Extending this relation to three dimensions, one arrives at the pressure gradient force in its general form

$$\vec{F}_{\nabla p}^* = -(\hat{x} \, \frac{\partial p}{\partial x} + \hat{y} \, \frac{\partial p}{\partial y} + \hat{z} \, \frac{\partial p}{\partial z}) = -\text{grad } p = -\nabla p \tag{2.44}$$

where $\hat{x}$, $\hat{y}$ and $\hat{z}$ are unit vectors, 'grad' denotes the gradient, and ∇ is the nabla (or del) operator.

Because the pressure corresponds to a momentum flux, the pressure gradient force for the case shown in Fig. 2.16 can also be written as

$$F_x^* = -\frac{\partial}{\partial x} \phi_{x,x}^{I(c)} \tag{2.45}$$

As such, it is clear that each pressure gradient force component corresponds to the divergence of a momentum flux. If the momentum flux $\phi_{x,x}^{I(c)}$ brings in more kinetic momentum than is taken away, it leads to an acceleration of the volume. This alternative interpretation is quite satisfying in that a derivation of the aerostatic equation is allowed purely on the basis of volumetric forces and without introducing artificial membranes. We mention for the sake of completeness that the divergence of the momentum flux, when extending this formalism to three dimensions, becomes the divergence of a momentum flux tensor (= pressure tensor), see Appendix A.6.

2.3.3 Barometric Law

The aerostatic equation expresses a relation between the two height dependent state parameters pressure and mass density. If the temperature profile is known, the height dependence of both parameters can be determined after first eliminating one of them with the help of the ideal gas law Eq. (2.28)

$$\rho = \overline{m}\, n = \overline{m}\, \frac{p}{k\, T}$$

Inserting this into the aerostatic equation yields

$$\frac{dp}{dz} = -\, \frac{\overline{m}\, g}{k\, T}\, p = -\, \frac{p}{H}$$

where we have introduced the *pressure scale height H*

$$H(h) = \frac{k\, T(h)}{\overline{m}(h)\, g(h)} \tag{2.46}$$

Separating the variables p and z and integrating over the height interval from h_0 to h, where h_0 is an arbitrary reference height and h is any given height, results in the following relation

$$\int_{p(h_0)}^{p(h)} \frac{dp}{p} = \ln \frac{p(h)}{p(h_0)} = -\int_{h_0}^{h} \frac{dz}{H(z)}$$

or

$$p(h) = p(h_0)\, \exp\left\{ -\int_{h_0}^{h} \frac{dz}{H(z)} \right\} \tag{2.47}$$

This last equation is the well-known *barometric law* for the vertical pressure profile in an atmosphere. An analogous barometric law for the density profile is obtained when the ideal gas law (2.28) is used to replace the pressure with the density

$$n(h) = n(h_0)\, \frac{T(h_0)}{T(h)}\, \exp\left\{ -\int_{h_0}^{h} \frac{dz}{H(z)} \right\} \tag{2.48}$$

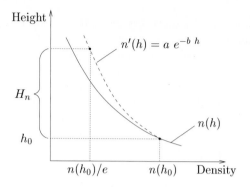

Fig. 2.17. Interpreting the density scale height H_n

Alternatively, the density profile may also be written in the form

$$n(h) = n(h_0) \; \exp\left\{ -\int_{h_0}^{h} \frac{dz}{H_n(z)} \right\}$$ (2.49)

where, in analogy with our definition of the pressure scale height $H = (|dp/dz|/p)^{-1}$, we have introduced the *density scale height* H_n

$$H_n = (|dn/dz|/n)^{-1}$$ (2.50)

Differentiating Eq. (2.48) with respect to h by means of the product rule, we obtain the following relation between the two scale heights

$$\frac{1}{H_n} = \left(\frac{1}{H} + \frac{1}{T}\frac{dT}{dh} \right)$$ (2.51)

Usually (and certainly above an altitude of 200 km) the second term of the sum on the right hand side is small compared to the first, so that the density scale height corresponds closely to the pressure scale height.

Since both scale heights play an important role in the description of the upper atmosphere, we need to have a clear understanding of their physical meaning. Considering the density scale height as an example, this may be obtained with the help of Fig. 2.17. We approximate the density profile $n(h)$ by an exponential function of the form $n'(h) = a\,e^{-bh}$, where a and b are constants. The fitting is done in such a way that the functional values and the slopes of the functions are equal at the reference height of interest h_0

$$n'(h_0) = n(h_0), \quad dn'/dz|_{h_0} = dn/dz|_{h_0}$$

This yields

$$n'(h) = n(h_0) \; \exp\left\{ -\left(\frac{1}{n}\left|\frac{dn}{dz}\right| \right)_{h_0} (h - h_0) \right\}$$

$$= n(h_0) \; \exp\{-(h - h_0)/H_n\}$$

The density scale height H_n thus describes a difference in height $h - h_0$, over which the density changes significantly, i.e. by a factor of e or $1/e$, respectively, when the actual profile is approximated by an exponential function. Correspondingly, the pressure scale height H describes a vertical distance over which the pressure (for any arbitrary pressure profile) varies approximately by a factor e or $1/e$. This interpretation, with the help of Eq. (2.50), may be applied generally to estimate 'typical density scale lengths', i.e. distances over which the density changes significantly. We will make use of this possibility on many occasions.

2.3.4 Heterospheric Density Distribution

The barometric law is valid whether we consider a single gas atmosphere, a well-mixed homosphere or a gravitationally separated heterosphere. However, it can only be explicitly evaluated in the first two cases, for which the mean particle mass is known. In the heterosphere, $\overline{m}(h)$ is an unknown function of height that depends on the density profiles of the individual gases. In order to determine these density profiles, we need a force balance relation similar to the aerostatic equation, but this time for a single gas in a gas mixture. Intuitively, one might be inclined to derive this by separating out selected terms from the aerostatic equation for the entire gas. After all, according to the Dalton-Stefan law, the total pressure of a gas mixture is equal to the sum of the partial pressures. This axiom follows immediately from the ideal gas law for a single gas $p_i = n_i \, k \, T_i$. Provided the individual gases are in thermodynamic equilibrium $(T_i = T)$, we can write

$$\sum_i p_i = \sum_i n_i \, k \, T = n \, k \, T = p$$

The aerostatic equation for a gas mixture can thus be written as follows

$$\sum_i \frac{dp_i}{dz} = -\sum_i n_i \, m_i \, g \qquad (2.52)$$

Unfortunately, this breakdown into various terms provides no help because we are not allowed to associate an individual term on the left hand side with a selected term on the right hand side by writing, for example, $dp_1/dz = n_1 m_1 g$. In this situation the only recourse is an *ab initio* derivation of an 'aerostatic equation' for a single gas in a gas mixture.

We consider a volume element containing a gas of species i as sketched in Fig. 2.18. Included among the forces that act on this volume element is the interaction force $F_{i,j}$, in addition to the previously described gravitational force $F_{g,i}$ and pressure gradient force $F_{\nabla p,i}$. $F_{i,j}$ is the force exerted by all the other gases on the single gas component of interest. In the case of charged gas particles, for example, this might be an electrical force, as described in more

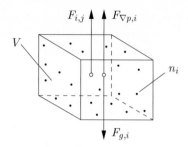

Fig. 2.18. Deriving the aerostatic equation for a single gas in a gas mixture

detail in Section 4.4.1. The only interactions available for neutral particles, however, are frictional forces, i.e. forces that accelerate the gas volume via collision-induced momentum transfer. The generalized form describing these frictional forces is derived in the following.

Frictional Forces. We consider a gas mixture consisting of gas particles of type 1 and type 2. The force exerted on a gas particle of type 1 from collisions with gas particles of type 2 can be expressed in the general form

$$\vec{F}_{1,2} = \vec{F}_{fr_1} = \frac{\sum \Delta \vec{I_1}}{\Delta t} = \langle \Delta \vec{I_1} \rangle \, \nu_{1,2}$$

where $\langle \Delta \vec{I_1} \rangle$ denotes the mean change in momentum suffered by a type 1 particle from a collision. The change in momentum is particularly simple to determine in the case of a head-on collision. Let u_1 be the mean velocity of a type 1 particle and u_2 be the same for a type 2 particle. Since we are interested only in forces acting in a specified direction (in our case the vertical direction), we need only consider the velocity components in this direction. To simplify the calculation, we choose a coordinate system that moves with the mean velocity of the type 2 particles. This results in the scenario sketched in Fig. 2.19. Assuming that the collisions are fully elastic and the total kinetic momentum and energy are conserved, we obtain

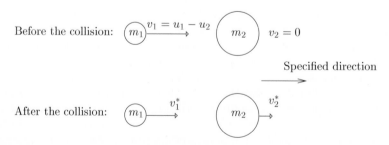

Fig. 2.19. Determining the change of momentum in the case of a head-on collision

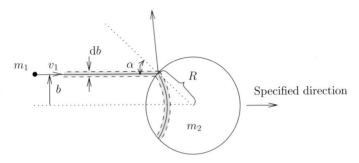

Fig. 2.20. Determining the mean change of momentum for collisions with arbitrary impact parameter

$$v_1\, m_1 = v_1^*\, m_1 + v_2^*\, m_2, \qquad v_1^2\, m_1/2 = (v_1^*)^2\, m_1/2 + (v_2^*)^2\, m_2/2$$

As a nontrivial solution of these equations, after some manipulation omitted here, we obtain

$$v_1^* = v_1(m_1 - m_2)/(m_1 + m_2)$$

so that the change in momentum of the type 1 particle becomes

$$\Delta I_1 = (v_1^* - v_1)\, m_1 = \frac{2\, m_1\, m_2}{m_1 + m_2}\, (u_2 - u_1) \tag{2.53}$$

The situation for the more general case of collisions with arbitrary impact parameter is shown in Fig. 2.20. The calculation has been simplified by reducing the type 1 particles (radius r_1) to point masses and letting the type 2 particles (radius r_2) assume a larger effective radius $R = r_1 + r_2$. When the impact parameter $b \neq 0$, the collision-induced change of momentum is reduced by the factor $\cos\alpha$, because only the velocity component perpendicular to the surface of the type 2 particle contributes to the momentum transfer. Further considering that we are interested here only in momentum transfer in a specified direction, the change of momentum is reduced upon projection onto this direction by another factor of $\cos\alpha$. Accordingly, we have

$$\Delta I_1(b) = \Delta I_1(b = 0)\ \cos^2\alpha$$

In order to determine the mean change in momentum, one must average over the momentum transfer associated with the various possible impact parameters

$$\langle \Delta I_1 \rangle = \int_{b=0}^{R} \Delta I_1(b = 0)\ \cos^2\alpha\ \frac{2\pi b\ db}{R^2\pi}$$

Here, the factor $2\pi b\ db/R^2\pi$ corresponds to that fraction of particles for which the change in momentum is just $\Delta I_1(b = 0)\cos^2\alpha$. Using the relation $b = R\sin\alpha$ (Fig. 2.20), it is found that $\cos^2\alpha = (R^2 - b^2)/R^2$, and one obtains

$$\langle \Delta I_1 \rangle = \Delta I_1(b=0) \, \frac{2}{R^4} \int_{b=0}^{R} (R^2 - b^2) b \, db = \frac{\Delta I_1(b=0)}{2}$$

The frictional force acting on a type 1 particle due to collisions with type 2 particles is thus determined to be $F_{fr_1} = \Delta I_1(b=0)\nu_{1,2}/2$. Accordingly, the associated *frictional force per unit volume* acting on an ensemble of type 1 particles is then $F^*_{fr_1} = n_1 F_{fr_1}$. More precise calculations (particularly more precise averaging procedures) yield a slightly larger value (by a factor 4/3). Incorporating this correction factor, the frictional force per unit volume becomes

$$F^*_{fr_1} = \frac{4}{3} \, n_1 \, \frac{m_1 \, m_2}{m_1 + m_2} \, \nu_{1,2} \, (u_2 - u_1) = m_1 \, n_1 \, \nu^*_{1,2} \, (u_2 - u_1) \qquad (2.54)$$

where we have formally introduced the *momentum transfer collision frequency*

$$\nu^*_{1,2} = \frac{4}{3} \, \frac{m_2}{m_1 + m_2} \, \nu_{1,2} \qquad (2.55)$$

Using Eq. (2.12), this can be written as follows

$$\nu^*_{1,2} = \sqrt{\frac{128 \, k}{9\pi}} \, \sigma_{1,2} \, n_2 \, \sqrt{\frac{m_2 \, T_{1,2}}{m_1 \, (m_1 + m_2)}} \qquad (2.56)$$

Upon extension to three dimensions one finally obtains

$$\vec{F}^*_{fr_1} = m_1 \, n_1 \, \nu^*_{1,2} \, (\vec{u}_2 - \vec{u}_1) \qquad (2.57)$$

As indicated, the frictional force is directly proportional to the difference in flow velocities of the interacting gases. This verifies the intuitively plausible suspicion, that neutral gases exert forces upon each other only if they are moving with different velocities. The interaction force vanishes for stationary gases ($u_1 = u_2 = 0$) and they are basically 'oblivious' to each other. This confirms the observation by Dalton, that stationary neutral gases in a mixture behave as if each were present alone.

Density Distribution. Since the interaction force on the species i, $F_{i,j} = F_{fr_i}$, is zero in the static situation depicted in Fig. 2.18, the aerostatic equation governing a single gas in a gas mixture is formally the same as that for the entire gas mixture. Thus we may write

$$\frac{dp_i}{dz} = - \, n_i \, m_i \, g \qquad (2.58)$$

The solution to this is the barometric law for a single heterospheric gas of species i in thermodynamic equilibrium with the gas mixture ($T_i = T$)

$$n_i(h) = n_i(h_0) \, \frac{T(h_0)}{T(h)} \, \exp \left\{ - \int_{h_0}^{h} \frac{dz}{H_i(z)} \right\} \qquad (2.59)$$

where the scale height is given by

$$H_i(h) = \frac{k\,T(h)}{m_i\,g(h)} \qquad (2.60)$$

The temperature in the upper thermosphere ($h \gtrsim 200$ km) can be taken as constant to a good approximation. Neglecting also the height dependence of the gravitational acceleration, the barometric law simplifies to the relation

$$n_i(h) = n_i(h_0)\,e^{-(h-h_0)/H_i} \qquad (2.61)$$

This corresponds to a straight line in a semi-logarithmic representation such as shown in Fig. 2.11. The density of the gas decreases by a factor $1/e$ for an increase in altitude by one scale height. The scale height, of course, is inversely proportional to the respective particle mass. As a result, light gases fall off with increasing altitude much more slowly than heavy gases. Comparing nitrogen and helium for example, we find $H_{He} = 7\,H_{N_2}$, i.e. the helium density decreases with height seven times more slowly than the nitrogen density (see Fig. 2.11). This explains why helium, a trace gas in the lower thermosphere, becomes the dominant gas constituent at altitudes above about 700 km. The same argumentation holds for the even lighter atomic hydrogen.

Later, the height range between roughly 200 and 600 km will be of particular interest. The transition to the exosphere takes place in this range and this is where the ionized gas density reaches its maximum. According to Fig. 2.9 and Fig. 2.11, $T \simeq T_\infty$ and $\overline{m} \simeq m_o$ are good approximations in this region. For a typical thermopause temperature of $T_\infty = 1000$ K and a mean gravitational acceleration of $\overline{g}(200-600\text{ km}) \simeq 8.8$ m/s^2, the scale height for atomic oxygen is found to be

$$H_o \simeq \frac{k\,T_\infty}{m_o\,\overline{g}} \simeq 60\text{ km} \qquad (2.62)$$

This means that the atomic oxygen in this region – and thus essentially the total density – decreases by a factor $1/e \simeq 0.37$ every 60 km.

In order to calculate the density profile in the lower thermosphere, one needs an explicit expression for the temperature variation in this region. A frequently used approximation is the so-called *Bates temperature profile*

$$T(h) = T_\infty - (T_\infty - T(h_0))\,e^{-s(h-h_0)} \qquad (2.63)$$

Typical values for the constants are $h_0 = 120$ km, $T(h_0) = 350$ K and $s = 0.021$ km^{-1}. This temperature profile has distinct advantages: it is in reasonably good agreement with the (admittedly sparse) observations, and admits an analytical solution of the integral in the barometric law. For these reasons it has found application in many atmospheric models.

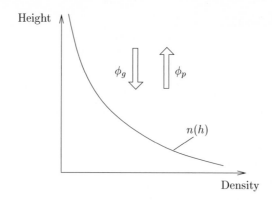

Fig. 2.21. Transport or diffusive equilibrium. ϕ_g and ϕ_p are the descent and expansion fluxes, respectively, and n is the density.

2.3.5 Gas Kinetic Interpretation of the Barometric Law

An alternative interpretation of the barometric density distribution is obtained when it is understood as the result of a transport equilibrium. A heterospheric gas is subject to two opposing tendencies. First it is accelerated toward the Earth because of gravity. This motion, slowed by the numerous collisions with other gas particles, is transformed into a locally constant descent velocity toward the Earth. Second it is located in a pressure gradient field that accelerates it in the direction of decreasing density, i.e. away from the Earth. This motion is also hindered by collisions such that a locally constant expansion velocity is attained. Both of these motions represent a transport of particles. A static density distribution can only be maintained under these circumstances when both fluxes, the descent flux from gravity ϕ_g and the outward expansion flux ϕ_p, are of equal magnitude as indicated in Fig. 2.21. This condition is known as *transport equilibrium* or also – for reasons to be explained later – as *diffusive equilibrium*.

It is easy to show that this alternative approach again leads to the aerostatic equation. The descent velocity u_g is derived from the requirement of force balance between gravity and internal friction

$$\vec{F}^*_{g_1} + \vec{F}^*_{fr_1} = n_1 m_1 \vec{g} - n_1 m_1 \nu^*_{1,2} \vec{u}_1 = \hat{z}(-n_1 m_1 g - n_1 m_1 \nu^*_{1,2} u_{g_1}) = 0$$

or

$$n_1\, m_1\, g = -n_1\, m_1\, \nu^*_{1,2}\, u_{g_1}$$

Here, other heterospheric gases (index 2) are considered to be at rest ($u_2 = 0$). The *descent flux* is therefore

$$\phi_{g_1} = n_1\, u_{g_1} = -n_1\, g/\nu^*_{1,2} \tag{2.64}$$

Similar to the descent velocity, we require another balance between the pressure gradient and frictional forces to obtain the expansion velocity

$$-\frac{\mathrm{d}p_1}{\mathrm{d}z} = n_1\, m_1\, \nu_{1,2}^{*}\, u_{p_1}$$

The *expansion flux* is thus given by

$$\phi_{p1} = n_1\, u_{p_1} = -\frac{1}{m_1\, \nu_{1,2}^{*}}\, \frac{\mathrm{d}p_1}{\mathrm{d}z} \qquad (2.65)$$

The aerostatic equation follows immediately from the condition $\phi_{g_1} + \phi_{p1} = 0$. Should this equilibrium not exist, either a descent flux or an expansion flux will be induced that strives to reestablish a density distribution compatible with the aerostatic equation.

Molecular Diffusion. While the descent process is evident from prior experience (recall here the case of a falling feather), the expansion velocity becomes considerably more obvious when understood as a diffusion process. Diffusion is known to be a mediating process by which density differences are smoothed out by the thermal motion of the gas particles. The exponential decrease of the gas density in the atmosphere represents a significant density gradient. Correspondingly, a diffusion flux will flow in the direction of decreasing density and thus increasing altitude.

In order to assess the effectiveness of this diffusion process, we consider an isothermal gas mixture consisting of a primary gas and a secondary gas. To simplify the calculation, we assume further that the gases have the same properties and differ only by an external label. Such a situation is also called self-diffusion. While the homogeneous, isothermal primary gas remains at rest, the secondary gas exhibits a density gradient in the negative z-direction; see Fig. 2.22. Let us determine the gradient-induced particle flux of the secondary gas through a surface oriented perpendicular to the density gradient. It should be noted that, since the particles can only move freely between two

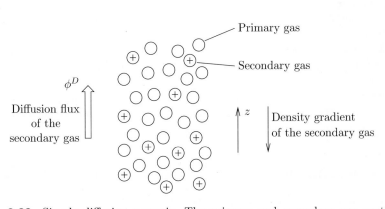

Fig. 2.22. Simple diffusion scenario. The primary and secondary gas particles are assumed to have the same properties (i.e. the same radius and mass) and are discriminated only by an external label.

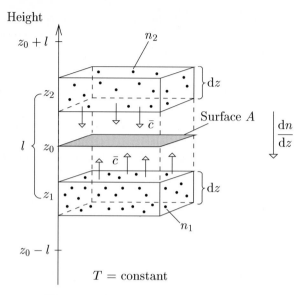

Fig. 2.23. Deriving the diffusion flux

consecutive collisions, the particle transport proceeds over scale lengths comparable to the mean free path. Consequently, we consider here the density exchange of the secondary gas between two volumes separated by just this distance. Volume element 1 lies below and volume element 2 above the surface of area A; see Fig. 2.23. The number of particles in each volume element amounts to

$$dN_1 = A \, dz \, n_1, \quad dN_2 = A \, dz \, n_2 = A \, dz \left(n_1 + \frac{dn}{dz} \, l + \cdots \right)$$

where all densities refer to the secondary gas and the mean free path $l_{1,1}$ is abbreviated as l.

Assuming a reduced velocity distribution, a fraction $1/6$ of the particles in each volume will have their velocity directed toward the surface A, penetrate the surface and enter the other volume without suffering a collision. The exchange of particles occurs over a time scale $\Delta t = l/\bar{c}$, where $\bar{c}$ is the mean random velocity of the secondary gas particles. The particle fluxes associated with this particle exchange are thus

$$d\phi_1 = \frac{1}{6} \frac{dN_1}{A \, \Delta t} = \frac{1}{6} \frac{n_1 \, \bar{c}}{l} \, dz, \quad d\phi_2 = \frac{1}{6} \frac{n_1 \, \bar{c}}{l} \, dz + \frac{1}{6} \bar{c} \left(\frac{dn}{dz}\right) dz + \cdots$$

so that the net flux through the surface A is

$$d\phi^D = d\phi_1 - d\phi_2 = -\frac{1}{6} \bar{c} \left(\frac{dn}{dz}\right) dz$$

In order to obtain the total diffusion flux, contributions from all volumes within a distance $\leq l$ of the surface A must be added together. Only those particles within these limits reach the surface within the time interval Δt. In our case we must thus integrate the net flux $d\phi^D$ over the distance l. This yields

$$\phi^D = -\frac{1}{6}\,\bar{c}\,\frac{dn}{dz}\int_{z_0-l}^{z_0} dz = -\frac{1}{6}\,\bar{c}\,\frac{dn}{dz}\int_{z_0}^{z_0+l} dz = -\frac{1}{6}\,\bar{c}\,l\,\frac{dn}{dz}$$

where we have assumed that the density gradient (not the density itself) is constant over the scale of a mean free path. Upon introducing the *diffusion coefficient* $D \sim \bar{c}\,l$ for simplicity, the diffusion flux assumes the following form (*Fick's law*)

$$\phi^D = -D\,\frac{dn}{dz} \tag{2.66}$$

Using the relations $l = \bar{c}/\nu$ and $3kT/2 = m\,\bar{c}^2/2$ (reduced velocity distribution!) the diffusion coefficient may be written as $D = \xi\,k\,T/m\,\nu$. While for our estimate the value of the constant $\xi = 1/2$, a more exact calculation based on a Maxwellian distribution yields $\xi = 3/2$. In this case we may write the *self-diffusion coefficient* as

$$D_{1,1} = \frac{3}{2}\,\frac{k\,T}{m\,\nu_{1,1}} = \frac{k\,T}{m\,\nu_{1,1}^*} \tag{2.67}$$

where, for clarity, we have introduced indices on the collision and frictional frequencies as well as the diffusion coefficient. The foregoing treatment is based entirely on the assumption that the main and secondary gases have the same properties. Correspondingly, when the gases are different

$$D_{1,2} = \frac{kT_1}{m_1\nu_{1,2}^*} \tag{2.68}$$

In order to show that the expression derived here for the diffusion flux is identical to that derived for the expansion flux, we recall that, for the isothermal case, changes in pressure are proportional to changes in density

$$\phi^D = -\frac{k\,T}{m\,\nu_{1,1}^*}\,\frac{dn}{dz} = -\frac{1}{m\,\nu_{1,1}^*}\,\frac{dp}{dz} = \phi_p \tag{2.69}$$

The barometric density distribution can thus be understood as an equilibrium state, for which the descent flux and the diffusion flux are equal. Extending the diffusion concept to all phenomena for which a particle flux occurs under the influence of an accelerating force (e.g. gravity force or pressure gradient force) in the presence of an opposing frictional force, the barometric density distribution represents not only a transport equilibrium, but also a state of *diffusive equilibrium*. A more detailed discussion of this extended diffusion concept may be found in Appendix A.5.

In closing, it is worthwhile questioning just why an effect based purely on the thermal motion of the particles, as diffusion is viewed in a strict sense, can be interpreted as a pressure gradient force. Recall here that every particle transport is inseparably associated with a momentum flux. It is the momentum gain or loss brought about by the diffusion that corresponds to the effect of a force.

2.3.6 Transition from Homosphere to Heterosphere

Molecular transport such as descent flux and molecular diffusion create a gravitationally separated atmosphere in the equilibrium state, by which each gas establishes its own height profile. Heavier gases have smaller scale heights and decrease more rapidly with height. Conversely, lighter gases with their large scale heights decrease more slowly with increasing altitude. A layered atmosphere (i.e. a literal *strato*sphere) is created in this way, by which the heavy gases dominate at lower heights and lighter gases are more prevalent at greater heights.

The gravitational separation would occur directly on the Earth's surface for an atmosphere governed exclusively by molecular transport processes. As we know, this is not the case. Instead, the atmosphere remains well mixed and of homogeneous composition up to an altitude of roughly 100 km. Mixing processes that counteract the effects of gravitational separation evidently dominate below this height. In order to understand why the homopause is located at 100 km altitude (and not, as suspected earlier, at the tropopause), we determine the effectiveness of the mixing and separation processes by means of their respective characteristic time constants. This requires that we first quantify the mixing process.

Eddy Diffusion. The primary reason for the mixing of the atmosphere is what is generally denoted as turbulence. Air parcels of varying composition are exchanged via irregular whirling motion, leading to a homogenization of the composition. This moderation process can be understood formally as a type of diffusion, by which compositional differences are smoothed out by associated particle fluxes. In this sense one also refers here to *eddy diffusion*. A quantitative analysis of this phenomenon may be undertaken with the help of Fig. 2.24.

Consider an atmosphere of total density n, within which a secondary gas constituent with density n_i exhibits a gradient in the mixing ratio given by $d(n_i/n)/dz$. Due to the uneven distribution of heat, a cylindrical turbulence eddy is created with a radius $l_\mathcal{E}$ and a circumferential velocity $u_\mathcal{E}$. As a consequence, air parcels in the lower part of the eddy are carried upward, and parcels in the upper part are transported downward. Because a complete rotation would merely recreate the original state, only incomplete eddy motions are of interest for the moderating process considered here. For simplicity, let us assume that the eddy motion is interrupted after a half rotation and then

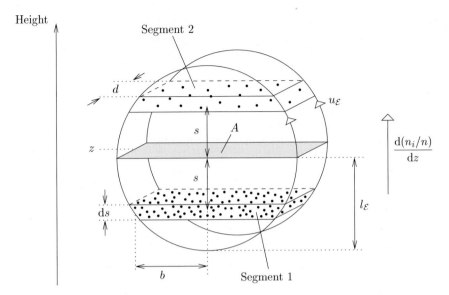

Fig. 2.24. Deriving the eddy diffusion flux

comes to rest. In this case, exactly the lower half of the cylinder will have turned to the top and the upper half will be at the bottom. Due to the gradient in the mixing ratio, this tumbling motion leads to a net flux of the secondary gas through the mid-plane of the cylinder. To calculate this particle flux, we consider two cylinder segments of thickness $\mathrm{d}s$, width $2b$ and depth d, located at the distance $\pm s$ from the mid-plane (see Fig. 2.24). The number of particles of gas constituent i in the segment of volume $\mathrm{d}V = d\,2b\,\mathrm{d}s$ is

$$(\mathrm{d}N_i)_1 = \mathrm{d}V\,n_1\,(n_i/n)_1$$

and

$$(\mathrm{d}N_i)_2 = \mathrm{d}V\,n_2\,(n_i/n)_2 = \mathrm{d}V\,n_2\left[(n_i/n)_1 + \frac{\mathrm{d}(n_i/n)}{\mathrm{d}z}\,2s + \cdots\right]$$

The cylinder segments move in the direction of the mid-plane in the course of the eddy rotation. The gas density is changed as a result and reaches the value $n(z) = n$ at the altitude of the mid-plane. The mixing ratio is not modified by this change in density, because the total gas and the secondary gas are rarefied or enhanced to the same degree by the pressure differences. The net transport of the secondary gas particles through the surface A resulting from the exchange of the cylinder segments is thus given by

$$(\mathrm{d}N_i)_A = [(\mathrm{d}N_i)_1 - (\mathrm{d}N_i)_2]_A \simeq -\mathrm{d}V\,n\,\frac{\mathrm{d}(n_i/n)}{\mathrm{d}z}\,2s$$

Noting that the exchange surface $A = 2l_\mathcal{E}\, d$ and the exchange time $\Delta t = \pi\, l_\mathcal{E}/u_\mathcal{E}$, we may write the differential flux of the secondary gas through the surface A as

$$\mathrm{d}\phi_i^\mathcal{E} = \frac{(\mathrm{d}N_i)_A}{A\,\Delta t} = -\frac{2}{\pi}\,\frac{u_\mathcal{E}}{l_\mathcal{E}^2}\,n\,\frac{\mathrm{d}(n_i/n)}{\mathrm{d}z}\,\sqrt{l_\mathcal{E}^2 - s^2}\;s\,\mathrm{d}s$$

where the width b has been replaced by the square root expression dependent on s. The total flux is obtained upon summing over the contributions from all cylinder segments

$$\phi_i^\mathcal{E} = -\frac{2}{\pi}\,\frac{u_\mathcal{E}}{l_\mathcal{E}^2}\,n\,\frac{\mathrm{d}(n_i/n)}{\mathrm{d}z}\int_{s=0}^{l_\mathcal{E}}\sqrt{l_\mathcal{E}^2 - s^2}\;s\,\mathrm{d}s = -\frac{2}{3\pi}\,u_\mathcal{E}\,l_\mathcal{E}\,n\,\frac{\mathrm{d}(n_i/n)}{\mathrm{d}z} \quad (2.70)$$

where the gradient in the mixing ratio is taken as constant over the scale length of the eddy and the integral was solved with the help of Eq. (A.5). Introducing the *eddy diffusion coefficient* K, the eddy diffusion flux may be written as follows

$$\phi_i^\mathcal{E} = -K\,n\,\frac{\mathrm{d}(n_i/n)}{\mathrm{d}z} \quad (2.71)$$

The formal similarity of this expression with Eq. (2.66), valid for the molecular diffusion flux, is obvious. The particle flux here is again directed such that it opposes heterogeneity. Moreover, a comparison of the relations (2.70) and (2.71) shows that the eddy diffusion coefficient is proportional to both the size of the eddy and its rotation velocity, $K \sim l_\mathcal{E}\,u_\mathcal{E}$. A formal similarity also exists here when we recall that the molecular diffusion coefficient is proportional to both the mean free path and the random velocity. Contrary to the situation with diffusion, however, no general theory of turbulence exists which allows a derivation of the eddy size and eddy velocity needed for the calculation of K. Rather, one must resort here to indirect measurements and their results tend to vary by orders of magnitude. Figure 2.26 shows a representative height profile of K, as found in the contemporary literature. Noteworthy is the fairly sudden decrease of the eddy diffusion coefficient above 100 km altitude. This is associated with the exponentially rising effectiveness of the viscosity, which increasingly suppresses the eddy motions. One also designates the atmospheric region of high turbulence as the *turbosphere* and its upper boundary as the *turbopause* (see overview in Fig. 2.13).

Time Constants for Transport Processes. In order to compare the effectiveness of the competing molecular and turbulence-induced transport processes, it is necessary to determine their respective time constants. These can be evaluated with the help of a very simple disturbance scenario. Consider an atmospheric layer of thickness $2H_{d(isturbance)}$, within which the density of a secondary gas deviates from its ambient values by an amount Δn_i. We want to estimate the time molecular and eddy diffusion need to eliminate this density disturbance. According to Fig. 2.25a, the deficit of secondary gas

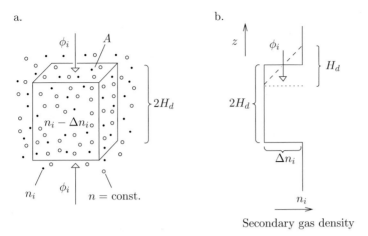

Fig. 2.25. Estimating the time constants for transport processes

particles in a volume element of the layer is $\Delta N_i = 2H_d\, A\, \Delta n_i$. This deficit will be compensated over time by particle fluxes which increase the number of particles by $\Delta N'_i(t) = 2\phi_i\, A\, t$. Equating these two expressions, we obtain directly the time constant for the smoothing process

$$\tau = H_d\, \Delta n_i/\phi_i$$

The particle flux required for evaluating this expression in the case of molecular diffusion is

$$\phi_i^D = D\,\left|\frac{dn_i}{dz}\right| \simeq D\,\frac{\Delta n_i}{H_d}$$

where the continually changing density gradient has been approximated by a crude mean value; see Fig. 2.25b. Combining the above, we obtain the *time constant for molecular diffusion*

$$\tau_D \simeq H_d^2/D \tag{2.72}$$

Because molecular diffusion and descent flux are of comparable magnitude in the atmosphere, this expression is valid for both of these processes.

An alternative derivation of this time constant may be obtained from the continuity equation (2.19), for which

$$\left(\frac{\partial n}{\partial t}\right)_{diff} = -\frac{\partial(n u_z)_{diff}}{\partial z} = \frac{\partial}{\partial z}\left(D\,\frac{\partial n}{\partial z}\right) \simeq D\,\frac{\partial^2 n}{\partial z^2} \tag{2.73}$$

Here, the diffusion coefficient is assumed to be independent of position to a first approximation. Equation (2.73) represents a simple form of the time-dependent diffusion equation, known also as *Fick's second law*. Assuming that the temporal and spatial variations of the density perturbation can be roughly

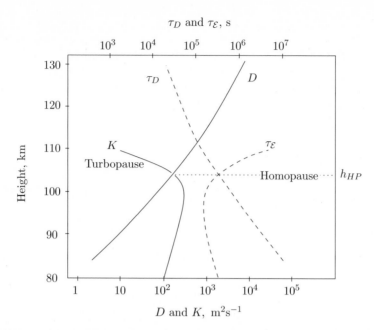

Fig. 2.26. Height profiles of the molecular diffusion coefficient D, the eddy diffusion coefficient K and the associated time constants for molecular diffusion τ_D and eddy diffusion $\tau_\mathcal{E}$. The value of H_d was set equal to the density scale height.

approximated by exponential functions, $n(t,z) \sim \exp(t/\tau_D)\exp(z/H_d)$, the time constant for molecular diffusion is derived directly from Eq. (2.73).

Correspondingly, for density compensation by eddy diffusion, we have

$$\phi_i^\mathcal{E} = K\,n\left|\frac{\mathrm{d}(n_i/n)}{\mathrm{d}z}\right| \simeq K\,n\left(\frac{n_i}{n} - \frac{n_i - \Delta n_i}{n}\right)\Big/H_d = K\,\frac{\Delta n_i}{H_d}$$

and

$$\tau_\mathcal{E} \simeq H_d^2/K \tag{2.74}$$

Homopause Height. Figure 2.26 shows a typical height profile for τ_D and $\tau_\mathcal{E}$. Because the molecular diffusion coefficient is inversely proportional to the collision frequency, which decreases exponentially with height, τ_D also decreases rapidly with increasing altitude. In contrast, $\tau_\mathcal{E}$ is relative constant below the turbopause and then rises swiftly toward higher altitudes. These two curves cross near 100 km altitude. Below this intersection $\tau_\mathcal{E} < \tau_D$ and the opposite is true above this height $\tau_\mathcal{E} > \tau_D$. Eddy diffusion thus dominates below this altitude – the atmosphere is well mixed; molecular transport processes are more effective above this height – a gravitationally separated heterosphere results. The intersection of the curves $\tau_D(h)$ and $\tau_\mathcal{E}(h)$ thus defines the height of the *homopause* h_{HP}

$$\tau_{\mathcal{E}}(h_{HP}) = \tau_D(h_{HP}) \quad \text{or} \quad K(h_{HP}) = D(h_{HP})$$

where a typical value for the terrestrial atmosphere is $h_{HP} \simeq 100$ km. Since the diffusion coefficient is inversely proportional to the particle mass, there will be, strictly speaking, different intersections for each gas constituent. These differences are of little consequence in view of the great variability of the eddy diffusion coefficient.

2.3.7 Atomic Oxygen and Hydrogen

Atomic oxygen, a highly reactive gas, is not a natural constituent of the lower atmosphere. Its prominence in the thermosphere is thus in need of explanation. Atomic oxygen is produced by the photodissociation of molecular oxygen

$$O_2 + \text{photon}(\lambda \le 242.4 \text{ nm}) \rightarrow O + O^{(*)} \tag{2.75}$$

where the dissociation products – depending on the photon energy – can occur in an excited state (*). Compared with the actually observed O densities, however, this production rate is surprisingly small. In fact, at this rate it would take weeks in constant daylight to accumulate the measured oxygen density at an altitude of 150 km. That significant O densities are attained in spite of this modest production rate may be attributed to the even less effective loss processes. Direct radiative recombination by the scheme O + O $\rightarrow$ O$_2$ + photon is very improbable because of energy and momentum conservation requirements. The only important loss process is the three body collisional recombination

$$O + O + M \rightarrow O_2^{(*)} + M \tag{2.76}$$

where the collision partner M (e.g. N_2) assumes some fraction of the recombination energy of 5 eV. Since three body collisions very seldom occur in tenuous gases, this loss process is extremely slow and the mean life expectancy of an oxygen atom, already a few months at a height of 100 km, becomes orders of magnitude longer at 150 km.

In view of such a slow loss process, it may be surprising that the molecular oxygen component in the upper atmosphere is not fully dissociated and converted into atomic oxygen. That this is not the case may be attributed to the effectiveness of transport processes. The molecular transport time constant for atomic oxygen above 100 km altitude is much shorter than the corresponding loss time constant (see Fig. 2.26). As a result, deviations from the barometric distribution produced by an excess of O and a deficit of O_2 are eliminated rather quickly. Surplus oxygen atoms are transported by a descent current toward the mesopause, where they recombine quickly because of the large density there. The resulting oxygen molecules move upward into the regions of the mid- and upper thermosphere. Besides conserving the molecular oxygen component, this exchange cycle has important consequences for

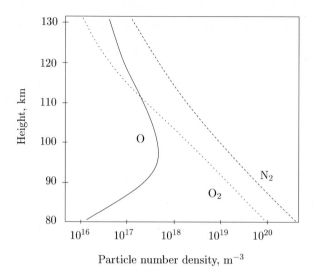

Fig. 2.27. Height profile of the atomic oxygen density in the lower thermosphere. The distribution below 100 km altitude depends strongly on the solar zenith angle and thus on the local time.

the energetics of the thermosphere because each atomic oxygen pair transports a recombination energy of 5 eV into the lower thermosphere/upper mesosphere. The question arises whether or not these additional descent and diffusion fluxes have any influence on the density distribution of O and O_2 above 100 km altitude. This is so only to a very small extent. The transport currents associated with the oxygen circulation constitute only a small fraction of the descent and diffusion fluxes already present in these regions. The equilibrium state is therefore only insignificantly disturbed.

A different situation exists below 100 km altitude. The loss rate of atomic oxygen grows rapidly with decreasing height (density scale height $\lesssim$ 6 km), and the mean life expectancy of these particles drops to only a few hours at an altitude of 80 km. The loss process thus becomes considerably more effective than the mixing process and the atomic oxygen density falls off rapidly; see Fig. 2.27. This decrease has no effect on the total density profile, however, because atomic oxygen is only a trace gas in this lower height region. How the density profile of an atmospheric gas is calculated when production and loss processes are important in addition to transport processes will be described later in Section 4.3.1.

If atomic oxygen is such an important constituent of the upper atmosphere, why doesn't the same hold for atomic nitrogen? This may be attributed to the small production rate and high loss rate for this species. Owing to the stronger bond of the N_2 molecule, photodissociation is much less frequent than for the O_2 molecule (factor 10^{-6}). Indeed, dissociative photo*ionization* processes and chemical reactions are considerably more important sources of N atoms than direct photodissociation. Furthermore, atomic nitrogen is a highly reactive gas that is especially rapidly depleted in the lower thermosphere (e.g. via reaction with oxygen to form nitric oxide NO). This all leads to distinctly smaller N densities. Even the peak density of this species at an altitude of 180-200 km is two orders of magnitude smaller than that of atomic oxygen.

As a highly reactive gas, atomic hydrogen is also not a natural constituent of the lower atmosphere. Like atomic oxygen, it is produced only at greater heights, in this case by the photodissociation of water molecules

$$H_2O + photon(\lambda \lesssim 240 \text{ nm}) \rightarrow H + OH$$

Once produced, hydrogen atoms possess similarly long life expectancies in the middle and upper thermosphere as oxygen atoms. These particles are lost either by chemical reactions or via evaporation effects in the upper thermosphere. This latter process, with its important consequences for the density distribution of the outermost atmosphere, will be discussed in more detail in the following section.

2.4 Exospheric Density Distribution

The exosphere is the outermost envelope of the terrestrial atmosphere. The densities are so small here that direct escape (evaporation) of gas particles is possible, a fact that provides the name for this region. In the following section we first determine that height at which the escape process becomes possible. This lower boundary of the exosphere is called the exobase. In order to overcome the force of gravity, the evaporating particles must possess a velocity greater than or equal to the escape velocity. Knowing the velocity distribution function at the exobase, we can determine how many of the gas particles reach or exceed this escape velocity. Particular attention is paid here to the Maxwellian distribution function and the distribution of velocity magnitudes based on it. This distribution function allows an estimate of the escaping particle flux, an important quantity for assessing the stability of the terrestrial atmosphere. The aerostatic equation, no longer valid in an evaporating atmosphere, cannot be used to determine the density profile. Instead of a force balance condition, we therefore consider the populations of exospheric particles in various orbital states. This leads to density models in satisfactory agreement with the observations.

2.4.1 Exobase Height

All gas particles moving outward in the lower thermosphere are prevented from escape. Sooner or later they will suffer a collision and are scattered

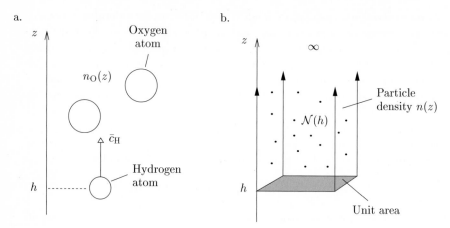

Fig. 2.28. (a) Calculating the exobase height and (b) defining the column density

backward in the direction of Earth. Since collisions become increasingly less frequent in the upper thermosphere, a particle moving away from Earth has a fair chance to avoid subsequent downward scatter and thus may escape into interplanetary space. We define the *exobase* as that height above which, on the average, a radially outward moving particle will suffer less than one (backscatter) collision. This height may be determined with the help of Fig. 2.28a. We assume for simplicity that only hydrogen atoms, because of their large random velocity ($\sim 1/\sqrt{m}$), have a distinct chance of overcoming Earth's gravity and escaping. Moreover, we assume that the exobase is located in an oxygen-dominated region of the thermosphere. The validity of both of these assumptions will be verified later.

The number of collisions dN_ν suffered by a radially outward moving hydrogen atom in the time interval dt is $\nu_{H,O}\, dt$. The particle moves a vertical distance $dz = \bar{c}_H\, dt$ during this time, so the number of collisions per height interval will be $dN_\nu = \nu_{H,O}\, dz/\bar{c}_H$. The total number of collisions encountered along the path from the height h up to an infinitely large distance is therefore

$$N_\nu(h) = \int_h^\infty (\nu_{H,O}(z)/\bar{c}_H)\, dz$$

$$= \sigma_{H,O}\sqrt{1 + m_H/m_O} \int_h^\infty n_O(z)\, dz \qquad (2.77)$$

where in the second step we have made use of the relation (2.13) with $T_O = T_H = T$. The integral in the above expression is an important atmospheric quantity called the *column density*

$$\mathcal{N}(h) = \int_h^\infty n(z)\, dz \qquad (2.78)$$

This corresponds to the number of particles in a vertical atmospheric column per column base area, i.e. the number of particles in a vertical column with a base area of 1; see Fig. 2.28b. For a base area of unity and without the density in the integrand, the integral just calculates the volume of the column. To obtain an explicit estimate of this quantity, we carry out a coordinate transformation with the help of the aerostatic equation

$$n \, dz = -dp/m \, g$$

and arrive at

$$\mathcal{N}(h) = - \int_{p(h)}^{0} \frac{dp}{m \, g} \simeq - \frac{1}{m \, g(h)} \int_{p(h)}^{0} dp = \frac{p(h)}{m \, g(h)}$$

or

$$\mathcal{N}(h) \simeq n(h) \, H(h) \tag{2.79}$$

In other words, the column density is given by the product of the density times the scale height, both evaluated at the base of the column. Since this derivation assumes that the particle mass can be taken out of the integral, Eq. (2.79) is valid only in the homosphere ($m = \overline{m}$) or for a single gas in the heterosphere ($m = m_i$). On the other hand, the total column density in the heterosphere (column densities are additive quantities!) is given by

$$\mathcal{N}_{heterosphere}(h) \simeq \sum_{i} n_i(h) \, H_i(h) \tag{2.80}$$

The above derivation also assumes the gravitational acceleration to be constant even though the integral extends over a large height interval. This is still possible, because practically the entire column content is concentrated in a very limited height interval near the base of the column. We demonstrate this with an example. For a column of atomic oxygen with a base located at an altitude of 400 km and a typical thermopause temperature of 1000 K, more than 99% of the particles in the column are found below 700 km. The acceleration due to gravity, however, changes by less than 10% over the height interval from 400 to 700 km. It is thus no serious error to approximate the gravitational acceleration g as constant and equal to its value at the base of the column ($= g(h)$). No assumptions are made, however, regarding the temperature profile in the given gas column.

Using Eqs. (2.77) and (2.79), we may write the defining equation for the exobase height h_{EB} as follows

$$N_\nu(h_{EB}) \simeq \sigma_{H,O} \sqrt{1 + m_H/m_O} \, n_O(h_{EB}) \, H_O(h_{EB})$$
$$= H_O(h_{EB})/l_{H,O}(h_{EB}) = 1 \tag{2.81}$$

This means that the (horizontal) mean free path of the hydrogen atoms at the exobase height becomes equal to the pressure scale height and thus also very close to the density scale height of the primary gas atomic oxygen. In gas physics the ratio of the mean free path to the typical density or system scale length is also denoted the *Knudsen number*. The Knudsen number near the exobase thus reaches the value $Kn(h_{EB}) \simeq 1$.

In order to obtain an explicit expression for the exobase height, let us assume that the temperature in the exobase region is nearly constant and may be approximated by the thermopause temperature ($T(h_{EB}) \simeq T_\infty$). With $H_O \simeq$ const. and $m_H/m_O \ll 1$, we obtain

$$\sigma_{H,O} \; H_O \; n_O(h_{Ref}) \; \exp\{-(h_{EB} - h_{Ref})/H_O\} \simeq 1$$

or

$$h_{EB} \simeq h_{Ref} + H_O \; \ln[\sigma_{H,O} \; H_O \; n_O(h_{Ref})] \tag{2.82}$$

For a collision cross section $\sigma_{H,O} \simeq 2 \cdot 10^{-19}$ m^2 ($r_H \simeq 1$, $r_O \simeq 1.5$, in 10^{-10}m), a thermopause temperature of 1000 K, a scale height $H_O(1000 \text{ K}) \simeq 60$ km, and a reference density of $n_O(250 \text{ km}, \; 1000 \text{ K}) = 1.5 \cdot 10^{15}$ m^{-3}, we obtain an exobase height of 420 km from Eq. (2.82). This result is obviously consistent with the original assumptions regarding the oxygen-dominated and isothermal exobase region. The exobase height shifts up and down for other thermopause temperatures, but these assumptions do not lose their validity.

2.4.2 Escape Velocity

In order to escape from the Earth's atmosphere, it is not enough that a gas particle above the exobase be simply moving outward. In addition to this prerequisite, it must also posses a sufficiently high velocity. The commonly used designation for that minimum velocity a body requires to overcome Earth's gravity is the *escape velocity*. All parameters associated with atmospheric escape in the following will carry the index 'es'. The escape velocity v_{es} is defined from the condition that the kinetic energy of the escaping body is equal to the work performed by the body against the Earth's gravitational field along its path outwards

$$\frac{1}{2} \; m \; v_{es}^2 = \int_r^\infty m \; g(r') \; \mathrm{d}r' = m \; g(r) \; r$$

where r is the geocentric distance and $g(r) \simeq G \; M_E/r^2$ the gravitational acceleration. The escape velocity is therefore given by

$$v_{es} = \sqrt{2 \; g(r) \; r} = \frac{v_{es}(h = 0)}{\sqrt{1 + h/R_E}} \tag{2.83}$$

where in the second step we have replaced the geocentric distance $r = R_E + h$ with the height. Inserting the indicated values yields an escape velocity $v_{es} \simeq 11$ km/s at heights near the exobase (see also Table 2.2).

The mean random velocity of gas particles, as taken from the definition for temperature, is $\bar{c} \simeq \sqrt{3\,k\,T/m}$. With a temperature of $T = 1000$ K this yields mean velocities for oxygen, helium and hydrogen atoms of $\bar{c}_O \simeq 1.25$ km/s, $\bar{c}_{He} \simeq 2.5$ km/s and $\bar{c}_H \simeq 5$ km/s, respectively. Evidently the mean velocity of even the light hydrogen particles is too small for escape! To make an estimate of the actual escape flux, it is necessary to know how many of the gas particles possess a velocity greater than the escape velocity. This information is contained in the velocity distribution functions of the gases.

2.4.3 Velocity Distribution in Gases

The distribution of the velocities of the individual particles in a gas is an issue of critical importance. Only if this velocity distribution is known, for example, is it possible to determine such quantities as the mean relative velocity (as needed for the calculation of the collision frequency) or the mean squared velocity (as used in the definition of the temperature). Specifically, the velocity distribution provides the information how many gas particles have velocities greater than the escape velocity and can thus evaporate from the atmosphere. Information about this velocity distribution can be presented in various forms. The general form is a so-called distribution function, the definition of which we explain in the following. Here, a simple but very important specific case of a distribution function, the Maxwellian distribution, is considered. A specialized distribution function for velocity magnitudes (speeds) can be derived from the Maxwellian distribution that provides estimates of the mean speed as well as the fraction of particles with velocities greater than the escape velocity.

Definition of the Distribution Function. Distribution functions describe not only the velocity distribution, but also the spatial distribution of the gas particles. Consider first a gas in ordinary space. The location of every gas particle in this *configuration space* at the time t is characterized by a position vector $\vec{r}_i$ – or more simply by a point at the tip of this vector. These points represent the instantaneous distribution of the gas particles in space at time t. We now consider a small volume element d^3r in this configuration space. The dimensions of this volume element are small compared with the local density scale lengths, but big enough to contain a statistically large number of gas particles. The position of the volume element is defined by the mean position vector $\vec{r}$ of the gas particles contained within the gas volume. A sketch of the situation in Cartesian coordinates is shown in Fig. 2.29a. Evidently, the total number of particles δN contained in d^3r is

$$\delta N = n(\vec{r}, t)\, d^3r \tag{2.84}$$

where n, as used previously, defines the particle number density, which can be dependent on position and time.

a. Configuration space

b. Velocity space

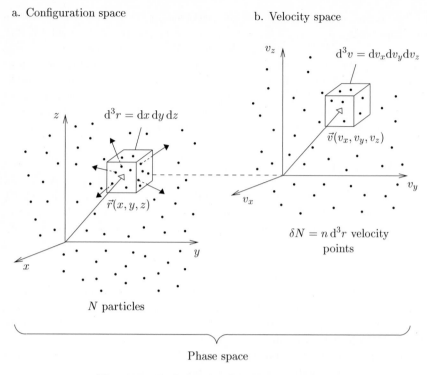

Phase space

Fig. 2.29. Defining the distribution function

As indicated in Fig. 2.29a, a distinct velocity may be ascribed to each of the δN particles. These velocities can be represented in a separate *velocity space*, either in the form of velocity vectors $\vec{v}_i$ or, more simply and analogous to the representation in configuration space, by a point at the tip of the velocity vector. There are a total of δN points in velocity space, whereby each of these points describes the velocity of one of the gas particles contained in the configuration space volume element d^3r.

Let us again consider a small volume element d^3v, now in velocity space. The dimensions of this volume should be small with respect to the scale lengths over which the density of the velocity points changes appreciably, but big enough to accommodate many velocity points. The position of the volume is given by the mean vector $\vec{v}$ of the velocities contained within d^3v. The situation in a Cartesian coordinate system is sketched in Fig. 2.29b. As in configuration space, one could now define a density in velocity space

$$dN = n_v(\vec{v}, \vec{r}, t) \, d^3v$$

where dN denotes the number of velocity points in d^3v. Since this velocity space density n_v would then depend on the number of particles in d^3r, it is more sensible to introduce a density normalized to this number δN by writing

$$dN/\delta N = g(\vec{v}, \vec{r}, t)\, d^3v \tag{2.85}$$

The quotient $dN/\delta N$ describes the fraction of the gas particles in d^3r that have a velocity within the volume d^3v. Noting that fractions can be interpreted as probabilities, the quotient $dN/\delta N$ also represents the probability that velocities are located within d^3v. The function g can thus be viewed as either a fractional density or a probability density in velocity space. It is usually referred to as the *velocity distribution function*.

The number of particles having position vectors in d^3r and, at the same time, velocity vectors in the volume d^3v is thus given by

$$dN = n(\vec{r}, t)\, g(\vec{v}, \vec{r}, t)\, d^3r\, d^3v \tag{2.86}$$

It is useful to combine the particle number density and the velocity distribution function together into a *distribution function f*

$$f(\vec{r}, \vec{v}, t) = n(\vec{r}, t)\, g(\vec{v}, \vec{r}, t) = \frac{dN}{d^3r\, d^3v} \tag{2.87}$$

The last step above implies that the distribution function can be formally interpreted as a density in a 6-dimensional *phase space*, depending on 3 positional and 3 velocity coordinates.

The distribution function contains essentially all necessary information for a statistical description of a gas. In principle, it can be determined with the help of the *Boltzmann equation*

$$\frac{\partial f}{\partial t} + \vec{v}\, \nabla f + \vec{a}\, \nabla_v f = \left(\frac{\delta f}{\delta t}\right)_{collisions} \tag{2.88}$$

where ∇ and ∇_v are the nabla operators in configuration and velocity space, respectively, $\vec{a}$ is the externally acting acceleration, and $(\delta f/\delta t)_{collisions}$ is the temporal change in the distribution function resulting from two-body elastic collisions. Unfortunately, no general, closed solution to this equation, introduced in 1872 by Ludwig Boltzmann, is known up to the present day. Indeed, it is not even known if such a solution exists! An approximate solution is obtained, for example, by expanding f into a truncated orthogonal series and then determining the coefficients of the series after substitution in the Boltzmann equation. This line of analysis will not be pursued further here; the interested reader is referred to the pertinent literature at the end of the chapter. We will instead content ourselves with introducing an extremely important solution of the time and position independent Boltzmann equation, the Maxwellian distribution function, also known simply as the Maxwellian distribution.

Maxwellian Distribution Function. If a gas is close to its equilibrium state, the distribution function will be subject only to slow temporal and spatial changes. To a first approximation, this allows one to neglect the

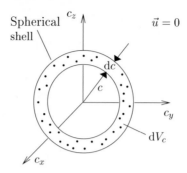

Fig. 2.30. Deriving the distribution function of the velocity magnitudes $h(c)$

collision-induced changes in the distribution function. This abbreviated form of the Boltzmann equation with $(\delta f/\delta t)_{collisions} = 0$ has a known solution, the *Maxwellian distribution function*

$$f_M(\vec{v}) = n \left(\frac{m}{2\pi\,k\,T}\right)^{3/2} e^{-\frac{m\,c^2}{2\,k\,T}} \tag{2.89}$$

where the factors following the density n correspond to the *Maxwellian velocity distribution function* g_M. Note that the distribution function depends on the actual velocity of the gas particles $\vec{v}$ via the relation $c^2 = (\vec{v} - \vec{u})^2$. Before we apply the Maxwellian distribution function for a determination of the escape flux from the atmosphere, we should get a better feel for the velocity distribution described by this function. In particular, we should further investigate the velocity distribution of the particles as compared to the escape velocity. We begin by considering only the *velocity magnitudes (speeds)*, independent of the direction taken by the particles. Under this limitation we thus ask ourselves how many gas particles $dN(c)$ in a gas at rest ($\vec{v} = \vec{c}$) have random speeds in the interval between c and $c + dc$.

Since $g_M = f(c^2)$ is a spherically symmetric function of c, the (differentially small) fraction $dN(c)/\delta N$ of gas particles with speeds between c and $c + dc$ corresponds to the amount of points in velocity space lying within a spherical shell of radius c, and thickness dc, i.e. within the volume dV_c; see Fig. 2.30. This means that

$$dN(c)/\delta N = g_M dV_c = g_M\, 4\pi\,c^2\,dc \tag{2.90}$$

In order to derive a generally applicable expression, independent of the size of the velocity interval dc, we divide both sides of the above equation by dc and obtain

$$h(c) = \frac{dN(c)/\delta N}{dc} = 4\pi \left(\frac{m}{2\pi\,k\,T}\right)^{3/2} c^2\, e^{-\frac{m\,c^2}{2\,k\,T}} \tag{2.91}$$

This so defined *distribution function of the velocity magnitudes* $h(c)$ denotes the fraction of gas particles with speeds between c and $c + dc$ under thermal

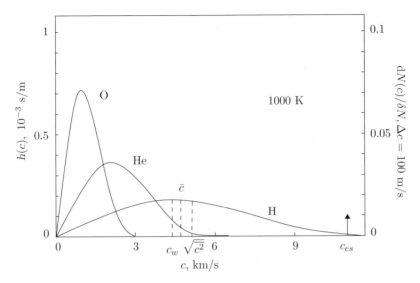

Fig. 2.31. Distribution functions of the velocity magnitudes $h(c)$ for atomic oxygen, helium and atomic hydrogen at a temperature of $T = 1000$ K. Also shown are the most probable velocity c_w, the arithmetic mean velocity $\bar{c}$, the root mean squared (RMS) velocity $\sqrt{\overline{c^2}}$, as well as the escape velocity at the exobase height c_{es}.

equilibrium conditions *normalized to the width of the given velocity interval* dc. Since fractions correspond to probabilities, $h(c)$ also represents a probability density dw/dc. As such, $h(c)$ gives the probability that a particle has a speed in the range c to $c + \mathrm{d}c$, again normalized to the width of the given velocity interval.

Regarding the graphic representation of this function, we note that $h(c)$ is always positive, possesses two zeroes at $c = 0$ and $c \to \infty$, increases monotonically proportional to c^2 at small c, decreases monotonically as $\exp(-c^2)$ at large c, and must therefore have a maximum somewhere in between. The position and amplitude of this maximum is dependent on m as well as on T. Figure 2.31 illustrates this behavior using examples for the three gases atomic oxygen, helium and atomic hydrogen at a temperature $T = 1000$ K.

The position of the maximum of each curve, which also gives the most frequently encountered and thus *most probable speed* c_w, may be determined by differentiating $h(c)$ (or, more easily, the logarithm of $h(c)$) with respect to c and setting the result to zero. One obtains

$$c_w = \sqrt{2\,k\,T/m} \tag{2.92}$$

Because of the asymmetric nature of $h(c)$, the mean speed $\bar{c}$ is not equal to the most probable speed c_w. By definition

$$\bar{c} = \int_0^\infty c \left(\frac{\mathrm{d}N(c)}{\delta N} \right) = \int_0^\infty c\, h(c)\, \mathrm{d}c \tag{2.93}$$

The mean value is calculated here in the usual way by multiplying each possible speed by its occurrence frequency (i.e. its probability) and then summing over all values of c. One obtains

$$\bar{c} = \frac{4}{\sqrt{\pi}} \frac{1}{c_w^3} \int_0^\infty c^3 \, e^{-(c/c_w)^2} \, dc$$

Making use of the substitution $x = (c/c_w)^2$ and then integrating by parts, or directly with the help of Eq. (A.11), the above integral is found to yield the value $c_w^4/2$. The *mean speed* is therefore given by

$$\bar{c} = \sqrt{8 \, k \, T / \pi \, m} \tag{2.94}$$

Analogously, and in agreement with our definition of temperature, we calculate the *root mean squared (RMS) speed* from the formula

$$\sqrt{\overline{c^2}} = \sqrt{\int_0^\infty c^2 \, h(c) \, dc} = \sqrt{3 \, k \, T / m} \tag{2.95}$$

In general

$$\sqrt{\overline{c^2}} = 1.085 \, \bar{c} = 1.224 \, c_w \tag{2.96}$$

and one makes only a small error when, as done earlier, $\bar{c}$ is approximated by $\sqrt{\overline{c^2}}$. In any case, even for hydrogen, the mean velocity is considerably smaller than the escape velocity, so only a small fraction of these particles will have enough energy to overcome the gravitational attraction. Denoting this fraction by b_{es}, it may be calculated from

$$b_{es} = \int_{c_{es}}^\infty dN(c)/\delta N = \int_{c_{es}}^\infty h(c) \, dc = \frac{4}{\sqrt{\pi}} \int_y^\infty x^2 \, e^{-x^2} \, dx$$

where we have introduced the variable $x = c/c_w$ and the constant $y = c_{es}/c_w$. Integrating by parts according to Eq. (A.4) with

$$u = -x/2 \,, \quad v' = -2x \, e^{-x^2} \,, \quad v = e^{-x^2}$$

yields

$$b_{es} = \frac{2}{\sqrt{\pi}} y \, e^{-y^2} + \frac{2}{\sqrt{\pi}} \int_0^\infty e^{-x^2} \, dx - \frac{2}{\sqrt{\pi}} \int_0^y e^{-x^2} \, dx$$

$$= \frac{2}{\sqrt{\pi}} y \, e^{-y^2} + 1 - \mathrm{erf}(y)$$

Here, the first integral is calculated with the help of Eq. (A.9) and the second integral corresponds by definition to the error function, see Eq. (A.10). Inserting typical values for c_{es} ($\simeq 11$ km/s) and c_w ($\simeq 4$ km/s for H at 1000 K) yields an estimate of the relative number of particles with speeds $\gtrsim c_{es}$ equal to $b_{es} \lesssim 2 \cdot 10^{-3}$, indeed a rather tiny fraction.

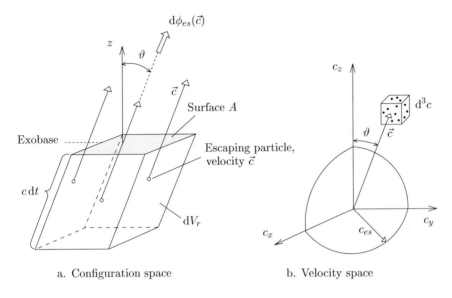

a. Configuration space   b. Velocity space

Fig. 2.32. Calculating the escape flux

2.4.4 Escape Flux and Stability of the Atmosphere

In spite of the low escape probability of the hydrogen atoms, a non-negligible fraction of the hydrogen atmosphere will be lost over an extended time. In order to estimate the length of time necessary for the entire available hydrogen atmosphere to evaporate, we must first determine the magnitude of the escape flux.

To simplify the calculation we assume that the exobase is a sharp boundary that separates the inner *collision dominated* atmosphere from the outer *collisionless* atmosphere. Since the collision frequency decreases exponentially with height, this assumption is not a substantial limitation of our estimate. Let the collision dominated region of the inner atmosphere be described by a Maxwellian distribution function and assume that particles can only escape if they penetrate the exobase in the upward direction (with arbitrary inclination) and possess a velocity greater than the escape velocity. To calculate the escape flux, we consider a surface A at the exobase height as sketched in Fig. 2.32a. Clearly, particles with velocities $\vec{c}$ in d^3c will only pass through the surface A in the time interval dt if they are located within the indicated parallelepiped, bounded at the top by the surface A. The direction and length of this parallelepiped are determined by the direction and magnitude of the velocity vector $\vec{c}$. Particles outside of this volume dV_r will either not pass through the surface A or not reach A in the time dt. The number of particles in the parallelepiped with velocities in d^3c is

$$dN(\vec{c}) = f_M \, dV_r \, d^3c = f_M \, A \, c \, dt \, \cos\vartheta \, d^3c$$

so that their contribution to the flux through the surface A is

$$d\phi(\vec{c}) = dN(\vec{c})/A \ dt = f_M \ c \ \cos\vartheta \ d^3c$$

In order to obtain the total flux through A, we must integrate over the different contributions of the velocity volumes d^3c in velocity space, restricting the integration to the space $c_z > 0$ and $c > c_{es}$ (see Fig. 2.32b). The form of the integrand and the hemispherical boundary for the range of integration suggest the use of spherical coordinates. With $d^3c = c^2 \sin\vartheta \ d\varphi \ d\vartheta \ dc$ we have

$$\phi_{es}(h_{EB}) = n(h_{EB}) \left(\sqrt{\pi} \ c_w\right)^{-3} \int_{\varphi=0}^{2\pi} \int_{\vartheta=0}^{\pi/2} \int_{c_{es}}^{\infty} c^3 \, e^{-(c/c_w)^2} \cos\vartheta \ \sin\vartheta \ d\varphi \ d\vartheta \ dc$$

The integration over the angles φ and ϑ are easily performed and yield the values 2π and $1/2$, respectively. The integral over the velocity c may be calculated with the help of the substitution $x = (c/c_w)^2$ and Eq. (A.8) and one obtains as a complete solution the so-called *Jeans escape flux*

$$\phi_{es}(h_{EB}) = n(h_{EB}) \frac{c_w}{2\sqrt{\pi}} \, e^{-X} \, (1 + X) \tag{2.97}$$

where we have introduced an abbreviation X called the *escape parameter*

$$X = \left(\frac{c_{es}}{c_w}\right)^2_{EB} = \frac{g_{EB} \, r_{EB} \, m}{k \, T_\infty} = \frac{r_{EB}}{H_{EB}} \tag{2.98}$$

X is an important characteristic parameter of planetary and lunar atmospheres. The larger the value of X, the smaller is the escape flux. As an example, for hydrogen at Jupiter $X \simeq 1400$ and the escape rate is correspondingly low. At Mercury, on the other hand, $X \simeq 2$ and the evaporation process is quite effective. For atomic hydrogen in the terrestrial atmosphere, assuming an exobase temperature of 1000 K and a scale height of 960 km, the escape parameter becomes $X \simeq 7$. This results in an escape flux of the order $\phi_{es}^{H}(1000 \text{ K}) \simeq 10^{12} \text{ m}^{-2}\text{s}^{-1}$, whereby we have taken a hydrogen density at the exobase height of n_H (420 km) $\simeq 10^{11}\text{m}^{-3}$ (see Appendix A.4).

How long would it take this escape flux to completely remove the hydrogen from the terrestrial atmosphere? To answer this question we equate the entire loss of hydrogen by evaporation over the time interval τ_{es} to the total hydrogen content in the Earth's atmosphere. This yields

$$\phi_{es} \ 4\pi \ r_{EB}^2 \tau_{es} \simeq N_H(h = 0) \ 4\pi \ R_E^2 \simeq n_H(h = 0) \ H(h = 0) \ 4\pi \ R_E^2$$

where we have considered the escape flux as constant to a first approximation. Neglecting the relatively small difference between r_{EB} and R_E, we obtain the *escape time* as

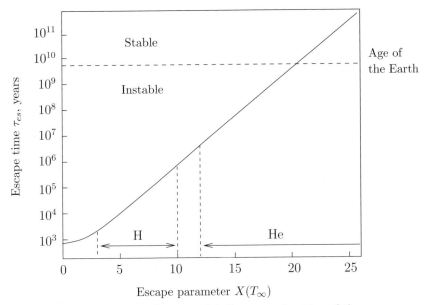

Fig. 2.33. Escape times of hydrogen and helium as a function of the escape parameter. The indicated ranges of X refer to different thermopause temperatures (adapted from Blum, personal communication).

$$\tau_{es} \simeq n_{\mathrm{H}}(h=0)\; H(h=0)\; /\; \phi_{es}$$

The entire available reserve of hydrogen in free and bound forms has a relative fraction of about 10^{-5} in the lower atmosphere. Assuming a scale height of 8 km and an escape flux of $10^{12}\mathrm{m}^{-2}\mathrm{s}^{-1}$, one obtains an escape time of $7 \cdot 10^4$ years, i.e. significantly less than the age of the Earth; see Fig. 2.33. Conservation of the hydrogensphere thus requires continual replenishment of hydrogen atoms from the Earth. On the other hand, the helium component of the Earth's atmosphere, with $X(1000\ \mathrm{K}) \simeq 28$, is quite stably bound.

It should be mentioned here that the Jeans theory of evaporation considerably underestimates the actually possible escape flux, which is determined essentially by nonthermal escape processes. On the other hand, only that number of particles can evaporate as can be supplied by the atmosphere below the exobase height via eddy and molecular diffusion. For hydrogen these processes deliver roughly 10^{12} atoms $\mathrm{m}^{-2}\mathrm{s}^{-1}$, and this represents a rigid upper bound on the possible escape flux.

2.4.5 Exospheric Density Distribution

At a minimum, when extrapolating the barometric density distribution into the exosphere, one must account for the height dependence of Earth's gravity. Replacing the height by the geocentric distance (the atmosphere can no longer be considered as planar in this range of distances), the aerostatic equation of a single gas assumes the following form

$$\frac{dp_i}{dr} = -\rho_i \frac{G\ M_E}{r^2} \tag{2.99}$$

Since the temperature in the exosphere corresponds approximately to the thermopause temperature, this relation may be written as follows

$$\frac{dn_i}{dr} = -\frac{m_i\ G\ M_E}{k\ T_\infty} \frac{n_i}{r^2} \tag{2.100}$$

Separation of the variables and subsequent integration leads to the density profile

$$n_i(r) = n_i(r_0)\ \exp\left\{\frac{m_i\ G\ M_E}{k\ T_\infty}\left(\frac{1}{r} - \frac{1}{r_0}\right)\right\} \tag{2.101}$$

Alternatively, this expression may be written as

$$n_i(r) = n_i(r_0)\ e^{-(r-r_0)/H_i(r)} \tag{2.102}$$

where the scale height $H_i(r)$ now depends on the geocentric distance

$$H_i(r) = \frac{k\ T_\infty\ r}{g(r_0)\ m_i\ r_0} = H_i(r_0)\ \frac{r}{r_0} \tag{2.103}$$

Evidently, the weakening gravitational attraction at larger distance results in an increasingly slower decrease in density. In a semi-logarithmic plot, as shown in Fig. 2.12 and more clearly in Fig. 2.36, this is revealed as a curvature in the density profile.

Equation (2.101) is a sufficiently good representation of the density profile for gases up to an altitude of a few 1000 km. This is especially true for the heavier gases atomic oxygen and helium, which have small escape rates. In the higher exosphere, however, the density profile is increasingly influenced by the evaporation of hydrogen atoms and the barometric law loses its validity. This follows immediately from its asymptotic behavior

$$n_i(r \to \infty) = n_i(r_0)\ e^{-r_0/H_i(r_0)} = \text{const.} \tag{2.104}$$

Since a constant density would correspond to an infinite column density and an infinitely extended atmosphere, this asymptotic value is surely unrealistic.

Why one would indeed expect the density in the exosphere to decrease more rapidly than predicted by the barometric law is illustrated by the following line of thought. Similar to a rocket thruster, the exosphere constantly propels material as an escape flux into interplanetary space. The resulting propulsion force produces an recoil acceleration of the remaining exospheric gases in the direction of Earth, which is added to the gravitational acceleration. This enhanced effective acceleration, however, leads to a decrease in the

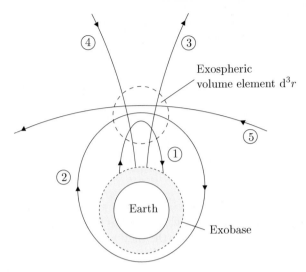

Fig. 2.34. Orbits of exospheric particles passing through a volume element d^3r in ordinary or configuration space

scale height and thus to a more rapid decrease in density. In order to obtain a more exact description of the expected deviations from the barometric law, we consider the partial densities associated with the different orbital characteristics of exospheric particles.

Orbits of exospheric particles and associated partial densities. The atmosphere can be considered collisionless to a good approximation at exospheric heights. At an altitude of 3000 km, for example, the mean free path of hydrogen atoms is greater than 200 000 km. The gas particles correspondingly move undisturbed on Keplerian orbits in the gravitational field of the Earth. These orbits are ellipses (circles) or hyperbolas (parabolas), depending on whether their velocity is smaller or larger than the escape velocity. Since $m \ll M_E$, the focus of these curves is located at the Earth's center of mass. Let us first consider particles with velocities less than the escape velocity. Suppose their closest approach to Earth along their elliptical orbit (i.e. their *perigee*) is located below the exobase within the dense atmosphere. These particles have evidently traversed the exobase on their upward trajectory, but lack sufficient energy to escape and thus fall back into the atmosphere. These particles, denoted with the index 1, are called ballistic particles because their trajectory is similar to that of a thrown ball or artillery shell; see Fig. 2.34 and the accompanying Table 2.3. On the other hand, if the perigee of their elliptical orbit is located in the exosphere, these must be particles which have experienced a rare collision after crossing the exobase and have been catapulted into an orbit around the Earth. These particles, denoted by the index 2, are thus called orbiting or satellite particles.

Table 2.3. Particle orbits in the exosphere and associated partial densities

No.	Orbit Type	Orbit Curve	Origin	Destination	Partial Density
1	Ballistic	Ellipse	Barosphere	Barosphere	n_1
2	Satellite	Ellipse	Exosphere	Exosphere	n_2
3	Escape	Hyperbola	Barosphere	Heliosphere	n_3
4	Capture	Hyperbola	Heliosphere	Barosphere	n_4
5	Transitory	Hyperbola	Heliosphere	Heliosphere	n_5

Of those particles with velocities equal to or greater than the escape velocity, we first consider those for which the orbits traverse the exobase. Should these be moving away from the Earth, they are clearly escaping (evaporating) particles and are denoted with the index 3. If they are moving toward the Earth, however, these must be interplanetary hydrogen atoms that become captured by the Earth and are denoted by the index 4. Finally, we may also find particles with velocities greater than or equal to the escape velocity, but for which the orbits do not intersect the exobase surface. These must be transitory particles from interplanetary space that fly through the outer atmospheric envelope and then vanish again into interplanetary space. These particles are assigned the index 5.

All of the above mentioned particle types will thus contribute to the density of any given volume element in the exosphere, each population associated with its particle density n_i. These contributions, however, will be of vastly different proportions. The hydrogen density in interplanetary space, for example, is so small that the partial densities associated with capture and transitory particles can be neglected to a very good approximation. Moreover, the production rate of satellite particles is small and their loss rate (e.g. by photoionization) is substantial, so their contribution to the density can also be neglected to a first approximation. The only remaining constituents are the ballistic particles, which deliver the main contribution to the exospheric density, and the escape (evaporating) particles, which gain in importance with increasing height. In order to estimate the sum of the partial densities of these two particle populations, we require knowledge of the distribution function valid in the exosphere.

Exospheric Distribution Function and Density Profile. In general, the sum of all velocity points in velocity space corresponds exactly to the number of particle in the corresponding volume element of configuration space

$$\delta N = \int_{VS} \mathrm{d}N = \mathrm{d}^3 r \int_{VS} f \, \mathrm{d}^3 v$$

where the integration is taken over the entire velocity space (VS). As such, it is obvious that the density of a gas is related to its distribution function by

$$n = \int_{VS} f\,\mathrm{d}^3 v \tag{2.105}$$

Within the collision dominated barosphere, the distribution function has the following form

$$f_{baro}(r,v) = n_{baro}(r)g_M(v)$$

where $n_{baro}(r)$ denotes the density profile arising from the barometric law under isothermal conditions, and $g_M(v)$ is the Maxwellian velocity distribution function. In contrast to the Maxwellian distribution, this special form of the *Maxwell-Boltzmann distribution function* is dependent on position. It may be easily shown that $f_{baro}(r,v)$ is a solution of the position dependent (but not time dependent) Boltzmann equation. Accounting for the spherical symmetry of both the density and the velocity distributions and assuming an atmosphere at rest ($\vec{v} = \vec{c}$), we obtain

$$\vec{v}\,\nabla f_{baro} + \vec{g}\,\nabla_v f_{baro}$$

$$= g_M\,c\,\frac{\partial}{\partial r}\left(n(r_0)\mathrm{e}^{-(r-r_0)/H}\right) + n_{baro}\,g\,\frac{\partial}{\partial c}\left(\left(\frac{m}{2\pi kT}\right)^{3/2}\mathrm{e}^{-mc^2/2kT}\right) = 0$$

A verification of Eq. (2.105) also yields the expected result

$$n(r) = n_{baro}(r)\int_{VS} g_M(v)\,\mathrm{d}^3 v$$

$$= n_{baro}(r)\int_{\vartheta=0}^{2\pi}\int_{\varphi=0}^{\pi}\int_{c=0}^{\infty}\left(\frac{m}{2\pi kT}\right)^{3/2}\mathrm{e}^{-\frac{m\,c^2}{2\,k\,T}}\,c^2\sin\vartheta\,\mathrm{d}\varphi\mathrm{d}\vartheta\mathrm{d}c = n_{baro}(r)$$

The situation is different for the case of the exosphere. Here we can no longer assume that the velocity distribution function corresponds to a complete Maxwellian distribution. Otherwise, for reasons of symmetry, just as many particles would have to be captured as those which escape. Accordingly, in the case of the exosphere there are some regions in velocity space that are practically empty and can thus be neglected during the integration over velocity space. These are the regions occupied by the capture particles, the transitory particles, and the satellite particles. Assuming further that the velocity space in the remaining regions (ballistic and escaping particles) can be described by a Maxwellian distribution (which implies that the density at the exobase height corresponds to a complete Maxwellian distribution), we obtain

$$n_{exo}(r) = n_{baro}(r)\int_{VS'(r)} g_M(v)\,\mathrm{d}^3 v \tag{2.106}$$

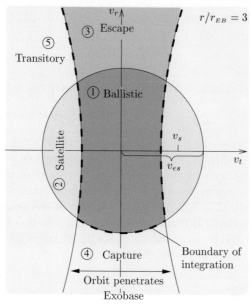

Fig. 2.35. Orbit types of exospheric particles and their corresponding ranges in velocity space for an exospheric volume element at a geocentric distance of $r = 3\,r_{EB}$ ($h \simeq 14\,000$ km). $v_s(= \sqrt{g\,r})$ is the velocity of a satellite particle in a circular orbit at this height and $v_{es}(= \sqrt{2g\,r})$ is the escape velocity at the same height. The dashed line delineates the range of integration $VS'(r)$ used for the calculation of the exospheric density. The complete regions in velocity space, actually three dimensional, are obtained by a rotation about the v_r-axis (adapted from Fahr and Shizgal, 1983; note that the references given here and in the following figure captions are summarized in Appendix B).

where $VS'(r)$ is the restricted volume of integration. The exospheric density evidently becomes smaller than the barospheric density because of this restriction. In order to carry out the integration, each orbit type (see Fig. 2.34) contributing to the density in the configuration space element d^3r must be associated with a corresponding region in velocity space. These associations are illustrated in Fig. 2.35 for an exospheric volume element at a distance of $r = 3\,r_{EB}$ (i.e. $h \simeq 14\,000$ km). Since we have a symmetrical situation in the horizontal direction, it is sufficient in this case to use a two-dimensional description, with one velocity component in the radial direction (v_r) and the other in the horizontal or transverse direction (v_t). Ballistic particles and satellite particles have velocity vectors in this velocity plane with coordinates inside a circle of radius $v = \sqrt{v_r^2 + v_t^2} = v_{es}$. The velocity coordinates of the other three particle populations lie outside this circle, whereby escaping particles are all in the upper half of the diagram (positive v_r) and capture particles are all in the lower half (negative v_r). Furthermore, it can be shown

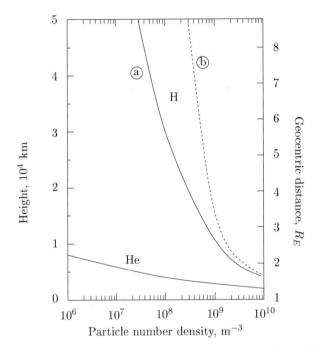

Fig. 2.36. Density profile in the exosphere for a temperature of 1000 K. Curve (a) denotes the sum of the ballistic and escape components ($n = n_1 + n_3$); curve (b) denotes the extrapolated barometric profile ($n = n_1 + n_2 + n_3 + n_4 + n_5$) (adapted from Banks and Kockarts, 1973).

that only particles with velocity coordinates located in a region bounded by two mirror-symmetric hyperbolas can traverse both the volume element d^3r as well as the exobase surface. The form and position of these two hyperbolas are dependent on the distance (height) of the exospheric volume element considered. At greater distances, for example, the trajectories of the particles traversing the volume element must be more closely directed toward Earth in order to penetrate the exobase.

With the help of these bounding curves the orbit types sketched in Fig. 2.34 can be associated with regions in the velocity plane shown in various degrees of shading in Fig. 2.35. Satellite particles, for example, all lie inside the circle, but outside the region bounded by the twin hyperbolas. Escaping particles all lie outside the circle, but inside the bounding hyperbolas and their velocities are directed away from the Earth (positive v_r component). The complete three-dimensional form of the regions is obtained by rotating the velocity plane about the v_r-axis.

This association also defines the boundary of the velocity space $VS'(r)$ over which the integral of Eq. (2.106) extends; see dashed line in Fig. 2.35. The integration yields the sum of the partial densities n_1 and n_3, which is plotted as curve (a) in Fig. 2.36 as a function of height. The extrapolated barometric hydrogen density is shown as curve (b) for comparison. It is seen that the barometric law describes the density profile quite well up to an altitude of 5000 km, but the deviation grows steadily toward greater heights, reaching a factor 10 at an altitude of 50 000 km.

Direct experimental determination of the hydrogen density in the exosphere is difficult because of the low density and energy of the particles. Indirect density information may be obtained from the radiation emitted by these particles. This is sunlight scattered from the hydrogen atoms, and is particularly intensive in the resonantly scattered Lyman-α radiation at 121.6 nm. Figure 7.16a shows an image of the exosphere at this wavelength, recorded in 1972 from the surface of the Moon by the APOLLO 16 team. The Earth seems to be embedded in a fluorescent shell, an apparition that explains the origin of the name for this region, the *geocorona*. An indirect determination of the hydrogen density from such images essentially confirms the density profile based on the ballistic and escape components $n_1 + n_2$ shown in Fig. 2.36.

Exercises

2.1 Atomic oxygen is the primary constituent in the thermosphere of Mars at heights above ca. 200 km (planetary data in Table A.3). The gas temperature there has almost reached its thermopause value of 200 K.

(a) Calculate the relative decrease in density in this region for an increase in altitude of 100 km. Assume the gravitational acceleration to be constant and equal to its value at 200 km.

(b) How large is the relative error if one were to calculate the oxygen density at a height of 3000 km without accounting for the height dependence of the gravitational acceleration? Assume again a reference height of 200 km.

2.2 Estimate the density scale height at an altitude of 150 km using the data given for the model atmosphere in Table A.4 and compare your result with the scale height listed in this table.

2.3 Calculate the heterospheric pressure scale height for thermodynamic equilibrium conditions ($T_i = T$).

2.4 A spherical satellite (1 m radius, 100 kg mass) orbits the Earth in the equatorial plane at a height of 300 km.

(a) Show that the Knudsen number ($\simeq$ mean free path divided by the dimension of the satellite) at this height is much larger than unity so that the flow of air ($T \simeq 1000$ K) onto the satellite corresponds to a free molecular flow (i.e. no shock wave is formed in front of the satellite).

(b) Calculate the velocity of the satellite (and thus approximately the flow velocity of the atmospheric gases), as well as its orbital period and its total (i.e. kinetic plus potential) energy.

(c) How large is the change in momentum incurred by an oxygen atom, on the average, during the impact with the ideally reflecting satellite? What is the corresponding change of momentum incurred by the satellite during its flight through the upper atmosphere and how large is the associated drag force?

(d) The temporal change in the satellite's total energy corresponds to the energy expended during the atmospheric deceleration (= drag force times satellite velocity). From this one may derive an expression for the temporal change in the satellite's orbital radius. What is the daily loss in altitude?

2.5 Saturn's moon Titan ($M = 135 \cdot 10^{21}$ kg, $R = 2575$ km) has an extraordinarily dense nitrogen atmosphere. The particle number density at an altitude of 1000 km reaches a value of nearly $n_{N_2} \simeq 10^{16}$ m^{-3}. An almost constant temperature of ca. 186 K prevails in this height region.

(a) Calculate the height of the exobase for molecular hydrogen ($\sigma_{H_2,N_2} \simeq 2 \cdot 10^{-19}$ m^2).

(b) Compare the random and escape velocities for H_2 molecules at this height. What is the fraction of escaping particles?

(c) Compute the Jeans escape flux and the escape time for the H_2 component of the Titan atmosphere. Let the surface temperature be 94 K, the surface pressure be 1496 mbar and the surface abundance of H_2 be 0.2% (primarily in bounded form).

References

Kinetic Theory of Gases

S. Chapman and T.G. Cowling, *The Mathematical Theory of Non-Uniform Gases*, Cambridge University Press, Cambridge, 1970

T.I. Gombosi, *Gaskinetic Theory*, Cambridge University Press, Cambridge, 1994

Upper Atmosphere

C.O. Hines, I. Paghis, T.R. Hartz and J.A. Fejer (eds.), *Physics of the Earth's Upper Atmosphere*, Prentice-Hall, Englewood Cliffs, NJ, 1965

R.M. Goody and J.C.G. Walker, *Atmospheres*, Prentice-Hall, Englewood Cliffs, NJ, 1972

P.M. Banks and G. Kockarts, *Aeronomy A / B*, Academic Press, New York, 1973

H.J. Fahr and B. Shizgal, Modern exospheric theories and their observational relevance, *Rev. Geophys. Space Phys.*, *21*, 75, 1983

J.W. Chamberlain and D.M. Hunten, *Theory of Planetary Atmospheres*, Academic Press, Orlando, 1987

M.H. Rees, *Physics and Chemistry of the Upper Atmosphere*, Cambridge University Press, Cambridge, 1989

T. Tohmatsu and T. Ogawa, *Compendium of Aeronomy*, Kluwer Academic Publishers, Dordrecht, 1990

See also the references in Chapter 1 and the figure references in Appendix B.

3. Absorption and Dissipation of Solar Radiation Energy

While the temperature profile of the upper atmosphere was assumed to be known in the previous chapter, its formation will be explained in the following. This requires that we understand the properties of the solar radiation, the primary source of energy for the upper atmosphere, and its absorption by atmospheric gases. The resulting heat input profile and a simple heat balance equation then allows us to understand the observed temperature variations. The spatial and temporal dependence of the solar irradiation leads to temperature, density and pressure differences that stimulate thermospheric winds. To calculate these, we introduce the momentum balance equation, which also serves to derive the fundamental characteristics of atmospheric waves.

3.1 Origin and Characteristics of Solar Radiation

In the following we will be talking about a rather average star. This star does not have a particularly striking size or luminosity, nor does it have an extraordinary surface temperature. Moreover, it is presently in that stage of evolution where stars spend most of their lifetimes (main sequence). Nevertheless this star is by far the most important one for us: it gives us the energy we need to exist and governs our life cycle. It also happens to be the only star we can study from relatively close range, thereby resolving the surface structure and enabling observations of processes that can only be surmised on other stars. We are referring here, of course, to the central celestial body of our solar system, the Sun. We begin by summarizing some fundamental characteristics of this object (solar physics in a nutshell!). Of particular interest beyond the interior structure is the atmosphere of the Sun, which determines the spectral characteristics of the solar radiation. Especially interesting for the terrestrial upper atmosphere are the ultraviolet and X-ray regions of the solar spectrum, which, as shown below, are subject to both systematic and irregular variations.

3.1.1 Interior Structure of the Sun

The Sun, a commonplace, middle-aged, main sequence star, is a gigantic gaseous ball with a radius of ca. 700 000 km; see Table 3.1. The gas ball

Table 3.1. Physical properties of the Sun. It should be noted that the entries for composition and rotation period are somewhat uncertain (adapted generally from Lang, 1992; see figure and table references in Appendix B).

Radius R_S		$6.96 \cdot 10^8$ m ($\simeq 109\ R_E$)
Mass M_S		$1.99 \cdot 10^{30}$ kg ($\simeq 333\,000\ M_E$)
Composition:	particle fraction	91% H, 8% He, 1% $\mathcal{M} > 4$
	mass fraction	72% H, 26% He, 2% $\mathcal{M} > 4$
Mass density:	mean	$1.41 \cdot 10^3$ kg/m^3 ($\simeq 0.25\ \rho_E$)
	at center	$1.5 \cdot 10^5$ kg/m^3
Energy production rate (luminosity)		$3.86 \cdot 10^{26}$ W
Effective radiation temperature		5780 K
Equatorial rotation period:	sidereal	ca. 24.8 days ($\simeq 2.14 \cdot 10^6$ s)
	synodical	ca. 26.6 days ($\simeq 2.30 \cdot 10^6$ s)
Equatorial inclination to the ecliptic		$7° 15'$
Distance from Earth:	mean	$149.6 \cdot 10^9$ m $= 1$ AU
	perihelion (January)	$147.1 \cdot 10^9$ m
	aphelion (July)	$152.1 \cdot 10^9$ m
Solar constant		$1.37 \cdot 10^3$ W/m^2 ($\pm 0.2\%$)
Age		$4.6 \cdot 10^9$ years
Life expectancy (total)		$10 \cdot 10^9$ years

is held intact by the extraordinarily strong gravitational force generated by the great mass of the object. This mass, roughly $2 \cdot 10^{30}$ kg, corresponds to about 300 000 times the mass of the Earth. In fact, 99.9% of the mass in the entire solar system is concentrated in the Sun. According to mass fraction, the Sun contains about 72% hydrogen, 26% helium, and 2% heavy elements such as oxygen, carbon and nitrogen. The dominance of hydrogen becomes more obvious when ordering by particle density fraction: 91% H with an 8% admixture of He. Because of the extremely high temperatures in the solar interior these constituents are all present in a completely ionized state. In other words we are dealing with a gas composed of electrons, protons and α-particles (helium nuclei, He^{++}).

What prevents this gas ball from collapsing into a superdense object under the influence of the powerful gravitational force and forming, for example, a white dwarf star with a diameter of a few 1000 km? Similar to the Earth's atmosphere, it is the gas pressure that opposes the force of gravity. The structure of the solar interior is thus determined by an equilibrium between gravitational and pressure gradient forces; other forces such as radiation pressure play only a secondary role.

In order to generate sufficiently large pressure forces in such massive objects, the gas must attain very high temperatures and densities, particularly in the center where extraordinarily strong compression occurs. Model calculations show that the temperature here reaches ca. $15 \cdot 10^6$ K. Associated with this is an astoundingly high particle density of roughly 10^{32} m^{-3}, which corresponds to a mass density of 150 g/cm^3, i.e., 150 times the density of water. The density does drop rapidly with distance from the center, however, so that the mean density of 1.4 g/cm^3 is only about one-fourth that of the Earth. Owing to the nearly exponential decrease in the density, roughly 75% of the Sun's mass is concentrated in a central core that occupies only 5% of the total volume.

The most important property of stars for us, of course, is that they produce a vast amount of radiative energy. Indeed, this is why we notice them at all at such large distances. Averaged over all wavelengths, the Sun (but by no means every star) radiates at a nearly constant energy rate. The Sun's output, roughly $4 \cdot 10^{26}$ W, is enough energy in one second to satisfy humankind's energy demand for untold ages. Where does this enormous energy come from and how is it produced? We have known since the 1930's that this energy is produced by nuclear fusion inside the Sun. By far the most important fusion process there is the conversion of four protons into one helium nucleus with the mass excess being transformed into free energy (*proton-proton cycle* or for short *pp-cycle* or *pp-chain*)

$$(1) \quad {}^1\mathrm{H} + {}^1\mathrm{H} \xrightarrow{10^{10}\mathrm{a}} {}^2\mathrm{H} + e^+ + \nu_e \ (0.25 \ \mathrm{MeV}), \quad 1.2 \ \mathrm{MeV}$$

$$(2) \quad {}^2\mathrm{H} + {}^1\mathrm{H} \xrightarrow{<10\mathrm{s}} {}^3\mathrm{He} + \gamma, \quad 5.5 \ \mathrm{MeV}$$

$$(3) \quad {}^3\mathrm{He} + {}^3\mathrm{He} \xrightarrow{10^6\mathrm{a}} {}^4\mathrm{He} + 2 \ {}^1\mathrm{H}, \quad 12.9 \ \mathrm{MeV}$$

The first two reactions above occur twice and create the input products for the third reaction. ${}^1\mathrm{H}$ designates a proton, ${}^2\mathrm{H}$ a deuterium or heavy hydrogen nucleus, ${}^3\mathrm{He}$ a helium isotope nucleus of mass number 3, ${}^4\mathrm{He}$ an α-particle, e^+ a positron, ν_e an electron neutrino, and γ electromagnetic radiation in the γ-range. The given reaction times correspond to the inverse collision frequencies and a(nni) stands for years. The energy released during these exothermic reactions is denoted on the far right. 'MeV' stands for megaelectronvolt, a unit of energy that is not officially a part of the *Système International d'Unités*, but is used here in recognition of its long tradition. By using the appropriate prefix (in this case 'Mega') one obtains easily comprehendible numbers and the physical meaning of this unit is also immediately obvious: it is the energy an electron (or any singly ionized particle) gains upon being accelerated through the given voltage difference. Conversion factors for SI units are given in the table of physical constants found on the inside back cover of the book.

The mass excess between the input and output products of the above reaction scheme amounts to about 0.7% of the four involved protons, corresponding to an energy of $E = \Delta m\, c_0^2 \simeq 26.7$ MeV (c_0 is the speed of light). The neutrinos carry away $2 \cdot 0.25$ MeV and we are left with roughly 26.2 MeV available in the form of free energy, most of which appears as γ-rays, i.e. photons, either directly from reaction (2) with ca. 11 MeV, or indirectly via electron–positron annihilation ($e^+ + e^- \rightarrow \gamma$).

It is evident that the speed of the fusion process is determined only by the first reaction, because the fusion of two protons to a deuterium nucleus is an extremely rare event at the prevailing temperature. Recall here that the protons must attain energies of 1 MeV to overcome their mutual repulsive force, but the mean energy of such particles at a temperature of $15 \cdot 10^6$ K is of the order of 2 keV. It is thus only the enormous number of particles in the solar core that renders reaction (1) sufficiently probable, thereby yielding the observed high energy production rate.

The well-known astrophysical problem of the missing neutrinos should also be mentioned. The Sun should produce some 10^{38} neutrinos per second according to the above reaction scheme. These neutrinos pass easily through matter because of their extremely small reaction cross sections. Only a tiny fraction of these solar neutrinos can be detected on Earth from the nuclear reactions they induce. Surprisingly, only about 60% of the theoretically predicted counting rate is measured and this has led to new hypotheses in neutrino physics (neutrino oscillations). Solar neutrinos are particularly interesting to us because they provide a direct link to the solar core.

Another interesting question is how the energy produced in the Sun's core is transported outward. Model calculations show that electromagnetic waves are responsible for this energy transport out to about 500 000 km from the Sun's center. This *radiative transfer* is very similar to the diffusion of particles in a gas. The photons propagate at a speed of c_0 along a zigzag path from one absorption and reemission to another, slowly working their way to the outside. Because of the high ambient density and associated short mean free path, it takes more than 100 000 years on the average before they reach the Sun's surface along their random journey. It is understandable that the numerous interactions also result in a degradation of the photon energy and a simultaneous increase in the number of photons. Relatively few, but highly energetic γ-rays start near the Sun's center and are absorbed by nuclear excitation processes. The subsequent reemission can proceed stepwise, however, distributing the excitation energy among several photons of lower energy.

Moving outward, the mean photon temperature and thus the gas temperature continually decrease, slowly at first but then increasingly faster because of unrestrained radiative escape at the Sun's surface. As the temperature

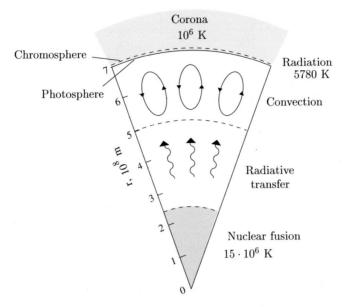

Fig. 3.1. Cross section through the Sun

gradient becomes too steep, the gas layer structure becomes instable. This instability can be explained as follows. Consider a parcel of gas that gets displaced from its state of rest toward the solar surface. The resultant decrease in pressure leads to expansion and cooling of the gas. If the ambient temperature decreases more slowly than that corresponding to the assumed adiabatic cooling, the gas parcel will now be heavier than its environment and will fall back downward to its original location (see also Section 3.5.3). Barometric density distributions are evidently stable with respect to disturbances when the temperature decreases slowly (and, of course, for positive temperature gradients). It is a different story when the rate of temperature decrease becomes more rapid than that corresponding to the adiabatic cooling rate. In this case the disturbed gas parcel remains warmer than its environment and the buoyancy drives it continually outward. At the same time, cooler gases will flow downward because they stay cooler than their environment in spite of the compressional heating. This leads to the formation of large-scale cyclic motions, or convection cells, as illustrated schematically in Fig. 3.1 (an entire spectrum of turbulent motions is excited in reality). Transport of energy occurs alongside these turbulent gas motions because the heat content of these gases is also moved from one location to another. This *convective heat transport* has a number of everyday analogues such as the hot air stream from a hair dryer or the heat transport of a fluid in a central heating system. Convective heat transport begins to assume control of the radial energy flux in the Sun at a distance of about 500 000 km (0.74 R_S) from the center and

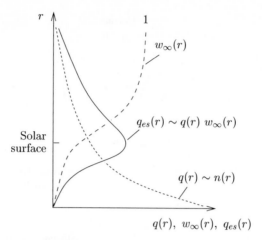

Fig. 3.2. Formation of the photospheric emission layer. $q(r)$ denotes the photon production rate and is assumed proportional to the gas density $n(r)$. $w_\infty(r)$ is the escape probability, for which the height profile is described by Eq. (3.4) in the simplest case. Finally, $q_{es}(r)$ is the production (emission) rate of *escaping* photons.

remains dominant out to the surface. Hot gases flow upward and cooler gases downward, much like the motion observed on a much smaller scale in the lower layers of our atmosphere.

Because the Sun is a gaseous object, its 'surface' is a matter of definition. It is entirely natural, however, to associate the surface with the limb of the visible solar disk. This boundary is very sharply defined, implying that the light must come from a very thin emission layer. This is denoted the *photosphere* and we consider its level of maximum radiative emission as the surface of the Sun. How such a photospheric emission layer is created may be understood as follows. It is immediately plausible that the intensity of the photon flux reaching our eyes is proportional to the production rate of these photons in the Sun. As a working hypothesis, we assume that this production rate is proportional to the density of the solar gas, i.e., every gas particle is considered to be a potential photon emitter. The photon production rate will thus drop rapidly with increasing distance from the Sun. On the other hand, the probability that a photon emitted in the solar interior can leave the Sun and reach our eye without reabsorption is essentially null. This probability grows toward unity only in the outer, increasingly thin gas shell. The production of *escaping* photons thus depends on both the photon production rate at the source as well as their probability of escape, resulting in the emission profile sketched in Fig. 3.2.

It can be shown that the thickness of this emission layer is only a few pressure scale heights (a few hundred kilometers), so that the photosphere is indeed only a very thin 'shell'. More exact calculations account for the fact that the primary absorber and emitter of visible light is not – as one might

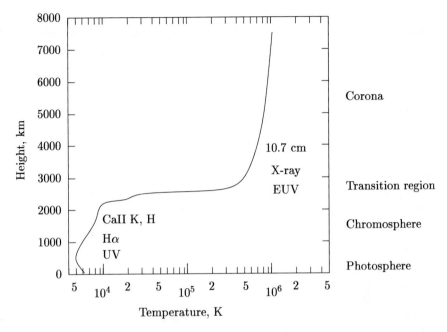

Fig. 3.3. Temperature profile in the solar atmosphere and heights of important radiative emissions. Emission at radio wavelengths is represented by the 10.7 cm radiation (adapted from Lean, 1988).

perhaps expect – neutral hydrogen, but rather the relatively rare (concentration only 1 ppm) negative hydrogen ion H^-. The variation of this rarefied gas with height, however, is also regulated by its production and loss rates.

3.1.2 Solar Atmosphere

The above definition of the solar surface serves equally well for defining the lower boundary of the solar atmosphere. The description of this atmosphere proceeds in a manner similar to that used for the Earth's atmosphere, i.e. by describing the height profiles of the various state parameters. The temperature variation serves again as a convenient guideline for subdividing the atmosphere into different layers. As shown in Fig. 3.3, the temperature characteristically decreases in the photosphere, rises slowly in the chromosphere, undergoes an extremely steep increase in the transition region, and reaches an almost constant value in the corona. Based on the temperature height profile, the photosphere thus corresponds to the terrestrial troposphere, the chromosphere to the stratosphere, and the transition region/corona to the thermosphere. This latter analogy implies that the transition region be considered a part of the corona.

Photosphere. According to our definition of the solar surface, the upper part of the photosphere is also the lowest layer of the solar atmosphere. This layer is about 500 km in thickness and characterized by a decrease in temperature from approximately 6400 to 4200 K. A direct indicator of this temperature decrease is the limb darkening seen in Fig. 3.4b. The lower intensity at the Sun's limb, where the radiation comes from higher layers in the photosphere because of the greater absorption along the longer ray path, can be explained only by a lower temperature, see Eq. (3.14). Analogous to the terrestrial troposphere, the temperature decrease is explained by radiative energy loss.

Just like the solar interior, the photosphere consists mostly of hydrogen and helium, except these gases now exist in their neutral form. The low degree of ionization ($\lesssim 0.1\%$) comes from the low thermal energy of the gas particles. A temperature of 6000 K corresponds to a particle energy of ca. 0.8 eV, quite small compared with the ionization energies of hydrogen (13.6 eV) and helium (24.6 eV). The density height profile can be determined using the barometric law, where the surface density is $n(R_S) \simeq 10^{23}$ m^{-3} and the scale height is $H = k\,T\,/\,\overline{m}\,g_S = k\,T\,R_S^2/\,\overline{m}\,G\,M_S \simeq 150$ km at a temperature of 6000 K.

Observing the Sun with a telescope through a haze (or through a dark filter), the solar disk appears featureless except for the remarkable *sunspots*. These amorphous spots, which often form groups and have typical diameters of 1000 to 40 000 km (i.e. up to 3 times the Earth's diameter), appear darker than the bright photosphere; see Fig. 3.4b.

Even as late as the 19th century, sunspots were thought to be part of a dark solar surface seen through gaps in brightly glowing clouds. Today we realize that they are relatively cool regions ($T \simeq 4000$ K) which emit significantly less radiation than their hot surroundings and thus appear dark. The reason for the lower temperature is thought to be a greatly reduced heat input due to strong magnetic fields that retard the convective heat transport below the sunspots. The strong magnetic fields within the sunspots (e.g. a few tenths of a tesla, which is up to ten thousand times the normal photospheric magnetic field) are direct evidence in support of this explanation.

Sunspots can also be used to determine the rotation period of the Sun. The rotation period at low latitudes, observed in a fixed coordinate system, is about 25 days (*sidereal* rotation period; see Table 3.1); the *synodic* rotation period (as observed from the Earth) is about two days longer. This is due to the motion of the Earth about the Sun, which is in the same direction as the solar rotation. It is interesting that the rotation period increases with latitude and is two full days longer at mid-latitudes, $T_{sid}(45°) \simeq 27$ days. This is known as the *differential* rotation of the photosphere. The inclination of the Sun's equator with respect to the ecliptic, i.e. the plane of Earth's orbit about the Sun, can also be determined from the motion of sunspots. It is about 7°.

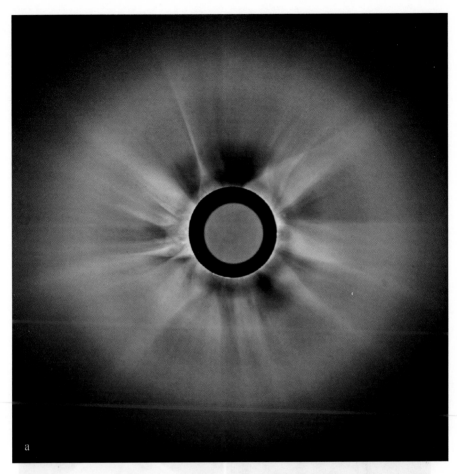

Fig. 3.4. The Sun observed at various wavelengths. (**a**) Corona in the visible, as observed during a solar eclipse. The corona has been occulted near the Sun because of the high radiation intensity there and observed at a longer exposure time in the outer regions. The bright disk in the center is the moon, illuminated by light reflected from Earth (J. Dürst and A. Zelenka, ETH Sternwarte, Zürich, 16 February 1980); (**b**) Photosphere with sunspots in visible light (Baader Planetarium, Mammendorf); (**c**) Chromosphere in the red Hα-Line at 656.3 nm (Kiepenheuer-Institut für Sonnenphysik, Freiburg); (**d**) Prominences (coronal arcs) in the red line of iron (Fe X) at 637.4 nm (National Solar Observatory, Sacramento Peak); (**e**) Corona in the EUV-line of 11-times ionized iron (Fe XII) at the wavelength 19.5 nm. Of course, this and the two following images are color coded representations of the recorded brightness (R. Schwenn, SOHO/EIT - Consortium, 11 September 1997); (**f**) Corona in X-rays at a wavelength of 6.35 nm (L. Golub, Smithsonian Astrophysical Observatory, 11 September 1989); (**g**) Corona at the radio wavelength 2.8 cm (E. Fürst and W. Hirth, Universität Bonn, 24 July 1973)

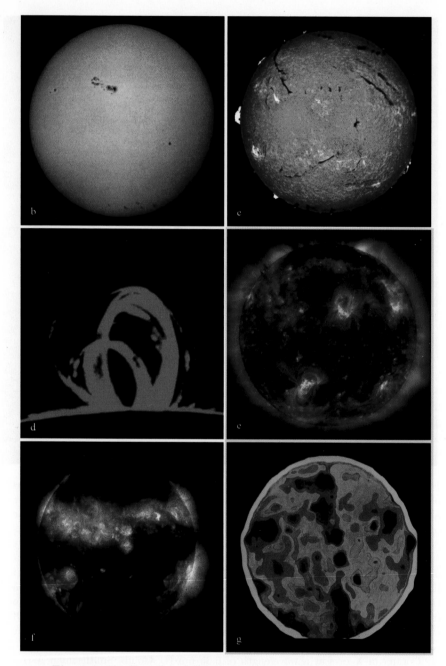

Fig. 3.4. The Sun observed at various wavelengths (continuation)

In order to account for the statistical appearance of sunspots, the relative sunspot number (also called the *Wolf* number) was introduced.

$$R = k \ (10 \ g + f) \tag{3.1}$$

The symbol g denotes the number of groups of spots, f is the number of individual spots, and k serves to standardize the observations (accounting for, among others, the seeing conditions!). As noted in Section 3.1.4, R displays systematic variations that provide information on the state of the Sun.

Chromosphere. The chromosphere begins just above the temperature minimum of the photosphere. The name (*chromos* means color) stems from the fact that it is visible for a short time (only a few seconds) on the solar limb as a thin magenta sliver during a solar eclipse. This roughly 2000 km thick gas layer is characterized by a moderate rise in temperature from about 4200 to 10 000 K. Similar to the stratosphere, this temperature inversion can only be explained by local heating. Absorption of photospheric UV radiation cannot explain this because the overall radiation balance of the chromosphere is positive, i.e. more energy is radiated than absorbed. The most popular scenario advanced today is that this region is heated by the dissipation of atmospheric waves. The photospheric convection excites a spectrum of acoustic waves that propagate outward. The energy associated with these waves is dissipated in the decreasing density of the ambient medium as soon as the wave propagation becomes nonlinear. As demonstrated in Section 3.5.2, the amplitude of an acoustic wave is inversely proportional to the square root of the gas density in the propagation medium. If the density is steadily decreasing, the amplitude of an outward propagating wave must continually grow in order to conserve energy. This is obviously only possible up to a certain limit, at which point nonlinearities set in and the wave is destroyed by viscosity and heat conduction. Directed kinetic energy is thus transformed into heat.

Knowing the temperature, one can determine the density height profile with the help of the barometric law, assuming a density at the lower boundary of $n(500 \text{ km}) \simeq 2 \cdot 10^{21}$ m^{-3}. The fraction of charged particles increases significantly upon approaching the upper boundary and leads to a distinct structuring of the ionized gas distribution by the magnetic fields. Classical examples of these are the *fibrils* and *spicules*, which brighten due to emission from hot ionized gas in magnetic flux tubes.

The chromosphere is observed not only on the Sun's limb, but also from above on the solar disk. This is only possible, however, in narrow-band observations of radiation produced predominantly in the chromosphere. Among these are the well-known red Hα (Balmer-α) radiation of atomic hydrogen at 656.3 nm and the violet lines of singly ionized calcium at 396.9 and 393.4 nm (CaII H-lines and K-lines). Why the Hα radiation is attributed to the chromosphere and not to the underlying photosphere can be understood as follows.

As shown in Fig. 3.2, the intensity of the radiation reaching us is proportional to the production rate of this radiation (and therefore assumed proportional to the density of the emitting gas) on the one hand, and to the escape probability on the other hand. For determining the escape probability we can make the intuitively plausible assumption that, on a differential scale, the probability of absorption is proportional to the length of the ray path. We note in passing that the probability of a photon being absorbed upon propagation through its associated mean free path is one, by definition. Hence the differential probability $dw_{col(lision)}(ds)$ that a photon be absorbed upon traversing the path segment ds is

$$dw_{col}(ds) = ds/l_{ph} \qquad (3.2)$$

where $l_{ph} = 1/\sigma^A n$ is the photon's mean free path and σ^A is its absorption cross section. An illustrative confirmation of this concept may be found in Section 3.2.2; see Eq. (3.21). Correspondingly, the probability $w_{f(ree)}(ds)$ that a photon traverses the path segment ds without absorption is

$$w_f(ds) = 1 - ds/l_{ph}$$

From this we can write the probability that a photon traverses first a segment s and then an additional segment ds without absorption

$$w_f(s + ds) = w_f(s)\, w_f(ds)$$

where we have accounted for the fact that the occurrence probability of two independent events is the product of the probabilities of each separate event. Alternatively, this probability can also be expressed approximately as a Taylor series

$$w_f(s + ds) = w_f(s) + \frac{dw_f}{ds}\, ds + \cdots$$

Equating the above two expressions leads to the differential equation

$$\frac{dw_f(s)}{ds} = -\frac{w_f(s)}{l_{ph}}$$

which can be easily separated and integrated. One obtains

$$w_f(s) = w_f(s_0)\, \exp\left(-\int_{s_0}^{s} ds'/l_{ph} \right) \qquad (3.3)$$

Utilizing this formula for a photon moving vertically upward from the starting height h $(w_f(h) = 1)$ and propagating through an infinitely long distance without absorption, we obtain

$$w_f(h \text{ to } \infty) = w_\infty(h) = \exp\left(-\int_{h}^{\infty} dz/l_{ph} \right)$$

or, using $l_{ph} = 1/\sigma^A n$ and recalling Eqs. 2.78 and (2.79)

$$w_\infty(h) = \exp(-H(h)/l_{ph}(h)) \tag{3.4}$$

Applying this result to the solar atmosphere, the production rate of escaping photons may be written as follows

$$q_{es}(r) = \text{const. } n(r) \, \exp(-H(r)/l_{ph}(r)) \tag{3.5}$$

The position of maximum emission can be obtained by setting the derivative of $q_{es}(r)$ (or more efficiently, the derivative of the logarithm of $q_{es}(r)$) equal to zero. This yields

$$\frac{d(\ln q_{es})}{dr} = \frac{1}{n}\frac{dn}{dr} - \frac{d}{dr}\left(\frac{H}{l_{ph}}\right) = 0$$

Assuming that the scale height is a constant within the relatively narrow emission layer, it follows that

$$\frac{1}{n}\frac{dn}{dr}\left(1 - \frac{H}{l_{ph}}\right) = 0$$

or

$$l_{ph} = H \tag{3.6}$$

The maximum emission thus occurs at the level where the mean free path of the photons becomes equal to the scale height of the absorbing (and reemitting) gas. Referring to Section 2.4.1, however, this condition also corresponds to the definition of the exobase, so the level of maximum radiative emission is essentially an exobase for photons. At the same time, it may be verified that the exobase is that height at which the probability of a vertically upward propagating particle or photon reaching infinite height (and thereby escaping) attains the value $1/e$, see Eq. (3.4).

The magnitude of the photon absorption cross section σ^A is dependent on the wavelength and attains particularly large values at the characteristic transitions of the absorbing gas. This is the case for $H\alpha$ radiation, for example, which excites the first Balmer (α) transition of atomic hydrogen. The hydrogen absorption is so large at the line center that the condition $\sigma^A n H = 1$ from Eq. (3.6) is fulfilled only for lower gas densities at greater heights. This explains why the maximum in the $H\alpha$ emission is shifted from the photosphere to the lower chromosphere.

Upon observing the chromosphere with the help of filters centered on the $H\alpha$ or CaII lines, it is found to be considerably more structured than the photosphere; see Fig. 3.4c. Particularly conspicuous are the brighter emission regions called *plage*(= beach) or *chromospheric faculae*. These are obviously regions of higher temperature. Similar to the sunspots, the activity of the plage regions has also been quantified using indices. Even more remarkable are the elongated structures of strongly reduced radiation called *filaments*. These are tubes of cooler and denser ionized gases confined by magnetic fields that extend high above the chromosphere. Situated between the glowing chromosphere and the observer, they absorb radiation and are thus relatively dark. Seen against the dark sky on the solar limb, their reemitted and scattered light appears as bright *prominences*; see Fig. 3.4d.

Corona. The corona was originally understood to be the radiant ring of brightness around the Sun that was visible for a few minutes during a total solar eclipse. Today we generally consider it to be the outermost, very tenuous and hot, gas envelope of the Sun, the source region of the brightness seen at eclipses. The corona extends from the transition region at a height of 2000 to 3000 km ($\simeq 1.004\,R_S$) up to several solar radii, the upper boundary usually being associated with the beginning of the solar wind regime. Since the transition from a collision dominated, quasi-static inner corona ($\lesssim 2\,R_S$) to a highly dynamic solar wind region above $6\,R_S$ occurs gradually, the location of this outer boundary is a matter of definition. Furthermore, the coronal magnetic field affects the density distribution and dynamics of this region in very different ways. In the following, we set the boundary at the coronal exobase, located at a heliocentric distance of about $3\,R_S$ ($1.5 \cdot 10^6$ km altitude), see Section 6.1.5.

Like the rest of the Sun, the corona is made up mostly of hydrogen with a small amount of helium and traces of heavier elements. The slow decrease of the particle density (large scale height!), the strong Doppler broadening of coronal emission, and the high degree of ionization inferred from the line emission of trace constituents all show, that these gases are very hot with temperatures from 1 to 2 $\cdot 10^6$ K (Alfvén, 1942). The thermal energy of the hydrogen and helium atoms is so much greater than their respective ionization energy that both gases exist in an almost fully ionized state (H^+, He^{++}). The corona is therefore a very hot gas mixture consisting essentially of electrons and protons, a small amount of α-particles, and trace concentrations of highly ionized heavier elements.

Our contemporary knowledge of the corona has been gleaned primarily from images of this gas envelope in various wavelength ranges. Particularly important among these are the images in the visible obtained during a solar eclipse, such as shown in Fig. 3.4a. Roughly 99% of the coronal brightness comes from photospheric light that is scattered from coronal electrons. The scatter process is similar to that occurring in the Earth's atmosphere. Indeed, the blue sky (or generally, the dayglow) is also a result of scattering from photospheric light, albeit in this instance from atmospheric neutral particles. The contributions of various contaminating components to a coronal image are shown in Fig. 3.5. Evidently, if the Sun doesn't just happen to be occulted, one is well advised to observe the corona from outside the Earth's atmosphere, thereby avoiding its strong 'light pollution'. This can be ac-

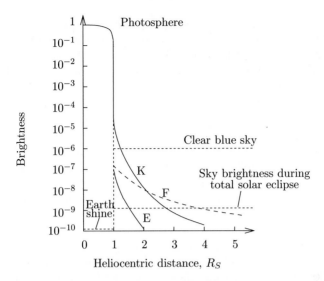

Fig. 3.5. Contributions of various radiation components to coronal brightness. K(ontinuum) stands for coronal scattered light, E(mission) comes from emission lines of coronal ions, and F(raunhofer) is the light scattered from interplanetary dust particles. This latter component can be separated from the K-corona on the basis of its dark absorption lines (Fraunhofer lines). These dark lines, an integral part of the photospheric spectrum, are attributed to absorption in the cool upper photosphere. Their radiation deficit, however, is compensated in the corona by the strongly Doppler-shifted scattering from fast electrons. The relative intensities given above are referred to a wavelength of 500 nm (adapted from van de Hulst, 1953).

complished with satellite borne *coronagraphs* that block out the intensive radiation of the photosphere and chromosphere with an external occulting disk in front of the telescope. Figure 8.24 is an example of a coronal image obtained with this technique.

Because the brightness of the scattered coronal light is proportional to the density of the absorbing and reemitting electrons, coronal images can be used to determine the electron density and its dependence on heliocentric distance. Typical results have the form

$$n_{\mathrm{e}}[\mathrm{m}^{-3}] \simeq \frac{10^{14}}{(r/R_S)^6} + \frac{6 \cdot 10^{11}}{(r/R_S)^2} \quad , \qquad r/R_S \gtrsim 1.3 \qquad (3.7)$$

where n_{e} is the electron density, r the heliocentric distance and R_S the Sun's radius. This must be considered as a mean value in view of the strong structure in the corona. We make use of this expression in Section 6.1.5.

Superposed on the scattered light from the corona are a series of emission lines (E(missions)-corona or L(ine)-*corona*), which at first remained unidentified. For a while one even introduced a new element, 'coronium', in order to explain their origin. We know today that the emission comes from trace gas atoms in very high states of ionization. Classic examples of these are the red line of 9-times ionized iron (Fe X) at 637.4 nm, the green line of 13-times ionized iron (Fe XIV) at 530.3 nm (neutral iron has a total number of 26 electrons) and the yellow line of 14-times ionized calcium (Ca XV) at 569.4 nm (neutral calcium has a total number of 20 electrons). These transition lines arise from the Zeeman splitting of metastable states with correspondingly small differences in energy. Emission from 'regular' transitions of these highly ionized gases lies completely outside the visible range.

The corona can be observed not only tangentially at the solar limb against the dark sky, but also from above against the solar disk. As in the case of the chromosphere, this is only possible in those wavelength ranges dominated by emission from the corona. Among these are the extreme ultraviolet (EUV) radiation, the X-ray radiation and radio emission. The latter is essentially thermal bremsstrahlung, i.e. radiation produced by the deceleration and deflection of thermal electrons as a result of collisions with other particles. Maps of the corona at radio wavelengths can be reconstructed from observations at a ground-based radio telescope. Images in the EUV and X-ray ranges can be obtained only from rockets and satellites, however, because of atmospheric absorption. Figure 3.4e,f and g show images in these three wavelength ranges.

Perhaps the most striking feature of these images is the strong spatial variability of the radiation intensity. We can loosely distinguish three different types of coronal regions: *active regions* with unusually high radiation intensity; the 'quiet' corona with average intensity; and *coronal holes* with below average brightness (by a factor 1/3) that thus appear dark.

Coronal holes are observed preferentially at high latitudes, see Fig. 3.4f, but can also sometimes extend a tongue down into equatorial regions, see Fig. 3.4g. Their low radiation intensity indicates that they are regions of reduced density that can also be detected in the visible, see Fig. 3.4a. Although we will get to know coronal holes later as source regions of the fast solar wind, we direct our attention here to the radiatively active regions.

The rise in temperature in the corona, similar to the earlier situation found in the chromosphere, requires an explanation. Heating by magneto-plasma waves is a much discussed, but possibly insufficient mechanism. Indeed, the energy source of coronal heating (and of stellar coronae in general) is a poorly understood phenomenon of astrophysics. It is worth noting that the amount of energy needed is not proportional to the observed increase in temperature. Much more energy is transported into the chromosphere, for example, than into the corona. Of importance for the energy budget are also the heat capacity of the respective gas layer and the effectiveness of energy loss processes. These interconnections will become clear in Section 3.3.5, where the temperature increase in the thermosphere (a proxy for the corona) is discussed.

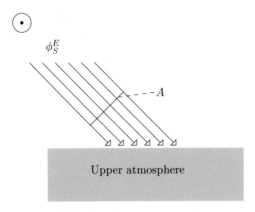

Fig. 3.6. Defining the energy flux of the solar radiation

3.1.3 Radiation Spectrum

The Sun's radiation is of imminent importance to the Earth and its inhabitants. As such, it is advisable to know the characteristics of the solar spectrum as described in the following section. After a short introduction, we focus on the short wavelength range because of its special consequences for the upper atmosphere.

Overview. The Sun's radiation impinges on the Earth to a good approximation in parallel rays. The angle between a ray from the center of the Sun's disk and another ray from the solar limb is only about 1/4 degree. The energy transported by the radiation is described by the radiative energy flux ϕ_S^E. This is the energy per unit time and unit area at the Earth's orbit, but above the atmosphere, which passes through a surface normal to the Sun's rays

$$\phi_S^E = \frac{E}{A\,t} \tag{3.8}$$

see Fig. 3.6. Written in this form, Eq. (3.8) expresses the fact that the radiation field is homogenous and constant in time when averaged over all wavelengths. Because of this temporal invariance, ϕ_S^E is denoted the *solar constant*. Table 3.1 lists this value as 1.37 kW/m². Integrated over a spherical surface with a radius r equal to the mean distance between Sun and Earth ($r = 1$ *astronomical unit* or 1 AU), the total radiated energy of the Sun per second, its luminosity, is calculated to be $L_S = \phi_S^E\,4\,\pi\,(1\,\text{AU})^2 = 3.86 \cdot 10^{26}$ W.

In many situations one is not interested in the entire energy flux, but rather only the energy flux in a certain wavelength interval. In this case it is useful to introduce the *spectral energy flux density*

$$S^E(\lambda) = \frac{\mathrm{d}\phi^E(\lambda)}{\mathrm{d}\lambda} = \frac{\mathrm{d}E}{A\,t\,\mathrm{d}\lambda} \tag{3.9}$$

As indicated by the definition, this describes the energy flux per wavelength interval $d\lambda$, measured at wavelength λ at the Earth's orbit. The fundamental unit of this quantity is thus watt per square meter (of area) per meter (of wavelength), but much smaller, more convenient wavelength units are commonly used in practice, depending on the particular spectral resolution. At radio wavelengths one frequently encounters a spectral energy flux density referred to a frequency interval $S^E(\nu)$, where $\nu = c_0/\lambda$. This may be converted to an energy flux density per wavelength interval using

$$S^E(\nu) = \left| \frac{d\phi^E(\nu)}{d\nu} \right| = \frac{d\phi^E(\nu = c_0/\lambda)}{d\lambda} \frac{\lambda^2}{c_0} = \left(S^E(\lambda) \frac{\lambda^2}{c_0} \right)_{\lambda = c_0/\nu} \tag{3.10}$$

whereby only the absolute magnitude of the frequency and/or wavelength interval is of interest.

If one plots the spectral energy flux density as a function of the wavelength or frequency, one obtains the *radiation spectrum* of the Sun. Figure 3.7 shows this spectrum in units of $Wm^{-2}\mu m^{-1}$, whereby the values again refer to the Earth's orbit at a solar distance of 1 AU. This double logarithmic representation dramatically documents the wide range of the Sun's spectral energy flux density, covering 24 orders of magnitude over 11 decades of wavelength. The usual designations of the wavelength ranges are denoted for better orientation at the top of the plot. We can thus estimate the relative intensities of the Sun's radiation in the X-ray, ultraviolet, visible, infrared and radio ranges. The visible band and part of the radio range have been shaded to indicate that this radiation can pass through the atmosphere and reach the Earth's surface. Radiation in all other wavelength ranges is absorbed at various heights in the Earth's atmosphere.

Figure 3.7 shows that the spectral energy flux density of the Sun attains its maximum of slightly more than $1~kW\,m^{-2}\mu m^{-1}$ in the visible range and decreases rapidly toward longer and (particularly) shorter wavelengths. A conspicuous interruption in this decrease is observed in the ultraviolet and X-ray regions, where the energy flux density stays roughly constant over two wavelength decades. Furthermore, the radiation in these regions undergoes significant time variations. The range of quasi-regular variations is roughly indicated here by the dotted area; the amplitude of short-term outbursts by the dashed line. The variability in these wavelength regions has no significant effect on the total solar flux (solar constant), because the energy flux density there is 4 to 5 orders of magnitude lower than that near the maximum in the visible range. A similar, but less strongly pronounced, weakening of the energy flux density falloff is observed at radio wavelengths. Temporal changes of the energy flux density, with quasi-regular variations of about the same order as those in the EUV range, are also observed at radio wavelengths.

Figure 3.7 also implies that the solar spectrum can be very well approximated over a wide spectral range by the spectrum of a classical 'black body' (or, in our case, a 'black sphere' with the radius of the Sun). 'Black' denotes

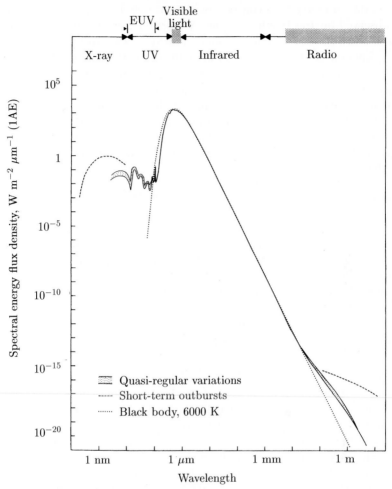

Fig. 3.7. Overview of the solar spectrum. The spectral energy flux density is referenced to a solar distance of 1 AU (adapted from Malitson, 1965).

here an object with ideal absorption and emission characteristics in radiative equilibrium. According to the *Planck radiation law*, radiation emitted from the surface of such a body follows a spectrum of the form

$$S_{BB}^{E}(\lambda) = 2\pi \, h_P \, c_0^2 \, \frac{1}{\lambda^5} \, \frac{1}{\exp(h_P \, c_0 \, / \lambda \, k \, T) - 1} \qquad (3.11)$$

In this expression h_P denotes the Planck constant, c_0 the speed of light, λ the wavelength, k the Boltzmann constant, and T the temperature of the black body. $S_{BB}^{E}(\lambda)$ is the amount of energy (in joules) radiated per second through a surface element of 1 m^2 at a wavelength λ per wavelength interval $d\lambda$ (in meters). In contrast to the strongly beamed energy flux density observed on the Earth, the energy emitted from a surface element of a black body (here, the Sun) is radiated away in all directions. Hence only a tiny fraction of this energy will actually reach the Earth. On the other hand, every nonobstructed surface element on the Sun will contribute to the energy flux at Earth. Rather than adding up all these individual contributions, we invoke instead the continuity of the radiated energy flux. All of the radiated energy from the surface must pass through a concentric sphere with a radius equal to the mean distance of the Earth's orbit, in other words $S_{BB}^E(\lambda)4\pi R_S^2 = \left(S_{BB}^E(\lambda)\right)_{1\,AU} 4\pi(1\,\mathrm{AU})^2$. Using the units of Fig. 3.7, it follows that

$$\left(S_{BB}^E(\lambda)\right)_{1AU} [\mathrm{Wm}^{-2}\mu\mathrm{m}^{-1}] = 10^{-6}(R_S/1\mathrm{AU})^2 S_{BB}^E(\lambda)[\mathrm{Wm}^{-2}\mathrm{m}^{-1}] \quad (3.12)$$

The theoretical spectrum described by Eq. (3.11) contains the temperature of the black body as a free parameter. This can be determined by fitting the curve (3.12) to the observed spectrum. For example, T is uniquely defined if one requires that the theoretical spectrum attains its maximum at the same wavelength λ_{max} as observed. In order to determine the position of the maximum, we differentiate the Planck spectrum (or even better, its logarithm) with respect to λ and set the derivative equal to zero. This yields an implicit equation, the numerical solution of which corresponds to the well-known *Wien's law*

$$\lambda_{max} = a_W/T \quad (3.13)$$

with $a_W = 0.002898$ K m. It is evident that the wavelength of maximum radiation shifts to lower values with increasing temperature. A black body temperature of 6400 K is obtained for the observed maximum radiation of the Sun at $\lambda_{max} \simeq 0.45$ μm (blue light). Alternatively, one could determine the black body temperature of the Sun by equating the total energy flux of the observed spectrum with that of the theoretical curve. Numerically integrating the Planck spectrum Eq. (3.11) over all wavelengths yields the *Stefan-Boltzmann law*

$$\phi_{BB}^E = \int_0^\infty S_{BB}^E \, d\lambda = a_{SB} \, T^4 \quad (3.14)$$

with $a_{SB} = 5.67 \cdot 10^{-8}$ W m^{-2}K^{-4}. Setting this relation equal to the Sun's luminosity divided by the area of the solar surface, one derives an *effective radiation temperature* of 5780 K.

A comparison of the black body spectrum with the observations allows us to verify some earlier assertions. The good agreement in the infrared and visible ranges demonstrates that the photosphere does indeed have a temperature of about 6000 K. Seen a different way, the radiation deficit in the near and far ultraviolet (as revealed by the shift of the spectral peak to larger wavelengths) means that the radiation must originate from cooler gas

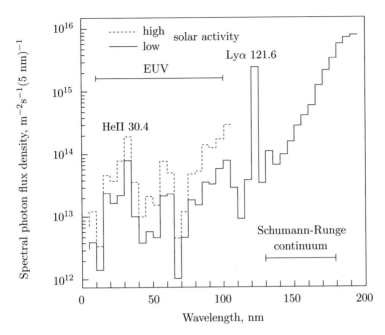

Fig. 3.8. The solar spectrum in the ultraviolet range. The spectral photon flux density is referred to a distance of 1 AU. Because the bin widths are 5 nm, the ordinate scale directly gives the photon flux density (in $m^{-2}s^{-1}$) for each spectral bin (adapted from Heroux and Hinteregger, 1978; Torr and Torr, 1985).

layers, as indicated in Fig. 3.3. On the other hand, the strongly enhanced radiation in the EUV and X-ray ranges implies that the gases emitting this radiation must be very hot. These gases, however, must also be relatively tenuous (and thus cannot represent a black body) because their radiative output lies well below that of the photosphere and is thus incompatible with the Stefan-Boltzmann law. Both of these conclusions agree with our previous conceptions of the corona. It is also clear that the EUV and X-ray radiation must be attributed to the corona, because the photospheric (black body) contribution is negligible. Just based on these few examples, it is obvious that a wealth of information is contained within the solar spectrum.

Ultraviolet and X-ray Radiation. As the most important energy source for the upper atmosphere, the radiation below about 200 nm, encompassing the ultraviolet and X-ray regions down to a wavelength of about 1 nm, has particular relevance for us. Figure 3.8 shows this spectral region with moderately high spectral resolution. In contrast to Fig. 3.7, this time we have plotted the spectral photon flux density $S^{ph}(\lambda)$ rather than the energy flux density. This quantity, which describes how many photons (per unit time, per unit area, per wavelength interval, at wavelength λ) pass through a surface oriented normal to the propagation direction, is obtained from the energy flux density upon dividing by the corresponding energy of the photons $E_{ph} = h_P\, \nu = h_P\, c_0/\lambda$

$$S^{ph}(\lambda) = \frac{\mathrm{d}\phi^{ph}(\lambda)}{\mathrm{d}\lambda} = S^E(\lambda)\Big/(h_P\, c_0/\lambda) \qquad (3.15)$$

Working with a photon flux density rather than energy flux density has practical reasons. Treatment of the interactions between photons (as particles) and gases, for example, can now be performed using the well-established gas kinetic approach. This considerably simplifies the calculation of the absorption of electromagnetic radiation in gases. We are thus interested in the number of photons per given wavelength interval that impinge upon the upper atmosphere.

As seen in Fig. 3.8, the decrease in the spectral photon flux density proceeds relatively smoothly in the far ultraviolet range. Considerable fluctuations are evident below ca. 130 nm, which result from superposition of a variable background continuum with numerous emission lines. Prominent strong emissions here are the Lyman-α radiation of hydrogen at 121.6 nm and the energetic radiation of singly ionized helium at 30.4 nm.

Particularly important wavelength ranges for heating the upper atmosphere are the so-called *Schumann-Runge continuum* (130 – 175 nm) and the *EUV (extreme ultraviolet)* radiation (10 – 100 nm). The energy flux $\phi_\infty^E(\Delta\lambda)$ incident on the upper atmosphere in each of these particular spectral ranges is

$$\phi_\infty^E(\Delta\lambda) = \sum_{\Delta\lambda} S^E(\lambda)\, \delta\lambda = \sum_{\Delta\lambda} (h_P c_0/\lambda) S^{ph}(\lambda)\, \delta\lambda$$

The index ∞ reminds us that this is the unattenuated energy flux impinging upon the Earth's upper atmosphere. Using $\delta\lambda = 5$ nm and the data presented in Fig. 3.8, one obtains

$$\begin{aligned}
\text{Schumann-Runge continuum:} \quad & \phi_\infty^E(SRC) \simeq 11 \pm 1\ [\mathrm{mW/m^2}] \\
\text{extreme ultraviolet radiation:} \quad & \phi_\infty^E(EUV) \simeq 4 \pm 2\ [\mathrm{mW/m^2}]
\end{aligned} \qquad (3.16)$$

where the bounds correspond to the naturally observed variations. Later, we refer back to these values for an estimate of the heating effects.

3.1.4 Variation of the Radiation Intensity

Within the spectral range of maximum energy flux, i.e. in the visible and neighboring infrared and ultraviolet ranges, the intensity of the solar radiation is practically constant, varying according to contemporary measurements by less than 0.3% (solar *constant*!). Even large sunspots are unable to influence the total intensity significantly, particularly because their radiation deficit is more than compensated by the surrounding brightly emitting

Table 3.2. Variations of radiation intensity in the EUV, X-ray and radio spectral regions, as observed from Earth.

1.	Intrinsic Variations	
	1.1	Growth and decline of emission centers (days to weeks)
	1.2	Solar activity cycle ($\simeq$ 11 years, quasi-regular)
	1.3.	Solar flares (minutes to hours)
2.	Geometric Effects	
	2.1	Solar rotation effect (e.g. 27–day, quasi-regular)
	2.2	Solar distance effect (1 year, regular)

regions (*photospheric faculae*). The situation is completely different in the EUV and X-ray regions, as well as at radio wavelengths. Distinct variations in the radiation intensity are observed here, brought about either by events on the Sun or by changes in the observation geometry; see Table 3.2. A more detailed description of the causes and characteristics of these variations is given in the following.

Growth and Decline of Emission Centers. A characteristic of the radiation emission in the EUV, X-ray and radio ranges is that it is not uniformly distributed in the corona, but mostly concentrated in localized emission centers; see Fig. 3.4e,f and g. The larger examples of these emission centers, which grow and decline with typical lifetimes of a few days to a few weeks, can have measurable repercussions for the total intensity.

Solar Activity Cycle. Emission centers of EUV, X-ray and radio radiation belong to a class of phenomena known collectively as *active regions*. Generally included in this category are any regions on the Sun with properties that deviate strongly from the average, in particular

- regions of abnormal brightness such as
  - coronal emission centers in the X-ray, EUV or radio spectral ranges
  - chromospheric plages or faculae (Hα, CaII lines)
  - photospheric radiation deficits in sunspots
- regions of strong photospheric magnetic fields (mostly associated with sunspots)
- regions with spectacular structures (filaments, prominences, etc.); and
- regions of explosive dynamics (solar flares, eruptive prominences, solar mass ejections)

The occurrences of these various anomalies are only loosely correlated in time and space. A common thread, however, is a lasting variation with respect to their occurrence frequency and intensity. There are times when the active

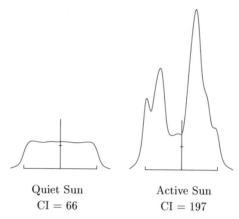

Quiet Sun

CI = 66

Active Sun

CI = 197

Fig. 3.9. Radio scan of the Sun at a wavelength of 10.7 cm during quiet and active conditions. The mean spectral energy flux density for the former case is $S^E(\nu) = 66 \cdot 10^{-22}$ Wm^{-2}Hz^{-1}, for the latter this value increases by a factor of three. The length of the horizontal axis defines the diameter of the photosphere (adapted from the National Research Council of Canada).

regions are rare or only weakly defined. Under these conditions the Sun is designated as 'quiet'. An 'active' Sun, on the contrary, is characterized by numerous strong and extended active regions. Figure 3.9 illustrates the difference between an active and quiet Sun on the basis of its radio emission. The two traces show east-to-west scans of the solar corona using a radio telescope of moderate resolution at a wavelength of 10.7 cm (2.8 GHz). During quiet conditions the corona appears as a smooth featureless surface, clearly different from the active Sun with its numerous intense emission centers. Various indices have been defined in order to describe the solar activity in a quantitative sense. The first such index used for this purpose is the relative sunspot number R (Wolf, 1847) defined by Eq. (3.1). Some 100 years later, following the discovery of solar radio radiation, the *10.7 cm radio flux index* or *Covington index* CI was introduced (Covington, 1946). This corresponds to the mean spectral energy flux density of the solar disk $S^E(\nu)$ at a frequency $\nu = 2.8$ GHz (wavelength: 10.7 cm) in units of 10^{-22} W m^{-2}Hz^{-1} (*solar flux units*). For example, the quiet Sun in Fig. 3.9 corresponds to a Covington index of CI=66; the active Sun to CI=197. The Covington index can reach values as high as CI=300 on extremely active days. Among the newer, although less often used indices, is the so-called *plage-index*, which attempts to quantify the intensity of the Hα and CaII radiation.

All of these indices display a clearly pronounced variation of the solar activity with an 11-year periodicity. This is documented in Fig. 3.10, again using the radio radiation as an example. These oscillations are designated the *solar cycle variations*. Both the period and the amplitude of these oscillations are themselves subject to irregular variations. Particularly striking are the

$S^E(\nu), 10^{-22}$ Wm^{-2}Hz^{-1}

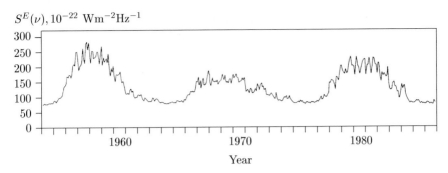

Fig. 3.10. Solar activity cycle in 10.7 cm radio radiation. Monthly mean values are plotted for an interval of 33 years. The energy flux density is referenced to 1 AU (adapted from the National Research Council of Canada).

strongly pronounced maximum in 1957/58 and the relatively weak maximum in 1967/70.

The solar cycle variations were first detected in 1843 by the pharmacist Heinrich Schwabe from sunspot observations. A satisfactory physical explanation of this phenomenon is still lacking up to the present day. Meanwhile, it has become clear that the fundamental period of the cycle must be 22 years, because the oscillations in activity are correlated with the dipole reversal of the solar magnetic field, which follows a 22-year cyclic variation on the average. Over a period of 22 years, for example, the heliographic north pole is first associated with the magnetic north pole, then, after about 11 years, with the magnetic south pole, and then, after another 11 years, again with the magnetic north pole. Maximum solar activity always occurs during the magnetic reversal phase. It is interesting that the reversal cycle not only regulates the intensity of the solar activity, but also the spatial distribution of the active regions on the solar disk. The so-called *butterfly* or *Spörer diagram* of sunspot distribution has become a popular representation of this latter effect. Remarkably, solar cycle variations can be tracked backwards in time over centuries using historically documented observations of sunspots (at least back to ca. 1610, the year of the first recorded observations of sunspots in Europe). Ironically, the solar cycle was very weakly developed exactly during the regency of the 'sun king' Louis XIV, because hardly any sunspots were observed during this interval (*Maunder minimum*, 1645–1715).

Solar Flares. Whereas solar flares are certainly among the most impressive spectacles on the Sun, they play only a subordinate role for the terrestrial upper atmosphere. This is mostly because of the short duration (often just minutes) of the observed emission outbursts. It thus seems appropriate to treat this phenomenon in detail later, within the context of solar-terrestrial relations (see Section 8.6.3).

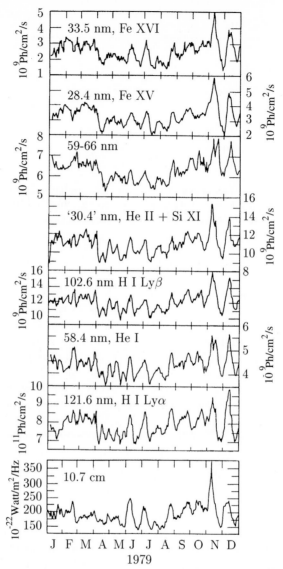

Fig. 3.11. Radiation intensity variations caused by solar rotation. The panels show the photon fluxes of various emission lines in the EUV and far ultraviolet ranges, as well as the energy flux density at the radio wavelength 10.7 cm, over a time interval of one year (adapted from Lean, 1988).

Solar Rotation Effect. As observed from Earth, the concentration of EUV, X-ray and radio radiation in emission centers is responsible for another characteristic modulation of the radiation intensity. A prominent emission center acts somewhat like the beam of a lighthouse that sweeps outward through interplanetary space as it rotates with the Sun. This emission center will attain its maximum illumination of Earth every 27 days, thereby producing a corresponding modulation of the radiation flux. As illustrated by the various intensity plots in Fig. 3.11, this modulation is best recognized if all the significant emission centers are situated on one side of the Sun: one then observes a well-pronounced, quasi-sinusoidal *27-day variation* of the solar photon fluxes. In addition to the 27-day variation (strongest in June/July and November/December), variations with shorter periods may be observed in this figure that are caused by other longitudinal distributions of the emission centers. For example, the 13.5 day variation seen in April/May can be attributed to two emission centers located about 180° apart in solar longitude.

It can also be seen in Fig. 3.11 that the variation of the radio flux density agrees to a certain extent with the variations of the UV line intensities. This correspondence suggests that the 10.7 cm radio flux may be used as a 'proxy' (substitute observation) for the EUV and X-ray intensities, which can only be obtained on an irregular basis from monitors above the Earth's atmosphere. This has now become standard practice.

Solar Distance Effect. The solar distance variation is a purely geometrical effect that is valid for all spectral ranges. It comes from the changing distance of the Earth from the Sun due to the tiny, but non-negligible eccentricity ($\varepsilon \simeq 0.017$) of the Earth's elliptical orbit. The Earth reaches its minimum distance to the Sun (perihelion) in January; its maximum distance (aphelion) in July. Because the radiation intensity is inversely proportional to the square of the distance, the solar flux decreases over the interval from January to July by ca. 7%. Evidently, the Southern Hemisphere will receive more radiation in summer and less radiation in winter than the Northern Hemisphere.

3.2 Extinction of Solar Radiation in the Upper Atmosphere

Radiation undergoes two fundamental interactions with a gas, emission and extinction, resulting in a radiation enhancement in the first case and a radiation attenuation in the second. Here, the extinction of solar radiation during its passage through the upper atmosphere is investigated. We first identify the important absorption processes that produce this extinction. Using a generalized approach, we then determine the extinction of monochromatic radiation in gases. In the next step we apply this approach to the special situation of the upper atmosphere, with particular emphasis on the absorption height. Finally, the energy deposition associated with the absorption of radiation is determined.

3.2.1 Absorption Processes

Solar radiation propagating in the upper atmosphere is absorbed and thus attenuated in various ways. Particularly important absorption processes are *photodissociation, photoionization,* and the combination of these two, *dissociative photoionization.* The absorption processes of interest here can be written formally in the following way

- Photodissociation ($\lambda \leq 242$ nm)

$$O_2 + \text{photon}(\lambda \leq 242 \text{ nm}) \rightarrow O + O , \qquad \sigma^D_{O_2} \qquad (3.17)$$

- Photoionization ($\lambda \leq 103$ nm)

$$\begin{aligned}
O \ + \text{photon}(\lambda \leq 91 \text{ nm}) \ &\rightarrow O^+ + e , & \sigma^I_O \\
N_2 + \text{photon}(\lambda \leq 80 \text{ nm}) \ &\rightarrow N_2^+ + e , & \sigma^I_{N_2} \\
O_2 + \text{photon}(\lambda \leq 103 \text{ nm}) &\rightarrow O_2^+ + e , & \sigma^I_{O_2}
\end{aligned} \qquad (3.18)$$

- Dissociative Photoionization ($\lambda \leq 72$ nm)

$$N_2 + \text{photon}(\lambda \leq 49 \text{ nm}) \rightarrow N^+ + N + e , \qquad \sigma^{DI}_{N_2} \qquad (3.19)$$

This notation makes use of the photon particle concept of electromagnetic radiation, even though the dissociation and ionization energies are given as upper bounds on the wavelength. According to this concept, the given interactions are understood as collisions between photons and gas particles, the probability of which is specified by the magnitude of an interaction or collision cross section σ^i_j. This collision cross section is associated only with the absorbing gas particle, the photon itself being considered as a massless point. The formal definition of σ^i_j follows from the approach used for calculating the extinction of radiation (see Section 3.2.2). As such, it is clear that the interaction cross sections of different absorption processes can be added together to yield a total absorption cross section σ^A_j, i.e. we require that

$$\sigma^A_j = \sigma^A_j(\varepsilon^D_j + \varepsilon^I_j + \varepsilon^{DI}_j + \cdots) = \sigma^D_j + \sigma^I_j + \sigma^{DI}_j + \cdots \qquad (3.20)$$

where ε^i_j is the probability that an absorption 'collision' leads to a dissociation, an ionization, a dissociative ionization or some other process.

It may be surprising at first glance that the often discussed radiation absorption by ozone is missing from the above interactions. Recall here that the ozone concentration in the upper atmosphere is extremely low. Sufficiently high densities are reached only in the middle atmosphere (20 – 30 km height), where ozone absorbs practically all UV-radiation at wavelengths above 242 nm. It may also come as a surprise that the dissociation of molecular nitrogen is virtually insignificant. This has to do with the relatively strong binding energy of this molecule, see Section 2.3.7. Finally, it may seem strange that excitation processes play only a minor role in the absorption. It can be

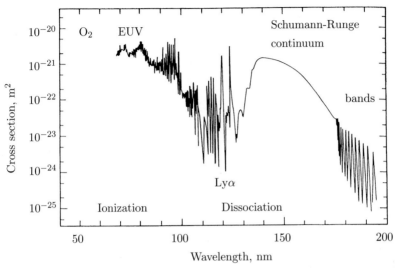

Fig. 3.12. Absorption cross section of molecular oxygen at wavelengths 70-195 nm. The absorption is dominated by ionization below $\lambda \simeq 103$ nm and by dissociation above this wavelength (adapted from Banks and Kockarts, 1973).

shown that the associated excitation cross sections are many times smaller than those of photodissociation or photoionization. An exception to this rule is the excitational absorption of molecular nitrogen in the wavelength range 80–100 nm (see Fig. 3.13).

In contrast to the collision cross sections for neutral gas particles, it is important that the value of the absorption cross section is strongly dependent on the energy of the incident photons. Indeed, a certain minimum of energy is needed to dissociate or ionize a gas particle and the absorption cross section above this borderline wavelength is virtually zero. Considerable variations can also occur at wavelengths below this boundary, however, as documented for the absorption cross section of molecular oxygen in Fig. 3.12. Particularly striking here are the quasi-periodic oscillations of the cross section by more than an order of magnitude observed in the range of the *Schumann-Runge bands* ($175 < \lambda < 195$ nm). Even more remarkable is the spectral band from 105 to 125 nm, where small changes in wavelength cause the cross section to vary by more than three orders of magnitude. Interestingly, the strong Lyman-α radiation of the Sun at 121.6 nm sits exactly in a minimum of the absorption cross section.

The ionization cross section for O_2 is nearly constant in the EUV range and relatively large compared with the dissociation cross section, hence dominating the total absorption cross section of this species. Figure 3.13 shows this behavior in the wavelength range between 5 and 103 nm, together with

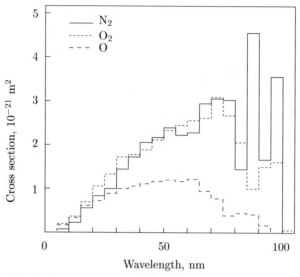

Fig. 3.13. Absorption cross sections of molecular nitrogen, molecular oxygen and atomic oxygen in the extreme ultraviolet range. Ionization dominates the absorption of radiation in this wavelength range. An exception to this rule is the range 80-100 nm, where absorption of molecular nitrogen occurs primarily by excitation into so-called pre-dissociation states (adapted from Torr et al., 1979).

the absorption cross sections of molecular nitrogen and atomic oxygen (essentially the ionization cross section for O). It should be noted that σ_j^A is plotted here on a linear scale rather than the logarithmic scale used in Fig. 3.12. It may be seen in Fig. 3.13 that the mean absorption cross section for atomic oxygen in the EUV range is $\sigma_O^A(EUV) \simeq 10^{-21}$ m^2 and is about twice as large for molecular nitrogen and oxygen. We will need these values later.

3.2.2 Extinction of Radiation in Gases

In order to calculate the extinction of electromagnetic radiation in a gas, we must first determine the probability that a photon undergoes a collision (and is thereby absorbed) upon propagating through the path length ds. This probability can be visualized intuitively from Fig. 3.14. The sum of the gas particle cross sections projected onto the area A is a measure of the probability for a collision over the path length ds. Conversely, the area remaining on A after subtracting this sum is a measure of the probability that a photon traverses this length without a collision. The collision probability $dw_{col}(ds)$ is thus the ratio of the area shaded by the projections to the total available area

$$dw_{col}(ds) = (\textstyle\sum \sigma^A)/A = (\sigma^A \; n(s) \; A \; ds)/A = \sigma^A \; n(s) \; ds \qquad (3.21)$$

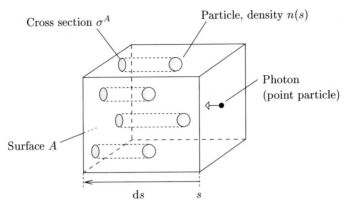

Fig. 3.14. Deriving the probability of an absorption for a photon traversing the path ds in a gas

Of course, this will only be valid as long as there is no overlapping between the projection areas. This condition is fulfilled when the distance between particles ($\simeq 1/\sqrt[3]{n}$) is large compared with their diameter ($\simeq \sqrt{\sigma^A}$) and ds is chosen to be sufficiently small.

Using the above collision probability, we can calculate the extinction of a photon flux in a gas volume. It is assumed that the photon flux has already passed through a certain path length in the gas and has a value of $\phi^{ph}(s)$ at the entrance surface of the gas volume. The number of photons (per unit area and time) suffering an absorption during their transit through the path length ds is thus $\phi^{ph}(s)\,\mathrm{d}w_{col}(\mathrm{d}s)$. The corresponding change in the photon flux is

$$\mathrm{d}\phi^{ph} = -\phi^{ph}(s)\,\mathrm{d}w_{col}(\mathrm{d}s) = -\phi^{ph}(s)\,\sigma^A\,n(s)\,\mathrm{d}s \qquad (3.22)$$

This expression is intuitively understandable. Indeed, the change in photon flux should be proportional to the initial photon flux, the size of the absorption cross section, the density of the absorbing gas particles and the length of the propagation path. The minus sign indicates that the photon flux decreases. This approach basically defines the value of the absorption cross section (and, by analogy, the other interaction cross sections). Separation of the variables s and ϕ^{ph} and subsequent integration leads to the *Lambert-Beer extinction law*

$$\phi^{ph}(s) = \phi^{ph}(s_0)\,\mathrm{e}^{-\tau(s)} \qquad (3.23)$$

where for brevity we have introduced the *optical depth* τ

$$\tau = \int_{s_0}^{s} \sigma^A\,n(s')\,\mathrm{d}s' \qquad (3.24)$$

Evidently, a large value of τ corresponds to a strongly absorbing and thus opaque, 'optically thick' gas volume, while a small value of τ indicates a

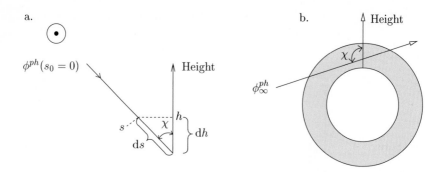

Fig. 3.15. (a) Relations between the variables for application of the Lambert-Beer extinction law to the upper atmosphere. **(b)** Modification of the absorption path at large zenith angle when accounting for the sphericity of the upper atmosphere.

weakly absorbing and thus transparent, 'optically thin' gas volume. Although originally introduced for the visible range, the term 'optical depth' is commonly used for all wavelength ranges. The exponential factor $\exp(-\tau) = \phi^{ph}(s)/\phi^{ph}(s_0)$ in the Lambert-Beer extinction law, for obvious reasons, is also called the *transmission factor*.

3.2.3 Extinction of Radiation in the Upper Atmosphere

In order to determine the extinction of solar radiation in the upper atmosphere, we first consider the simple case of a planar (parallel plane layered) single gas atmosphere. The Lambert-Beer extinction law is adjusted to this situation using the relations sketched in Fig. 3.15a

$$s \mathrel{\hat{=}} h; \qquad s_0 = 0 \mathrel{\hat{=}} h \to \infty$$

$$\phi^{ph}(s) \mathrel{\hat{=}} \phi^{ph}(h); \qquad \phi^{ph}(s_0 = 0) \mathrel{\hat{=}} \phi^{ph}(h \to \infty) = \phi^{ph}_{\infty} \qquad (3.25)$$

$$ds = -dh/\cos\chi = -dh \ \sec\chi$$

Here, ϕ^{ph}_{∞} denotes the photon flux of the Sun outside the Earth's atmosphere and χ is the angle of incidence, or *zenith angle*. Substitution into the Lambert-Beer extinction law and exchanging the upper and lower integration limits yields the desired expression for the photon flux at the height h

$$\phi^{ph}(h) = \phi^{ph}_{\infty} \, e^{-\tau(h)} \qquad (3.26)$$

where, with the help of Eqs. (2.78) and (2.79), the optical depth takes the following simple form

$$\tau(h) = \int_{h}^{\infty} \sec\chi \, \sigma^A \, n(h') \, dh' = \sec\chi \, \sigma^A \, \mathcal{N}(h) \simeq \sec\chi \, \sigma^A \, n(h) \, H(h) \quad (3.27)$$

Equation (3.27) must be modified in two ways when considering the real upper atmosphere, which consists of several gases and also cannot be considered planar at large zenith angles. It follows from the approach used for deriving the collision probability in Eq. (3.21) that the contributions of the different gas components to the absorption may be added together. It is thus adequate to replace the gas specific product of absorption cross section, density and scale height in Eq. (3.27) with the sum of these products for the gas mixture. Much more effort is required to account properly for the spherical shell form of the atmosphere (see Fig. 3.15b). This can be accomplished, without going into details, by introducing the so-called *Chapman function for grazing incidence* $ch(\chi, h)$. It is sufficient to know that $ch(\chi, h) \simeq \sec \chi$ at zenith angles smaller than ca. $80°$ and that the Chapman function remains positive and finite for zenith angles $\chi \geq 90°$ (in contrast to $\sec \chi$). It is also understandable that the Chapman function must depend on the height. The sphericity is much more important in the upper atmosphere than in the lower atmosphere, where the absorption is much stronger anyway. The optical thickness thus assumes the form

$$\tau(h) \simeq ch(\chi, h) \sum_{j=O,N_2,O_2} \sigma_j^A \, n_j(h) \, H_j(h) \qquad (3.28)$$

where only the most important gases are considered.

The height at which the absorption reduces the incident solar radiation by a factor $1/e$ is of special interest. This so-called *absorption height h_A* is obtained – whenever the radiation is not incident at grazing angles – from the condition

$$\tau(h_A) \simeq \sec \chi \sum_{j=O,N_2,O_2} \sigma_j^A \, n_j(h_A) \, H_j(h_A) = 1 \qquad (3.29)$$

Because the expression from Eq. (2.59) must be used for the densities, this implicit equation for h_A can only be solved numerically. Two representative results are shown in Fig. 3.16. The absorption height is plotted as a function of the wavelength ($\sigma_j^A = f(\lambda)!$) for two different conditions of solar zenith angle and solar activity. A general result is that all radiation in the wavelength range 5 to 175 nm is absorbed above a height of about 100 km. While the far ultraviolet radiation barely reaches down to this altitude, the extreme ultraviolet radiation is absorbed above 150 km (and even above 200 km at times of high solar activity or for large zenith angle) because of the larger EUV absorption cross sections. The dependence of the absorption height on the solar zenith angle is indicated directly in Eq. (3.29). However, there are also indirect dependences on the solar zenith angle and the solar activity, via the density and scale height of the absorbing gases. Smaller zenith angles and higher solar activity effectively increase the atmospheric temperature, thus leading to greater scale heights and densities and, as a consequence, higher absorption heights. The dependence on the zenith angle is therefore

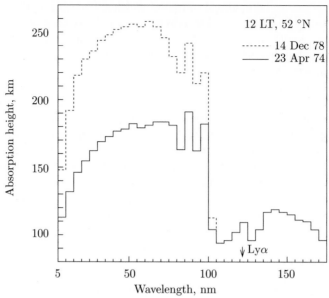

Fig. 3.16. Absorption heights in the far and extreme ultraviolet range for noon conditions at mid-latitudes. The solid line refers to a spring day during low solar activity (CI=73); the dashed line is a winter day during high solar activity (CI=203). An arrow denotes the relatively large penetration depth of the Lyman-α radiation (adapted from Kockarts, 1981).

more complicated than indicated explicitly by the function $\sec \chi$. The lack of variability at longer wavelengths is due mainly to the steep density increase in the lower thermosphere ($H_n \simeq 6\,\mathrm{km}$).

3.2.4 Energy Deposition from Radiation Absorption

In the following we determine the energy deposition associated with the absorption of radiation, which, of course, is responsible for the heating of the upper atmosphere. The first step is to calculate the number of photons absorbed in a given volume element of the atmosphere. For this purpose we consider a cylinder with a base surface area $\mathrm{d}A$ and length $\mathrm{d}s$, aligned parallel to the incident radiation; see Fig. 3.17. The number of photons entering the cylinder on the upper end is $\mathrm{d}N_{ph}^{+} = \phi^{ph}\,\mathrm{d}A\,\mathrm{d}t$; the number leaving at the bottom is $\mathrm{d}N_{ph}^{-} = (\phi^{ph} - |\mathrm{d}\phi^{ph}|)\,\mathrm{d}A\,\mathrm{d}t$. The difference is the number of absorbed photons $\mathrm{d}N_{ph} = |\mathrm{d}\phi^{ph}|\,\mathrm{d}A\,\mathrm{d}t$. The number of photons absorbed at the height h per volume element $\mathrm{d}V = \mathrm{d}A\,\mathrm{d}s$ and per time interval $\mathrm{d}t$ is thus given by

$$
\left.\frac{\mathrm{d}N_{ph}}{\mathrm{d}V\,\mathrm{d}t}\right|_{h} = \left.\left|\frac{\mathrm{d}\phi^{ph}}{\mathrm{d}s}\right|\right|_{h} = \left.\sigma^{A}\,n(s)\,\phi^{ph}(s)\right|_{h} = \sigma^{A}\,n(h)\,\phi^{ph}(h) \qquad (3.30)
$$

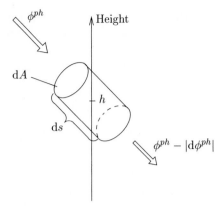

Fig. 3.17. Determining the number of photons absorbed at height h per volume element and time interval

where we have made use of the expressions (3.22) and (3.25). It is easy to verify that this result is independent of the form of the volume element used in the derivation. Moreover, no assumptions were made concerning the solar zenith angle, so Eq. (3.30) must also be valid for grazing incidence radiation. The only restriction comes from the fact that we have considered here a single gas atmosphere. Clearly, the contributions must be summed for an atmosphere with multiple constituents.

Alternatively, the number of photons absorbed per unit volume and unit time can be calculated from the divergence of the photon flux. According to Eq. (2.18), we have

$$\frac{dN_{ph}}{dV\,dt} = \frac{dn_{ph}}{dt} = -\mathrm{div}\vec{\phi}^{ph} = -\frac{d\phi_s^{ph}}{ds} = \sigma^A n(s)\phi^{ph}(s) = \sigma^A n(h)\phi^{ph}(h)$$

where the path length coordinate s, as before, is along the direction of the photon flux.

Because the energy of the photons is also transferred at each absorption event, the extinction of radiation is always accompanied by an energy deposition into the atmosphere. With $E_{ph} = h_P c_0/\lambda$ the energy deposition rate per unit volume q^E at height h can be written as

$$q^E(h) = \frac{dN_{ph}\,E_{ph}}{dV\,dt} = (h_P\,c_0/\lambda)\,\sigma^A\,n(h)\,\phi^{ph}(h)$$

or

$$q^E(h) = \sigma^A\,n(h)\,\phi_\infty^E\,e^{-\tau(h)} \qquad (3.31)$$

where we have introduced the energy flux $\phi_\infty^E = (h_P c_0/\lambda)\,\phi_\infty^{ph}$ associated with the unattenuated photon flux ϕ_∞^{ph}. The height profile of this energy deposition rate per volume can be understood from the following consideration. Because the density at great heights is very low and the transmission factor

$e^{-\tau(h)}$ cannot exceed 1, the energy deposition there will drop to very low values. There are simply not enough gas particles at large heights to cause appreciable absorption of the incident radiation. At lower heights the density is high enough, but the transmission factor goes to zero as a double exponential $(\tau(h) \sim n(h) \sim e^{-h})$, so the energy deposition rate here also quickly becomes quite small. There are simply no photons left at lower altitudes to be absorbed by the gas particles. The maximum energy deposition must be attained at that intermediate height, where the density of the gases and the incident photon flux are both sufficiently large to produce intense absorption of the radiation.

In order to derive the height dependence of the energy deposition rate explicitly, we consider the very idealistic case of a planar isothermal single gas atmosphere. Equation (3.31) can be written as follows

$$q^E(h) = \left(\phi_\infty^E \, \cos\chi/H\right) \, \tau(h) \, e^{-\tau(h)} \qquad (3.32)$$

The maximum of this curve may be obtained from $dq^E/dh = 0$ or, more readily, from $d(\ln q^E)/dh = 0$. This latter condition leads to an equation of the form

$$\frac{d\tau}{dh}\left(\frac{1}{\tau} - 1\right) = 0$$

Noting that $d\tau/dh \sim \exp[-(h - h_0)/H]$ vanishes only for $h \to \infty$, it may be concluded that

$$\tau(h_{max}) = 1 \qquad (3.33)$$

This corresponds exactly to the definition of the absorption height. In other words, the maximal energy deposition occurs at that altitude where the intensity of the incident radiation has decreased from its original value by a factor $1/e$. The energy deposition at the maximum is calculated to be

$$q^E(h_{max}) = q_{max}^E = \frac{\phi_\infty^E \, \cos\chi}{H \, e} \qquad (3.34)$$

where e is the base of the natural logarithm.

It is useful at this point to introduce the height of maximum energy deposition at normal incidence $h_{max}^* = h_{max}(\chi = 0)$ as a reference height in the barometric density height profile. According to Eq. (3.27) and with $h_{max}^* = h_A^* = h_A(\chi = 0)$, we obtain $n(h_{max}^*) = 1/\sigma^A H$, so the optical depth can now be written as follows

$$\tau(h) = \sec\chi \, e^{-(h - h_{max}^*)/H}$$

Introducing the maximum energy deposition rate for normal incidence $(q_{max}^E)^* = q_{max}^E(\chi = 0)$, Eq. (3.32) can be written as

$$q^E(h) = (q_{max}^E)^* \, \exp\left\{1 - \frac{h - h_{max}^*}{H} - \sec\chi \, e^{-(h - h_{max}^*)/H}\right\} \qquad (3.35)$$

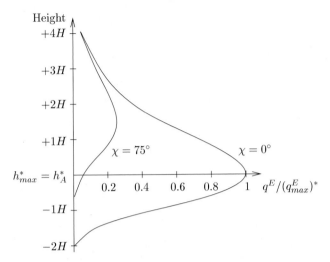

Fig. 3.18. Chapman production function for two different solar zenith angles. h_{max}^* and h_A^* denote the height of maximum energy deposition rate and absorption height, respectively, at normal incidence. H is the scale height of the absorbing gas. The height $+1H$ stands for $h_{max}^* + 1H$, etc. q^E is the energy deposition rate per volume and $(q_{max}^E)^*$ is the maximum of this quantity for normal incidence.

Although previously discussed by others, this type of height profile has become known as the *Chapman production function*. Figure 3.18 illustrates the height dependence of this function for two different solar zenith angles. As may be seen, the maximum energy deposition rate is decreased and shifted to higher altitudes when the direction of the incident radiation is slanted. The half-thickness of the layer remains unchanged at about 2.5 scale heights. Note that the energy deposition rate above about $h_{max}^* + 2H$ becomes independent of the solar zenith angle. This can be attributed to the transmission factor at great heights, which is close to unity even for large solar zenith angles. As a result, the energy deposition rate here is only dependent upon the density of the absorbing gas.

In order to determine the height profile of the energy deposition rate in the real upper atmosphere, we must retract a number of the simplifying assumptions made so far. The upper atmosphere is certainly not isothermal, thereby considerably complicating the analytical calculation. Because we are dealing with a multiple component gas mixture rather than a single gas species, we must also replace the product of absorption cross section and density appearing in Eq. (3.31) with a sum of the products accounting for the contributions from the various gas constituents. The same correction must be made for the optical depth. These modifications alone force us to resort to numerical calculations. Furthermore, because the incident solar radiation is polychromatic rather than monochromatic, we will also have to add up the

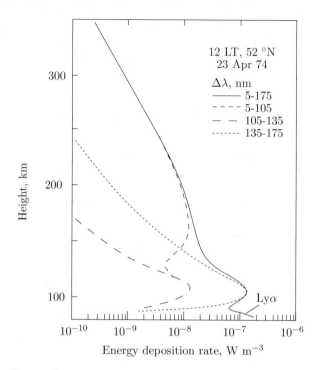

Fig. 3.19. Energy deposition rate per volume for noontime conditions at mid-latitudes. The solar activity on this particular day, 23 April 1974, was low (CI=73). Note that the energy deposition rate of the EUV-radiation rises again at low altitudes. This can be attributed to absorption of long and short wavelength radiation in this region; see also Fig. 4.6 (adapted from Kockarts, 1981)

different contributions from the individual wavelength regions. Accordingly, Eq. (3.31) assumes the following form for the real upper atmosphere

$$q^E(h) \simeq \sum_\lambda \sum_i n_i(h) \, \sigma_i^A(\lambda) \, \phi_\infty^E(\lambda) \, \exp\left\{ -ch(\chi, h) \sum_j \sigma_j^A(\lambda) \, n_j(h) \, H_j(h) \right\}$$

(3.36)

where $\sum_\lambda$ indicates a summation over the different wavelengths of interest and $\sum_i$, $\sum_j$ are summations over the gas components of the upper atmosphere. The Chapman function again accounts for the sphericity of the upper atmosphere.

As an example, Eq. (3.36) has been evaluated for local noon on a spring day at mid-latitude during a time of low solar activity. The resulting energy deposition profile is shown in Fig. 3.19. This height profile (given by the solid curve) obviously shows very little resemblance to the Chapman production function. Examining the contributions from the various wavelength regions separately, however, one may discern a qualitative agreement. The solar EUV range does indeed produce an energy deposition layer with a maximum very close to the absorption height appropriate for this radiation (see Fig. 3.16). Moreover, the thickness of this layer is quite consistent with the relatively large scale heights in this region. The energy deposition layer produced by the Schumann-Runge continuum radiation, with its maximum near the absorption height for this wavelength band and its moderate thickness (in agreement with the small scale heights at 100 km altitude), also bears similarity to the Chapman production function. Concerning the actual value of the maximal energy deposition, it is seen in Eq. (3.34) that this should be dependent on the scale height, but also on the intensity of the incident energy (photon) flux. Indeed, the scale height is smaller and the energy flux is greater for the case of the Schumann-Runge continuum radiation, thereby leading to a considerable enhancement of the maximum energy deposition rate. Remarkable in our example is the absorption of the very intense and deeply penetrating solar Lyman-α radiation, which produces a separate energy deposition layer of even higher maximum intensity. Generally, the exact form of the energy deposition profile will depend strongly on the solar zenith angle and the corresponding radiation intensity. The particular height profile shown in Fig. 3.19 should thus only be considered as an example.

3.3 Heating and Temperature Profile

As the preceeding discussion has shown, absorption of solar radiation has a definitive influence on the properties of the upper atmosphere. Atomic oxygen is produced, changing the chemical composition of the upper atmosphere and hence its absorption characteristics. Photoionization transforms the upper atmosphere into a conducting medium. Heat is generated during the radiation absorption process, bringing important changes for the temperature in this region. This latter process is investigated in the present section. We first determine the heat production rate and estimate the temperature increase induced by this heating. It turns out that effective heat loss processes are needed in order to prevent excessive heating of the upper atmosphere. It is shown that key roles are played here by radiative cooling in the lower, and molecular heat conduction in the upper thermosphere. A simple form for the temperature profile in the upper atmosphere can be obtained in the case when the heat production is just compensated by the loss via heat conduction. Finally, systematic variations in temperature and density induced by the changes in heat input, as well as airglow phenomena, are described.

3.3.1 Heat Production

Only a fraction of the absorbed radiative energy is transformed into heat. Determining the heating efficiency is not a trivial task, as shown by a glance at the energetics of photoionization. For example, the interaction

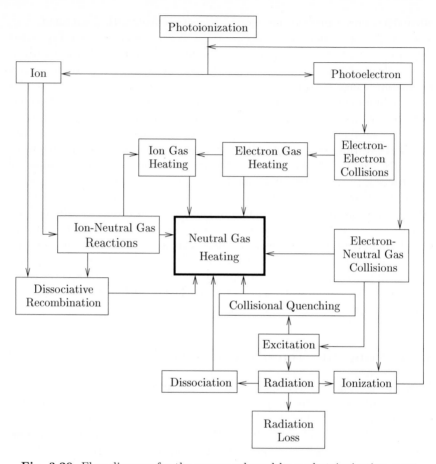

Fig. 3.20. Flow diagram for the energy released by a photoionization event

$$O + \underbrace{photon(\lambda = 30.4 \text{ nm})}_{41 \text{ eV}} \rightarrow \underbrace{O^+}_{} + e + \underbrace{excess \text{ energy}}_{27 \text{ eV}}$$
$$\underbrace{}_{14 \text{ eV}}$$

does not produce direct heating of the neutral upper atmosphere. The ionization energy of ca. 14 eV is first transformed into potential chemical energy and the excess energy of 27 eV appears in the form of kinetic energy of the *photoelectron* released in the process. The further development of this energy transfer is sketched in Fig. 3.20. On one hand, the fast photoelectron is decelerated by collisions with the ambient thermal electrons, leading to heating of the electron gas. Subsequent heat exchange of this hot electron gas with the cooler neutral gas produces a first contribution to heating of the neutral upper atmosphere. A corresponding heat exchange also occurs with the cooler ion gas, a large part of the transferred heat going directly into the neutral gas. In this context, it should be noted that the electron gas temperature in the sunlit upper atmosphere is higher than the ion gas temperature, which, in turn, is higher than or about the same as, the neutral gas temperature (see Fig. 4.3). On the other hand, the photoelectron is slowed down by collisions with the neutral gas particles. Should these collisions occur elastically, this will lead to heating of the neutral upper atmosphere. For inelastic interactions one must distinguish between collisional ionization and collisional excitation. Whereas the energy transfer cycle begins anew for the former case, the second case is followed either by *collisional quenching* of the excited state or emission of radiation. During the collisional quenching process the excitation energy is transformed into kinetic energy of the participating collision partners and thus contributes to the heating of the neutral upper atmosphere. Emission of radiation can have different consequences. The emitted photon could be lost to the upper atmosphere if it escapes into interplanetary space or the middle atmosphere without being reabsorbed. Should it have sufficient energy, the photon could trigger a secondary photoionization, thereby inducing a further energy transfer cycle from the beginning. Finally, the emitted radiation could produce a photodissociation and the associated heat production would have to be investigated in a separate energy flow diagram. The following example

$$O_2 + \underbrace{photon(\lambda = 150 \text{ nm})}_{8 \text{ eV}} \rightarrow \underbrace{O + O^{(*)}}_{5 \text{ eV}} + \underbrace{\text{excess energy}}_{3 \text{ eV}}$$

shows, however, that this process will lead to direct heating of the neutral upper atmosphere because the excess energy will be at least partly converted to kinetic energy of the two resulting oxygen atoms (the asterisk in parenthesis indicates that the particles may be created in an excited state during the dissociation process). Even so, a large part of the dissociation energy (roughly 5 eV) is lost to the upper atmosphere, because it is transported with the oxygen atoms into the middle atmosphere and released there via recombination (see Section 2.3.7).

As indicated in the left part of Fig. 3.20, the potential chemical energy of the ion is mostly converted into heat and thus contributes to the heating of the neutral gas. The primary reactions responsible for this process, treated in more detail in Chapter 4, can be skipped for the present. Only a small fraction of the available energy goes toward heating the ion gas, and even most of this energy is ultimately transferred to the neutral gas via heat exchange.

The numerous processes presented in Fig. 3.20 indicate that calculating the heating of the neutral upper atmosphere is a rather complicated assignment that can really only be handled with the help of numerical models. This is all the more so because Fig. 3.20 only shows the most important, by no means all, energy transfer channels. Moreover, additional processes beginning with a primary photodissociation event must be considered when determining the total heat production. In view of this complexity, it suffices here to review some of the main results of the presently available model calculations.

One of these results deals with the effectiveness of the various heating mechanisms. It turns out that the upper atmosphere is heated primarily by

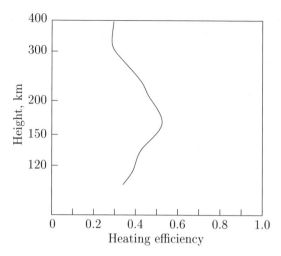

Fig. 3.21. Representative heating efficiency height profile (adapted from Roble et al., 1987)

the photodissociation of molecular oxygen below about 150 km altitude; the principal heat source is exothermal chemical reactions in the range between 150 and 250 km altitude (e.g. $O^+ + O_2 \rightarrow O_2^+ + O + 1.56$ eV; $O_2^+ + e \rightarrow O + O^{(*)} + 12$ eV); and heat exchange with the hotter electron and ion gases is mostly responsible for the heating at altitudes above ca. 250 km. The concept of *heating efficiency* has been introduced in order to describe the total effectiveness of these heating processes

$$\eta^W = q^W / q^E \tag{3.37}$$

As before, q^E is the radiation energy absorbed per unit volume and time and q^W is the heat produced by this energy deposition. The dependence of this parameter on height is shown in Fig. 3.21. It may be seen that only 30 to 55% of the absorbed radiation energy is converted into heat. The rest is channeled primarily into potential chemical energy or into radiation out of the given volume element.

Knowing the heating efficiency, it is possible to determine the heat production profile from a given energy deposition profile; see Fig. 3.22. It is important here to distinguish between the heat production rate per volume and the heat production rate per particle. As shown, this latter quantity is essentially constant above 200 km altitude. Indeed, the volumetric heat rate decreases exponentially with height, but this is also the case for the number of particles, which divide the heat input among themselves. Figure 3.22 also indicates that extreme ultraviolet radiation is mostly responsible for heating the upper atmosphere above ca. 150 km and the far ultraviolet range (particularly the Schumann-Runge continuum) is the main agent below this height.

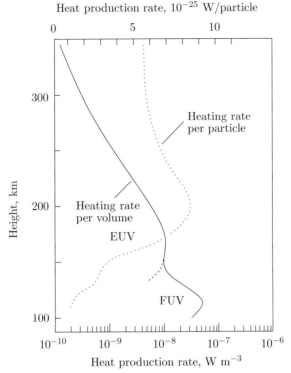

Heat production rate, 10^{-25} W/particle

Fig. 3.22. Height profile of the heat production rate. The conditions correspond to those of Fig. 3.19.

3.3.2 Temperature Increase from Heating

In order to get a feeling for the temperature increase produced by the generation of heat, let us consider specifically the upper thermosphere. The optical depth here is small and the transmission factor near unity, so that the unattenuated incident solar radiation is available. This will be the case above 300 km altitude for average solar activity and moderately slanted incident radiation. Under these conditions Eqs. (3.31) and (3.37) may be combined as

$$q^W (h \gtrsim 300 \text{ km}) \simeq \eta^W \; \sigma^A \; n \; \phi^E_\infty \qquad (3.38)$$

This heat input increases the internal energy and hence the temperature, but it also leads to an expansion of the gas. Considering a gas volume that expands with the atmosphere such that the weight of the gas column supported by this volume remains the same, then the expansion occurs at constant pressure. In this case Eq. (2.32) is valid and the heat input per mass $\Delta Q'$ will produce an increase in temperature $\Delta T = \Delta Q'/c_p$. We can obtain the rate of temperature increase resulting from a given heat input per gas volume and unit time if we multiply $\Delta Q'$ by the mass density ρ and divide by the time interval Δt. This yields

$$\frac{\Delta T}{\Delta t} = \frac{1}{\rho\,c_p}\left(\frac{\rho\,\Delta Q'}{\Delta t}\right) = \frac{q^W}{\rho\,c_p} = \frac{q^W}{n\,k\,(1+f/2)} \qquad (3.39)$$

or with Eq. (3.38)

$$\frac{\Delta T}{\Delta t} = \frac{\eta^W\,\sigma^A\,\phi_\infty^E}{k\,(1+f/2)} \qquad (3.40)$$

Substituting some numerical values into the above formula ($\eta^W = 0.3$, $\sigma_O^A(EUV) \simeq 10^{-21}$ m^2, $\phi_\infty^E(EUV) \simeq 4$ mW/m^2, $f_o = 3$), we obtain the astoundingly high temperature increase rate of ca. 125 K/h. This is many times higher than what is actually observed. Under the conditions given in Fig. 3.25, for example, the entire daily temperature increase is only about 160 K. This value would be exceeded in slightly more than one hour if only the heating process is considered. The increase in temperature is evidently moderated very effectively by heat loss processes, which will be described in more detail in the following.

3.3.3 Heat Losses by Radiative Cooling

One must distinguish between the loss processes by which the heat is really lost and those by which the heat is merely redistributed. Radiative cooling, by which the heat is converted to electromagnetic radiation and irreplaceably lost to the heat budget, belongs to the first category. Heat convection and molecular heat conduction are examples from the second category, because the heat is extracted locally but not destroyed. Whereas convection redistributes the heat by transport of the entire gas volume, including its heat content, molecular heat conduction relies on the thermal motion of individual gas particles in a gas at rest. All of these heat loss processes are important for the upper atmosphere, heat conduction being dominant in the upper thermosphere. Radiative cooling, the main loss process in the lower thermosphere, will be discussed first.

Collisions between gas particles moving with their thermal velocities can induce these to emit radiation. Checking the possible excitation states of the main thermospheric gases (O, N$_2$, O$_2$, He), it is found, however, that such thermal excitation is quite rare at the predominant temperatures of 500 to 2500 K ($\hat{=}$ 0.06 to 0.32 eV). Only fine structure transitions are excited more frequently, because their energy level differences are relatively small. One of the more prominent examples of such fine structure is the ground state triplet splitting of atomic oxygen, for which the excited energy levels are at 0.02 and 0.028 eV; see Fig. 3.30. Excitation of the first of these two transitions does yield infrared emission at 63 μm, but calculations show that the upper atmosphere loses only a few per cent of its heat content because of this process. As molecular oxygen and nitrogen are known to be even less effective in this regard, it may be concluded that radiative losses from the major thermospheric gases are small.

Radiative cooling is important, however, for trace gases like nitrogen oxide (NO) and carbon dioxide (CO_2), because their vibrational and rotational transitions are effectively excited at thermospheric temperatures. Examples are the 5.3 μm (0.23 eV) infrared radiation of NO and the 15 μm (0.08 eV) infrared line of CO_2. This radiation only becomes more intense in the lower thermosphere ($h \lesssim 150$ km), however, where these trace gases attain sufficiently large densities. Radiative cooling then dominates the heat loss of the upper atmosphere.

3.3.4 Heat Losses by Molecular Heat Conduction

Convection plays only a subordinate role in the *vertical* transport of heat in the upper atmosphere because turbulence is suppressed by the large viscosity, and the positive temperature gradient maintains a stable stratification of the gas. This mechanism, just like radiative cooling, can thus be discarded as a heat loss process in the upper thermosphere. Molecular heat conduction is essentially the only remaining alternative capable of keeping the heating of this region within its observed limits. (More precisely, this is valid only for a horizontally uniform or a global mean atmosphere, within which *horizontal* heat convection by winds can be neglected.) Because molecular heat conduction is also important for other planetary and stellar atmospheres, it will be treated here in some detail. We first examine the effectiveness of the heat conduction and then determine the associated heat loss.

Molecular Heat Conduction. Thermal gradients in a gas are associated with a heat flux due to the thermal motion of the particles that strives to eliminate the existing inhomogeneities. An assessment of this effect proceeds in a manner similar to that used for diffusion (compare with Section 2.3.5). We consider a gas at rest of uniform density that has a temperature gradient in the positive z direction. In the following we determine the heat flux induced by the temperature gradient that passes through a reference surface oriented perpendicular to the z direction. Remember here that the transport process due to the thermal motion of the gas particles takes place at the random thermal velocity of the particles and over distances comparable with the mean free path. We are therefore dealing with the heat exchange of two volumes separated by just this distance, one of which is above the reference surface and the other one below; see Fig. 3.23. Again assuming a reduced velocity distribution, 1/6 of the particles in each volume will move toward the reference surface A, penetrate this surface and enter the other volume without suffering a collision. The exchange time for this amounts to $\Delta t = l/\bar{c}$. According to Eq. (2.25), each particle carries an amount of heat equal to

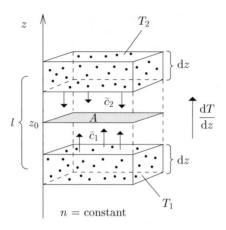

Fig. 3.23. Deriving the molecular heat flux

$f(kT/2)$, thereby establishing a heat transport through the reference surface. The heat flux emanating from each of the two volumes is

$$(\mathrm{d}\phi_z^W)_1 = \left[\frac{1}{6} \, n \, A \, \mathrm{d}z \left(\frac{f \, k \, T_1}{2}\right)\right] \Big/ \left(A \, \frac{l}{\bar{c}_1}\right) = \frac{1}{12} \, k \, f \, n \, \bar{c}_1 \, T_1 \, \frac{\mathrm{d}z}{l} \quad (3.41)$$

and

$$(\mathrm{d}\phi_z^W)_2 = \frac{1}{12} \, k \, f \, n \, \bar{c}_2 \, T_2 \, \frac{\mathrm{d}z}{l} = \frac{1}{12} \, k \, f \, n \left(\bar{c}_1 \, T_1 + \frac{\mathrm{d}(\bar{c} \, T)}{\mathrm{d}z} \, l + \cdots\right) \frac{\mathrm{d}z}{l} \tag{3.42}$$

The net heat flux through the surface A resulting from the heat exchange of the two volumes is thus

$$\mathrm{d}\phi_z^W = (\mathrm{d}\phi_z^W)_1 - (\mathrm{d}\phi_z^W)_2 \simeq -\frac{1}{12} \, k \, f \, n \, \frac{\mathrm{d}(\bar{c} \, T)}{\mathrm{d}z} \, \mathrm{d}z = -\frac{1}{8} \, k \, f \, n \, \bar{c} \, \frac{\mathrm{d}T}{\mathrm{d}z} \mathrm{d}z$$

where in the last step we have made use of the expression for the mean speed $\bar{c} = \sqrt{8kT/\pi m}$. When adding these contributions to obtain the total heat flux, we assume that the temperature gradient (not the temperature itself) is constant over the distance of a mean free path, and that the random velocity can be approximated by its value at the reference surface A. In this case we have

$$\phi_z^W = -\frac{1}{8} \, k \, f \, n \, \bar{c} \, \frac{\mathrm{d}T}{\mathrm{d}z} \int_{z_0-l}^{z_0} \mathrm{d}z = -\frac{1}{8} \, k \, f \, n \, \bar{c} \, l \, \frac{\mathrm{d}T}{\mathrm{d}z} \tag{3.43}$$

As an abbreviation we introduce the *heat conductivity*, which is proportional to the density, the mean random speed and the mean free path

$$\kappa = \xi' \, k \, f \, n \, \bar{c} \, l_{1,1} = \xi \, k^{3/2} \, f \, \sqrt{T}/(\sqrt{m} \, \sigma_{1,1}) = a_\kappa \, \sqrt{T} \tag{3.44}$$

where ξ' and ξ are numerical constants and we have placed indices on the mean free path and collision cross section for clarity. The relation for the heat flux now assumes the following simple form (*Fourier's law*)

$$\phi_z^W = -\kappa \frac{dT}{dz} \tag{3.45}$$

or, when extending to three dimensions

$$\vec{\phi}^W = -\kappa \text{ grad } T \quad (= -\kappa \nabla T) \tag{3.46}$$

The factor ξ in Eq. (3.44) is about 0.14 in our estimate. More exact calculations yield the considerably larger value $\xi = 25 \sqrt{\pi}/64 \simeq 0.7$. More important, of course, is that our estimate correctly reflects the dependence of the heat flux on the various gas parameters.

For later reference, we write here from Eq. (3.44) the explicit relation for the heat conductivity of molecular nitrogen

$$\kappa_{N_2} \text{ [W/K m]} \simeq 2 \cdot 10^{-3} \sqrt{T \text{ [K]}} \tag{3.47}$$

to sufficiently good agreement with experimental results. Remarkably, the heat conductivity does not depend on the density. High density does correspond to a larger number of 'heat carriers', but at the same time it also strongly inhibits their mobility. The heat conductivity thus has a dependence on the temperature alone, which, in turn, leads to a weak dependence on the altitude.

Heat Losses by Heat Conduction. The heat losses caused by heat conduction, of course, do not depend exclusively on the magnitude of the heat flux. A large, but spatially constant heat flux will bring the same amount of heat into a given volume as it takes out – the net effect is zero. Evidently, it is the divergence of the heat flux that determines the net gain or loss of heat. Formally, and completely analogous to the derivation of the continuity equation (2.17), this can be demonstrated using the following balance approach.

We consider a gas volume with base area A and height Δz, as sketched in Fig. 3.24. A heat flux transports the amount of heat ΔQ^+ through the upper boundary into the volume during the time Δt. Simultaneously, an amount of heat ΔQ^- is transported out through the lower boundary. These changes in the heat budget are given by

$$\Delta Q^+ = -\phi_z^W(z+\Delta z/2) \, A \, \Delta t \simeq -\left(\phi_z^W(z) + \frac{d\phi_z^W}{dz}\frac{\Delta z}{2} + \cdots\right) A \, \Delta t \tag{3.48}$$

and

$$\Delta Q^- = -\phi_z^W(z-\Delta z/2) \, A \, \Delta t \simeq -\left(\phi_z^W(z) - \frac{d\phi_z^W}{dz}\frac{\Delta z}{2} + \cdots\right) A \, \Delta t \tag{3.49}$$

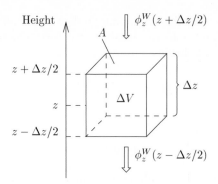

Fig. 3.24. Determining the change in heat content of a gas volume per volume and per time caused by an inhomogeneous heat flux

where the minus sign on ϕ_z^W designates the flux direction (negative z direction). The net change in the heat content of our volume is therefore

$$\Delta Q = \Delta Q^+ - \Delta Q^- = -\frac{d\phi_z^W}{dz}\, \Delta V\, \Delta t$$

Accordingly, the change in heat content per volume and per time induced by the heat flux can be written

$$d_z^W = \frac{\Delta Q}{\Delta V\, \Delta t} = -\frac{d\phi_z^W}{dz} \tag{3.50}$$

or, when extending to three dimensions

$$d^W = -\operatorname{div}\vec{\phi}^W \quad (= -\nabla\vec{\phi}^W) \tag{3.51}$$

It should be noted here that d^W (d again coming from *divergence*) is negative and the volume under consideration loses heat (a heat sink) if the divergence is positive, i.e. if a heat flux in the positive z direction increases with z, or (as in Fig. 3.24) a heat flux in the negative z direction decreases with z. Conversely, the case when d^W is positive corresponds to a heat source.

3.3.5 Heat Balance Equation and Temperature Profile

Heat loss must be considered along with the heat input in order to obtain a temperature increase rate commensurate with the observations. A corresponding extension of the estimate (3.39) leads to the following heat balance equation

$$\rho\, c_p\, \frac{\partial T}{\partial t} \simeq q^W - l^W + d^W \tag{3.52}$$

Evidently, the temporal change of the heat content of a gas volume is determined by its *effective* heating rate. This is composed of the heat production

q^W, the 'real' heat losses by radiative cooling l^W, and the heat gain or loss from transport processes d^W. In the case of the one-dimensional, horizontally uniform upper atmosphere considered here, d^W is limited essentially to the heat inflow or outflow caused by molecular heat conduction.

Apparently, the heat balance equation above is a partial, nonlinear differential equation that can only be solved numerically. Just recall here the complicated form of the heat production term given by Eqs. (3.36) and (3.37). A general idea of the temperature profile in the upper atmosphere can be obtained, however, when a greatly simplified form of the heat balance equation is considered. Radiative cooling, for example, can be neglected to a good approximation within the upper thermosphere. Furthermore, for times during the day and well removed from sunrise or sunset, the left side of the heat balance equation can be neglected with respect to the production and transport terms on the right side. This is surely the case in the vicinity of the temperature maximum ($\partial T/\partial t \simeq 0$), but this approximation is even justified during the increasing temperature phase because the increase rate stays moderate (according to Fig. 3.25 it is of the order of 17 K/h, i.e. negligible compared with the production rate of 125 K/h). We are left with an equilibrium situation, in which the heat production is just compensated by the heat conduction losses

$$q^W \simeq -d^W = \text{div}\vec{\phi}^W = \frac{d\phi_z^W}{dz} \qquad (3.53)$$

where the last step above assumes again that we have a planar, horizontally uniform atmosphere. Integrating this relation over the height interval from h to ∞ yields

$$\int_h^\infty q^W(z)\,dz \simeq \int_{\phi_z^W(h)}^0 d\phi_z^W = -\phi_z^W(h) = \kappa(h)\,\frac{dT}{dz}(h) \qquad (3.54)$$

where the heat flux at infinity is set equal to zero. Evidently, for this particular equilibrium situation, the total heat produced in a gas column with a base area of unity (left side of Eq. (3.54)) is converted into a heat flux passing downward through the base of the column. For an explicit calculation of the integrated heat production rate we consider a single gas atmosphere with average properties (σ^A, $\eta^W \neq f(h)$) and introduce the optical depth as a new variable. Using

$$\tau(z) = \sec\chi\,\sigma^A \int_z^\infty n(z')\,dz' \quad \text{or} \quad dz = -d\tau/(\sec\chi\,\sigma^A\,n(z)) \qquad (3.55)$$

and Eqs. (3.31) and (3.37), we obtain

$$\int_h^\infty q^W(z)\,dz = -\eta^W\,\cos\chi\,\phi_\infty^E \int_{\tau(h)}^0 e^{-\tau}d\tau = \eta^W\,\cos\chi\,\phi_\infty^E\,(1 - e^{-\tau(h)}) \qquad (3.56)$$

Substituting this into Eq. (3.54) finally yields

$$\frac{\mathrm{d}T}{\mathrm{d}z}(h) \simeq \eta^W \cos\chi \, \phi_\infty^E \, (1 - \mathrm{e}^{-\tau(h)})/\kappa(h) \tag{3.57}$$

This equation for the temperature gradient allows us to make some important statements about the temperature profile in the upper atmosphere. For example, if we let the height go to infinity, the optical depth goes to zero and therefore also the temperature gradient. It follows that the temperature at great heights must be nearly constant, in agreement with the observations. Furthermore, because the optical depth becomes quite small at a point 2-3 scale heights above the absorption height, the asymptotic temperature (i.e. the thermopause temperature) should be reached already at an altitude of 250 to 350 km, again in agreement with the observations. The isothermal behavior of the upper thermosphere is immediately evident from a physical point of view. The heat content of this region is very small because of the low densities, but the heat conductivity stays large. This means that even the slightest temperature gradients are sufficient to induce the heat fluxes necessary for equalizing the temperature.

The temperature gradient must not disappear entirely, however. In fact, it must be large enough at any given height to enable downward flow of all heat produced above this height. This requirement implies that the temperature gradient must become increasingly steeper with the increasing heat production in the middle thermosphere (i.e. below ca. 250 km altitude), in order to support a sufficiently large heat flux. According to Eq. (3.57), the maximum temperature gradient reached this way is

$$\left(\frac{\mathrm{d}T}{\mathrm{d}z}\right)_{max} \simeq \frac{\eta^W \cos\chi \, \phi_\infty^E}{\kappa} \tag{3.58}$$

Assuming that this value is reached at an altitude of about 150 km (most of the EUV radiation is absorbed above this height) and using $\eta^W \simeq 0.4$, $\phi_\infty^E(\mathrm{EUV}) \simeq 4$ mW/m^2, $\kappa_{N_2}(150 \text{ km}) \simeq 2 \cdot 10^{-3} \sqrt{T(150 \text{ km})} \simeq 52$ mW/K m and $\cos\chi \simeq 0.5$, one obtains a temperature gradient of ca. 15 K/km, in rather good agreement with the temperature gradient actually measured there.

The explanation outlined here for the temperature gradient in the terrestrial thermosphere is also applicable to other planetary thermospheres. In any case, the outer regions of these upper atmospheres are nearly isothermal because of their low thermal capacity and high heat conductivity. Moreover, since these upper atmospheres are composed of infrared inactive gases, molecular heat conductivity is the dominant loss process. This further implies that the outer gas envelopes will continue to warm up until the growing temperature gradient suffices to discharge the heat produced at all higher altitudes via heat conduction. Heat production and heat loss via conduction thus compensate each other in the equilibrium case. These principles are also valid for the upper atmosphere of the Sun, as shown by the temperature profile in the transition region and corona. Some differences in the details arise here from the fact that we are dealing with an ionized, rather than a neutral gas. It thus follows that magnetic fields play an important role in channeling the heat fluxes.

3.3.6 Estimate of the Thermopause Temperature

In order to derive the thermopause temperature explicitly, we again consider a planar, single gas atmosphere with average properties, for which neither the absorption cross section, the heating efficiency, nor the mean molecular mass depends on the height. We also account for heating effects of the EUV radiation only. Combining Eqs. (3.44) and (3.57) we obtain

$$T^{1/2} \, dT \simeq \eta^W \, \cos \chi \, \phi_\infty^E \, (1 - e^{-\tau}) \, dz \, / \, a_\kappa \tag{3.59}$$

Introducing again the optical depth as an independent variable from Eqs. (3.55) and (3.27)

$$dz = -d\tau/(\sec \chi \, \sigma^A \, n) \simeq -d\tau \, H/\tau$$

we can write Eq. (3.59) in the following way

$$T^{-1/2} \, dT \simeq -\frac{\eta^W \, \cos \chi \, \phi_\infty^E \, k}{a_\kappa \, m \, g} \left(\frac{1}{\tau} - \frac{e^{-\tau}}{\tau} \right) \, d\tau \tag{3.60}$$

Integrating this relation from a lower boundary h_0, where $T(h_0) = T_0$ and $\tau(h_0) = \tau_0$, up to a height where the temperature approaches the thermopause temperature and the optical thickness goes to zero, the left side of the above equation goes to $2(T_\infty^{1/2} - T_0^{1/2})$. The right side, not integrable in closed form, must be either numerically integrated or described by a (convergent) series

$$-\int_{\tau_0}^0 \left(\frac{1}{\tau} - \frac{e^{-\tau}}{\tau} \right) \, d\tau = \int_0^{\tau_0} \left(\frac{1}{\tau} - \left(\frac{1}{\tau} - 1 + \frac{\tau}{2!} - \frac{\tau^2}{3!} + \frac{\tau^3}{4!} - \cdots \right) \right) \, d\tau$$

$$= \tau_0 - \frac{\tau_0^2}{2 \cdot 2!} + \frac{\tau_0^3}{3 \cdot 3!} - \frac{\tau_0^4}{4 \cdot 4!} + \cdots = J(\tau_0) \tag{3.61}$$

see also Eq. (A.14). The thermopause temperature can thus be expressed as

$$T_\infty \simeq \left[\sqrt{T_0} + C \cos \chi \, \phi_\infty^E \right]^2 \tag{3.62}$$

where the constant C is a product of the time independent parameters

$$C = \eta^W \, k \, J(\tau_0)/(2 \, a_\kappa \, m \, g) \tag{3.63}$$

In agreement with our assumptions, we choose the lower boundary height as $h_0 = 150$ km and the corresponding boundary temperature $T_0 = 670$ K,

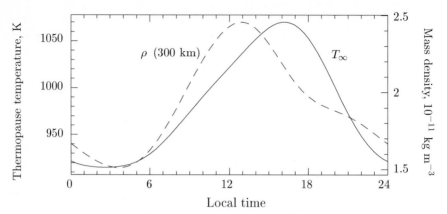

Fig. 3.25. Daily variation of the thermopause temperature T_∞ and the mass density ρ at a height of 300 km for mid-latitudes (51°N, 7°E) during spring equinox (21 March). The solar activity is taken as moderate (CI = 120); the geomagnetic activity as weak (Kp = 2). The variations were calculated using the empirical (i.e. observationally based) MSIS 86 model (Hedin, 1987).

see Appendix A.4. Using $\chi = 60°$, $\sigma^A(\text{EUV}) \simeq 10^{-21}$ m^2, $n(h_0) \simeq 5 \cdot 10^{16}$ m^{-3}, $\mathcal{M}(h_0) = 24$ and $H(h_0) \simeq 25$ km, we obtain an optical depth of $\tau_0 = \sec\chi \, \sigma^A n H = 2.5$. A calculation of the series (3.61) for this value yields $J(\tau_0) \simeq 1.5$. Further selecting $\eta^W \simeq 0.4$, $a_\kappa \simeq (a_\kappa)_{\text{N}_2} \simeq 2 \cdot 10^{-3}$ W/m K$^{3/2}$ and $\phi_\infty^E(\text{EUV}) \simeq 4$ mW/m^2, we obtain $C \, \cos\chi \, \phi_\infty^E \simeq 11$ and a thermopause temperature of $T_\infty \simeq 1400$ K, in sufficiently good agreement with actually measured values, particularly in view of the many simplifications and uncertainties associated with this estimate. Among these simplifications is especially the one dimensional atmosphere, for which any heat exchange by horizontal winds remains excluded (see also Section 3.4 and Appendix A.7).

3.3.7 Temperature and Density Variations

As seen in Eq. (3.62), the thermopause temperature depends on the incidence angle χ and the radiation intensity ϕ_∞^E. It should therefore display daily and seasonal variations as well as a sensitivity to solar rotation and the solar cycle. This is indeed the case and Fig. 3.25 documents the day-night variations at middle latitudes. The amplitude of the diurnal variation during the spring equinox is seen to be about 160 K. One striking feature is that the temperature maximum is not reached at midday, but rather during the afternoon. Among other reasons, this phase shift is due to the previously neglected horizontal heat transport from winds. Another factor is that a considerable fraction of the heat input during the morning warm-up phase must be invested in the work needed for the expansion of the gas. The accumulated heat can be expended primarily for the internal energy and thus for

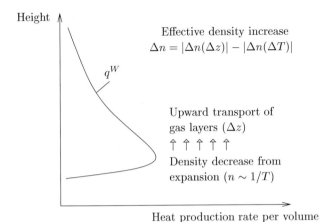

Fig. 3.26. Illustrating the density increase associated with a rise in temperature in the upper thermosphere. $\Delta n(\Delta z)$ and $\Delta n(\Delta T)$ denote the transport-induced and the thermally-induced changes in density.

increasing the gas temperature only after the gas expansion declines in the afternoon hours.

A direct indicator of the expansion and contraction of the gases (sometimes called *atmospheric breathing*) is the variation in density at 300 km altitude, also shown in Fig. 3.25. Given the rise in temperature, the increase in density observed during the day can only be explained by upward transport of denser gases from lower altitudes induced by expansion; see Fig. 3.26. A simple estimate confirms this effect. Let $n_1(h)$ be the density at the height h at a temperature $T_1(h)$ and $n_2(h)$ the density at the same height at a different temperature $T_2(h) > T_1(h)$. Using Eq. (2.48) and assuming constant lower boundary conditions, the ratio of the two densities is

$$\frac{n_2}{n_1}(h) = \underbrace{\frac{T_1}{T_2}(h)}_{<1} \ \underbrace{\exp\left\{\int_{h_0}^{h}\left(1 - \frac{T_1}{T_2}(z)\right)\frac{\mathrm{d}z}{H_1(z)}\right\}}_{>1} \tag{3.64}$$

where the first factor describes the temperature-induced decline and the second represents the transport-induced increase of the density. Evaluating the integral, it is found that the exponential factor already begins to dominate at a point two scale heights above the heat input altitude, thereby resulting in a net increase in density at greater heights. It may be noted that the relative increase of the density in the upper thermosphere is considerably larger than that of the temperature (see Fig. 3.25).

In addition to the diurnal oscillation, distinct variations of the temperature and density are observed that depend on the season, solar rotation, and solar cycle. Examples of these are shown in Figs. 3.27 and 3.28. In contrast,

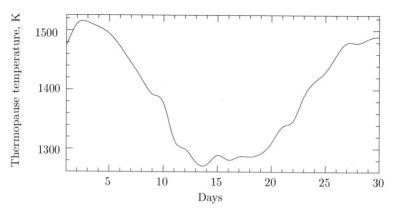

Fig. 3.27. Variation of the thermopause temperature caused by solar rotation in the time interval 20 June to 19 July 1980. The maximum Covington index during this period was CI=255; the minimum was CI=147. The calculation is based on the MSIS 86 model (Hedin, 1987) for a mid-latitude location (51°N, 7°E) at a local time of 1400 hours.

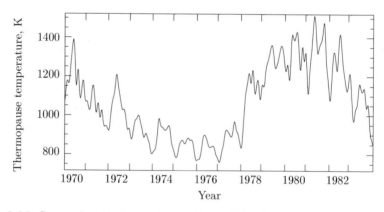

Fig. 3.28. Seasonal and solar cycle variations of the thermopause temperature in the years 1970 to 1983. The calculation, based on the MSIS 86 model (Hedin, 1987), is carried out for the 15th day of each month for a local time of 1400 hours at a mid-latitude location (51°N, 7°E). Semi-annual oscillations, which are a result of global thermospheric dynamics, are clearly seen in addition to the annual variations.

because of the relatively long reaction time of the thermosphere, solar flare effects are very small.

3.3.8 Airglow

The upper atmosphere is not only heated by the absorption of solar radiation but also stimulated to fluoresce. This *airglow* is below the level of visibility

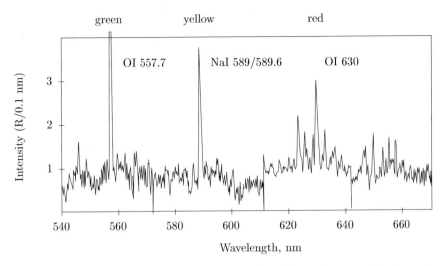

Fig. 3.29. Spectrum of the nightglow in the visible. The ordinate scale is intensity per wavelength interval (adapted from Broadfoot and Kendall, 1968).

from the ground and therefore rather unspectacular. However, it does contain a wealth of information about the thermosphere and is thus a subject of intensive research. We will be content here with a few remarks on this topic.

Similar to what we learned for the lower atmosphere (key: 'blue sky'), a large fraction of the dayglow in the upper atmosphere consists of scattered solar radiation. For example, solar Lyman-α radiation at 121.6 nm excites the Earth's hydrogen envelope, the geocorona, to fluoresce (see Fig. 7.16a). The intense glow of the sunlit upper atmosphere in Fig. 7.16d is also partly attributable to resonantly scattered sunlight. The UV radiation seen in this image is dominated by emission from atomic oxygen at 130.4 nm (more precisely the triplet at 130.2, 130.5 and 130.6 nm, see Fig. 3.30). These prominent emission lines are also excited by inelastic collisions with photoelectrons.

Another component of the airglow is created by chemical reactions. This chemical fluorescence is responsible for the *nightglow* of the upper atmosphere. As shown in Fig. 3.29, the nightglow spectrum in the visible range is dominated by the green line of atomic oxygen at 557.7 nm. This radiation, which attains its maximum at a height of 95-100 km, arises from the recombination of atomic oxygen

$$
\begin{aligned}
O + O + M &\rightarrow O_2{}^* + M \\
O_2{}^* + O &\rightarrow O_2 + O(^1S) \\
O(^1S) &\xrightarrow{1s} O(^1D) + \text{photon}(557.7 \text{ nm})
\end{aligned}
\tag{3.65}
$$

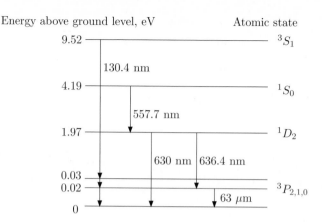

Fig. 3.30. Some excitation states and corresponding emission lines of atomic oxygen important for the thermosphere

The symbol M stands for an arbitrary three-particle collision partner (usually N_2) and the asterisk means that the oxygen molecule resulting from the recombination is often created in an excited state. The energy diagram of the transition responsible for the emission of the 557.7 nm radiation is displayed in Fig. 3.30. Noting the long lifetime of the ^1S-state, it is clear that this metastable state is de-excited by a 'forbidden' transition. For comparison, the 'normal' life expectancy of an excited state is only of the order of 10^{-8}s!

The intensity of the 557.7 nm radiation, as with all other airglow emissions, is given in units of rayleighs (R, named after the son of the more famous Lord Rayleigh, a pioneer of airglow research in the 1920s). One rayleigh corresponds to the emission rate of 10^6 photons per second radiated isotropically from an atmospheric column with a base area of 1 cm^2. The form of this unit is adapted to measurement conditions: a photometer or spectrometer is able to record only the integrated luminosity within a column along the viewing direction. It is sufficient to state here that 1 kilorayleigh (kR) corresponds approximately to the luminosity of the Milky Way. For the typical intensities of a few hundred rayleighs measured from the ground, it is obvious that the fluorescence of the 557.7 nm line is hardly visible. Observed from an orbiting space station in a direction tangential to the horizon, however, the emission layer becomes clearly perceptible because of the considerably larger column density (see Fig. 7.16c).

Another conspicuous emission of the nightglow is the yellow doublet (unresolved) D lines of sodium at 589 and 589.6 nm. The sodium atoms responsible for this emission were originally released from meteorites that disintegrated in the atmosphere at altitudes of 80–100 km.

Finally, a third prominent emission is the red line of atomic oxygen at 630 nm, which originates essentially from the dissociative recombination of molecular oxygen ions

$$O_2^+ + e \rightarrow O + O(^1D)$$

$$O(^1D) \xrightarrow{110s} O(^3P) + photon(630 \text{ nm})$$

(3.66)

The extraordinarily long lifetime of the metastable ^1D state basically restricts the emission of the 630nm line to heights above 200 km. At lower altitudes the metastable state is de-excited by frequent collisions with other gas particles (collisional quenching). The longevity of the excited state also means that the emitting oxygen atoms are thermally (and certainly also dynamically) integrated into their environment. Measurements of the 630 nm line from satellites or from ground can thus be utilized to determine the temperature (from the line broadening) and wind velocity (from the line shift) of the upper atmosphere.

3.4 Thermospheric Winds

Our previous considerations may have left the impression that the thermosphere is a more or less static phenomenon. This impression is deceiving: the thermosphere is actually in motion on all spatial and temporal scales. The dominant mode of motion is corotation with the Earth. This corotation is evidently transferred to the outer gas envelope first by friction from the Earth's surface and subsequently by frictional forces between the individual gas layers. The resultant *corotation velocity* is given by

$$u_{corotation} = \Omega_E(R_E + h) \cos \varphi$$

(3.67)

where Ω_E and R_E, as earlier, denote the angular velocity and radius of the Earth and φ is the geographic latitude. For heights near 300 km, corotation velocities of 500 m/s are reached at the equator and still a good 50% of this value at a latitude of 60°. Superposed on this corotation are numerous additional motions, among which the most important are the global wind circulation and atmospheric waves. The global wind circulation, taking the diurnal winds as an example, is documented in the following. The momentum balance equation needed for the calculation of such winds is then presented.

3.4.1 Diurnal Wind Circulation: Observations

As indicated in Fig. 3.25 of the preceeding section, considerable temperature and density differences exist in the thermosphere between the day and night sides. The resulting pressure differences are shown in Fig. 3.31 for a mid-latitude location and in Fig. 3.32 on a global scale. The latter figure shows

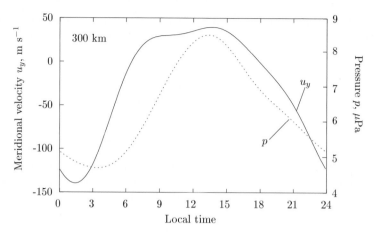

Fig. 3.31. Diurnal variation of the pressure p and the meridional wind velocity u_y for an altitude of 300 km at mid-latitudes (51°N, 7°E) during the spring equinox (21 March). The solar activity is moderate (CI=100); the geomagnetic activity weak (Kp=2). Positive velocities correspond to northward directed winds (see Fig. 3.33). The displayed variations were calculated with the help of the empirical (observationally based) models MSIS 86 (Hedin, 1987) and HWM 93 (Hedin, 1996).

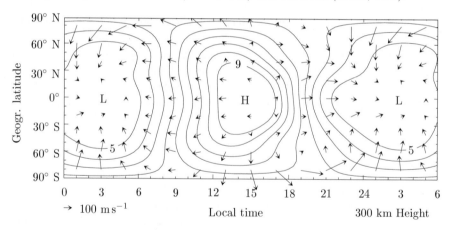

Fig. 3.32. Global pressure and wind distribution for an altitude of 300 km during the spring equinox (21 March). The distribution was computed for the time 12 UT, moderate solar activity (CI=100) and weak geomagnetic activity (Kp=2). The contours are lines of constant pressure (isobars). The highest plotted pressure level of the high pressure region (H) corresponds to a value of 9 μPa; the lowest level of the low pressure region (L) to a value of 5 μPa. The difference between isobars is thus 0.5μPa. Wind speed and direction at a given point are denoted by arrows drawn away from the point in question. The wind speed scale is shown below the abscissa at lower left. The displayed distribution was calculated using the empirical models MSIS 86 (Hedin, 1987) and HWM 93 (Hedin, 1996).

lines of constant pressure, isobars, as observed at an altitude of 300 km during the spring equinox and moderate solar activity. Under these conditions there exists a well-defined zone of high pressure in the afternoon sector at equatorial latitudes. Because of its shape, this is sometimes called the *pressure bulge*. By comparison, the low pressure region shifted by 12 hours is considerably flatter and the position of its minimum is thus less sharply defined. The total pressure difference between the high and low pressure regions, amounting to somewhat more than 4 μPa, is sufficient to generate significant balancing winds. This airflow, denoted the *tidal wind* because of its 24-hour periodicity, is also displayed in Fig. 3.32. It generally follows the pressure gradient, attaining velocities as high as 200 m/s ($\cong$ 700 km/h!) at mid-latitudes. It should be noted that this diagram holds for an observer moving with the rotating Earth. Otherwise the corotation of the upper atmosphere must be superposed onto the winds. In the following we address the issue of how these winds can be calculated. We begin by considering the various forces acting on a moving gas and then sum these together in a force equilibrium relation.

3.4.2 Inventory of Relevant Forces

Consider a stationary volume element in an inertial system, through which various gases are flowing at different velocities. We are interested in those forces which act on any of these gases within the given volume element. Applying the condition that the sum of all forces must be zero, one can derive information about the dynamical state of the gas. When conducting an inventory of forces, it is convenient to discriminate between so-called internal forces, forces acting from outside or external forces, frictional forces and the inertial force. In addition, we must consider the apparent forces associated with the transformation from an inertial system to a coordinate system rotating with the Earth. In order to keep the qualitative description of these forces as simple as possible, we restrict our attention here to the important case of horizontal gas motion. The x-coordinate points toward east, the y-coordinate toward north and the z-coordinate is directed upward; see Fig. 3.33. Choosing the x-coordinate (east-west direction) as representative for the horizontal, we first consider only the x-component of the relevant forces.

Internal Forces. Internal forces arise from the momentum transport carried by the motion of the gas particles. The gas in the volume element considered can gain or lose a quantity of momentum I due to the motion of the gas particles. In either case, the temporal change of momentum of the gas volume corresponds to an internal force acting on this gas volume given by $\partial I/\partial t$. The volumetric force generated by an inhomogeneous momentum flux can be determined with the help of Fig. 3.34a. Independent of the particular form of the momentum, a quantity of momentum ΔI_x^+ will be transported into the volume through the surface A on the left side (the notation is the same as introduced in Section 2.1.2). Simultaneously, a quantity of momentum ΔI_x^- is transported out of the volume on the right side. One obtains

$$\Delta I_x^+ = \phi_{x,x}^I(x - \Delta x/2) \, A \, \Delta t = \left(\phi_{x,x}^I(x) - \frac{\partial \phi_{x,x}^I}{\partial x} \frac{\Delta x}{2} + \cdots \right) A \, \Delta t$$

and

$$\Delta I_x^- = \phi_{x,x}^I(x + \Delta x/2) \, A \, \Delta t = \left(\phi_{x,x}^I(x) + \frac{\partial \phi_{x,x}^I}{\partial x} \frac{\Delta x}{2} + \cdots \right) A \, \Delta t$$

so that the net change in momentum contained within the volume element ΔV may be written as $\Delta I_x = \Delta I_x^+ - \Delta I_x^- \simeq -(\partial \phi_{x,x}^I/\partial x) \Delta V \Delta t$. Recalling that the temporal change of the momentum contained within the volume corresponds to a force acting on the volume and dividing this by the size of the volume element considered, one obtains the force per unit volume

$$F_x^* = \frac{\Delta I_x/\Delta t}{\Delta V} = -\frac{\partial \phi_{x,x}^I}{\partial x} \tag{3.68}$$

Depending upon whether we consider the momentum associated with the bulk or random velocity, or the momentum transport caused by bulk or random motion, we distinguish in the following between the flow gradient force, the pressure gradient force and the viscosity; see Fig. 3.34.

Flow Gradient Force. The flow gradient force accounts for the flow-induced transport of momentum associated with the flow velocity, for which $\phi_{x,x}^{I(u)} = (mu_x)nu_x$, see also Eq. (2.16). The flow gradient force is therefore given by

$$(F_x^*)_{flow} = -\frac{\partial \phi_{x,x}^{I(u)}}{\partial x} = -\frac{\partial(\rho u_x^2)}{\partial x} \tag{3.69}$$

Pressure Gradient Force. The pressure gradient force is generated by the momentum transport arising from the random motion of the gas particles.

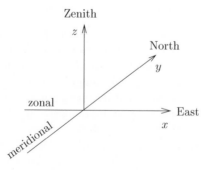

Fig. 3.33. Coordinate system used for the description of thermospheric winds

a. Flow gradient force

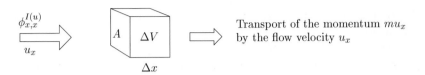

b. Pressure gradient force

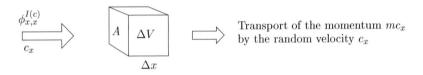

c. Viscosity (force)

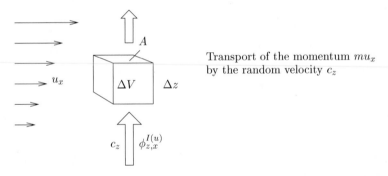

Fig. 3.34. Internal forces generated by momentum transport. For simplification, only the x-component of the momentum, interesting for zonal winds, is considered. (**a**) Transport of the momentum mu_x by the flow velocity u_x. (**b**) Transport of the momentum mc_x by the random velocity c_x. (**c**) Transport of the momentum mu_x by the random velocity c_z.

Thereby, only the momentum component perpendicular to a given surface is considered. The associated net momentum flux has already been determined in Section 2.1.2 as $\phi_{x,x}^{I(c)} = m\,n\,\overline{c^2}/3 = p$, see Eqs. (2.22) and (2.23). Substituting this relation into Eq. (3.68), one obtains the already familiar expression for the pressure gradient force

$$(F_x^*)_{pressure} = -\frac{\partial \phi_{x,x}^{I(c)}}{\partial x} = -\frac{\partial p}{\partial x} \tag{3.70}$$

Viscosity. The situation is somewhat more complicated for the viscosity. Here again, the random velocity of the particles is responsible for the momentum transport, but this time it is not the momentum component perpendicular to a reference surface – as for the pressure – but rather the component parallel to this surface and aligned with the momentum along the given flow direction that is considered. Taken as a statistical average, this corresponds to the momentum associated with the flow motion mu_x. As such, we are dealing here with the transport of a quantity, described by a macroscopic parameter, which is caused by the thermal motion of the particles. The situation corresponds to those encountered earlier in connection with diffusion or molecular heat conduction.

For a better understanding, let us consider the volume element shown in Fig. 3.34c. Gas particles will enter the lower boundary surface of this volume element as a result of their random motion. However, because of the flow speed gradient in the z-direction, these particles carry a smaller momentum in the x-direction than those particles simultaneously leaving the volume through the lower boundary surface. The result is a net transport of momentum mu_x in the negative z-direction, i.e. a net transport that strives to eliminate the existing velocity differences.

For a quantitative treatment of this effect we refer back to the approach used when dealing with diffusion or molecular heat conduction. For example, replacing the heat content of a particle $fkT/2$ in Eqs. (3.41) and (3.42) by its mean momentum mu_x, one obtains (with $n, \bar{c} \neq f(z)$)

$$(d\phi_{z,x}^{I(u)})_1 = \frac{1}{6} m \, n \, \bar{c} \, (u_x)_1 \, dz/l$$

and

$$(d\phi_{z,x}^{I(u)})_2 = \frac{1}{6} m \, n \, \bar{c} \, (u_x)_2 \, dz/l = \frac{1}{6} m \, n \, \bar{c} \, ((u_x)_1 + (du_x/dz) \, l + \cdots) dz/l$$

The net momentum flux through the lower boundary surface is thus

$$\phi_{z,x}^{I(u)} \simeq -\frac{1}{6} m \, n \, \bar{c} \, l \, \frac{du_x}{dz} = -\eta \frac{du_x}{dz} \tag{3.71}$$

where we have assumed that the flow speed gradient (not the flow speed itself) can be considered constant over scales comparable with the mean free path. The *viscosity coefficient* η (also known as the *coefficient of internal friction, dynamic viscosity* or simply *viscosity*) is by definition

$$\eta = \xi' \, m \, n \, \bar{c} \, l = \xi \sqrt{mkT}/\sigma_{1,1} = a_\eta \sqrt{T} \tag{3.72}$$

Similar to the heat conductivity, η is proportional to the product of the density, the mean free path and the mean random velocity. Whereas the factor ξ is equal to $1/(3\sqrt{\pi})$ in our estimate, more accurate calculations yield a distinctly larger value $\xi = 5\sqrt{\pi}/16$.

In order to determine the volume force produced by the momentum fluxes through the upper and lower boundary surfaces, we substitute $\phi_{z,x}^{I(u)}$ into an equation of the form (3.68), but this time corresponding to the z-direction, and obtain

$$(F_x^*)_{viscosity} = -\frac{\partial \phi_{z,x}^{I(u)}}{\partial z} = \frac{\partial}{\partial z}\left(\eta \frac{\partial u_x}{\partial z}\right) \simeq \eta \frac{\partial^2 u_x}{\partial z^2} \qquad (3.73)$$

where we assume in the last step that the viscosity coefficient is independent of z to a first approximation.

It is important to note that viscosity, similar to diffusion and heat conduction, is an equalization process. While diffusion and heat conduction smooth out inhomogeneities in density and temperature, viscosity acts to reduce bulk flow velocity differences. Of course, such velocity differences can also be present in the horizontal direction, but the associated viscosity forces are comparatively weak. This is a consequence of the large horizontal scales over which such velocity changes occur. In the vertical direction, however, the wind velocity can increase from zero to its maximum over a height range of only 100 km. Considerable nonlinearities are also observed in this case. It is thus sufficient to account for only the vertical velocity variations in Eq. (3.73).

External Forces. Gravity is the only external force acting on the neutral gases of the upper atmosphere

$$(\vec{F}^*)_{gravity} = \rho \vec{g} \qquad (3.74)$$

Because this force is directed vertically, it plays no role for the horizontal motion under consideration here.

Frictional Forces: Ion Drag Force. The *ion drag force* is a frictional force between neutral gas particles and ions. A certain fraction of the gas particles in the upper atmosphere, as mentioned earlier, is ionized. When these charged particles move through the Earth's magnetic field, they experience a magnetic force. Here we are not interested in the details of this interaction, as they are discussed at length in Chapter 5. For the present, it is sufficient to know that the charge carriers can move freely only along the Earth's magnetic field and not perpendicular to it. Since the terrestrial magnetic field is tilted to the horizon in the meridional direction, these charge carriers cannot move at all in the zonal direction and can move only conditionally in the meridional direction. For the zonal direction, the ionized particles therefore behave like a stationary gas, an obstacle through which the thermospheric neutral gas must move. The frictional force between two gases of different flow velocity has already been determined in Section 2.3.4. Applying Eq. (2.57) to a zonal wind flow, the ion drag force with $(u_x)_{ions} = 0$ is found to be

$$(F_x^*)_{ion\ drag} = -\rho\,\nu_{n,i}^*\,u_x \qquad (3.75)$$

where ρ is the mass density and $\nu_{n,i}^*$ is the momentum transfer collision frequency between neutrals and ions. Because $\nu_{n,e}^* \ll \nu_{n,i}^*$ (see Eq. (2.56)), the corresponding electron drag force is negligible.

Inertial Force. The sum of all forces acting on a gas volume will produce an acceleration which is opposed by the inertial force. According to Newton's law, this acceleration is equal to the temporal change of the momentum of the body. When the body is a gas, in our example a volume element with mass density ρ and velocity u_x, then the x-component of this inertial force per unit volume is given by

$$(F_x^*)_{inertial} = -\frac{\partial(\rho u_x)}{\partial t} \qquad (3.76)$$

Apparent Forces: Coriolis Force. The above form for the inertial force is valid only for a nonaccelerated coordinate system, i.e. an inertial system. Here we are interested, however, in the wind flow at a given point on Earth (see, for example, Fig. 3.31), i.e. in the wind flow in a coordinate system that rotates with the Earth. Accordingly, Newton's law must be transformed into this corotating system. This yields two additional inertial force terms, which are denoted *apparent forces* because they arise only as a consequence of the transformation. These are the *centrifugal force* and the *Coriolis force*. The centrifugal force has already been discussed in Section 2.3.1. Its vertical component can either be neglected or combined with gravity as part of a reduced gravitational acceleration for more accurate calculations. While the centrifugal force always acts on a gas volume rotating with the Earth, Coriolis forces appear only for gases that are moving relative to the rotating coordinate system. As an example, the form of the Coriolis force for the zonal x-direction is derived here using simple argumentation.

We consider an airflow directed from north to south; see Fig. 3.35. Viewed from outside the Earth, this will follow a straight line trajectory. Observed from the rotating Earth, however, it appears to be deflected toward the west. For the corotating observer this behavior can only be understood as an acceleration toward the west, the magnitude of which is determined as follows. Let the velocity of the airflow in the inertial system be u_y. At this speed it will cover the distance $\overline{AB} = y$ over the time t. A stationary observer of this airflow on the Earth's surface will move because of the Earth's rotation from point B to point C during the same length of time. The distance covered by the observer is given by $x = \Delta v_x\, t$, where Δv_x is the difference between the corotation velocities at points A and B. With $v_x(A) = \Omega_E\, r'$ and $v_x(B) = \Omega_E(r' + \Delta r')$ the difference in velocity is found to be $\Delta v_x = \Omega_E\, \Delta r'$. Finally, noting that $\Delta r' \simeq y\,\sin\varphi$ and $y = u_y\, t$, we arrive at the relation $x = u_y \Omega_E \sin\varphi\, t^2$. The observer on the Earth attributes this deflection x to an acceleration perpendicular to the flow velocity u_y. The fact that the deflection arising from this acceleration is proportional to the square of the time

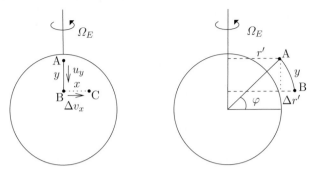

Fig. 3.35. Deriving the Coriolis acceleration

indicates that the acceleration a is constant (for which we have an expression of the form $x = a\,t^2/2$). Comparing these, we obtain a Coriolis acceleration of $a_{Coriolis} = 2u_y \Omega_E \sin\varphi$, and for the x-component of the Coriolis force per unit volume

$$(F_x^*)_{Coriolis} = 2\,\rho\,u_y\,\Omega_E\,\sin\varphi \qquad (3.77)$$

The flow is deflected toward the right in the Northern Hemisphere and to the left in the Southern Hemisphere. An analogous expression can be derived for the Coriolis force in the y-direction associated with the zonal wind flow. In this case, however, the deflection is caused by the horizontal component of the modified centrifugal force.

3.4.3 Momentum Balance Equation

In the following the forces introduced in the preceeding section will be combined in the form of a *force equilibrium relation*. As previously, we consider only forces in the east-west direction (x-component)

$$-\frac{\partial(\rho u_x^2)}{\partial x} - \frac{\partial p}{\partial x} + \eta\,\frac{\partial^2 u_x}{\partial z^2} - \rho\,\nu_{n,i}^*\,u_x - \frac{\partial(\rho u_x)}{\partial t} + 2\,\rho\,u_y\,\Omega_E \sin\varphi = 0 \quad (3.78)$$

Since the left side of this equation can be considered as a tally of momentum gains and losses, this relation is also known as the *momentum balance equation*. An alternative form of this relation is obtained if the first and fifth terms on the left side are differentiated (product rule) and then suitably combined

$$-\left(m\,u_x\,\frac{\partial(n u_x)}{\partial x} + m\,n\,u_x\,\frac{\partial u_x}{\partial x} + m\,u_x\,\frac{\partial n}{\partial t} + m\,n\,\frac{\partial u_x}{\partial t} \right)$$

$$= -\rho\left(\frac{\partial u_x}{\partial t} + u_x\,\frac{\partial u_x}{\partial x} \right)$$

where, in the second step, we have made use of the continuity equation $\partial n/\partial t + \partial(n u_x)/\partial x = 0$. Rearranging, we can write the momentum balance equation in the following way

$$\rho \left(\frac{\partial u_x}{\partial t} + u_x \frac{\partial u_x}{\partial x} \right) = \rho \left(\frac{\partial}{\partial t} + u_x \frac{\partial}{\partial x} \right) u_x$$

$$= -\frac{\partial p}{\partial x} + \eta \frac{\partial^2 u_x}{\partial z^2} - \rho\, \nu_{n,i}^*\, u_x + 2\, \rho\, u_y\, \Omega_E\, \sin\varphi \quad (3.79)$$

We now extend this relation to three dimensions

$$\rho \frac{D\vec{u}}{Dt} = -\nabla p + \eta \frac{\partial^2 \vec{u}_h}{\partial z^2} + \rho\, \vec{g} + \rho\, \nu_{n,i}^*\, (\vec{u}_i - \vec{u}) + 2\, \rho\, \vec{u} \times \vec{\Omega}_E \quad (3.80)$$

Here, for the sake of brevity, we have introduced the horizontal velocity $\vec{u}_h = \hat{x}u_x + \hat{y}u_y$ and the *convective derivative*

$$\frac{D}{Dt} = \frac{\partial}{\partial t} + (\vec{u}\,\nabla) \quad (3.81)$$

The details of this extension will not be explained further here. Whereas with some terms it is immediately obvious, with others it is less intuitive. As an example, for the second term on the left side, $u_x \partial u_x / \partial x \rightarrow (\vec{u}\nabla)\vec{u}$, it should be recalled that the flow velocities u_j and u_k (for three-dimensional flow in the i-, j- and k-directions) must also account for the transport of momentum mu_i if this flow has gradients in the j- and/or k-directions. This is elucidated by the various components of the quantity $(\vec{u}\nabla)\vec{u}$ in Eq. (A.39). Moreover, restrictions apply to the extension of the viscosity term. To a good approximation, vertical winds are neglected with respect to horizontal winds and horizontal variations are neglected with respect to vertical variations. More general expressions for the viscosity term may be found in Appendix A.6. One must account for the Earth's acceleration, of course, when extending the treatment to three dimensions. Finally, it should be explicitly stated that the above relation is valid for a single gas (or a completely homogeneous gas mixture with average properties), as indicated sometimes by indices on the various parameters.

Equation (3.80) is denoted as the force equilibrium or momentum balance equation, sometimes also simply as the *transport equation* or *equation of motion*. These latter designations are appropriate because the equation is utilized to determine the flow velocity and therefore the transport and motion of a gas. Before attempting such a determination, it is highly instructive to explain the origin of the designation 'convective derivative' for D/Dt. For this, rather than the previously considered stationary (or *Euler*) volume element, we now work with a volume of gas that moves along with the flow (*Langrange* volume element). In contrast to the Euler case, the same gas particles are always contained within the Lagrange volume element. The acceleration acting on these particles can be resolved into two types. In the first case, one can have a purely temporal change of the flow velocity. A second possibility is that the gas particle ensemble can be accelerated because of spatial variations of the flow field. These two sources of acceleration are illustrated in Fig. 3.36,

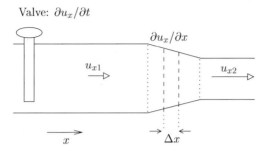

Fig. 3.36. Gas flow through a constricted pipe system

an example showing airflow through a constriction in a pipe system. In order to maintain continuity of the airflow through the system, the velocity in the narrower pipe must be greater than that in the wider pipe, and the gas volume must be accelerated in the conical interface piece connecting the two pipe sizes. The velocity increases by an amount $\Delta u_x = (\partial u_x/\partial x)\Delta x$ as the gas flows through the distance Δx. The time required to move through this distance is $\Delta t \simeq \Delta x/u_x$, so the change in velocity per unit time (i.e. the acceleration) is

$$\frac{\Delta u_x}{\Delta t} = \frac{(\partial u_x/\partial x)\,\Delta x}{\Delta x/u_x} = u_x\,\frac{\partial u_x}{\partial x}$$

Another type of acceleration occurs when the flow velocity is regulated by opening and closing the valve. The change in flow velocity in the entire pipe system resulting from this action is purely temporal and is described by the partial derivative of the velocity with respect to time $\partial u_x/\partial t$. The total acceleration of our moving (*convected*) gas volume is thus given by

$$\left(\frac{\partial}{\partial t} + u_x\frac{\partial}{\partial x}\right) u_x$$

where the expression in parentheses corresponds to the convective derivative. Note that we do not incur any temporal and spatial changes of the gas mass with the Lagrange approach: the volume element always contains the same gas particles. Furthermore, there is evidently no flow gradient force acting on the volume element, because the volume moves along with the flow. The same momentum balance equation holds accordingly for every point of our flow field, irrespective of whether we choose the Euler or Lagrange approach. Alternatively, the momentum balance equation (like all other balance equations) can also be derived from the Boltzmann equation (2.88). This derivation has the advantage that the mathematical structure and validity regime of the momentum balance equation can be more clearly recognized. The disadvantage is that the derivation is less evident and mathematically more elaborate. We thus content ourselves here by outlining the derivation in Appendix A.6.

3.4.4 Calculation of Thermospheric Winds

The momentum balance equation (3.80) is a fairly complicated system of coupled, nonlinear, partial differential equations. It is a 'system' because the vector relation contains three separate component equations (the x-, y- and z- components in a Cartesian coordinate system), and it is 'coupled' because the velocities of the other components appear in the viscosity and Coriolis terms. Furthermore, the velocity of the ion component $\vec{u}_i$ must generally be determined self-consistently. We will therefore limit ourselves in the following to a discussion of greatly simplified forms of this equation. These abridged equations have analytical solutions that can be compared with observations. As a control, consider first the case of a stationary atmosphere. Setting $\vec{u}$, $\vec{u}_i = 0$, Eq. (3.80) is reduced to the aerostatic equation. Not surprisingly, this is a special form of the momentum balance equation.

Next we consider the case of horizontal wind flow in the lower atmosphere. When examining weather maps, it is striking that the wind flow moves along isobars rather than following the pressure gradient. The air just circulates around the high and low pressure zones without directly inducing a pressure compensation. In the Northern Hemisphere, the circulation moves clockwise for a high pressure zone and counterclockwise for a low pressure region. The sense of circulation is opposite to this in the Southern Hemisphere. This observation implies that the Coriolis force plays a dominant role in the lower atmosphere. Neglecting all terms except the pressure gradient and Coriolis forces in Eq. (3.80), one obtains the so-called *geostrophic approximation* for the horizontal equation of motion

$$u_x \simeq -\frac{1}{(2\,\rho\,\Omega_E \sin\varphi)}\frac{\partial p}{\partial y}\,, \qquad u_y \simeq \frac{1}{2\,\rho\,\Omega_E \sin\varphi}\frac{\partial p}{\partial x} \qquad (3.82)$$

which is valid for mid-latitudes. For a given density and pressure distribution, these equations yield horizontal winds that are often in remarkably good agreement with the observations.

One look at Fig. 3.32 clearly shows that the geostrophic approximation cannot be extended to the upper atmosphere. To a good approximation, wind currents at an altitude of 300 km follow the existing pressure gradients, and direct pressure compensation occurs between the warmer dayside and cooler nightside. Coriolis forces evidently play only a subordinate role in the thermosphere. We will thus neglect these and, as a trial, all other terms in the momentum balance equation except the pressure gradient and ion drag forces. In this case the zonal wind component becomes

$$u_x \simeq -\frac{1}{\rho\,\nu_{n,i}^*}\frac{\partial p}{\partial x} \qquad (3.83)$$

We examine this solution with the help of the model calculations shown in Fig. 3.32, noting that atomic oxygen is the dominant neutral gas and ion constituent at a height of 300 km. Using $\sigma_{O,O^+} \simeq 8 \cdot 10^{-19}\,\mathrm{m^2}$ (σ_{O,O^+} is about twice as large as $\sigma_{O,O}$ due to resonance effects), n_{O^+} (300 km) $\simeq 5 \cdot 10^{11}\,\mathrm{m^{-3}}$, $m_{O^+} \simeq m_O = 16\,m_u$ and $T_{O^+}(300\,\mathrm{km}) \simeq T_O(300\,\mathrm{km}) \simeq 1000$ K, the momentum transfer collision frequency computed from Eq. (2.56) becomes $\nu^*_{O,O^+} \simeq 4 \cdot 10^{-4}\,\mathrm{s^{-1}}$. Further, we take the oxygen density from Appendix A.4 to be $n_O(300\,\mathrm{km}) \simeq 6 \cdot 10^{14}\,\mathrm{m^{-3}}$, and estimate from Fig. 3.32 a value for the pressure gradient in the dawn sector at equatorial latitudes: $\Delta p/\Delta x \simeq 4 \cdot 10^{-13}$ Pa/m. Putting these values into Eq. (3.83), the zonal wind velocity in this region is calculated as $u_x \simeq 60$ m/s, in good agreement with the winds actually observed there. Evidently, consideration of just the pressure gradient and ion drag forces yields wind directions and wind speeds in rough agreement with the observations. That ion drag is indeed a dominating force in the upper atmosphere may be inferred on the one hand from an order of magnitude comparison with the competing forces and on the other hand from the fact that much higher wind velocities are observed in the night sector, where the ion density is smaller, than on the dayside, (see Figs. 3.31 and 3.32).

The variation of the thermospheric winds with height is determined essentially by the viscosity. The viscous forces, which strive to smooth out velocity differences, become continually more successful in this effort with increasing height. The height dependence of the viscosity coefficient is primarily determined by the temperature profile, even though strongly attenuated by the square root, see Eq. (3.72). At any rate, η goes to a constant maximum value upon approaching the thermopause temperature. This means that a momentum flux $\phi^{I(u)}_{z,x}$ caused by a given velocity difference will be greater in the upper thermosphere than for the same velocity difference in the lower thermosphere. This is true even though the smallest momentum fluxes are capable of smoothing out velocity differences in the upper thermosphere because of the low density there. The velocity assimilation caused by the viscosity is therefore much more effective in the upper than in the lower thermosphere. Accordingly, the wind velocity in the upper thermosphere, similar to the temperature, tends to approach a constant limiting value. This is documented in Fig. 3.37, which shows the meridional wind currents in a cross section through the midnight sector. The limiting velocity is obviously attained in this example even below the altitude 300 km.

It may also be seen in Fig. 3.37 that the mass flow directed toward the equator in the upper thermosphere is balanced by a poleward directed mass flow in the lower thermosphere. Of course, the return flow velocities in the lower thermosphere are much smaller because of the greater densities there. All things considered, however, the wind circulation resembles the flow field of a Hadley cell. Superposed on this flow, easily recognized in Figure 3.37, is a wind circulation from the summer to the winter hemisphere (seasonal winds). This is evidently driven by the pressure gradient that exists between the hotter summer and cooler winter hemispheres.

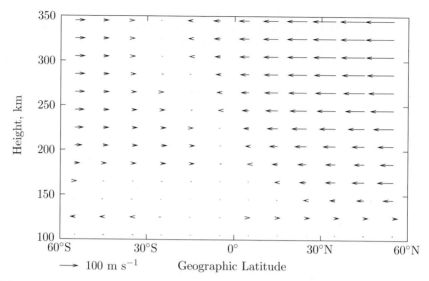

Fig. 3.37. Cross section through the meridional wind u_y in the midnight sector (0 hours local time, 0° geographic longitude). The distribution holds for the northern summer solstice (21 June) at moderate solar activity (CI=100) and weak geomagnetic activity (Kp=2). The winds were calculated using the model HWM 93 (Hedin, 1996).

It is worth noting that ion drag can produce not only deceleration, but also acceleration. This is the case in the polar upper atmosphere where intensive electric fields can accelerate the ions to high velocities. Upon transferring a fraction of this kinetic energy to the neutral gas via collisional friction, substantial wind velocities (> 1000 m/s) can be achieved. This additional wind source will be addressed in more detail in Section 7.5.1.

3.5 Atmospheric Waves

Waves are omnipresent in the terrestrial upper atmosphere and, as such, classic examples of thermospheric dynamics. A short introduction to this broad field of study is given here. We first refresh our understanding of the formal description of waves. The parameters necessary for such a description are then derived using acoustic waves as an example. This is followed by a study of atmospheric buoyancy oscillations. Finally, albeit only qualitatively, we discuss atmospheric gravity waves, i.e. a phenomenon that unites elements of both acoustic waves and buoyancy oscillations.

3.5.1 Wave Parameters

Various parameters are necessary in order to completely describe a wave. Among these are the wave amplitude a_0, which gives the maximum deviation of a quantity a from its equilibrium value due to the wave; another is the angular frequency $\omega = 2\pi/\tau$ (τ is the wave period), which determines the number of oscillations per unit time the wave generates at a given point in space; and finally the wavenumber $k = 2\pi/\lambda$ (λ is the wavelength), which defines the spatial density of the wave crests and troughs. Additional parameters are used to describe the phase, polarization and direction of propagation of the wave. It is sufficient in the following to consider the case of a plane harmonic wave. Letting the wave propagate in the x-direction, we have

$$a(t, x) = a_0 \, \sin(\omega t - kx) \tag{3.84}$$

An essential quantity is the velocity with which the various phases of the wave, e.g. its maxima and minima, propagate. This *phase velocity*, which may be determined from the condition $\omega t - kx = $ const., is given by

$$v_{ph} = \omega/k \tag{3.85}$$

This must be distinguished from the *group velocity* v_{gr}, which gives the propagation velocity applicable to modulation on a carrier wave or impulsive disturbances formed by a superposition of waves. The group velocity is given by

$$v_{gr} = \frac{\partial \omega}{\partial k} \tag{3.86}$$

3.5.2 Acoustic Waves

Plane acoustic waves are a particularly simple form of atmospheric waves. Although less important for the thermosphere, they are nevertheless well suited for an introduction into the more complicated physics of the dominant wave type, the atmospheric gravity waves. Furthermore, acoustic waves display similarities to an important type of plasma waves.

The physics of acoustic waves is explained in Fig. 3.38. Alternating zones of density enhancement and depletion are produced in a gas by the rhythmic motion of a membrane. Pressure gradient forces $-\partial p/\partial x$ act at the edge of a newly formed enhancement zone, accelerating the gas in the x-direction. This creates a new enhancement zone at a larger distance from the membrane and the disturbance travels in the x-direction at the speed of sound (acoustic speed) v_S, as the phase velocity is denoted. Following each enhancement zone is a depletion zone with its associated pressure gradient that pushes the gas in the negative x-direction. The gas moves back and forth during the transit of the enhancement and depletion zones generated by the alternating acceleration, but it remains stationary when averaged over time. We denote the spatial deflection of the gas by ξ_x and the velocity of this deflection by u_x.

Clearly, waves must obey the fundamental equations describing their propagation medium in order to exist at all. For propagation in the thermosphere these are the density, momentum, and energy balance equations.

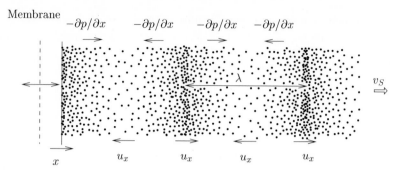

Fig. 3.38. Density distribution in acoustic waves and the positions of maximum pressure gradient and gas velocity

Here we assume that the energy equation can be approximated by the adiabatic law (see Appendix A.7). We also use a greatly simplified form for the equation of motion

$$\rho \frac{\partial u_x}{\partial t} + \rho\, u_x\, \frac{\partial u_x}{\partial x} = -\frac{\partial p}{\partial x} \tag{3.87}$$

Comparing this with Eq. (3.79), it is seen that the Coriolis force (to a very good approximation), as well as the viscosity and frictional forces (to a first approximation), are neglected here. Moreover, gravity also remains unconsidered.

Evidently, Eq. (3.87) represents a relation between the three atmospheric parameters u_x, ρ and p. In order to calculate one of these quantities, the other two must be eliminated. This is accomplished with the help of the continuity equation and adiabatic law. The former relation, of course, is a special form of the density balance equation, and the latter relation assumes that exchange of energy via heat conduction does not occur with these waves. We use these two equations to derive an additional relation between p and u_x. Differentiating the continuity equation (2.19) for the x-direction

$$\frac{\partial n}{\partial t} + u_x\, \frac{\partial n}{\partial x} + n\, \frac{\partial u_x}{\partial x} = 0$$

and differentiating the adiabatic law (2.36) with respect to the variable i yields

$$\frac{\partial n}{\partial i} = \text{const.}\ \frac{1}{\gamma}\, p^{1/\gamma - 1}\, \frac{\partial p}{\partial i} = \frac{n}{\gamma p}\, \frac{\partial p}{\partial i}$$

This expression can be used in the continuity equation with $i = t$ and $i = x$. The result is the following relation between p and u_x

$$\frac{\partial p}{\partial t} + u_x\, \frac{\partial p}{\partial x} + \gamma\, p\, \frac{\partial u_x}{\partial x} = 0 \tag{3.88}$$

Simplifying further, let us assume that the pressure and density deviations caused by the wave, p_1 and ρ_1, are small compared to their equilibrium values p_0 and ρ_0, and that the background atmosphere is homogeneous and at rest

$$p(x,t) = p_0 + p_1(x,t) \quad \text{with} \quad p_1 \ll p_0, \; p_0 \neq f(x,t)$$
$$\rho(x,t) = \rho_0 + \rho_1(x,t) \quad \text{with} \quad \rho_1 \ll \rho_0, \; \rho_0 \neq f(x,t) \tag{3.89}$$
$$u_x(x,t) = u_{1x}(x,t) \quad \text{and} \quad u_0 = 0$$

Using this small perturbation approach, Eqs. (3.87) and (3.88) can be approximated by

$$\frac{\partial u_{1x}}{\partial t} + u_{1x}\frac{\partial u_{1x}}{\partial x} + \frac{1}{\rho_0}\frac{\partial p_1}{\partial x} = 0 \tag{3.90}$$

$$\frac{\partial p_1}{\partial t} + u_{1x}\frac{\partial p_1}{\partial x} + \gamma\, p_0 \frac{\partial u_{1x}}{\partial x} = 0 \tag{3.91}$$

It should be mentioned that the disturbance variables $\rho_1 \ll \rho_0$ and $p_1 \ll p_0$, but not necessarily their derivatives, can be neglected in the above expressions. We now make the additional assumption that the nonlinear terms $u_{1x}\partial u_{1x}/\partial x$ and $u_{1x}\partial p_1/\partial x$ can be neglected with respect to the other terms of their respective equations. The conditions under which this is possible will be investigated later with the help of the solutions we obtain. Two linear equations for the disturbance quantities u_{1x} and p_1 are found under this approximation. Differentiating the upper equation with respect to t and the lower equation with respect to x and eliminating p_1 by substitution, we obtain

$$\frac{\partial^2 u_{1x}}{\partial t^2} - \frac{\gamma p_0}{\rho_0}\frac{\partial^2 u_{1x}}{\partial x^2} = 0 \tag{3.92}$$

This corresponds to a wave equation for the velocity u_{1x}, for which we can anticipate a solution of the form

$$u_{1x} = (u_{10})_x \; \sin(\omega t - kx) \tag{3.93}$$

where $(u_{10})_x$ is the amplitude of the velocity oscillation. Using this in Eq. (3.92), we arrive at the following dispersion relation between the frequency ω and the wavenumber k

$$-\omega^2 + k^2\gamma\, p_0/\rho_0 = 0 \tag{3.94}$$

Combining the equations (3.85), (3.94), (2.28) and (2.34), one obtains the following expression for the speed of sound

$$v_S = \omega/k = \sqrt{\gamma\, p/\rho} = \sqrt{(1 + 2/f)\, kT/m} \tag{3.95}$$

where we have now dropped the index '0' on the background pressure, density and temperature. The speed of sound is evidently independent of the frequency and is thus also equal to the group velocity. Using the appropriate mean degree of freedom and the mean gas particle mass, Eq. (3.95) yields typical values for v_S near the ground and in the upper thermosphere of $v_S(h = 0) \simeq 340$ m/s and $v_S(h = 300$ km, $T = 1000$ K $) \simeq 860$ m/s, respectively. A comparison with Table 2.1 shows that the magnitude of these values is consistent in each case with the random velocities of the gas particles.

Our derivation was based on the assumption, among others, that the nonlinear term $|u_{1x}\,\partial u_{1x}/\partial x|$ is small compared with the other two terms $|\partial u_{1x}/\partial t|$ and $|(1/\rho_0)\partial p_1/\partial x|$, which are of about equal magnitude under these conditions. Substituting the solution (3.93) into Eq. (3.90) and comparing terms, it is found that our assumption evidently requires $\omega \gg (u_{10})_x\, k$, or $(u_{10})_x \ll v_S$. In other words, our approach is valid only if the deflection velocity is small with respect to the speed of sound. An equivalent condition is that the deflection amplitude of the gas volume $(\xi_{10})_x$ must be small compared with the wavelength λ. This may be easily shown by integrating Eq. (3.93), $(u_{10})_x = (\xi_{10})_x\, \omega$.

While acoustic waves play only a subordinate role in the thermosphere (triggered, for example, by earthquakes or by the supersonic motion of the aurora), they are of great importance for the solar atmosphere. Acoustic waves, which are continually being excited by vertical photospheric convective motions, travel outward into the chromosphere and contribute there to the heating of the gas (see Section 3.1.2). The dissipation of the wave energy is brought about, for example, by the strong growth of the deflection amplitude and velocity, a process that can be understood from the following. The time-averaged kinetic energy density of an upward propagating acoustic wave is given by

$$\langle E^*_{kin}\rangle = \frac{1}{2}\,\rho\,\langle u_{1z}^2\rangle = \frac{1}{2}\,\rho\,(u_{10})_z^2\,\langle \sin^2(\omega t - kz)\rangle = \frac{1}{4}\,\rho\,(u_{10})_z^2 \qquad (3.96)$$

where ρ is the density associated with the deflection velocity u_{1z} and the temporal mean is calculated with the help of the integral (A.6). As with all wavelike processes, the mean potential energy density can be taken as equal to the kinetic energy density. The total energy density of an acoustic wave is thus given by $\langle E^*\rangle = \rho\,(u_{10})_z^2/2 = \rho\,\omega^2\,(\xi_{10})_z^2/2$. Because this energy density is transported at the velocity v_S, the resulting energy flux (i.e. the *wave intensity*) is found to be $\phi_S^E = \langle E^*\rangle A(v_S\Delta t)/A\Delta t = \langle E^*\rangle v_S$, where $A(v_S\Delta t)$ is the volume of the sound wave that passes through the surface A in the time Δt. Assuming that this energy flux and the speed of sound stay roughly constant during the propagation of the wave into higher atmospheric layers, this requires that the deflection amplitude and velocity, which are inversely proportional to the square root of the density, must increase, i.e. $(u_{10})_z, (\xi_{10})_z \sim 1/\sqrt{\rho}$. This increase can easily exceed a factor 100 as the wave progresses through the upper photosphere and lower chromosphere. It is thus not difficult to visualize that this must sooner or later lead to non-linearities $((u_{10})_z \gtrsim v_S,\ (\xi_{10})_z \gtrsim \lambda)$ and dissipation of coherent wave energy. Furthermore, viscosity and heat conduction effects become more important with increasing height, inducing additional dissipation of wave energy. In summary, the heating by acoustic waves provides an important contribution to the heat budget of the chromosphere.

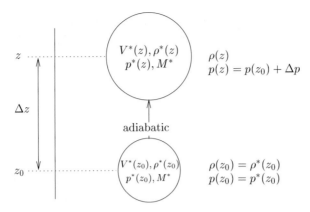

Fig. 3.39. Derivation of the buoyancy oscillation frequency. The quantities with asterisk refer to the gas volume under consideration with constant mass M^*. The acceleration of gravity g is taken as constant.

3.5.3 Buoyancy Oscillations

Whereas acoustic waves play only a subordinate role in the thermosphere, a variation of this wave type modified by buoyancy effects is of great importance. In order to understand this more complicated wave type better, it is appropriate to first explain the origin of buoyancy oscillations. We consider an air parcel in a stable layered atmosphere that is raised from its equilibrium location at z_0 to a height z; see Fig. 3.39. Because of the lower pressure in its new surroundings, the gas volume will expand and perform work. This happens at the cost of its internal energy and the gas must cool off. For an atmosphere in equilibrium, the temperature in the gas volume is now less than outside and the gas parcel becomes heavier than its environment. As a result, it sinks back down in the direction of its original position. The gas parcel will overshoot its equilibrium position, however, because of the kinetic energy gained by its downward motion. At lower heights the gas parcel, compressed and heated with respect to its equilibrium state and thus lighter than its new environment, will be accelerated by an upward directed buoyancy force back to its original position. In this way, the gas volume evidently bobs up and down about its equilibrium height, a phenomenon denoted the *buoyancy oscillation*. We determine the frequency of this oscillation in the following.

A well-known result from elementary mechanics is that the restoring force of an oscillating system F_{res} is proportional to the deviation (deflection) from the equilibrium position. Denoting M^* the mass of the oscillating body, z_0 its equilibrium position and $\Delta z = z - z_0$ the deflection from this position, then Newton's law requires

$$M^* \frac{\mathrm{d}^2 z}{\mathrm{d}t^2} = M^* \frac{\mathrm{d}^2 \Delta z}{\mathrm{d}t^2} = F_{res} = - K \, \Delta z \qquad (3.97)$$

where K is a proportionality constant (e.g. the spring constant). A solution of this differential equation is a simple harmonic oscillation of the form

$$\Delta z = (\Delta z)_0 \; \sin(\omega t) \tag{3.98}$$

where $(\Delta z)_0$ is the deflection amplitude and ω is the frequency of this oscillation. Inserting this solution into the differential equation, we obtain an expression for the oscillation frequency

$$\omega = \sqrt{K/M^*} \tag{3.99}$$

In the situation sketched in Fig. 3.39, the restoring force is the difference between the pressure gradient and gravity forces for the given deflection

$$F_{res} = -V^*(z) \; \partial p/\partial z|_z - M^* \; g$$

With $V^*(z) = M^*/\rho^*(z)$ and $\partial p/\partial z|_z = -\rho(z)g$ we can write this expression in the following way

$$F_{res} = M^* \; g \; (\rho(z) - \rho^*(z))/\rho^*(z) \tag{3.100}$$

The strength of this restoring force is controlled primarily by the difference $\rho(z) - \rho^*(z)$, rather than by the particular value of the denominator. For the small deflections considered here, the quantity $\rho^*(z)$ in the denominator can thus be replaced without further ado by the somewhat larger value $\rho^*(z_0) = \rho(z_0)$.

Furthermore, the density $\rho(z)$ in the numerator can be approximated by the following Taylor series, truncated at two terms

$$\rho(z) \simeq \rho(z_0) + \partial \rho/\partial z|_{z_0} \; \Delta z$$

The same approximation can be used for the density $\rho^*(z)$, whereby we assume that the density variation in the gas volume proceeds adiabatically, i.e., with no exchange of thermal energy

$$\rho^*(z) \simeq \rho^*(z_0) + \left.\frac{\partial \rho^*}{\partial p}\right|_{z_0}^{ad} \Delta p \simeq \rho(z_0) + \left(\frac{\rho}{\gamma p} \frac{\partial p}{\partial z}\right)_{z_0} \Delta z = \rho(z_0) - \frac{\rho(z_0)g}{v_S^2(z_0)} \Delta z$$

In the second step we have made use of the adiabatic law (2.36) and the approximation $\Delta p \simeq (\partial p/\partial z)\Delta z$ as well as the aerostatic equation (2.38). In the third step we retrieved the expression for the speed of sound (3.95). The restoring force can thus be written as follows

$$F_{res} = \frac{M^* g}{\rho(z_0)} \left[\left.\frac{\partial \rho}{\partial z}\right|_{z_0} + \frac{\rho(z_0)g}{v_S^2(z_0)} \right] \Delta z = -K \; \Delta z$$

This relation defines the value of the proportionality constant K and, using Eq. (3.99), the value of the desired *buoyancy oscillation frequency*

$$\omega_g = \sqrt{\frac{K}{M^*}} = \sqrt{-g \left(\frac{1}{\rho(z_0)} \frac{\partial \rho}{\partial z} \bigg|_{z_0} + \frac{g}{v_S^2(z_0)} \right)} \tag{3.101}$$

The quantity ω_g is also denoted the *Brunt-Väisälä frequency* in honor of its 'discoverers'. In order to simplify the above expression, the term $(1/\rho)\partial\rho/\partial z$ will be rewritten as follows. From the expression in Eq. (3.95) for the speed of sound we obtain $\ln\rho = \ln\gamma + \ln p - \ln v_S^2$. This allows us to write

$$\frac{1}{\rho}\frac{\partial\rho}{\partial z} = \frac{\partial(\ln\rho)}{\partial z} = \frac{\partial(\ln p)}{\partial z} - \frac{\partial(\ln v_S^2)}{\partial z}$$

$$= -\frac{\rho g}{p} - \frac{1}{v_S^2}\frac{\partial v_S^2}{\partial z} = -\frac{\gamma g}{v_S^2} - \frac{1}{v_S^2}\frac{\partial v_S^2}{\partial z}$$

which we can then substitute into Eq. (3.101) and obtain

$$\omega_g = \sqrt{\frac{g^2}{v_S^2}(\gamma - 1) + \frac{g}{v_S^2}\frac{\partial v_S^2}{\partial z}} = \sqrt{\frac{g}{T}\left(\frac{g}{c_p} + \frac{\partial T}{\partial z}\right)} \tag{3.102}$$

where all quantities refer to the given equilibrium height z_0. If the speed of sound (or the temperature) can be assumed constant to a first approximation within the range of oscillation, the expression for the buoyancy oscillation frequency reduces to the simple form

$$\omega_g \simeq g\sqrt{\gamma - 1}/v_S = g/\sqrt{c_p T} \tag{3.103}$$

For the upper thermosphere (300 km) at a temperature of 1000 K, the buoyancy oscillation frequency is found from this formula to be $\tau_g = 2\pi/\omega_g \simeq 13$ minutes.

3.5.4 Gravity Waves

In order to dispel any possible misunderstandings: gravity waves have nothing to do with the fascinating, but hitherto undetected gravitational waves. Rather, we are dealing here with a special form of atmospheric wave, for which buoyancy plays an important role alongside compressional effects. The treatment follows essentially that of the acoustic waves, except that now, in addition to the pressure gradient, we account for the force of gravity in the momentum balance equation – thereby also providing the name for this wave type. Equation (3.87) is replaced by the two-dimensional equation of motion

$$\rho\frac{\partial u_x}{\partial t} = -\frac{\partial p}{\partial x} \tag{3.104}$$

and

$$\rho\frac{\partial u_z}{\partial t} = -\frac{\partial p}{\partial z} - \rho g \tag{3.105}$$

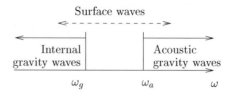

Fig. 3.40. Frequency range for the three types of atmospheric gravity waves

where the nonlinear terms $\rho u_x \partial u_x / \partial x$ and $\rho u_z \partial u_z / \partial z$ have already been neglected. Together with the continuity equation and the adiabatic law, we thus have four equations available for determining the four unknowns u_x, u_z, ρ and p. The restoring force for gravity waves is provided by both the pressure and buoyancy accelerations, the latter, however, acting only in the vertical direction. The wave propagation thus becomes anisotropic with the waves in the horizontal direction behaving differently from those in the vertical direction. This circumstance calls for an approach using different wavenumbers in the horizontal and vertical directions, generally expressed as follows

$$a(x, z, t) = a_0 \sin(\omega t - k_x x - k_z z) \tag{3.106}$$

'a' again stands for one of the four unknowns. Inserting this into the simplified equation set and applying the assumption of small perturbations (3.89), we can derive (after a somewhat extensive calculation that is omitted here) the following dispersion relation

$$\omega^4 - [v_S^2 \, (k_x^2 + k_z^2) + (\gamma g / 2 v_S)^2] \, \omega^2 + (v_S \omega_g)^2 \, k_x^2 = 0 \tag{3.107}$$

Obviously, this condition for the propagation of gravity waves is considerably more complicated than the corresponding relation (3.94) for acoustic waves. It is a quartic equation for ω, yielding solutions only for the cases

$$\omega \geq \omega_a = \gamma g / 2 v_S \tag{3.108}$$

or

$$\omega \leq \omega_g \tag{3.109}$$

where ω_a denotes the *acoustic resonance frequency*. Waves of the first case are referred to as *acoustic gravity waves*; those of the second case are called *internal gravity waves*. The frequency range for both wave types is given in Fig. 3.40. As the name implies, acoustic gravity waves are a slightly modified variety of acoustic waves, which, because of their relative high frequency, are only weakly affected by buoyancy forces. Like acoustic waves, for which the gas motion occurs primarily parallel to the propagation direction, they are quasi-longitudinal waves. The internal gravity waves are propagating buoyancy oscillations that are modified by density compression effects. They are quasi-transverse waves, because the gas motion is essentially perpendicular

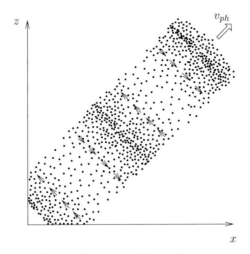

Fig. 3.41. Density and velocity distribution (arrows) in an idealized, plane internal gravity wave propagating upward along a tilted path

to the propagation direction. Figure 3.41 illustrates the density and velocity distribution in a plane gravity wave propagating upward along a tilted path.

Equation (3.107) is already a special form of the dispersion relation. The general form allows for an additional wave type called the *external gravity wave* or *surface wave*. These waves are characterized by a propagation in the horizontal direction, a height-independent phase, and an amplitude that decreases exponentially with distance from their defining surface. As such, this type corresponds to commonly observed water surface waves or seismic waves on the Earth's surface. Because the atmosphere has no well-defined surfaces, however, this wave type plays only a minor role.

Gravity waves are most often detected by their ionospheric signature and are held responsible for propagating, quasi-periodic fluctuations of the electron density and layer height. These effects can be detected by radio sounding methods (compare with Section 4.7.3). Such perturbations have also been designated as *traveling ionospheric disturbances*, or TIDs for short. A topic of special interest in Chapter 8 will be the signatures of large-scale internal gravity waves. Large-scale in this sense means wavelengths of a few 1000 km and periods of the order of one hour. Such waves have their origin at polar latitudes, where the upper atmosphere is repeatedly heated by energy injection during disturbed conditions. Triggered by the sudden expansion of the atmospheric gas, these waves (or, more properly, their impulse-like superposition) propagate equatorward with velocities of up to 800 m/s. Among other effects, they produce a significant increase in the electron density (i.e. a *positive ionospheric storm*) at mid-latitudes and a considerable disturbance of the neutral gas density at equatorial latitudes (see Sections 8.4.2 and 8.5.2).

Exercises

3.1 Determine the mass of hydrogen converted to helium in the Sun per second. How many solar neutrinos pass through the average human body in this time $(A_{body} \simeq 1 \text{ m}^2)$?

3.2 Using Fig. 3.8, compare the magnitudes of the photon fluxes and particularly the radiated energy fluxes in the neighborhood (5-nm intervals) of the Lyman-α and HeII lines.

3.3 Deduce the present phase of the solar cycle with the help of the 10.7 cm radio flux data reported on the internet (http://www.drao.nrc.ca).

3.4 A monochromatic photon flux penetrates into a homogeneous gas of density n. The absorption cross section is σ^A. Calculate the mean penetration depth of the photons.

3.5 The intense Lyman-α radiation of the Sun is absorbed mostly by molecular oxygen $(\sigma^A_{O_2}(\text{Ly}\alpha) \simeq 10^{-24} \text{ m}^2)$ at heights near the mesopause $(T \simeq 160 \text{ K}, n_{total}(85 \text{ km}) \simeq 2 \cdot 10^{20} \text{ m}^{-3})$. Calculate the absorption height for this radiation and the maximum energy deposition rate for an incidence angle of $50°$. How large would the rate of change of temperature be at this height if heat losses were neglected $(\eta^W \simeq 0.3)$?

3.6 As a result of 'atmospheric breathing', a gas volume of mass M in an isothermal atmosphere is raised from the height h_1 to the height h_2. Show that the work expended in this process corresponds exactly to the adiabatic work of expansion produced by the gas volume during its upward motion. How long must the Sun shine until this amount of energy is available to the gas volume in the form of heat $(h_1 = 300 \text{ km}, h_2 = 350 \text{ km}, \tau \simeq 1)$?

3.7 An earthquake produces a seismic wave with a period of 10 minutes, causing the Earth's surface to move up and down with a displacement amplitude of ± 5 mm.

(a) Assuming a constant speed of sound and neglecting losses, how large is the amplitude of the resulting acoustic wave at an altitude of 300 km?

(b) A radio wave at a frequency of 5 MHz is reflected from the ionosphere, which moves up and down with the amplitude calculated in (a) at the frequency of the seismic (acoustic) wave. How large is the frequency shift of the reflected wave caused by the Doppler effect?

References

The Sun

C.J. Durrant, *The Atmosphere of the Sun*, Adam Hilger, Bristol and Philadelphia, 1988

H. Zirin, *Astrophysics of the Sun*, Cambridge University Press, Cambridge, 1988

M. Stix, *The Sun*, Springer-Verlag, Berlin, 1989

Popularizations: the Sun

R. Giovanelli, *Secrets of the Sun*, Cambridge University Press, Cambridge, 1984

H. Friedman, *Sun and Earth*, Scientific American Books, Inc., New York, 1986.

L. Golub and J.M. Pasachoff, *Nearest Star*, Harvard University Press, Cambridge, MA/USA, and London, 2001

K.R. Lang, *The Cambridge Encyclopedia of the Sun*, Cambridge University Press, Cambridge, 2001.

Dynamics of the Upper Atmosphere, Airglow

R.W. Schunk, Mathematical structure of transport equations for multispecies flows, *Rev. Geophys. Space Phys.*, *15*, 429, 1977

P. Stubbe, Interaction of neutral and plasma motions in the ionosphere, in *Handbuch der Physik* (S. Flügge, ed.), *Geophysik III, Teil VI*, 247, Springer-Verlag, Berlin 1982

H. Volland, *Atmospheric Tidal and Planetary Waves*, Kluwer Academic Publishers, Dordrecht, 1988

S.N. Ghosh, *The Neutral Upper Atmosphere*, Kluwer Academic Publishers, Dordrecht, 2002

See also the references in the two preceeding chapters and the figure and table references in Appendix B.

4. Ionosphere

By ionosphere we mean the *ionized* component of the upper atmosphere. This definition implies that we are dealing with a mixture of thermal and gravitationally bound charge carrier gases. Although only present as a trace constituent, this charge carrying component has important consequences. For example, it enables the flow of electric currents, thereby leading to magnetic field perturbations and electrodynamic heating effects. It also influences the dynamics of the upper atmosphere by producing or slowing down thermospheric winds. Finally it modifies electromagnetic waves, whether it be by refraction or reflection, attenuation or rotation of the plane of polarization.

There were a number of early speculations on the existence of a conducting layer in the upper atmosphere similar to what we now call the ionosphere. Gauß (1839; again note that, when given in the text, the year denotes the timing of an event and does not identify a reference) and later Kelvin (1860) theorized that fluctuations of the geomagnetic field were produced by upper atmospheric currents. This hypothesis was elaborated by Stewart (1883), who attributed the regular daily variations of the geomagnetic field to dynamo currents caused by tidal winds. The physical nature of these upper atmospheric currents remained an enigma, however, because the existence of free electrons and ions was not generally accepted until about 1900. The notion of a conducting layer in the upper atmosphere was rejuvenated in 1901. That was the year Marconi first succeeded in transmitting radio waves across the Atlantic, a feat that was explained independently by Kennelly, Heaviside and Lodge (1902) as reflection of the waves from free charge carriers in the upper atmosphere. This explanation was supported by the work of Taylor (1903) and Fleming (1906), who, following Lodge, suggested that solar UV radiation is responsible for the formation of these charge carriers. The existence of the *Kennelly-Heaviside layer* nevertheless remained controversial (an 'academic myth') until the year 1924, when two independent groups, Breit and Tuve in the USA and Appleton and Barnett in England, were able to prove the reality of the ionosphere with specially contrived radio wave experiments. Appleton was later awarded the Nobel prize for his work in the field of ionospheric physics. The experiments of these two groups mark the beginning of experimental exploration of the ionosphere as well as the beginning of active investigation of space.

Monitoring the ionosphere with radio waves is still an active branch of research today. There are, for example, some 100 ionosonde stations around the world that investigate the structure and variability of the ionosphere with the help of reflected radio waves (echo sounding). There are also powerful radar facilities that record the incoherent backscattering from this layer. In addition to these ground-based observations, space technology has provided the possibility of *in situ* exploration of the ionosphere. This chapter first describes the height profile of several important ionospheric quantities determined from these various investigative techniques. This is followed by sections explaining the observed density distribution. The concluding section is devoted to radio wave propagation in the ionosphere.

4.1 Height Profile of Ionospheric State Parameters

Generally, all of the quantities introduced for the description of a neutral gas are also applicable to a description of an ion or electron gas. This holds for macroscopic state parameters such as mass density, pressure or temperature as well as for gas kinetic quantities such as the mean free path or the collision frequency. In particular, all of the defining equations summarized in Section 2.1 retain their full validity.

Figure 4.1 shows the height profile of the ionospheric electron density, as observed during the day at mid-latitudes and for low solar activity. The layered structure of the charge carrier distribution is obvious. For this particular case, the layer is located at a height of about 240 km, peaks at an ion density of about $5 \cdot 10^{11}$ particles per m^3, has a thickness of about 120 km, and contains a column density of a few 10^{17} particles per m^2. These layer parameters are subject to strong fluctuations and the ranges of typical daily values are

Maximum ionization density:	$n_\mathrm{m} \simeq 1 - 30 \cdot 10^{11}$ m^{-3}
Height of the maximum:	$h_\mathrm{m} \simeq 220 - 400$ km
Layer thickness:	$\simeq 100 - 400$ km
Column density:	$\mathcal{N}_\mathrm{e} \simeq 1 - 10 \cdot 10^{17}$ m^{-2}

When trying to infer the ion density from the electron density shown in Fig. 4.1, it should be noted that the ionosphere everywhere represents a quasi-neutral mixture of charge carrier gases, for which the sum of the positive ions is always equal to the sum of the electrons and negative ions. A good approximation above an altitude of roughly 90 km is

$$h \gtrsim 90 \text{ km}: \qquad n_\mathrm{i} = \sum_j n_j \simeq n_\mathrm{e} = n$$

where n_i is the total density of positive ions, n_j the partial density of the ion species j, n_e the electron density and n (no index) the density of the

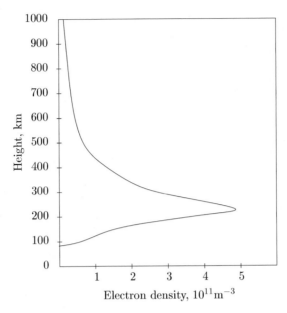

Fig. 4.1. Representative density profile of the electron density during the day at mid-latitudes and for low solar activity

electrons *or* the total density of the positive ions (but not the total density of all charge carriers). The contribution of negative ions above 90 km altitude is thus negligibly small and the total density of positive ions is about equal to the electron density.

The ionic composition of the ionosphere is shown in Fig. 4.2. The electron density profile (e^-) corresponds to that of Fig. 4.1, except that now the layer structure of the distribution is not as easy to recognize with the logarithmic representation. It is seen that the molecular ions O_2^+ and NO^+ dominate in the lower ionosphere, which is certainly surprising for the case of the NO^+ ions. In the region of the maximum and in the upper ionosphere, the primary ion is O^+, which seems reasonable in view of the dominance of thermospheric atomic oxygen in this region. Again surprising are the absence of larger He^+ densities and the direct transition from O^+ to H^+ as primary ion in the upper ionosphere.

We will use the ion composition described above as a guideline for a classification of the ionosphere; see Table 4.1. The lower region of the ionosphere dominated by O_2^+ and NO^+ ions is denoted the *E region*; the atomic oxygen ion region above this is called the *F region*. Continuing upward, we come to the region of dominant H^+ ions, the *protonosphere* or *plasmasphere*, which, perhaps somewhat arbitrarily, is not considered as part of the ionosphere. Finally, following the alphabet, the lowermost region of the ionosphere, where cluster ions and negative ions play an important role, is denoted the *D region*.

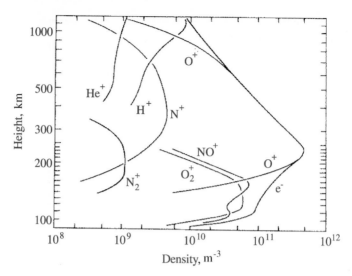

Fig. 4.2. The composition of the ionosphere as observed during the day at mid-latitudes for low solar activity (adapted from Johnson, 1966)

Different and more detailed classifications are possible (e.g., one discriminates between an F1 and an F2 region in the scientific literature), but these need not be discussed here. Note that the heights given in Table 4.1 should only be considered as typical values. This is particularly evident for the transition from O^+ to H^+ ions, which can occur at an altitude anywhere between 600 and 2000 km.

As seen from the magnitudes of the densities plotted in Fig. 4.1 and 4.2, electrons and ions represent trace gases and the upper atmosphere is only weakly ionized ($n/n_n \ll 1$, where n_n is the neutral gas density). Typical values for the ratio of electron to neutral gas density are 10^{-2} at the top of the ionosphere at a height of about 1000 km, 10^{-3} at the layer peak, and 10^{-8} at the bottom of the layer near 100 km altitude. In spite of these small relative densities, the ionosphere is nevertheless the largest concentration of charge carriers in the Earth's space environment.

Typical height profiles for the ion and electron temperature under noon-time conditions are shown in Fig. 4.3. The corresponding height profile for

Table 4.1. Classification of the ionosphere according to composition

Ionosphere	D region	$h \lesssim 90$ km	e.g.	$H_3O^+ \cdot (H_2O)_n$, NO_3^-
	E region	$90 \lesssim h \lesssim 170$ km		O_2^+, NO^+
	F region	$170 \lesssim h \lesssim 1000$ km		O^+
Plasmasphere		$h \gtrsim 1000$ km		H^+

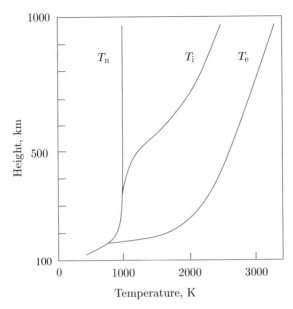

Fig. 4.3. Representative height profiles for the neutral gas, ion, and electron temperature in the noon meridian at mid-latitudes and for low solar activity (adapted from Köhnlein, 1984)

the neutral gas temperature is also shown for comparison. As indicated, the electron component is in thermal equilibrium with the neutral gas only in the lower ionosphere. A decoupling of the two components already begins at a height of 150 km with a distinctly stronger increase in the electron temperature. Furthermore, in contrast to the neutral gas temperature, the electron temperature does not approach a constant limiting value, but steadily increases toward greater heights. This implies that a heat source must exist in the plasmasphere that maintains a downward-directed heat current.

Owing to the much larger interaction cross section, the ion component stays in thermal equilibrium with the neutral gas up to a height of about 350 km. The ion temperature above this height displays an increase in the direction of the electron temperature, however, without ever reaching it. The different temperatures of the three gas components result in a continual heat flux from the electrons to the ions and from the ions to the neutral gas. Indeed, the ion and electron gases form the primary heat source for the neutral gas above about 250 km (see Section 3.3.1). This is true, however, only as long as the Sun is shining. At night the temperatures decline steeply and assimilate so that, to a good approximation, $T_e \simeq T_i \simeq T_n$ below roughly 500 km altitude.

4.2 Ionization Production and Loss

One of our main goals is to explain the observed height profile of the ionization density and this is best accomplished using a step-by-step approach. First the production and then the loss of ionization are described.

4.2.1 Ionization Production

Various processes contribute to the production of charge carriers. The primary process to be considered is photoionization of thermospheric gases by solar EUV and X-ray radiation. Secondary processes include ionization by photoelectrons and scattered or reemitted radiation, but particularly charge exchange reactions. Finally, precipitation of energetic particles plays an essential role in the ionization production at polar latitudes.

Primary Photoionization. Simple (i.e. nondissociative) photoionization processes have the general form

$$X + \text{photon}(\lambda \lesssim 100\,\text{nm}) \longrightarrow X^+ + e \qquad (4.1)$$

where X is an atom or molecule in the upper atmosphere. Of particular importance for the ionosphere is photoionization of the predominant gases O, N_2 and O_2

$$\begin{aligned}
\text{O} \ + \text{photon}(\lambda \leq 91\,\text{nm}) \ &\rightarrow \text{O}^+ + e \\
\text{N}_2 + \text{photon}(\lambda \leq 80\,\text{nm}) \ &\rightarrow \text{N}_2^+ + e \\
\text{O}_2 + \text{photon}(\lambda \leq 103\,\text{nm}) &\rightarrow \text{O}_2^+ + e
\end{aligned}$$

In order to obtain a quantitative estimate of the number of ions (and electrons) created, we again consider the absorption of a monochromatic photon flux by a single gas atmosphere of species X. According to Eq. (3.30), the number of photons absorbed by this atmosphere per unit volume and time is

$$\frac{\mathrm{d}N_{ph}}{\mathrm{d}V\,\mathrm{d}t} = \sigma_x^A \, n_x \, \phi^{ph}$$

Once absorbed, these photons can produce ionization, but also dissociation or excitation of the gas particles. Denoting the fraction of absorbed photons leading to an ionization by ε_x^I, the number of ions produced by primary photoionization (superscript PI) per unit volume and time is given by

$$q_{x+}^{PI} = \varepsilon_x^I \, \sigma_x^A \, n_x \, \phi^{ph} \qquad (4.2)$$

The parameter ε_x^I is usually referred to as the *ionization efficiency*. Note that $\varepsilon_x^I = 0$ for wavelengths greater than the ionization limit of the given species. Moreover, $\varepsilon_x^I \simeq 1$ for atomic gases and for wavelengths less than the ionization limit, because photoionization is the main absorption process in this case. When considering the dissociative ionization process, it should

Table 4.2. Ionization frequencies averaged over all relevant wavelengths. The range of values reflects the variations in solar activity, $J_X \sim \phi_\infty^{ph}$ (adapted mainly from Torr and Torr, 1985).

Ion species X^+	$\overline{J}_X \; [10^{-7} s^{-1}]$
O^+	$2 - 7$
N_2^+	$3 - 9$
O_2^+	$5 - 14$
He^+	$0.4 - 1$
H^+	$0.8 - 3$

be noted that not only photons absorbed by the mother gas, but also those absorbed by other relevant gases may provide a significant contribution (e.g. N_2 for the case of N^+ production, see Eq. (3.19)). Finally, be aware of the principal difference between the ionization efficiency ε_X^I introduced here and the heating efficiency η^W considered earlier. Whereas ε_X^I involves only primary photoionization, η^W was defined under consideration of both primary and secondary heating processes.

Alternative forms of the above relation are obtained by replacing the product $\varepsilon_X^I \, \sigma_X^A$ by the *ionization cross section* σ_X^I or, going a step further by replacing the product $\varepsilon_X^I \, \sigma_X^A \phi_\infty^{ph}$ by the *ionization frequency* J_X

$$q_{X^+}^{PI} = \sigma_X^I \, n_X \, \phi^{ph} = J_X \, n_X(h) \, e^{-\tau(h)} \qquad (4.3)$$

where we have made use of Eq. (3.26). The form on the right side of Eq. (4.3) emphasizes that the ionization frequency J_X, in contrast to the density n_X and the optical depth τ, does not depend on the height. Typical values of J_X, averaged over all wavelengths of interest here, are summarized in Table 4.2.

As indicated in Eq. (4.3), the height variation of the primary ion production rate is governed by the interplay between the decreasing mother gas density and the increasing radiation intensity with increasing height. This leads to the formation of a production layer, the maximum of which occurs at a height where the radiation is still sufficiently intense to yield a large amount of ionization and the gas density becomes large enough to absorb a significant amount of the incident radiation; see Fig. 4.4. Indeed, for the case considered here, absorption of monochromatic radiation in a planar, isothermal, single gas atmosphere, this profile corresponds exactly to the Chapman production function described by Eq. (3.35) and sketched in Fig. 3.18. In contrast to this very simplified situation, the real upper atmosphere involves absorption of polychromatic radiation by a heterogeneous gas mixture. In order to obtain the total production rate of an ion species, we must therefore sum over all wavelengths λ and over all possible gas constituents j. This last

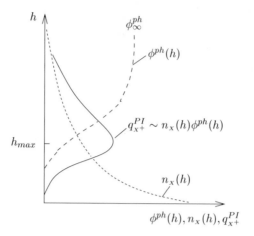

Fig. 4.4. Formation of ionization production layers

summation is necessary because not only the mother gas, but also all other gases, attenuate the incident radiation. We obtain

$$q_{x+}^{PI}(h) = n_x(h) \sum_\lambda J_x(\lambda) \exp\{-\sum_j \tau_j(h,\lambda)\} \qquad (4.4)$$

where we have neglected dissociative ionization to a first approximation. Figure 4.5 shows representative production profiles for the three ion species N_2^+, O_2^+ and O^+. A striking feature of these profiles, deviating from the simple Chapman production function with a single maximum near the absorption height for EUV radiation (ca. 170 km), is an additional maximum in the lower thermosphere. This may be explained by breaking down the production rates according to wavelength range. Figure 4.6 shows the total production rate

$$q^{PI}(h, \Delta\lambda_i) = q_{N_2^+}^{PI}(h, \Delta\lambda_i) + q_{O_2^+}^{PI}(h, \Delta\lambda_i) + q_{O^+}^{PI}(h, \Delta\lambda_i)$$

$$= \sum_s n_s(h) \, J_s(\Delta\lambda_i) \exp\{-\sum_j \tau_j(h, \Delta\lambda_i)\} \qquad (4.5)$$

for ten different wavelength ranges $\Delta\lambda_i$ ($s, j = N_2$, O_2 and O). As indicated, the maximum in the lower thermosphere forms as a result of the absorption of the longest and shortest wavelength ranges of the EUV radiation (intervals 1 and 10). The large penetration depth of this radiation arises from the relative small absorption cross sections in these wavelength ranges (see Fig. 3.13 and 3.16). Generally, the exact form of the production profile is strongly dependent on the actual solar zenith angle and the solar activity. The profiles shown above thus only serve as examples.

Secondary Ionization Processes. As mentioned in Section 3.3.1, the photoelectrons released as a result of the primary ionization process can attain

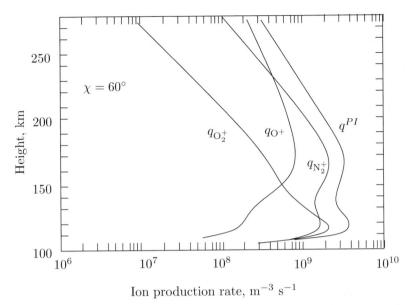

Fig. 4.5. Representative production profiles for the most important ions from primary photoionization and their sum q^{PI}. As earlier, χ denotes the zenith angle of the incident radiation (adapted from Matuura, 1966).

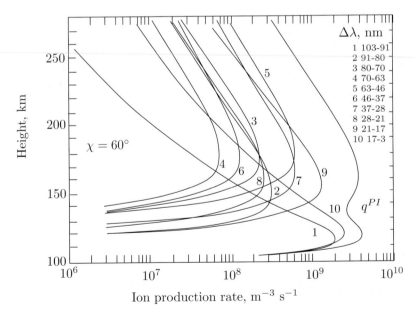

Fig. 4.6. Ionization production profiles for various wavelength ranges (see also Fig. 4.5, adapted from Matuura, 1966)

sufficiently high energies to ionize neutral gas particles themselves. For example, a photon of wavelength $\lambda \simeq 30.4$ nm can produce a photoelectron from an oxygen atom with a kinetic energy of almost 27 eV, which is larger than the ionization energy of atmospheric gases. Another possibility is that energetic photoelectrons can induce neutral gas particles to emit EUV radiation, thereby triggering additional photoionization events. Both of these processes are included in Fig. 3.20. It is estimated that the ionization rate due to these secondary ionization processes, when averaged over all wavelength ranges and constituents, contributes about 20 % of the primary production rate in the upper ionosphere. The secondary production rate in the E region can even be as large as that from the primary photoionization processes.

Charge Exchange. Simple (i.e. nondissociative) charge exchange reactions have the general form

$$X + Y^+ \xrightarrow{k^{CE}_{X,Y^+}} X^+ + Y \tag{4.6}$$

Although ions of the species X^+ are produced by these reactions, ions of the species Y^+ are lost, so that the total ionization density remains the same. Important examples for this type of production process are the reactions (2) to (5) in Table 4.3. The reactions (2) and (4) are an important source of O_2^+ ions in the lower ionosphere, and reaction (3) is even the main source of H^+ ions in the plasmasphere. Somewhat more complicated are the charge exchange reactions accompanied by dissociation and formation of new molecules, e.g. the reactions (1), (6) and (7). It turns out that these reactions are the main source of NO^+ ions in the lower ionosphere.

In order to describe the effectiveness of the charge exchange processes, we introduce the reaction constant $k^{CE}_{s,t}$. Using the nomenclature of Eq. (4.6), the reaction constant is defined by the following relation

$$q^{CE}_{X^+} = k^{CE}_{X,Y^+} \, n_X \, n_{Y^+} \tag{4.7}$$

The production of X^+ ions per unit volume and time by charge exchange is thus assumed to be proportional to the densities of the mutually interacting gases X and Y^+. The corresponding constant of proportionality k^{CE}_{X,Y^+} implicitly contains the temperature dependence of the given reaction; see Table 4.3.

A gas kinetic interpretation of the reaction constant may be formulated as follows. Every charge exchange reaction corresponds to an interaction or an 'inelastic collision' between the participating gas particles. Referring again to Eq. (4.6), a particle of type X will travel one mean free path l_{X,Y^+} before it suffers a charge exchange collision with a particle of type Y^+. The mean time required to travel this distance is thus

$$\Delta t = l_{X,Y^+}/\bar{c}_X = 1/\nu_{X,Y^+}$$

Table 4.3. Important chemical reactions in the ionosphere (adapted from Schunk, 1983)

(1)	$O^+ + N_2$	$\longrightarrow$	$NO^+ + N,$

$$k_1 = 1.533 \cdot 10^{-18} - 5.92 \cdot 10^{-19} \, (T/300) + 8.60 \cdot 10^{-20} \, (T/300)^2 \; ;$$

$$300 \leq T \leq 1700 \text{ K}$$

$$k_1 = 2.73 \cdot 10^{-18} - 1.155 \cdot 10^{-18} \, (T/300) + 1.483 \cdot 10^{-19} \, (T/300)^2 \; ;$$

$$1700 < T \leq 6000 \text{ K}$$

(2) $\quad O^+ + O_2 \quad \longrightarrow \quad O_2^+ + O,$

$$k_2 = 2.82 \cdot 10^{-17} - 7.74 \cdot 10^{-18} \, (T/300) + 1.073 \cdot 10^{-18} \, (T/300)^2$$
$$- 5.17 \cdot 10^{-20} \, (T/300)^3 + 9.65 \cdot 10^{-22} \, (T/300)^4; \; 300 \leq T \leq 6000 \text{ K}$$

(3) $\quad O^+ + H \quad \rightleftharpoons \quad H^+ + O, \qquad \overrightarrow{k}_3 = 2.5 \cdot 10^{-17} \sqrt{T_n}$

$$\overleftarrow{k}_3 = 2.2 \cdot 10^{-17} \sqrt{T_i}$$

(4) $\quad N_2^+ + O_2 \quad \longrightarrow \quad O_2^+ + N_2, \qquad k_4 = 5 \cdot 10^{-17} \, (300/T)$

(5) $\quad N_2^+ + O \quad \longrightarrow \quad O^+ + N_2, \qquad k_5 = 1 \cdot 10^{-17} \, (300/T)^{0.23} \; ;$

$$T \leq 1500 \text{ K}$$

(6) $\quad N_2^+ + O \quad \longrightarrow \quad NO^+ + N, \qquad k_6 = 1.4 \cdot 10^{-16} \, (300/T)^{0.44} \; ;$

$$T \leq 1500 \text{ K}$$

(7) $\quad N^+ + O_2 \quad \longrightarrow \quad NO^+ + O, \qquad k_7 = 2.6 \cdot 10^{-16}$

(8) $\quad N^+ + O_2 \quad \longrightarrow \quad O_2^+ + N, \qquad k_8 = 3.1 \cdot 10^{-16}$

(9) $\quad He^+ + N_2 \quad \longrightarrow \quad N^+ + He + N, \quad k_9 = 9.6 \cdot 10^{-16}$

(10) $\quad He^+ + N_2 \quad \longrightarrow \quad N_2^+ + He, \qquad k_{10} = 6.4 \cdot 10^{-16}$

(11) $\quad He^+ + O_2 \quad \longrightarrow \quad O^+ + He + O, \quad k_{11} = 1.1 \cdot 10^{-15}$

(12) $\quad N_2^+ + e \quad \longrightarrow \quad N + N, \qquad k_{12} = 1.8 \cdot 10^{-13} \, (300/T_e)^{0.39}$

(13) $\quad O_2^+ + e \quad \longrightarrow \quad O + O, \qquad k_{13} = 1.6 \cdot 10^{-13} \, (300/T_e)^{0.55}$

(14) $\quad NO^+ + e \quad \longrightarrow \quad N + O, \qquad k_{14} = 4.2 \cdot 10^{-13} \, (300/T_e)^{0.85}$

(15) $\quad O^+ + e \quad \longrightarrow \quad O^{(*)} + h\nu, \qquad k_{15} \simeq 1.4 \cdot 10^{-18} \, (1160/T_e)^{0.5}$

where the reaction constants k_i are in $[\text{m}^3\text{s}^{-1}]$, $T \simeq T_n$ for small ion drift velocities and $T_i \simeq T_n$. In the presence of polar electric fields, the temperature in the F region increases as $T[\text{K}] \simeq T_n[\text{K}] + 0.33 \, \mathcal{E}_{eff}^2[\text{mV/m}]$, where $\vec{\mathcal{E}}_{eff} = \vec{\mathcal{E}}_\perp + \vec{u}_n \times \vec{\mathcal{B}}$ ($\vec{\mathcal{E}}_\perp = $ externally applied electric field component perpendicular to the magnetic field $\vec{\mathcal{B}}$, $\vec{u}_n = $ neutral gas velocity and $\vec{\mathcal{B}} = $ geomagnetic field vector; see also Section 7.5.2).

Accordingly, all n_X particles in a unit volume will be transformed into X^+ ions within the time span Δt. The corresponding production rate must thus be given by

$$q_{X^+}^{CE} = n_X/\Delta t = n_X\, \nu_{X,Y^+} \tag{4.8}$$

Comparing this relation with Eq. (4.7), we arrive at the following gas kinetic interpretation of the reaction constants (generalized form)

$$k_{s,t} = \nu_{s,t}/n_t \tag{4.9}$$

One might be tempted to use Eq. (4.8) directly, since ν_{X,Y^+}, in fact, is the number of charge exchange collisions per second. For the situation under consideration here, it should be noted that each X particle suffers only one collision before it is transformed. As such, it is more appropriate to work with the mean free path, a quantity that may be defined as the average value of travel distances for one-time collisions.

Particle Precipitation. In addition to photons, the upper atmosphere can be ionized by incident energetic particles, and this process plays a particularly important role at high latitudes. For precipitating energetic electrons this process may be written as

$$X + e_{primary}(E \gtrsim 12 \text{ eV}) \rightarrow X^+ + e_{secondary} + e_{primary} \tag{4.10}$$

where the index 'primary' denotes the precipitating electron and the index 'secondary' the newly produced secondary electron. In contrast to photons, energetic particles can induce many primary ionization events (auroral electrons have energies from a few 100 up to many 1000 eV), thereby considerably complicating a calculation of the total effect. Furthermore, the secondary electrons are mostly energetic enough to induce ionization events themselves, and this holds as well for any tertiary electrons and all other electrons of higher order. One possible approach simulates the penetration of energetic particles into the thick atmosphere with the help of a Monte-Carlo technique. Another method is to solve a complex electron transport equation. Some results of such calculations are summarized in Section 7.4.2.

4.2.2 Ionization Losses

Were we to account only for ionization production, the actually observed ionization density would be far exceeded in a very short time. Let us take the E region as an example. With $n(130 \text{ km}) \simeq 10^{11}$ m^{-3} and $q(130 \text{ km}) \simeq 3 \cdot 10^9$ m^{-3}s^{-1} (compare Fig. 4.2 and Fig. 4.5), one obtains the time required to build up the observed ionization density $\tau_q = n/q \simeq 30$ s. Were it not for very effective loss processes with similarly short time constants, the ionization density of this region would grow to unrealistically high values within minutes. Here we identify these ionization loss processes and describe their height dependence.

Dissociative Recombination of Molecular Ions. By far the most important loss process for *molecular* ions is the so-called dissociative recombination. As implied by the name, the molecular ion involved in this reaction divides into its constituents

$$XY^+ + e \xrightarrow{k_{XY^+}^{DR}} X^{(*)} + Y^{(*)} \tag{4.11}$$

The asterisks again denote the possible excited states of the recombination products. Specific examples for this type of recombination are the reactions (12) to (14) of Table 4.3. The number of ions lost by this process per unit volume and time is

$$l_{XY^+}^{DR} = k_{XY^+}^{DR} \, n_{XY^+} \, n_e \tag{4.12}$$

Note that the symbol l (from *loss*) in Eq. (4.12) denotes a loss rate, i.e. a loss of charge carriers per unit volume and time, and should not be confused with the mean free path. Since the two particles involved in a dissociative recombination event are again transformed into two particles, there is no problem in conserving both momentum and energy. The associated reaction constants are correspondingly large. For the reactions (12) to (14) listed in Table 4.3 at typical electron gas temperatures, these are

$$k_{N_2^+}^{DR} \simeq k_{O_2^+}^{DR} \simeq k_{NO^+}^{DR} \simeq 10^{-13} \text{ m}^3\text{s}^{-1} \tag{4.13}$$

and this value is much larger than all other reaction constants in the table.

Radiative Recombination of Atomic Ions. The most obvious loss process for *atomic* ions is direct recombination, whereby the excess energy is either entirely or partly expended as radiation

$$X^+ + e \xrightarrow{k_{X^+}^{RR}} X^{(*)} + \text{photon} \tag{4.14}$$

A concrete example is the reaction (15) given in Table 4.3. Since this type of recombination creates one particle from two input particles, it is relatively rare. Conservation of energy and momentum requires exact expenditure of the recombination energy in excited states, but this storage is only possible in precisely defined quanta. The radiative recombination time constant for reaction (15) is correspondingly small

$$k_{O^+}^{RR} \simeq 10^{-18} \text{ m}^3\text{s}^{-1} \tag{4.15}$$

Charge Exchange. In contrast to radiative recombination, charge exchange is a very important loss process. Ions of a particular species are lost here, but not the ionization density as a whole

$$X^+ + Y \xrightarrow{k_{X^+Y}^{CE}} X^{(*)} + Y^+ \tag{4.16}$$

where in this case X^+ and Y can be either an atomic or molecular particle. Table 4.3 summarizes an entire suite of such reactions. Particularly important for the loss of atomic oxygen are the reactions (1) and (2), for which the reaction constants at a temperature $T = 1000$ K are

$$k^{CE}_{O^+,N_2} \simeq 5 \cdot 10^{-19} \text{ m}^3\text{s}^{-1} \;, \quad k^{CE}_{O^+,O_2} \simeq 125 \cdot 10^{-19} \text{ m}^3\text{s}^{-1} \qquad (4.17)$$

These are not very impressive values compared with the reaction constants for radiative recombination. However, comparing the associated loss rates in the lower F region (i.e. below the ionization peak), we find

$$l^{CE}_{O^+} = k^{CE}_{O^+,N_2} \, n_{O^+} \, n_{N_2} + k^{CE}_{O^+,O_2} \, n_{O^+} \, n_{O_2} \gg l^{RR}_{O^+} = k^{RR}_{O^+} \, n_{O^+} \, n_e \qquad (4.18)$$

The densities of the neutral gas constituents N_2 and O_2 in this region are much larger than the electron density. At a height of 200 km, for example, $n_{N_2}/n_e \simeq 10^4$ and $n_{O_2}/n_e \simeq 700$ (see Fig. 4.2 and Appendix A.4). Only above the ionization density maximum do the densities of the molecular constituents decrease to a value that renders charge exchange reactions unimportant. Since transport-induced loss processes are already dominant at these heights, however, radiative recombination is relegated to a subordinate role in the entire ionosphere. The molecular ions NO^+ and O_2^+ created by the charge exchange reactions of atomic oxygen with N_2 and O_2 decay very quickly via dissociative recombination. Considering the overall ionization loss in the lower F region, it is found that this is controlled almost exclusively by the efficiency of the charge exchange reactions (1) and (2).

Height Profile of the Loss rates

E Region (NO^+, O_2^+). Dissociative recombination is by far the quickest process responsible for the removal of the dominant molecular ions NO^+ and O_2^+ in the E region. As such, the total loss rate is given by

$$l_{\text{E region}} \simeq k^{DR}_{NO^+} \, n_{NO^+} \, n_e + k^{DR}_{O_2^+} \, n_{O_2^+} \, n_e$$

Using

$$k^{DR}_{NO^+} \simeq k^{DR}_{O_2^+} = \alpha \; (\simeq 10^{-13} \text{ m}^3\text{s}^{-1})$$

and

$$n_{NO^+} + n_{O_2^+} \simeq n_e = n$$

the loss rate can be rewritten as follows

$$l_{\text{E region}}(h) \simeq \alpha \, n^2(h) \qquad (4.19)$$

Since the loss coefficient α varies only weakly with height via its temperature dependence, the loss rate variation with height is essentially determined by the ionization density. Accordingly, the loss rate in the E region grows quadratically with increasing ionization density.

F Region (O^+). Destruction of atomic oxygen ions in the lower F region progresses essentially via the charge exchange reactions (1) and (2) in Table 4.3. The corresponding loss rate is

$$l_{\text{F-Region}} \simeq k^{CE}_{O^+,N_2} \, n_{N_2} \, n_{O^+} + k^{CE}_{O^+,O_2} \, n_{O_2} \, n_{O^+}$$

With

$$n_{O^+} \simeq n_e = n$$

we can write this relation in the following form

$$l_{\text{F region}}(h) \simeq \beta(h) \, n(h) \tag{4.20}$$

The height dependence of the loss term is determined by both the ionization density and, to a greater extent, the loss coefficient $\beta(h)$

$$\beta(h) = k^{CE}_{O^+,N_2} \, n_{N_2}(h) + k^{CE}_{O^+,O_2} \, n_{O_2}(h) \tag{4.21}$$

Using Eq. (4.17) (for a temperature of 1000 K), this yields

$$\beta[\text{s}^{-1}] \simeq 5 \cdot 10^{-19} \, (n_{N_2}[\text{m}^{-3}] + 25 \, n_{O_2}[\text{m}^{-3}]) \tag{4.22}$$

4.2.3 Chemical Composition

Knowledge of the production and loss processes described in the two previous subsections provides the basis for understanding the chemical composition of the ionosphere. The essential features are summarized here once again. For simplicity, reference to the various chemical reactions will be according to their numbers in Table 4.3.

- The prominent role played by the NO^+ ions in the E region may be attributed to the very effective production of this species by the charge exchange reactions (1), (6) and (7). Direct photoionization plays a subordinate role, because the mother gas NO exists only in trace concentrations in the lower thermosphere.
- The absence of N_2^+ ions in the E region, in spite of a copious production rate (see Fig. 4.5), occurs because the losses of this species via the charge exchange reactions (4)–(6) ($\sim n_{O_2,O}$) are much larger than those of the competing ions O_2^+ and NO^+ via dissociative recombination ($\sim n_e$).
- The lack of O^+ ions in the E region may be explained by the strongly suppressed production of this species in this region (see Fig. 4.6) and by the exponential growth of the loss coefficient $\beta(h)$ with decreasing height.
- The absence of the NO^+ and O_2^+ ions in the F region, on the other hand, is a consequence of the decreasing production of these species with height via reactions (1), (6), (7) and photoionization. Another factor, as indicated in Eq. (4.19), is the sharply increasing loss rate driven by the increasing electron density.

- The dominance of H^+ ions, which can occasionally extend down to relatively low heights (e.g. 600 km), is attributed to the effective production
  of this species by means of the charge exchange reaction (3). At the same
  time, the dominant ion O^+ is removed by this process.
- The dominance of the H^+ ions also explains why the He^+ ions play only
  a subordinate role in the terrestrial ionosphere. This latter species is produced primarily by photoionization – a slow process compared to the
  charge exchange reaction (3). Furthermore, the charge exchange reactions
  (9)–(11) are effective loss processes for this species.

One remark on this topic should not go unmentioned. The initial identification of the reactions so concisely summarized here was achieved only with
great difficulty. Even today, the measurement or calculation of the applicable reaction constants is by no means a simple task and the results are thus
subject to some uncertainty.

4.3 Density Profile in the Lower Ionosphere ($h < h_m$)

The approach we use to determine the density profile in the lower ionosphere
(i.e. below the density peak) is fundamentally different from that used to
calculate the density variation in the thermosphere. The reason for this is
that production and loss processes, in contrast to the thermosphere, play a
decisive role in the lower ionosphere. We thus first introduce an equation for
the determination of the density that accounts for this feature. Simplified
forms of this equation are then used to determine the density profile in the
E and lower F regions.

4.3.1 Density Balance Equation

To determine the density of a gas in the presence of production and loss
processes, we consider a stationary volume element embedded in the gas.
The temporal variation of the prevailing gas density in this volume element
depends upon the specific production of gas particles, the specific loss of gas
particles, and the net effect of the gas particles being transported into and out
of this volume element. In general: the change in density with time is equal
to the density gain by production, minus the density loss by destruction,
plus/minus the density gain/loss by transport. Denoting the density of a gas
species s with n_s, the production term with q_s, the loss term with l_s and the
transport term (which, in contrast to q_s and l_s, can assume either positive
or negative values) with d_s, the *density balance equation* may be written as
follows

$$\frac{\partial n_s}{\partial t} = q_s - l_s + d_s \tag{4.23}$$

This relation is formally analogous to the heat balance equation introduced
in Section 3.3.5, whereby q_s and l_s are again 'true' gains and losses and d_s

describes the changes brought about by redistribution. Moreover, this relation corresponds to the continuity equation (2.18), extended to include production and loss terms. As shown there, the transport-induced density gain or loss rates are equal to the negative divergence of the particle flux, $d_s = -\mathrm{div}\vec{\phi}_s$, so the density balance equation may also be written in the following way

$$\frac{\partial n_s}{\partial t} = q_s - l_s - \mathrm{div}\vec{\phi}_s = q_s - l_s - \mathrm{div}(n_s \vec{u}_s) \tag{4.24}$$

This relation is a partial, nonlinear differential equation that can only be solved numerically in the general case. Furthermore, self-consistently derived density distributions of the other charge carrier gases are required, particularly for the calculation of the production and loss terms. Just recall the charge exchange reactions and their importance for the ionosphere. Taken together, this means that we must solve a system of coupled differential equations. Here we will limit our discussion to strongly simplified forms of the density balance equation. These shortened equations possess analytical solutions and are adequate for understanding the ionospheric density distribution, at least for the salient features.

In the lower ionospheric regime of interest here, transport-induced density variations are negligibly small. The reason for this is the strong friction in the dense lower thermosphere that prevents any independent motion of the charge carriers. Moreover, during the day the rate of change of the density $\partial n/\partial t$ is also small with respect to the production and loss terms. This is especially true near local noon when the ionization density reaches its diurnal maximum and $\partial n/\partial t \to 0$, but the observed rate of density change is also comparatively small during the morning and afternoon hours (as discussed in Section 4.6). Under these conditions the density balance equation reduces to a simple *production-loss equilibrium*

Daytime: $$q_s \simeq l_s \tag{4.25}$$

If the production proceeds via photoionization and the loss is due to chemical reactions, as is the case for the lower ionosphere, the approximation given by Eq. (4.25) is also denoted as *photo-chemical equilibrium*.

During the night when the production rate is negligible we obtain

Nighttime: $$\frac{\partial n_s}{\partial t} \simeq -l_s \tag{4.26}$$

The range of validity for these approximations is investigated more closely in Section 4.5. First let us apply them to the E region and the lower F region.

4.3.2 Density Profile in the E Region

According to Eq. (4.19), the loss term in the E region assumes the general form

$$l \simeq \alpha \, n^2$$

The ionization density profile during the day under the condition of photo-chemical equilibrium is therefore given by

$$n(h,t) \simeq \sqrt{q(h,t)/\alpha} \qquad (4.27)$$

The height and time variations in the density thus follow those of the production, the rate of change being moderated by the square root. This is shown schematically in Fig. 4.7a. In order to obtain an explicit expression for the density profile, we approximate $q(h,t)$ by the Chapman production function Eq. (3.35). The result is a strongly idealized density profile, the so-called *Chapman (α)-layer*

$$n(h,t) = \sqrt{(q_{max}^{PI})^{*}/\alpha} \, \exp\left\{ \frac{1}{2} \left[1 - \frac{h - h_{max}^{*}}{H} - \sec\chi \, e^{-(h-h_{max}^{*})/H} \right] \right\} \qquad (4.28)$$

where $(q_{max}^{PI})^{*}$ denotes the maximum ionization production rate at normal radiation incidence (i.e. $\chi = 0$) and h_{max}^{*} is the height at which this maximum occurs.

During the night we have

$$\frac{\partial n}{\partial t} \simeq -\alpha \, n^2$$

Separation of variables and integration leads to the following height and time dependent density distribution

$$n(h,t) \simeq \frac{n(h,t_0)}{1 + \alpha \, n(h,t_0) \, (t - t_0)} \qquad (4.29)$$

Evidently high ionization densities decay faster than low densities after sunset so that the altitude profile becomes flatter.

4.3.3 Density Profile in the Lower F Region

In the lower F region the loss term from Eq. (4.20) takes the general form

$$l \simeq \beta \, n$$

so that the ionization density profile during the day becomes

$$n(h,t) \simeq q(h,t) \, / \, \beta(h,t) \qquad (4.30)$$

Since the production rate q decreases more slowly with height than the loss coefficient β, the equilibrium density effectively increases with increasing height. This is shown schematically in Fig. 4.7b. To obtain an explicit expression for

a. E region (O_2^+ , NO^+) : $n \simeq \sqrt{q/\alpha}$

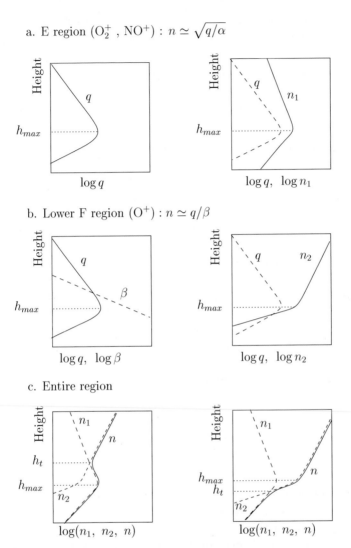

b. Lower F region (O^+) : $n \simeq q/\beta$

c. Entire region

Fig. 4.7. Schematic height profile of the ionization density in the lower ionosphere under the condition of photo-chemical equilibrium. h_t denotes the transition height from the E region to the F region (adapted from Ratcliffe, 1972).

the height dependence of the density, we consider the region above the maximum of ionization production ($h > h_{max}$ in Fig. 4.4). Here the transmission factor $e^{-\tau}$ quickly approaches unity, so that the height variation of the O^+ production rate corresponds to a first approximation to that of the mother gas n_O, see Eq. (4.3). Neglecting also the height dependence of the temperature (this has reached roughly 70 % of its thermopause value at the height of the production maximum), the variation of the production rate with altitude may be written as

$$q_{O^+}(h) \sim n_O(h) \sim e^{-(h-h_0)/H_O}$$

Furthermore, considering that molecular oxygen and molecular nitrogen have nearly the same weight, the height dependence of the loss coefficient β may be described, to a good approximation, using a combined scale height H_{N_2,O_2} for both gases

$$\beta(h) \simeq k^{CE}_{O^+,N_2}\, n_{N_2}(h_0)e^{-(h-h_0)/H_{N_2}} + k^{CE}_{O^+,O_2}\, n_{O_2}(h_0)e^{-(h-h_0)/H_{O_2}}$$
$$\sim e^{-(h-h_0)/H_{N_2,O_2}}$$

We thus obtain

$$n(h) \simeq \frac{q_{O^+}(h)}{\beta(h)} \sim \frac{\exp[-(h-h_0)/H_O]}{\exp[-(h-h_0)/H_{N_2,O_2}]} = \exp[(h-h_0)/H_{ql}]$$

where we have introduced the scale height

$$H_{ql} = (1/H_{N_2,O_2} - 1/H_O)^{-1} = \frac{k\,T}{(\mathcal{M}_{N_2,O_2} - \mathcal{M}_O)\,m_u\,g} \tag{4.31}$$

The ionization density evidently grows exponentially with increasing height and the effective scale height H_{ql} with $\mathcal{M}_{N_2,O_2} \simeq 30$ corresponds closely to that of atomic oxygen.

For nighttime conditions we have

$$\frac{\partial n}{\partial t} \simeq -\beta\, n$$

which leads to the height and time dependent density distribution

$$n(h,t) \simeq n(h,t_0)\, e^{-\beta(h)\,(t-t_0)} \tag{4.32}$$

where, to sufficient accuracy, we have neglected the time dependence of the loss coefficient. Since β increases toward lower heights, the density decrease proceeds much faster there.

The results for the E and lower F region must be combined to obtain the density profile in the entire lower ionosphere. This is shown in Fig. 4.7c. Evidently, the details of the combined density profile depend on whether the transition height from E to F region, $h_{t(ransition)}$, lies above or below the height of maximum production h_{max}.

When applying this superposition procedure to the real ionosphere, it should be noted that the production maximum at a height of about 110 km lies completely within the E region (see Fig. 4.5). The density profile in this region will thus follow that of the production profile to a first approximation. In contrast, the secondary production peak at about 170 km altitude is located close to the transition height h_t, so that either a secondary maximum or a ledge will be formed in the density profile.

4.4 Density Profile in the Upper Ionosphere ($h > h_m$)

Since the neutral gas densities decrease exponentially with height, the production and loss terms become increasingly less important. Considering quasi-stationary conditions, this leads to the following approximations

$$\frac{\partial n_s}{\partial t}, \ q_s, \ l_s \to 0 \tag{4.33}$$

Accordingly

$$\mathrm{div}\vec{\phi}_s \to 0 \tag{4.34}$$

For a horizontally uniform ionosphere this implies that

$$\phi_z = n \ u_z \to \text{constant} \tag{4.35}$$

or, for the case of interest here, namely a static ionosphere that is gravitationally bound to the Earth

$$u_z \to 0 \tag{4.36}$$

This condition can only be used indirectly for the determination of the ionization density profile. The velocity of a gas volume vanishes only if all of the forces acting on the volume are in equilibrium. We are clearly abandoning the density balance equation with this requirement and return to the force balance relation or momentum balance equation for a determination of the density profile.

4.4.1 Barometric Density Distribution

Similar to the determination of the density profile for thermospheric neutral gases, for which Eq. (4.33) is certainly fulfilled to good approximation, we consider the case where the pressure gradient and gravitational forces acting on an ion gas and its associated electron gas are in equilibrium. This leads immediately to the aerostatic equation

$$\frac{dp_s}{dz} = -\rho_s \ g$$

where s stands for either the ion gas (in our case the O^+ gas) or the electron gas. Assuming (to sufficient accuracy for our purposes) that the ion and electron temperatures are constant, this relation yields the familiar barometric law

$$n_s(h) = n_s(h_0) \ e^{-(h-h_0)/H_s}$$

where the scale height of the ions is considerably smaller than that of the lighter electrons

$$H_i = \frac{k \ T_i}{m_i \ g} \quad \ll \quad H_e = \frac{k \ T_e}{m_e \ g} \tag{4.37}$$

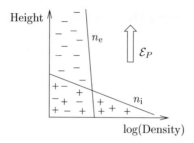

Fig. 4.8. Charge separation and build up of a polarization field by the different density distributions of the ion and electron gases

This results in the density distributions sketched in Fig. 4.8. These distributions are obviously unrealistic because they lead to a strong separation of the charge carriers. The positive dominated charge density at lower heights and the predominant negative charge density at greater heights combine to produce a strong electric polarization field. This in turn exerts an enormous force on the charge carrier gases and destroys the above imposed equilibrium of forces. In order to obtain a realistic density distribution, we must consider the electric force caused by the polarization field in addition to the pressure gradient and gravitational forces. The implied extension to the force balance relation is written as follows

$$\frac{\mathrm{d}p_i}{\mathrm{d}z} = -\rho_i\, g + n_i\, e\, \mathcal{E}_P \tag{4.38}$$

$$\frac{\mathrm{d}p_e}{\mathrm{d}z} = -\rho_e\, g - n_e\, e\, \mathcal{E}_P \tag{4.39}$$

where e is the elementary electric charge and $\mathcal{E}_P$ is the electric polarization field strength. Assuming that the electric polarization field is strong enough to enforce charge neutrality everywhere, i.e. $n_i(z) \simeq n_e(z) = n(z)$, in good agreement with the observations, then the above equations may be added together to obtain

$$\frac{\mathrm{d}(p_i + p_e)}{\mathrm{d}z} \simeq -n\, m_i\, g \tag{4.40}$$

where we have neglected the electron mass with respect to the ion mass. We may again assume to a sufficiently good approximation that T_i and T_e are nearly constant in the upper ionosphere. These two quantities vary by a factor less than three in the region between 400 and 1000 km (see Fig. 4.3), corresponding to a relatively small temperature gradient. Under this approximation the variables can be separated and the subsequent integration yields the barometric law, valid for both the ion as well as the electron gas

$$n(h) = n(h_0)\, \mathrm{e}^{-(h-h_0)/H_P} \tag{4.41}$$

where we have introduced the so-called *plasma scale height*

$$H_P = k \, (T_{\mathrm{i}} + T_{\mathrm{e}}) \, / \, m_{\mathrm{i}} \, g \tag{4.42}$$

As can be seen, the scale height of the ion gas in Eq. (4.37) is more than doubled by the polarization field. In contrast, the electron gas scale height is reduced by more than a factor of 10^4. Evidently, the light electron gas uses the polarization field to 'pull' the ions upward. Vice versa, in striving to attain an equilibrium state of complete charge neutrality, the heavy ion gas drags the electrons downward.

4.4.2 Polarization Field

The electric polarization field (also called the *Pannekoek-Rosseland field*) required to establish the above plasma density distribution is very small. Substituting Eq. (4.41) into Eq. (4.39) and using the fact that $m_{\mathrm{e}} \ll m_{\mathrm{i}} \, T_{\mathrm{e}} \, /(T_{\mathrm{e}}{+}T_{\mathrm{i}})$, we arrive at the relation

$$\mathcal{E}_P \simeq \frac{m_{\mathrm{i}} \, g}{e} \, \frac{T_{\mathrm{e}}}{T_{\mathrm{e}} + T_{\mathrm{i}}} \tag{4.43}$$

so that we calculate $\mathcal{E}_P < 1\mu$V/m for the oxygen ions of interest here. In order to produce a polarization field of this magnitude, it is sufficient to displace the electron gas only a tiny distance with respect to the ion gas. For example, considering the upper ionosphere to a first approximation as a homogenous slab of moderate density, an upward displacement of the electron gas gives rise to an electric field similar to that in a plate capacitor; see Fig. 4.9. The field strength is given by

$$\mathcal{E}_P = \frac{Q}{\varepsilon_0 \, A} = \frac{e \, n}{\varepsilon_0} \, \Delta z \tag{4.44}$$

where Q is the charge stored in either the upper or lower layer and ε_0 is the dielectric constant. The separation distance required to generate the given polarization field is thus

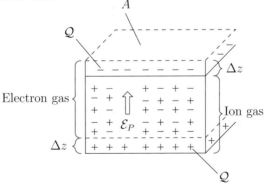

Fig. 4.9. Estimating the charge carrier separation distance needed for creating a specific electric polarization field

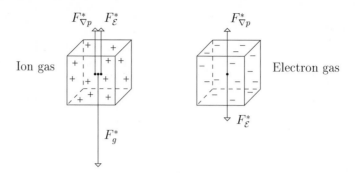

Fig. 4.10. The important volume forces acting on the ion and electron gases in the upper ionosphere ($T_i \simeq T_e$)

$$\Delta z = \frac{\varepsilon_0 \, \mathcal{E}_P}{e \, n} = \frac{\varepsilon_0 \, m_i \, g}{n \, e^2} \, \frac{T_e}{T_e + T_i} \tag{4.45}$$

Using the approximation $T_i \simeq T_e$ (absolutely adequate for this estimate) and taking a mean density in the upper ionosphere of $5 \cdot 10^{10}$ m^{-3}, we calculate a separation distance of $\Delta z \simeq 10^{-9}$ m. This corresponds to only a few ion diameters. Our assumption $n_i(h) \simeq n_e(h)$ is thus fulfilled to a very good approximation.

The continual striving toward charge neutrality, as demonstrated here, is one of the fundamental properties of every *plasma*. This concept has come to mean any partially or completely ionized mixture of gases that exhibits charge neutrality as a whole. In this sense it is justified and common practice to speak of ionospheric plasma instead of ionospheric charge carrier gases. In general, it is the particular behavior of partly or fully ionized gases that is emphasized by the designation 'plasma'. Because of the electromagnetic interaction of the charge carrier gases among themselves or with external electromagnetic fields, plasmas display an entire suite of collective phenomena that are not found in neutral gases (see, for example, Sections 4.7.1 and 6.3).

Inserting the polarization field strength into the momentum balance equations (4.38) and (4.39), one can estimate the relative magnitudes of the volume forces. The resultant force equilibrium situations, sketched in Fig. 4.10, are different for the ion gas and electron gas.

4.4.3 Transport Equilibrium

As explained in Section 2.3.5, a barometric density distribution can always be viewed as a transport or diffusive equilibrium. The downward current caused by the gravitational force and the upward current induced by the pressure gradient force are just balanced so that the net current vanishes. The validity of this interpretation should be examined here for the case of the upper ionosphere. We consider an ion-electron gas mixture for which

- $n_i \simeq n_e = n$
- $\vec{u}_i \simeq \vec{u}_e = \vec{u}$

In addition to quasi-neutrality (implying the presence of a Pannekoek-Rosseland field), we also assume that the ion and electron gases move together (i.e. *ambipolar*) with the same velocity. Velocity differences correspond to relative charge carrier motions and thus electric currents. For the purely vertical layering and motion considered here, such currents would lead inevitably to charge build-up at the upper and lower boundaries of the ionosphere. The charge separation, in turn, generates an additional electric polarization field that will quickly stifle any further relative motion of the charge carriers.

As shown earlier for the case of the thermosphere, the value of the downward current can be determined by equating the gravitational and frictional forces. The mass density of our charge carrier gas mixture (or plasma) is now equal to the sum of the single gas mass densities; the total frictional force is the sum of the frictional forces acting on the single gases. Since the ions and electrons are moving with the same velocity, only collisions with neutral gas particles need be considered. For a neutral gas atmosphere at rest ($u_n = 0$) we have

$$n \left(m_i + m_e \right) g \;=\; -n \left(m_i \, \nu^*_{i,n} + m_e \, \nu^*_{e,n} \right) u_g \qquad (4.46)$$

The common downward velocity of ions and electrons is denoted here by u_g. Alternatively, this relation can be derived by adding the momentum balance equations for the ions and electrons, provided only gravity and frictional forces are considered, see Eq. (3.80). Noting that $m_i \gg m_e$ and $m_i \, \nu^*_{i,n} \gg m_e \, \nu^*_{e,n}$, we may rewrite Eq. (4.46) as

$$\phi_{g,s} \;=\; n \, u_g \;\simeq\; -\frac{1}{m_i \, \nu^*_{i,n}} \, m_i \, n \, g \qquad (4.47)$$

where $\phi_{g,s}$ designates the descent flux of either the ion or electron component. To demonstrate that $m_i \, \nu^*_{i,n}$ is really large with respect to $m_e \, \nu^*_{e,n}$, we refer back to Eq. (2.56). With $\sigma_{e,n} = (r_e + r_n)^2 \pi \simeq \sigma_{i,n}/4$, $m_e \ll m_i \simeq m_n \simeq m_o(= 16 \, m_u)$ and $T_e \simeq T_i$, we find that the momentum transfer collision frequency of the electron component is a factor 60 greater than that of the ion component. Nevertheless, with $m_i \simeq 30\,000 \, m_e$ it is clear that the product of momentum transfer collision frequency and mass $m_i \, \nu^*_{i,n} \gg m_e \, \nu^*_{e,n}$.

The ambipolar upward flux may be determined by equating the frictional force to the pressure gradient force

$$\frac{\mathrm{d}(p_i + p_e)}{\mathrm{d}z} \;=\; -n \left(m_i \, \nu^*_{i,n} + m_e \, \nu^*_{e,n} \right) u_p \qquad (4.48)$$

from which one obtains

$$\phi_{p,s} \;\simeq\; -\frac{1}{m_i \, \nu^*_{i,n}} \frac{\mathrm{d}(p_i + p_e)}{\mathrm{d}z} \qquad (4.49)$$

The total flux of the ion and electron components is thus found to be

$$\phi_s = \phi_{g,s} + \phi_{p,s} \simeq -\frac{1}{m_i \, \nu_{i,n}^*} \frac{d(p_i + p_e)}{dz} - \frac{m_i \, n \, g}{m_i \, \nu_{i,n}^*}$$

$$= -D_P \left(\frac{dn}{dz} + \frac{n}{H_P} \right) \tag{4.50}$$

where in the final expression we have assumed constant ion and electron temperatures and have introduced the *ambipolar* or *plasma diffusion coefficient*

$$D_P = k \, (T_i + T_e)/m_i \, \nu_{i,n}^* \tag{4.51}$$

Equation (4.50) formally corresponds to the diffusion equation (A.45) derived in Appendix A.5 as long as the temperature gradients can be assumed negligibly small. Representative values of the momentum transfer collision frequency $\nu_{O^+,n}^*$ and the corresponding plasma diffusion coefficient can be seen in Fig. 4.11. Setting the total flux $\phi_s = 0$, as required in a static ionosphere, we arrive immediately at the aerostatic equation and thus, what we intended to prove, the barometric density distribution for the ionospheric charge carrier gases.

A complication we have not considered up to now comes from the fact that the Earth's magnetic field considerably limits the freedom of motion of

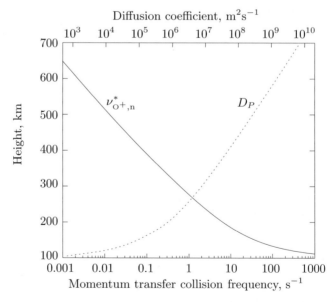

Fig. 4.11. Typical height profile of the momentum transfer collision frequency and the plasma diffusion coefficient for O^+ ions in the thermosphere for daytime conditions. Note that $\sigma_{O^+,O}$ is greater than $\sigma_{O,O}$ because of resonance effects.

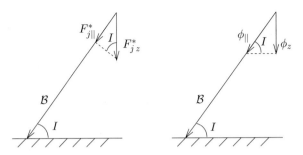

Fig. 4.12. Influence of the Earth's magnetic field on the ionization transport

the ionospheric charge carriers. The details of this interaction are discussed in Section 5.3. It suffices for now to realize that the charge carriers are only free to move parallel along the terrestrial magnetic field. It follows that magnetic forces may only be ignored for the components of the force balance equations (4.46) and (4.48) parallel to the magnetic field lines. This means that

$$\phi_{j\parallel} \simeq \frac{1}{m_i \, \nu_{i,n}^*} \, F_{j\parallel}^* \qquad (4.52)$$

where $\phi_{j\parallel}$ is the plasma flux parallel to the field and $F_{j\parallel}^*$ is the parallel component of each considered volume force. The motion of plasma in the Earth's magnetic field is sketched in Fig. 4.12. Denoting the angle between the Earth's magnetic field and the Earth's surface by the inclination I, the parallel volume forces $F_{j\parallel}^*$ are obtained from the vertical volume forces by multiplying them with the factor $\sin I$. Furthermore, the vertical plasma flux of interest here ϕ_{jz} is related to the field parallel plasma flux $\phi_{j\parallel}$ by the same factor $\sin I$. As a result, we obtain

$$\phi_{jz} = \phi_{j\parallel} \, \sin I \sim F_{j\parallel}^* \, \sin I = F_{jz}^* \, \sin^2 I \qquad (4.53)$$

and Eq. (4.50) may now be written

$$(\phi_s)_z = -D_P \, \sin^2 I \left(\frac{dn}{dz} + \frac{n}{H_P} \right) \qquad (4.54)$$

This modified vertical plasma flux clearly also vanishes for a barometric density distribution, which is evidently unaffected by the presence of a magnetic field.

4.4.4 Production-Generated Downward Current

A rather modest, but non-negligible amount of ionization production occurs in the upper ionosphere. Insufficiently countered by the still smaller chemical loss rates, this production works to create an ionization density that increases with height (see Section 4.3.3). In order to maintain the density distribution decreasing with height according to the barometric law, this production-generated surplus density must be continually removed by a downward current into the lower ionosphere where very effective chemical loss rates dominate. The situation is reminiscent of the one encountered in the thermosphere, where, in the equilibrium case, newly produced heat was continually removed by a downward-directed heat flux (see Section 3.3.5). We can estimate the magnitude of the production-generated ionospheric downward flux in a manner similar to this earlier exercise. Consider the entire amount of ionization produced by primary photoionization in an atmospheric column of base area A in the time interval Δt

$$N_{O^+}^{PI}(h) = A \, \Delta t \int_h^\infty q_{O^+}^{PI}(z) \, \mathrm{d}z \simeq A \, \Delta t \, J_o \int_h^\infty n_o \, \mathrm{d}z = A \, \Delta t \, J_o \, n_o(h) \, H_o$$

where the transmission factor $e^{-\tau}$ appearing in Eq. (4.3) has been set equal to unity for the upper ionosphere. Neglecting chemical losses and assuming the equilibrium case, this ionization amount must be transported downward through the area A in the time interval Δt. This requires an ambipolar downward flux given by

$$\phi_{g,o^+}^q(h) \simeq N_{O^+}^{PI}(h)/A \, \Delta t \simeq J_o \, n_o(h) \, H_o \tag{4.55}$$

As an example, using Table 4.2 and Appendix A.4, and assuming low solar activity, we calculate a production-generated downward flux of $\phi_{g,o^+}^q(400 \text{ km}) \simeq 10^{12} \text{ m}^{-2}\text{s}^{-1}$ at an altitude of 400 km. This should be compared with the 'regular' downward flux associated with transport equilibrium: $\phi_{g,o^+}(400 \text{ km}) = g \, n/\nu_{o^+,n}^* \simeq 10^{13} \text{ m}^{-2}\text{s}^{-1}$, where the ionization density and momentum transfer collision frequency were taken as $n(400 \text{ km}) \simeq 10^{11} \text{ m}^{-3}$ and $\nu_{o^+,n}^*(400 \text{ km}) \simeq 0.09$, respectively (see Figs. 4.2 and 4.11). The production-generated downward flux amounts to only about 10 % of the 'regular' downward flux and thus represents a relatively small disturbance of the diffusive equilibrium and the barometric density distribution. It is sufficient that the density decrease with height becomes somewhat less rapid than predicted by the barometric density distribution, so that the upward flux becomes somewhat smaller than the descent flux. Then a production-generated downward flux of the order $\phi_g^q = \phi_g + \phi_p \neq 0$ flows toward Earth. It is the divergence of this O^+ flux, increasing toward lower altitudes, that is by far the most important ionization loss for the upper ionosphere. It is noteworthy that the upward flux completely vanishes at the height of the ionization maximum because $\mathrm{d}n/\mathrm{d}z = 0$. The production-generated downward flux thus corresponds here to the regular downward flux, $\phi_g^q(h_m) = \phi_g(h_m) = g \, n(h_m)/\nu_{o^+,n}^*(h_m)$.

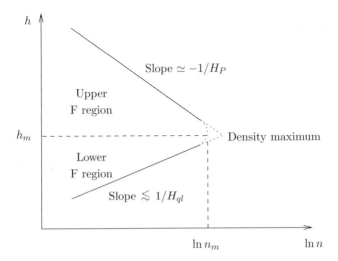

Fig. 4.13. Formation of the ionization density maximum

4.5 Density Maximum and Ionospheric Time Constants

The density profiles in the lower and upper ionosphere described in the previous sections elucidate the reason for the formation of the ionization density
maximum in the F region. While the lower F region is characterized by an
exponential increase in ionization density with a scale height H_{ql}, the upper
F region exhibits an exponential decrease in density with a scale height H_P.
Continuity of the combined density profile requires a smooth transition between both regions – a transition which must also define the maximum of the
ionization density. This is shown schematically in Fig. 4.13. The remaining
issues to be settled are the height at which this maximum forms and what is
the peak ionization density there.

As explained earlier, the increase of the ionization density in the lower
F region is attributed to the fact that the loss rate, dominated by chemical
reactions, decreases faster with height than the production rate. This increase should thus continue upward as long as chemical reactions remain the
dominant loss process. On the other hand, the density decrease in the upper
ionosphere is maintained as long as the divergence of the downward current
is capable of removing the ionization produced at these heights. The respective effectiveness of these two loss processes evidently determines whether the
ionization density increases or decreases. The transition from the lower to the
upper ionosphere will thus occur where chemical reactions are replaced by
ionization transport as the primary loss process, and the ionization maximum
will form at that height where the effectiveness of the two loss processes reach
the same magnitude. A quantitative measure of effectiveness is provided by
the associated time constants. These are defined and interpreted here, fol-

lowed by a description of their height profiles. These height profiles are then used to estimate the location and magnitude of the density maximum.

4.5.1 Ionospheric Time Constants

In order to estimate the time constants of interest, we write the density balance equation (without indices) in the following form

$$
\frac{\partial n}{\partial t} = q - l + d = \left|\frac{\partial n}{\partial t}\right|_q - \left|\frac{\partial n}{\partial t}\right|_l \pm \left|\frac{\partial n}{\partial t}\right|_d
$$

$$
= \frac{n}{\tau_q} - \frac{n}{\tau_l} \pm \frac{n}{\tau_d} \tag{4.56}
$$

The production, loss, and transport time constants are thus defined by the following relations

$$
\tau_q = \left(\frac{1}{n}\left|\frac{\partial n}{\partial t}\right|_q\right)^{-1} = \frac{n}{q} \tag{4.57}
$$

$$
\tau_l = \left(\frac{1}{n}\left|\frac{\partial n}{\partial t}\right|_l\right)^{-1} = \frac{n}{l} \tag{4.58}
$$

$$
\tau_d = \left(\frac{1}{n}\left|\frac{\partial n}{\partial t}\right|_d\right)^{-1} = \frac{n}{|d|} \tag{4.59}
$$

They evidently estimate the time required for the observed ionization density to be either built-up by production, destroyed by chemical reaction losses, or moved into or out of a given volume element by transport. This will be illustrated by an example involving the loss time constant.

Consider the nighttime decrease of the ionization density as sketched schematically in Fig. 4.14a. Starting at an arbitrarily chosen time t_0, the loss rate observed at this time is *linearly* extrapolated. The intersection of the associated tangent to the curve (straight line of slope $\partial n/\partial t$) with the abscissa $n = 0$ yields an estimate of the time required to completely deplete the ionization density present at the time t_0. This is expressed as

$$
\left.\frac{\partial n}{\partial t}\right|_{t_0} = \tan \alpha = \frac{n(t_0)}{\tau_l}
$$

so that

$$
\tau_l = \left(\frac{1}{n}\left|\frac{\partial n}{\partial t}\right|\right)^{-1}
$$

where we have suppressed the time variation of $\tau_l = f(t_0)$.

Another interpretation that comes closer to our conception of a time constant is obtained if we approximate the actually observed density decrease

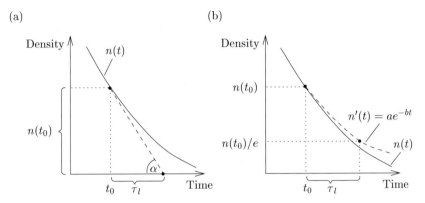

Fig. 4.14. Interpreting the loss time constant

$n(t)$ by an exponential decrease $n'(t)$; see Fig. 4.14b. The two curves are fitted at the starting point t_0 by equating the values of the functions and their derivatives, $n'(t_0) = n(t_0)$ and $(dn'/dt)_{t_0} = (dn/dt)_{t_0}$. This yields

$$n'(t) = n(t_0) \exp\left\{ -\left(\frac{1}{n} \left| \frac{\partial n}{\partial t} \right| \right)_{t_0} (t - t_0) \right\}$$

Defining the loss time constant as that time in which the approximation function $n'(t)$ – and thus the original function to a good approximation – decreases by a factor $1/e$ with respect to its initial value, one again obtains

$$\tau_l = \left(\frac{1}{n} \left| \frac{\partial n}{\partial t} \right| \right)^{-1} \tag{4.60}$$

These interpretations are valid, of course, independently of the type of loss process considered.

In order to obtain an explicit expression for the time constants, we use the second form of their defining equations. The chemical loss time constant in the F region below the ionization maximum is then given by

$$\tau_l(h < h_m) = n/l \simeq 1/\beta(h) \tag{4.61}$$

see Eqs. (4.20) and (4.58). This time constant evidently increases rapidly with height and reaches a value of roughly 30 minutes at a height of 270 km under the conditions valid for Fig. 4.15.

An estimate of the time constant associated with transport processes may be obtained by considering the rate of density change produced by a downward flux in the upper ionosphere. Neglecting the Earth's magnetic field and using Eqs. (2.17) and (4.47), we have

$$d_z = -\frac{d\phi_g}{dz} = g \frac{d(n/\nu_{i,n}^*)}{dz} \tag{4.62}$$

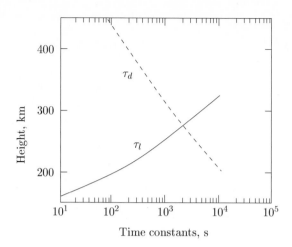

Fig. 4.15. Typical height profile of the ionospheric loss and transport time constants for daytime conditions

Assuming T_n, T_e and T_i to be constant to first approximation, the height variations of the ionization density and the momentum transfer collision frequency may be described by exponential functions

$$n(z) \sim e^{-(z-z_0)/H_P}$$

$$\frac{1}{\nu^*_{i,n}} \sim \frac{1}{n_n(z)} \sim e^{(z-z_0)/H_n}$$

where n_n and H_n denote the density and scale height of the neutral gas component, i.e. atomic oxygen for the upper ionosphere. Writing out the derivative contained in Eq. (4.62), we obtain

$$d_z \simeq \frac{g\,n}{\nu^*_{i,n}}\left(\frac{1}{H_n} - \frac{1}{H_P}\right) \simeq \frac{g\,n}{\nu^*_{i,n}H_P}$$

where, for the expression on the right, we have made use of the very good approximation $m_i \simeq m_{O^+} \simeq m_O \simeq m_n$, as well as $T_i \simeq T_e \simeq T_n$, also a sufficiently accurate approximation for our purposes. The transport time constant may thus be written as

$$\tau_d(h > h_m) = n/|d| \simeq \nu^*_{i,n}H_P/g = H_P^2/D_P \tag{4.63}$$

Formally, this corresponds exactly to the diffusion time constant Eq. (2.72) derived in Section 2.3.6 for a simple disturbance scenario. Since $\nu^*_{i,n}$ increases exponentially with decreasing altitude, τ_d quickly grows and reaches values of about 30 minutes at an altitude of 270 km under the conditions valid for Fig. 4.15.

4.5.2 Ionization Density Maximum

As indicated in Fig. 4.15, chemical loss processes predominate at heights below roughly 270 km ($\tau_d < \tau_l$). Accordingly, the ionization density increases with height in this region. In contrast, transport-induced losses dominate above this height ($\tau_d < \tau_l$), leading to a decrease of the ionization density according to the barometric law. Separating these regions is a transition layer where both time constants are about equal, and this transition also defines the height of maximum ionization density h_m

$$\tau_l(h_m) \simeq \tau_d(h_m) \tag{4.64}$$

Under the conditions valid for Fig. 4.15 this equation is fulfilled at $h_m \simeq$ 270 km, in basic agreement with observed values. The actual height of the maximum is a function of local time, season and solar activity, as reflected in its dependence on the neutral gas density (τ_l), as well as the plasma scale height and plasma diffusion coefficient (τ_d). An important feature of the ionization density maximum is that it is *not* located at the same height as the maximum of ionization production, but rather about 100 km higher.

The magnitude of the peak ionization density can be estimated from extrapolation of the photo-chemical equilibrium conditions into the height range of the maximum (see Fig. 4.13). This yields

$$n_m \lesssim \frac{q(h_m)}{\beta(h_m)} \simeq \frac{J_{\mathrm{O}}\, n_{\mathrm{O}}(h_m)}{k^{CE}_{\mathrm{O}^+,\mathrm{N}_2}\, n_{\mathrm{N}_2}(h_m) + k^{CE}_{\mathrm{O}^+,\mathrm{O}_2}\, n_{\mathrm{O}_2}(h_m)} \tag{4.65}$$

where we have set the transmission factor $e^{-\tau}$ equal to unity in the height region of interest here. Taking the maximum height $h_m \simeq 270$ km derived above and assuming moderate solar activity, we obtain a peak ionization density of $n_m \lesssim 10^{12}$ m^{-3}, again in good agreement with the observations.

Figure 4.16 summarizes the essential points of the present discussion. More precise calculations accounting for the effects of the Earth's magnetic field replace the plasma diffusion coefficient in Eq. (4.63) by the product $D_P \cdot \sin^2 I$ from Eq. (4.54). Furthermore, electric fields and particularly thermospheric winds can trigger plasma fluxes that are capable of considerably modifying the height and magnitude of the ionization density maximum (see also Section 8.5.2).

4.5.3 Ionoexosphere

Does the ionosphere possess a quasi-collisionless outer envelope, comparable to the neutral exosphere, from which particles can evaporate? The basic answer to this question is affirmative, albeit only for the polar ionosphere. Only in these regions are there so-called 'open' magnetic field lines (see Section 7.6.2), along which the particles can escape. In contrast, the terrestrial magnetic field assumes a closed (dipolar) configuration at mid-latitudes, in which

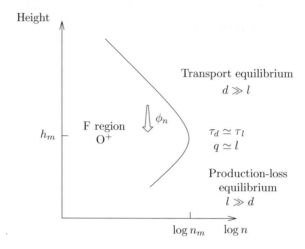

Fig. 4.16. Physical processes in the vicinity of the ionization density maximum

particles are captured like in a 'magnetic bottle' and prevented from escaping. As indicated earlier, this outer region is denoted the plasmasphere and is treated in detail in Section 5.4.3. The escape flux at high latitudes is appropriately named the *polar wind* after its place of origin. The approach used to describe this phenomenon theoretically is similar to that used for modeling the solar wind. The interested reader is thus referred to Section 6.1.

4.6 Systematic Variations of the Ionization Density

Similar to the magnitude of the energy and heat deposition, the actual amount of ionization production is also dependent on the solar zenith angle and radiation intensity. According to Eq. (4.3), this dependence is given by

$$q_{x+}^{PI} \sim J_x \ (= f(\phi_{\infty}^{ph})) \ e^{-\tau(=f(\sec \chi))}$$

Correspondingly, clear signatures of daily and seasonal variations, as well as solar rotation and solar cycle effects, are observed, especially in the lower ionosphere where the ionization density is proportional to the square root of the production rate. The situation is more complicated in the upper ionosphere, for example at the height of the ionization density peak. While the above described variations are observed, they sometimes appear in strongly modified, in some cases even inverse, forms. First we illustrate 'normal' daily variations as observed in winter at mid-latitudes during a period of high solar activity; see Fig. 4.17. Both the maximum ionization density and the ionospheric column density are plotted. Striking is the relatively sudden increase in the ionization density after sunrise. We can use the steep rise in density to estimate the size of the previously neglected term containing the rate of density change and obtain $\partial n / \partial t \simeq 10^8 \ \mathrm{m^{-3} s^{-1}}$. This must be compared with a

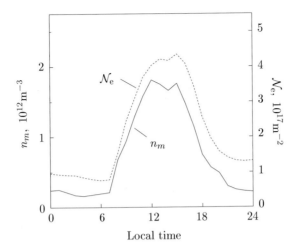

Fig. 4.17. Daily variation of the maximum electron density n_m and ionospheric column density $\mathcal{N}_e$. The data are for a mid-latitude location (Ottawa, 45°N), winter conditions (6 December 1982) and a period of high solar activity (CI=210).

production rate of $q_{O^+} \simeq J_O\, n_O(h_m) \simeq 10^9$ m^{-3}s^{-1}, where we have used $J_O = 7\cdot 10^{-7}$ s^{-1} and $n_O = 1.4\cdot 10^{15}$ m^{-3} (at $h_m = 270$ km, $T_\infty \simeq 1200$ K). Evidently, the rate of change of density is indeed small compared to the production term and thus also small with respect to the loss or transport terms.

The dependence of a typical ionization density profile on solar activity and season is illustrated in Fig. 4.18. Whereas the variations arising from solar activity agree with our expectations, the clearly smaller ionization densities in summer are surprising. This phenomenon, the *seasonal anomaly*, is caused by the neutral upper atmosphere, more specifically by the seasonal variations of the thermospheric composition and global wind circulation.

Spatial variations are observed in addition to temporal changes. Hence, considerable differences exist between the ionospheric conditions at low, middle and high latitudes. Among the peculiarities of the low-latitude ionosphere, for example, is the so-called *equatorial anomaly*. This feature is distinguished by higher ionization density on both sides of the equator, rather than at the equator itself; see Fig. 4.19. The cause of this phenomenon is the *equatorial plasma fountain* sketched in the lower part of this figure. This fountain is driven by an electric field observed in the immediate vicinity of the magnetic equator, which is directed from west to east during the day. In combination with the Earth's magnetic field, directed from south to north at the equator, this results in an upward directed $\vec{\mathcal{E}} \times \vec{B}$ drift in the F region (see Section 5.3.3) that transports the ionospheric plasma to great heights. Density differences are thus formed with respect to the neighboring latitudinal regions unaffected by the drift. Accordingly, the plasma drifts away from the high

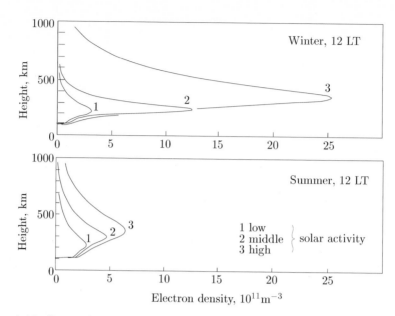

Fig. 4.18. Seasonal and solar activity induced variations of the electron density profile for noontime conditions at mid-latitudes (adapted from Wright, 1962)

density region along the magnetic field lines, supported in its motion by the parallel component of the gravitational force, which becomes effective when the field lines are no longer horizontal. This lateral plasma flux is responsible for the depletion of the ionization density at the equator and concurrent enhancement at subequatorial latitudes. Another anomaly of the low-latitude ionosphere is the occurrence of spectacular plasma instabilities as described in more detail in Appendix A.16.

The polar ionosphere also displays a number of striking features. Responsible for these, among others, are

- additional ionization production by precipitating energetic particles
- additional ionization loss from plasma escape along open magnetic field lines
- changes in the chemical reaction rates due to electric fields and particle precipitation
- horizontal ionization transport by electric fields
- heating by electric currents and
- heating by magnetospheric plasma via heat conduction

Since these phenomena exhibit complex spatial and temporal variations, it is not surprising that a systematic description of the polar ionosphere is difficult and would require detailed discussions beyond the scope of this book.

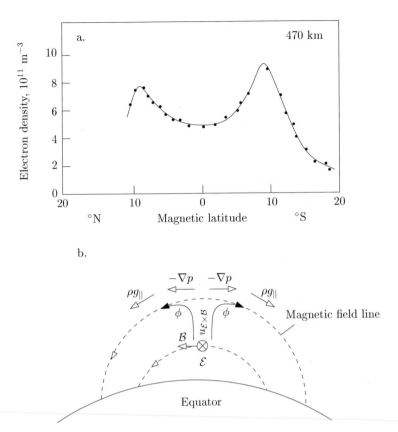

Fig. 4.19. The equatorial anomaly and its formation. (**a**) Latitudinal variation of the electron density at a height of 470 km along the 110 °E meridian at noon. The data were obtained using radio sounding from the ALOUETTE I satellite (adapted from Eccles and King, 1969). (**b**) The principle of the equatorial plasma fountain. $\mathcal{E}$ denotes the eastward directed electric field strength, $\mathcal{B}$ is the northward directed flux density of the terrestrial magnetic field, $u_{\mathcal{E}\times\mathcal{B}}$ is the upward directed plasma drift, $-\nabla p$ is the pressure gradient force acting in the meridional direction, $\rho g_{\parallel}$ is the component of the gravitational acceleration along the magnetic field, and ϕ is the ambipolar flux of the ionospheric plasma.

4.7 Radio waves in the Ionosphere

Every electromagnetic wave is altered during its passage through a plasma. Among the characteristics subject to change are the propagation direction, the amplitude and the wave velocity. This can be exploited to extract information about the properties of the plasma medium from the type of changes observed. On the other hand, one can make concerted use of available plasmas to induce desired changes (e.g. in the propagation direction). Both applications play an important role in the propagation of radio waves in the

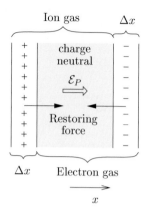

Fig. 4.20. Deriving the plasma frequency

ionosphere. Radio waves, of course, not only serve as a diagnostic tool for ionospheric research, they are also the longstanding carriers of classical radio communication in the long, middle, and short wave ranges (30 kHz – 30 MHz).

The key to understanding the wave-plasma interaction is the oscillation of the plasma electrons induced by the electric field of the wave. Oscillating charges themselves emit electromagnetic radiation. This secondary radiation is superposed on the primary wave and leads to the observed changes in the characteristics and propagation behavior of the wave. A generalized treatment of these interactions requires considerable mathematical manipulation and will not be pursued further here. Instead, we constrain our consideration to greatly simplified interaction scenarios, for which the result of the complex superposition of primary and secondary wave fields is described with the help of macroscopic material constants such as conductivity and refraction index. We first investigate the natural resonant oscillation of a plasma. Following this we show how a plasma undergoing forced oscillations can sometimes behave like a dielectric and other times like a conductor. The section is concluded with a discussion of the complications that arise upon applying a magnetic field.

4.7.1 Natural and Forced Oscillations of a Plasma

Consider a charge neutral plasma sheet of density $n = n_i = n_e$. Displacing the electron gas with respect to the ion gas by applying an external electric field, we produce a polarization field given by $\mathcal{E}_P = (e\, n/\varepsilon_0)\Delta x$; see Fig. 4.20 and Eq. (4.44). Turning the external field off again, the electron gas will now feel a restoring force proportional to the displacement, given by $F^*_{\mathcal{E}_P} = -e\, n\, \mathcal{E}_P = -(e^2\, n^2/\varepsilon_0)\Delta x$, that accelerates the electrons in the direction of the ion gas. The electrons gain kinetic energy in this process, causing them to overshoot their desired position of equilibrium. This then leads to the build up of a polarization field and its associated restoring force in the opposite direction, causing the electrons to decelerate, turn around, and accelerate back again toward their equilibrium position. The electron gas evidently becomes excited in this way to oscillate about its position of rest. The same holds for the ions, but because their mass is so much larger than that of the electrons, their motion is practically imperceptible. In other words, to a very good approximation, it is the electron gas that oscillates while the ions stay put in the mutual position of equilibrium.

Similar to our derivation of the buoyancy oscillation frequency, we can determine the electron-ion oscillation frequency by considering a balance of forces for which the restoring force is opposed solely by the inertial force

$$n\,m_e\,\frac{d^2(\Delta x)}{dt^2} = -\,\frac{e^2 n^2}{\varepsilon_0}\,\Delta x \tag{4.66}$$

The solution to this differential equation is a harmonic oscillation of the form

$$\Delta x = (\Delta x)_0 \sin(\omega_p t) \tag{4.67}$$

where $(\Delta x)_0$ denotes the displacement amplitude and ω_p is the angular frequency of the oscillation, generally known as the *plasma frequency*. The value of the plasma frequency, obtained by inserting the solution into the differential equation, is

$$\omega_p = \sqrt{\frac{e^2 n}{\varepsilon_0 m_e}} \tag{4.68}$$

or, expressed as a handy rule-of-thumb

$$\omega_p[\mathrm{s}^{-1}] \simeq 56.4\sqrt{n[\mathrm{m}^{-3}]}\,, \quad f_p[\mathrm{Hz}] \simeq 9\sqrt{n[\mathrm{m}^{-3}]} \tag{4.69}$$

A more precise description of this plasma oscillation requires the inclusion of damping processes. Among the damping processes in the ionosphere are collisions of the oscillating electrons with the ions and neutral gas particles, assumed to be at rest. The force balance equation (4.66) must therefore be extended with a friction term of the form $F_{fr}^* = -n\,m_e\nu_{e,s}^* u_e = -n\,m_e\nu_{e,s}^* d(\Delta x)/dt$, see Eq. (2.57), where s stands for the neutral gas and ion components. Moreover, there are radiative energy losses for which the associated damping, in many cases, can be formally described by an expression similar to the friction force $F_{rd}^* \simeq -n\,m_e\,\nu_{rd}^* d(\Delta x)/dt$. In this case ν_{rd}^* represents a type of momentum transfer collision frequency that summarizes the effects of radiative damping. The differential equation for damped plasma oscillations thus assumes the following form

$$n\,m_e\,\frac{d^2(\Delta x)}{dt^2} + n\,m_e\nu^*\,\frac{d(\Delta x)}{dt} + \frac{e^2 n^2}{\varepsilon_0}\,\Delta x = 0 \tag{4.70}$$

where we have made use of the abbreviation $\nu^* = \nu^*_{e,n} + \nu^*_{e,i} + \nu^*_{rd}$. The solution to this equation is an exponentially damped oscillation at the plasma frequency ω_p.

Finally, we consider the case when the plasma sheet shown in Fig. 4.20 is excited to undergo *forced oscillations* by an external alternating electric field. The corresponding force balance equation may now be written as

$$n\, m_e \frac{d^2(\Delta x)}{dt^2} + n\, m_e \nu^* \frac{d(\Delta x)}{dt} + \frac{e^2 n^2}{\varepsilon_0} \Delta x = -e\, n\, \mathcal{E}_0 \sin(\omega t) \quad (4.71)$$

where $\mathcal{E}_0$ and ω are the amplitude and frequency of the applied AC electric field and the minus sign comes from the negative charge of the electrons. As well known from mechanics, an externally excited oscillator will oscillate at the excitation frequency, perhaps phase shifted and with smaller amplitude, following an initial synchronization period. The time dependent displacement of the electron gas thus takes the form

$$\Delta x = (\Delta x)_0 \sin(\omega t - \varphi) \quad (4.72)$$

Inserting this into the differential equation (4.71), we obtain a defining equation for the amplitude and phase of this oscillation

$$-\omega^2 (\Delta x)_0 \sin(\omega t - \varphi) + \omega\, \nu^* (\Delta x)_0 \cos(\omega t - \varphi)$$
$$+ \omega_p^2 (\Delta x)_0 \sin(\omega t - \varphi) = -(e/m_e)\, \mathcal{E}_0 \sin(\omega t) \quad (4.73)$$

Solutions of special forms of this equation are discussed in the following.

4.7.2 The Ionosphere as a Dielectric

As will be shown, the type of interaction between a radio wave and the ionosphere depends on the frequency of the wave. Sometimes the ionosphere behaves like a dielectric, sometimes like a metallic reflector. Let us illustrate the first case with a concrete example. Consider a radio wave in the short wave range at a frequency of $f = 5$ MHz transmitted vertically upward in the direction of the ionosphere. This can be taken as a plane parallel wave at great distance from the transmitter. Assume further that the electric field oscillates in the $\pm$ x-direction and the associated magnetic field in the $\pm$ y-direction; see Fig. 4.21. Since the air molecules are only weakly polarizable and the ionization density in the lower and middle atmosphere is negligibly small, this region can be considered a vacuum from an electromagnetic point of view. The phase velocity of the radio wave is thus equal to the speed of light

$$v_{ph} = c_0 = 1/\sqrt{\mu_0 \varepsilon_0} \quad (4.74)$$

and the index of refraction, defined by

$$n_{ref} = c_0/v_{ph} = \sqrt{\varepsilon_r} \quad (4.75)$$

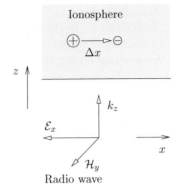

Radio wave

Fig. 4.21. Configuration of a plane radio wave impinging perpendicularly onto a horizontally-layered ionosphere. $\mathcal{E}_x$ and $\mathcal{H}_y$ denote the electric and magnetic field strengths of the radio wave and k_z its wavenumber. Δx indicates the magnitude of the displacement of the electron gas with respect to the ion gas.

has the value 1. The other quantities appearing in the above equations are the magnetic permeability μ_0 and electric permittivity ε_0 of free space, and the dielectric constant ε_r.

The propagation characteristics of this wave are changed when it meets the rapidly increasing electron density of the lower ionosphere. Let us first consider a range of altitudes for which the radio frequency is large compared with the plasma frequency and at the same time large compared with the momentum transfer collision frequency. The first condition applies well into the E region for the transmitted frequency $f = 5$ MHz. For estimating the lower boundary of the region, we assume that radiative damping occurs only for frequencies near the plasma frequency and that Coulomb collisions with ions are important only above the E region. In this case the second condition may be written as follows: $\omega \gg \nu_{e,n}^*$. We use Eq. (2.56) with $m_e \ll m_n$ and $r_e \ll r_n$ for a determination of the momentum transfer collision frequency

$$\nu_{e,n}^* \simeq \sqrt{\frac{8k}{9\pi}}\, \sigma_{n,n}\, n_n \sqrt{\frac{T_e}{m_e}} \tag{4.76}$$

For $\sigma_{n,n} \simeq 3 \cdot 10^{-19}$ m^2 and for $n_n \simeq 4 \cdot 10^{20}$ m^{-3}, $T_e \simeq T_n \simeq 200$ K (typical values at an altitude of 80 km), we compute a momentum transfer collision frequency of $\nu_{e,n}^*(80\ \text{km}) \simeq 3 \cdot 10^6$ s^{-1}. This is, however, a factor of ten less than the wave frequency $\omega \simeq 3 \cdot 10^7$ s^{-1}. We are thus assured that our subsequent treatment is certainly valid for the upper D region and lower E region at this wave frequency.

With $\omega^2 \gg \omega_p^2$ and $\omega \gg \nu^* \simeq \nu_{e,n}^*$, and noting that the sine and cosine functions are never greater than unity, Eq. (4.73) assumes the following simple form

$$-\omega^2 (\Delta x)_0 \sin(\omega t - \varphi) \simeq -(e/m_e)\, \mathcal{E}_0 \sin(\omega t) \tag{4.77}$$

This physical situation corresponds to a high frequency regime, in which the friction and restoring forces can be neglected because of the small displacement amplitudes and velocities, and only the inertial force opposes the electric force of the radio wave acting on the electron gas. Equation (4.77) is fulfilled for $\varphi = 0$ and $(\Delta x)_0 = e\,\mathcal{E}_0/m_e\omega^2$, so that

$$\Delta x = \frac{e}{m_e\omega^2}\,\mathcal{E}_0\sin(\omega t) \tag{4.78}$$

We use this expression to determine the conductivity of the plasma. Jumping ahead to Eq. (5.4) for a moment, the current density may be expressed as follows

$$j = \sum_s q_s\,n_s\,u_s = -e\,n\,u_e = -e\,n\,\frac{d(\Delta x)}{dt} = -\frac{e^2 n}{m_e\omega}\,\mathcal{E}_0\cos(\omega t) \tag{4.79}$$

from which the conductivity is given by

$$\sigma = \frac{j}{\mathcal{E}} = -\frac{\varepsilon_0\omega_p^2}{\omega}\,\cot(\omega t) \tag{4.80}$$

The temporal average of this expression, however, is equal to zero, so the ionosphere behaves like an insulator at this frequency for the range of altitudes considered here. Strictly speaking, it behaves like a dielectric because it is a polarizable medium as a result of the induced charge separation. Similar to the more familiar usage in optics, we characterize the properties of this dielectric with an index of refraction. From classical electrodynamics we utilize the following relation between the index of refraction n_{ref} and the dielectric polarization $\vec{P}$ (= electric dipole moment per unit volume)

$$n_{ref} = \sqrt{1 + \vec{P}/\varepsilon_0\vec{\mathcal{E}}} \tag{4.81}$$

In our case, the dielectric polarization (positive in the direction of the positive charge) is given by

$$P = -e\,n\,\Delta x = -\frac{e^2 n}{m_e\omega^2}\,\mathcal{E}_0\sin(\omega t) \tag{4.82}$$

so that the index of refraction becomes

$$n_{ref} = \sqrt{1 - \left(\frac{\omega_p}{\omega}\right)^2} \tag{4.83}$$

The index of refraction becomes smaller than unity when $\omega > \omega_p$. As a consequence, the *phase* velocity of the radio wave $v_{ph} = c_0/n_{ref}$ actually exceeds the speed of light. Evidently, the phase of the secondary wave radiated by the oscillating electrons runs ahead of the phase of the primary wave such that the superposition of both wave fields yield an effectively higher propagation velocity. Furthermore, if $n_{ref} < 1$, a wave impinging obliquely onto the ionosphere will be refracted according to the familiar *Snell's law*

$$\sin \vartheta_2 = \frac{\sin \vartheta_1}{n_{ref}} \qquad (4.84)$$

with $\vartheta_2 > \vartheta_1$.

In addition to the refractive characteristics we are interested in the losses in our dielectric. Important for this issue is the phase shift between the current density and the electric field strength. The power absorbed per unit volume is given by

$$P^* = j\,\mathcal{E} \sim -\cos(\omega t)\,\sin(\omega t)$$

This means that power is transferred from the radio wave to the oscillating electron gas during the second and fourth quadrants of the oscillation cycle ($P^* > 0$). The power is returned to the radio wave again in the first and third quadrants ($P^* < 0$). The returned power is incomplete, however, because of the always present frictional damping. Denoting the friction force per unit volume by F_{fr}^*, the dissipated power per unit volume is given by

$$P_{fr}^* = \left| F_{fr}^* \frac{d(\Delta x)}{dt} \right| = n\, m_e \nu_{e,n}^* \left(\frac{d(\Delta x)}{dt} \right)^2 \qquad (4.85)$$

For Δx we use the expression given in Eq. (4.78) and, after averaging over time, obtain

$$\langle P_{fr}^* \rangle = \frac{e^2 \mathcal{E}_0^2}{2\, m_e} \frac{n\, \nu_{e,n}^*}{\omega^2} \qquad (4.86)$$

The losses are thus inversely proportional to the square of the radio frequency, implying that one should work with the highest possible frequencies for improving radio communications. The losses are also directly proportional to the ionization density and to the momentum transfer collision frequency (and thus also to the neutral gas density). This leads to the following height dependence for the frictional losses. Variations of the functions $n(h)$ and $n_n(h)$ oppose each other below the ionization density peak and their product reaches a maximum at that height where, on the one hand, enough electrons are available to extract significant power from the radio wave and, on the other hand, enough neutral gas particles are present to provide significant friction. This is typically the case below an altitude of 100 km in the upper D region and lower E region; see Fig. 4.22. Nevertheless, even here the damping is relatively weak so that our radio wave can traverse the lower ionosphere without severe losses. Only in exceptional cases (e.g. enhanced X-ray radiation or intense precipitation of energetic particles) does the ionization density grow to such high values that radio waves are significantly or even completely absorbed (see Sections 8.6.3 and 8.7).

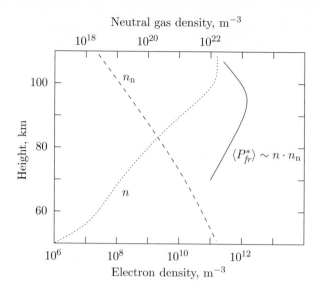

Fig. 4.22. Formation of the absorption maximum in the lower ionosphere. Here n denotes the electron density and n_n is the neutral gas density. The absorption is proportional to the product of these two quantities. The electron density profile corresponds to local noon conditions at mid-latitudes for a period of low solar activity.

4.7.3 The Ionosphere as a Conducting Reflector

The ionization density and its associated plasma frequency continue to increase above the previously considered D and E regions. If the plasma frequency reaches the transmission frequency of our radio wave, i.e. $\omega_p(h) = \omega$, then the propagation conditions of the ionosphere undergo a fundamental change. As shown in Eq. (4.73), the restoring force and inertial force compensate each other in this case and the amplitude of the oscillation is limited only by the damping. To emphasize the uniqueness of this situation, we equate the transmission frequency to the plasma frequency and write Eq. (4.73) in the form

$$\omega_p \nu^* (\Delta x)_0 \cos(\omega_p t - \varphi) = -(e/m_e)\,\mathcal{E}_0 \sin(\omega_p t) \qquad (4.87)$$

With $(\Delta x)_0 = e\,\mathcal{E}_0/m_e \nu^* \omega_p$ and $\varphi = -\pi/2$, it follows that

$$\Delta x = \frac{e}{m_e \nu^* \omega_p}\,\mathcal{E}_0 \sin(\omega_p t + \pi/2) = \frac{e}{m_e \nu^* \omega_p}\,\mathcal{E}_0 \cos(\omega_p t) \qquad (4.88)$$

Considering at first only the damping by neutral gas collisions, the amplitude of the oscillation becomes a factor $\omega/\nu_{e,n}^*$ larger than for the dielectric case. The electrons thus possess a much greater mobility, thereby significantly increasing the current density and conductivity

$$j(\omega_p) = \frac{e^2 n}{m_e \nu^*_{e,n}} \, \mathcal{E}_0 \sin(\omega_p t) \tag{4.89}$$

and

$$\sigma(\omega_p) = \frac{\varepsilon_0 \, \omega_p^2}{\nu^*_{e,n}} \tag{4.90}$$

This can be demonstrated by the following numerical example. Taking the previously used transmission frequency $f = 5$ MHz, the electron density must increase to $n \simeq 3 \cdot 10^{11}$ m^{-3} in order that the plasma frequency reaches the wave frequency. Such densities are typically observed in the lower F region at an altitude of about 200 km where the momentum transfer collision frequency, according to Eq. (4.76), is less than $\nu^*_{e,n} \simeq 200$ s^{-1}. The actual value of $\nu^*_{e,n+i}$ is about twice as large because collisions between electrons and ions are also important in this height region. Even when accounting for these Coulomb collisions (see Section 5.3.5), the conductivity reaches values of about 20 S/m (S = siemens = Ω^{-1}), which is equivalent to that of an electric conductor. It is well known, however, that electromagnetic waves are reflected at electric conductors, and this is exactly what happens in the case of the ionosphere. Radio waves from the ground penetrate into the ionosphere up to that height where the local plasma frequency reaches the wave frequency. At this point they are reflected back toward Earth.

This ionospheric property is exploited using the technique of *radio sounding*. A wave train at frequency ω is transmitted upward toward the ionosphere and the elapsed time Δt is recorded when the reflected signal is received by the *ionosonde*. Assuming that the average group velocity of the wave $v_{gr} = c_0 \, n_{ref}$ does not deviate greatly from the speed of light, the reflection height may be estimated from $h_{refl} \simeq c_0 \Delta t / 2$. One then immediately knows the electron density at the reflection height from the relation $\omega_p(h_{refl}) = \omega$. By varying the transmission frequency, this method can be used to determine the entire ionization density profile below the ionization density peak; see Fig. 4.23. Radio sounding from orbiting satellites, in fact, are not only capable of determining the density profile of the upper ionosphere, but can also investigate the density profile of magnetospheric plasmas, e.g. the plasma sheet of the magnetotail or the magnetospheric boundary layer, two regions which will be introduced in the next chapter.

Comparing the relations given in Eqs. (4.87) and (4.89), we note that at resonance ($f_p = f$) the current density is in phase with the electric wave field. This means that the radio wave must be continually losing energy, $P^* = j \, \mathcal{E} > 0$. If only frictional damping were present, this energy would be transformed into frictional heat and the wave would be strongly absorbed. This is only partly the case, however. The electrons, with their large displacement amplitudes and associated large accelerations, become very effective dipole radiators. Absorption and subsequent coherent (i.e. equally phased) reemission of the impinging radio wave is thus the primary damping process

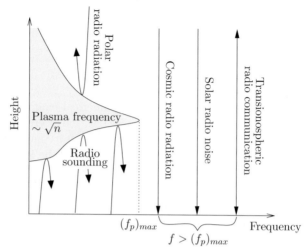

Fig. 4.23. Examples of reflection and transmission of artificial and natural radio waves. $(f_p)_{max}$ corresponds to the critical frequency f_0F2.

in the case of resonance. The reemitted waves are radiated in both the forward and backward directions. Indeed, if one approximates the ionospheric reflection layer by an areal array consisting of innumerable tiny dipoles, then upward and downward are the main directions of radiation. Concerning the phase of the reemitted radiation, it may be verified that the electric field is 180 degrees out of phase with the current density, $\mathcal{E}_{secondary} \sim -j_{dipole}$. Since the primary wave field is in phase with the current density, however, the primary and secondary waves cancel each other in the forward direction and are superposed to form a standing wave in the reverse direction (i.e. back toward Earth) in the case of continuous transmission. The reflection and superposition characteristics become more complicated for inclined incidence. The primary and secondary waves again interfere destructively in the forward direction and constructively in the backward direction, but now the reflected wave does not return to its source transmitter. The familiar reflection law holds in this case: the angle of incidence is equal to the angle of reflection. This is exploited for global radio communications to overcome the curvature of the Earth's surface; see Fig. 4.24.

In order to avoid a misunderstanding: radio waves with frequencies above the maximum plasma frequency of the ionosphere (also called the *critical frequency* and denoted for the 'ordinary' wave by f_0F2) are not reflected for normal incidence and pass with only moderate damping through the ionosphere. This holds for radio waves in the VHF and UHF bands (30 – 3000 MHz), for example, which, among other uses, are reserved for transmission of television signals. Of course, solar radio noise at 2.8 GHz ($\lambda = 10.7$ cm), used for the previously described Covington index, is also not reflected. The same is true

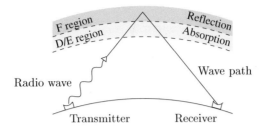

Fig. 4.24. Principle of subionospheric radio communications in the short wave regime (3-30 MHz). Note (in lieu of a detailed explanation) that the reflection frequency increases to $f_{refl} = f_p/\cos\vartheta$ for inclined incidence at the angle ϑ.

of the cosmic radio noise at 30 MHz, used for the determination of absorption effects (*riometer*). Clearly, no reflection occurs for the microwave signals at S and X band ($1.6 \le f \le 10.9$ GHz), which are standard radio communication ranges for satellites within or above the ionosphere. Polar kilometric radiation in the 100 to 400 kHz range (see Section 7.4.3), on the other hand, is reflected from the topside of the ionosphere and cannot penetrate down to the Earth's surface.

It remains to be mentioned that every radio wave, whether it is reflected or not, is subject to (even if only very weak) *incoherent* (differently phased) backscattering. This may be attributed essentially to the ubiquitous small scale irregularities in the ionospheric density distribution and its associated index of refraction. Because this incoherent backscatter radiation contains valuable information about the physical state of the ionosphere at the scatter location (e.g. the local ionization density and velocity as well as the local ion and electron temperatures), radar stations have been built to measure these quantities. The frequencies used for this purpose are in the VHF and UHF ranges, so the radio waves penetrate through the ionosphere and only their incoherently backscattered signal is recorded. Large antennas and powerful transmitters are needed here because of the very modest degree of backscattering.

4.7.4 Magnetic Field Influence

Completely ignored up to now is that magnetic forces also act upon moving charges. As explained in more detail in Section 5.1, the magnetic force per unit volume may be written as

$$\vec{F}_B^* = \vec{j} \times \vec{B}$$

where $\vec{j} = \sum_s q_s n_s \vec{u}_s$ is the current density and $\vec{B}$ is the magnetic flux density. Consider first the force exerted by the magnetic field of the radio wave. For a plane electromagnetic wave in a lossless dielectric, we utilize from electrodynamics a relation between the magnetic and electric field strengths

$$\frac{\mathcal{E}}{\mathcal{H}} = \sqrt{\frac{\mu_0}{\varepsilon_r \varepsilon_0}} \tag{4.91}$$

from which, making use of the relations (4.74) and (4.75), we obtain an expression for the magnetic flux density of the wave

$$\mathcal{B} = \mu_0 \mathcal{H} = \sqrt{\varepsilon_r} \sqrt{\mu_0 \varepsilon_0} \, \mathcal{E} = n_{ref} \, \mathcal{E}/c_0 \tag{4.92}$$

The resulting ratio of magnetic to electrical volume force is thus

$$\frac{F_{\mathcal{B}}^*}{F_{\mathcal{E}}^*} = \frac{-e \, n \, u_e \, n_{ref} \, \mathcal{E}/c_0}{-e \, n \, \mathcal{E}} = \frac{n_{ref} u_e}{c_0} < \frac{u_e}{c_0} \ll 1 \tag{4.93}$$

The inequality at the far right may be verified from the resonant case considered in the previous section, where we inferred that $u_e(\omega_p) \le e \, \mathcal{E}_0/(m_e \nu_{e,n}^*)$. Even for a relatively strong electric field strength of $\mathcal{E}_0(200 \text{ km}) {\simeq} 10 \text{ mV/m}$ and a collision frequency of $\nu_{e,n}^*(200 \text{ km}) \simeq 200 \text{ s}^{-1}$, this yields a rather small ratio of velocities $u_e/c_0 \simeq 0.03$. The influence of the wave magnetic field is thus negligibly small, even in the resonance case with its large current densities.

The much stronger magnetic field of the Earth, however, cannot be neglected. We account for it by including a term in the (now required) *vector* form of the force balance equation given by $\vec{F}_{\mathcal{B}}^* = -e \, n \, (\text{d}(\Delta \vec{x})/\text{d}t) \times \vec{B}_E$. Clearly, an additional term of this type considerably complicates the description of the wave propagation, which now depends on the angle between the propagation direction of the wave and the Earth's magnetic field. Furthermore, the ionosphere now becomes birefringent, i.e. the incident radio wave is split into an *ordinary* and an *extraordinary* component. These two components have differing propagation velocities, which, among others, leads to a rotation of the plane of polarization of a linearly polarized radio wave (*Faraday effect*). To a good approximation, the *Faraday rotation* of a radio wave traversing the entire ionosphere is proportional to the column density $\mathcal{N}_e$ and can be used to measure this physical quantity.

Exercises

4.1 An atomic oxygen ion is produced at an altitude of 200 km. Following its creation, it moves through the ambient neutral oxygen gas as a result of its thermal motion. Statistically, because of its random walk, one predicts that it will move a distance of $d = s\sqrt{N}$ from its place of origin, where s corresponds to the step size (= mean free path) and N the number of steps (= number of collisions). How large is this distance over the expected lifetime of the oxygen ion and by what factor is this distance decreased at a height of 150 km?

4.2 An electron density of 10^{11} m^{-3} is measured at an altitude of 500 km for noontime conditions at mid-latitudes during average solar activity.

(a) Using the production time constant, estimate the time needed to build up this electron density by photoionization.

(b) How large is the chemical loss rate at this height upon accounting for both charge exchange and radiative recombination?

(c) By what factor does the photo-chemical equilibrium density exceed the observed value? Knowing the transport time constant, estimate how long it takes until density disturbances caused by photo-chemical processes are eliminated.

(d) How large are the regular and production-induced downward currents at this height?

4.3 Figure 4.16 suggests that the transport term is negligibly small because $q \simeq l$. Indeed, measurements show that this term has a null at the position of the maximum with negative (positive) values above (below) the maximum. On the other hand, according to Eqs. (4.64), (4.58) and (4.59), $\tau_d = n/|d| \simeq \tau_l = n/l$ or $|d| \simeq l$. Discuss the reason for this apparent contradiction.

4.4 The planet Venus, like the Earth, possesses an ionosphere. For heights above about 200 km, this consists of oxygen ions and associated electrons in static equilibrium (barometric density distribution). The ion density at 200 km can be taken as $n_{o+} \simeq 10^{11}$ m^{-3}. Assume the ion and electron temperatures in this region are independent of height and of the same order of magnitude $T_i \simeq 2000$ K, $T_e \simeq 5000$ K. Contrary to the terrestrial ionosphere, the Venusian ionosphere has a relatively sharp upper boundary, the ionopause. This is located at the height where the pressure of the ionospheric plasma is equal to the pressure of the incident solar wind. It should be noted that Venus has no magnetosphere that could intercept the solar wind far above the ionosphere and that the magnetized solar wind plasma is unable to penetrate into the ionosphere. Calculate the stand-off distance to the ionopause for the case when the solar wind ram pressure takes the value $p_{sw} \simeq 4 \cdot 10^{-9}$ N m^{-2}. To a first approximation, consider the gravitational acceleration at Venus to be independent of height and use its value at 200 km.

References

H. Rishbeth and O.K. Garriot, *Introduction to Ionospheric Physics*, Academic Press, New York, 1969

J.A. Ratcliffe, *An Introduction to the Ionosphere and Magnetosphere*, Cambridge University Press, Cambridge, 1972

S.J. Bauer, *Physics of Planetary Ionospheres*, Springer-Verlag,Berlin, 1973

A. Giraud and M. Petit, *Ionospheric Techniques and Phenomena*, Reidel Publishing Company, Dordrecht, 1978

K.G. Budden, *The Propagation of Radio Waves*, Cambridge University Press, Cambridge, 1985

G.S. Ivanov-Kholodny and A.V. Mikhailov, *The Prediction of Ionospheric Conditions*, Reidel Publishing Company, Dordrecht, 1986

M.C. Kelley, *The Earth's Ionosphere*, Academic Press, San Diego, 1989

K. Rawer, *Wave Propagation in the Ionosphere*, Kluwer Academic Publishers, Dordrecht, 1993

H. Kohl, R. Rüster and K. Schlegel (eds.), *Modern Ionospheric Science*, European Geophysical Society, Katlenburg-Lindau, 1996

R. W. Schunk and A. F. Nagy, *Ionospheres*, Cambridge University Press, Cambridge, 2000

See also the references in Chapters 1 and 2 and the figure and table references in Appendix B.

5. Magnetosphere

The Earth's external magnetic field plays a key role in space physics. This arises from its strong interaction with charged particles, resulting in a number of fascinating phenomena. Among these are the capture of energetic particles in the Earth's radiation belt and the striated configuration of the aurora. Furthermore, the conductivity of the ionosphere is substantially modified by the magnetic field and important solar-terrestrial relations are based on the interaction of the solar wind with the magnetosphere. The large-scale structure of the geomagnetic field and its associated plasma populations are described in the following. In this context it is appropriate to distinguish between the near-Earth and the distant geomagnetic field. As an introduction to the topic, we first review some fundamentals of magnetostatic theory.

5.1 Fundamentals

The primary quantity for describing a magnetic field is the magnetic flux density or magnetic induction $\vec{B}$. This determines the strength and direction of the magnetic force acting on a charge q moving at the velocity $\vec{v}$

$$\vec{F_B} = q\,\vec{v} \times \vec{B} \tag{5.1}$$

Although first presented in this form by Heaviside, $\vec{F_B}$ is also known as the Lorentz force. An equivalent macroscopic definition of the magnetic flux density is obtained by considering an ensemble of charge carriers moving in a given direction $\vec{l}$ at the velocity $\vec{u}$; see Fig. 5.1. According to Eq. (5.1), the force acting on this ensemble of particles of density n is given by

$$d\vec{F_B} = q\,n\,A\,dl\,\vec{u} \times \vec{B}$$

where $d\vec{F_B}$ and dl are differentially small quantities, not in an absolute sense, but relative to the particular macroscopic dimensions. Introducing the electric current associated with these charge carriers (i.e. the net charge transport through a reference surface per unit time caused by their motion)

$$\vec{I} = q\,n\,\vec{u}\,A$$

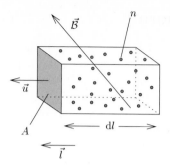

Fig. 5.1. Definition of the magnetic flux density using macroscopic quantities

one obtains the well-known relation

$$d\vec{F}_{\mathcal{B}} = dl \, \vec{\mathcal{I}} \times \vec{\mathcal{B}} = \mathcal{I} \, d\vec{l} \times \vec{\mathcal{B}} \tag{5.2}$$

This describes the differentially small force acting on a directed current element of length dl in a magnetic field. The force per unit volume dV acting on the charge carrier ensemble is thus found to be

$$\vec{F}_{\mathcal{B}}^* = \frac{d\vec{F}_{\mathcal{B}}}{dV} = \frac{dl \, \vec{\mathcal{I}} \times \vec{\mathcal{B}}}{dl \, A} = \vec{j} \times \vec{\mathcal{B}} \tag{5.3}$$

where we have introduced the current density $\vec{j} = \vec{\mathcal{I}}/A$. Note that the current density corresponds to a charge carrier flux, i.e. charge carrier transport per unit area and time, and that each charge carrier species s is associated with its own current density

$$\vec{j}_s = q_s \, n_s \, \vec{u}_s \tag{5.4}$$

The definition of the magnetic flux density according to Eq. (5.1) is formally similar to that of the electric field

$$\vec{F}_{\mathcal{E}} = q \, \vec{\mathcal{E}} \tag{5.5}$$

It is thus more appropriate to designate the quantity $\vec{\mathcal{B}}$ (rather than the quantity $\vec{\mathcal{H}} = \vec{\mathcal{B}}/\mu$ – as usually done –) as the magnetic field. With few exceptions, this convention is followed throughout this book.

The unit for the magnetic flux density (here magnetic field strength) is the tesla (symbol: T)

$$\vec{\mathcal{B}} \text{ in tesla, } T = \frac{Wb}{m^2} = \frac{V \, s}{m^2} = \frac{J \, s}{C \, m^2} = \frac{N \, s}{C \, m}$$

Two other non-SI units are often used in geophysical applications. These are the gauss (Γ) and the gamma (γ), derived from the Gaussian system of units, both of which pose no conversion problems

$$1 \; \Gamma \; \hat{=} \; 0.1 \; \text{mT}, \qquad 1\gamma \; \hat{=} \; 1 \; \text{nT}$$

The familiar field lines, the most common pictorial representation of a magnetic field, show the direction of the magnetic field at all locations in space. From a practical standpoint, this is the directional alignment assumed by the north pole of a test magnet (compass needle). The intensity of the field is described either by the density of field lines or by a contour map of isodynamic lines, i.e. loci of equal field strength $|\vec{B}|$.

5.2 The Geomagnetic Field Near the Earth

In this context 'near the Earth' is understood to mean anywhere within roughly 6 Earth radii of the Earth's center or ca. 30 000 km above the Earth's surface. This boundary can vary, depending on the particular geophysical conditions and the particular direction one moves away from the Earth. The general configuration of the near-Earth magnetic field has been known for some time. William Gilbert, whose book 'De Magnete' appeared in 1600, realized that the Earth itself was a giant magnet ('Magnus magnes ipse est globus terrestris'). He modeled the Earth's field with a small magnetic globe – a 'terrella' – and was able to bring some order to the observations available at the time.

Our contemporary view of the structure of the near-Earth magnetic field is shown in Fig. 5.2. This field is characterized by a magnetic field direction parallel to the Earth's surface (i.e. horizontal) for regions at low geographic latitude and perpendicular to the Earth's surface (i.e. vertical) for regions of high geographic latitude. The position where the field direction is exactly horizontal, i.e. where the *inclination* or *dip angle* with respect to the Earth's surface goes to zero, is designated the magnetic *inclination* or *dip equator*. Correspondingly, the two points where the magnetic field is exactly perpendicular to the Earth's surface are designated the magnetic *inclination* or *dip poles*. One distinguishes here between the *boreal* pole (BP) in the Northern Hemisphere (which, since the field points into the Earth, is magnetically a south pole!) and the *austral* pole (AP) in the Southern Hemisphere (which, since the field points away from the Earth, is magnetically a north pole).

The positions of the magnetic inclination poles are subject to secular variations. In the year 1965, for example, the following values were applicable, BP(1965): (75.6 °N, 259 °E); AP(1965): (66.3 °S, 141 °E). A recent determination of the southern inclination pole by ship in the year 2000 revealed that it had been displaced in the interim by about 3° to the northwest, AP (2000): 64.7 °S, 138.1 °E. A follow-up determination of the northern inclination pole also showed a considerable displacement to the northwest, BP (2001): 81.3 °N, 249.2 °E. The magnetic dip equator is fairly close to the geographic equator, but deviations of up to −17.5° are observed near the *South Atlantic anomaly* ($\simeq$ 293 °E); see also Fig. 8.5.

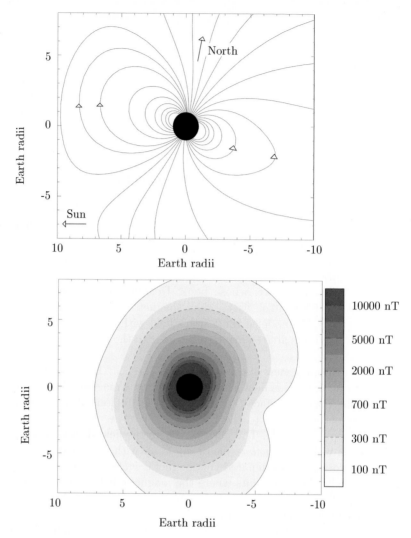

Fig. 5.2. The near-Earth geomagnetic field for 6 December 1989, 11 UT, during moderate magnetic activity (AE = 250-400 nT), according to the semiempirical model of Tsyganenko, 1990. Upper panel: field lines; lower panel: isodynamic lines. Arrows show the direction to the Sun and geographic north, i.e. along the Earth's rotation axis.

The isodynamic lines display an elliptical form with stronger field intensities near the poles and weaker field strengths around the equator. A remarkable feature is the rapid decrease in field intensity with distance from the Earth, thereby leading to considerable variations in the field strength along the magnetic field lines.

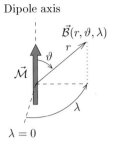

Fig. 5.3. Spherical coordinate system for describing a dipole field

The semiempirical model used to construct Fig. 5.2 uses a large number of coefficients and is much too complex for the analytic calculations to be performed in the following. Even at the expense of losing some precision, it is much more instructive to work with a simple, manipulable description of the geomagnetic field. A promising approach for such an approximation is suggested by the configuration of the field shown in Fig. 5.2: the structure of the near-Earth field corresponds quite closely to that of the familiar dipole field. In spherical coordinates (see Fig. 5.3), such a dipole field may be described by the following simple expressions

$$\mathcal{B}_\vartheta = \frac{\mu_0 \mathcal{M}}{4\pi} \frac{1}{r^3} \sin \vartheta \qquad (5.6)$$

$$\mathcal{B}_r = \frac{2\mu_0 \mathcal{M}}{4\pi} \frac{1}{r^3} \cos \vartheta \qquad (5.7)$$

$$\mathcal{B}_\lambda = 0$$

where $\mathcal{M}$ denotes the magnetic dipole moment and $\mu_0 = 4\pi \cdot 10^{-7}\,\mathrm{Tm/A}$ is the magnetic permeability in vacuum. The dipole moment evidently characterizes the intensity of the dipolar magnetic field.

Rather than the fictitious, infinitesimally small magnetic dipole, one can portray the origin of the magnetic field as a current loop or a bar magnet: in both cases the *far* field of these sources closely approaches the dipole field configuration. This can be checked for a current loop using the Biot-Savart law. As it turns out, the dipole moment of a current loop is given by the product of the current strength $\mathcal{I}$ times the area A enclosed by the circuit

$$\vec{\mathcal{M}} = A\,\mathcal{I}\,\hat{r}_{cl} \times \hat{l} \qquad (5.8)$$

As sketched in Fig. 5.4, $\hat{r}_{cl}$ and $\hat{l}$ are unit vectors in the direction of the circular loop radius and the current flow, respectively. The dipole moment for the case of a bar magnet is given by the product of the pole strength times the distance between poles, $\vec{\mathcal{M}} = |\mathcal{P}_\mathcal{B}|\,\vec{d}$.

Fitting a dipole field to the Earth's magnetic field is a quite satisfactory approximation of the true field configuration in near-Earth space. Placing

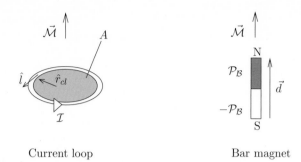

Current loop   Bar magnet

Fig. 5.4. Defining the magnetic dipole moment $\vec{\mathcal{M}}$ for a circular current loop and a bar magnet

the dipole at the Earth's center, the remaining free parameters of the dipole approximation assume the following values

$$\mathcal{M}_E \simeq 7.7 \cdot 10^{22} \text{ A m}^2, \quad \delta \simeq 11°, \quad \lambda_g^{BP} \simeq 290°\text{E (70°W)}$$

where $\mathcal{M}_E$ denotes the magnetic dipole moment of the Earth, δ is the dipole axis tilt with respect to the Earth's rotation axis, and λ_g^{BP} is the geographic longitude of the northern intersection point of the dipole axis with the Earth's surface; see Fig. 5.5. In combination with Eqs. (5.6) and (5.7), these values yield the *centered dipole approximation* of the geomagnetic field. A summary of the components of this model is presented in Fig. 5.6. Position is given in terms of the geocentric distance r and the *(geo)magnetic latitude* φ. The

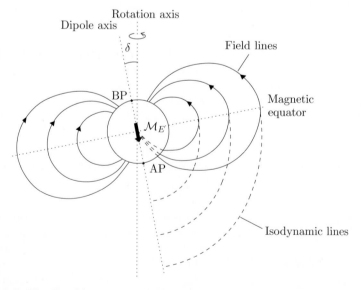

Fig. 5.5. The Earth's magnetic field approximated as a geocentric dipole field

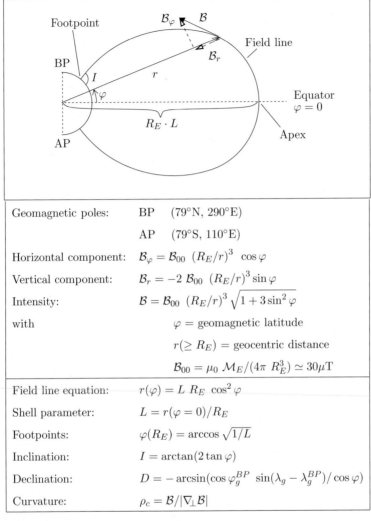

Geomagnetic poles:	BP (79°N, 290°E)
	AP (79°S, 110°E)
Horizontal component:	$\mathcal{B}_\varphi = \mathcal{B}_{00} \ (R_E/r)^3 \ \cos\varphi$
Vertical component:	$\mathcal{B}_r = -2 \ \mathcal{B}_{00} \ (R_E/r)^3 \sin\varphi$
Intensity:	$\mathcal{B} = \mathcal{B}_{00} \ (R_E/r)^3 \sqrt{1 + 3\sin^2\varphi}$
with	$\varphi = $ geomagnetic latitude
	$r(\geq R_E) = $ geocentric distance
	$\mathcal{B}_{00} = \mu_0 \ \mathcal{M}_E/(4\pi \ R_E^3) \simeq 30\mu\text{T}$
Field line equation:	$r(\varphi) = L \ R_E \ \cos^2\varphi$
Shell parameter:	$L = r(\varphi = 0)/R_E$
Footpoints:	$\varphi(R_E) = \arccos\sqrt{1/L}$
Inclination:	$I = \arctan(2\tan\varphi)$
Declination:	$D = -\arcsin(\cos\varphi_g^{BP} \ \sin(\lambda_g - \lambda_g^{BP})/\cos\varphi)$
Curvature:	$\rho_c = \mathcal{B}/|\nabla_\perp\mathcal{B}|$

Fig. 5.6. Geocentric dipole approximation of the geomagnetic field

latter is measured relative to the *(geo)magnetic equator*, i.e. relative to a plane oriented perpendicular to the dipole axis and passing through the center of the dipole. For the centered dipole approximation, of course, this is the Earth's center.

A calculation of particle trajectories in the geomagnetic field requires an equation for a dipole field line. This may be derived from the condition that magnetic field lines are defined at all points to be co-aligned with the direction of the magnetic field. According to Fig. 5.7, this leads to the following relation

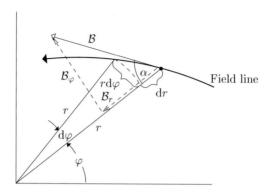

Fig. 5.7. Relation between the radius vector and the latitude of a field line

between the geocentric distance and the magnetic latitude of any point on a
field line

$$\frac{r\,d\varphi}{dr} \simeq \tan\alpha = \frac{B_\varphi}{B_r} = -\frac{1}{2}\frac{\cos\varphi}{\sin\varphi}$$

Separation of variables and using the substitution $x = \cos\varphi$, we obtain the
differential equation

$$\frac{dr}{r} = 2\frac{dx}{x}$$

with the solution

$$r = \frac{r_1}{\cos^2\varphi_1}\cos^2\varphi$$

Introducing the *shell parameter* L, defined as the geocentric distance of a field
line in the geomagnetic equatorial plane measured in units of Earth radii, i.e.
$r_0 = r(\varphi = 0) = L\,R_E$, the equation for a field line assumes the following
form

$$r = L\,R_E\,\cos^2\varphi \tag{5.9}$$

Graphical instructions for constructing field lines from this equation are presented in Fig. 5.8.

Of further interest are the *footpoints* of a field line and the tilt or inclination of these field lines at the footpoint. The magnetic latitude of a footpoint
at a height h above the Earth's surface ($r = R_E + h$) may be calculated from

$$\varphi(R_E + h) = \arccos\sqrt{\frac{R_E + h}{R_E\,L}} \tag{5.10}$$

Since, for most of the cases of interest here, $h \ll R_E$, Eq. (5.10) reduces to
the following simple form

$$\varphi(R_E) = \arccos\sqrt{1/L} \tag{5.11}$$

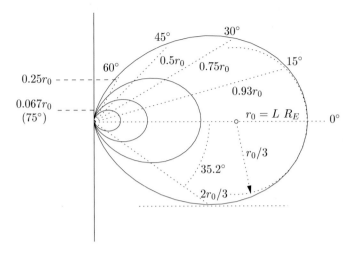

Fig. 5.8. Guide for constructing dipole field lines (adapted from Kertz, 1969)

In order to determine the tilt of a field line, we introduce the characteristic parameters sketched in Fig. 5.9. The parameter H (not to be confused with the magnetic field strength $\mathcal{H}$) denotes the projection of $\mathcal{B}$ onto the *H*orizontal plane, and the angle between $\mathcal{B}$ and H is defined as the *I*nclination I. For the centered dipole approximation considered here, the horizontal component corresponds to the φ-component of the magnetic field so that

$$I = -\arctan\left(\mathcal{B}_r/\mathcal{B}_\varphi\right) = \arctan(2\tan\varphi) \qquad (5.12)$$

This angle is independent of the height and, as indicated by the minus sign, is taken as positive in the Northern Hemisphere. Figure 5.10 illustrates the steepness of the magnetic inclination for various magnetic latitudes.

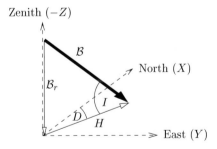

Fig. 5.9. Definitions of the horizontal component H, the inclination I and the declination D of the geomagnetic field. As an alternative to these three parameters, Cartesian coordinates X, Y and Z can also be used to describe the magnetic field vector using the above coordinate system.

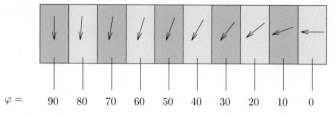

Fig. 5.10. Inclination of the geocentric dipole field in the Northern Hemisphere at various magnetic latitudes

The deviation of the horizontal component H from geographic north, denoted the *Declination*, is obtained from

$$D(\varphi, \lambda_g) = -\arcsin(\cos \varphi_g^{BP} \, \sin(\lambda_g - \lambda_g^{BP})/\cos \varphi) \tag{5.13}$$
$$\simeq -\arcsin(0.19 \, \sin(\lambda_g + 70°)/\cos \varphi)$$

where λ_g^{BP}, φ_g^{BP} are the geographic longitude and latitude of the geomagnetic north pole, and λ_g, φ are the geographic longitude and latitude of the observer. As indicated by the minus sign in the above equation, a deviation to the east is considered to be positive.

Another noteworthy property of a field line is its curvature. As shown in Appendix A.11, the radius of curvature for a planar field line is given by the expression

$$\rho_c = \frac{\mathcal{B}}{|\nabla_\perp \mathcal{B}|} \tag{5.14}$$

where $\nabla_\perp \mathcal{B}$ is the magnetic field strength gradient in the direction perpendicular to the magnetic field. In the equatorial plane we thus have

$$|\nabla_\perp \mathcal{B}| = \left|\frac{\partial}{\partial r}\mathcal{B}_\varphi\right| = \left|-\frac{3\mathcal{B}_\varphi}{r}\right| \tag{5.15}$$

and hence

$$\rho_c(\varphi = 0) = L \, R_E/3 \tag{5.16}$$

The radius of curvature of a field line in the equatorial plane is thus exactly one third of the geocentric distance of the field line. This result can be useful when constructing dipole field lines (see Fig. 5.8).

Finally, we consider the variation of field strength along a given field line. This is obtained by inserting Eq. (5.9) into the expression for the magnetic field intensity given in Fig. 5.6

$$\mathcal{B}(\text{field line}) = \frac{\mathcal{B}_{00}}{L^3} \frac{\sqrt{1 + 3\sin^2 \varphi}}{\cos^6 \varphi} \tag{5.17}$$

An *eccentric dipole approximation* is often used instead of the centered dipole. For this approximation the magnetic dipole is displaced about 500 km in the direction of the Mariana Islands in the Pacific Ocean, but without changing the orientation of the dipole. A better fit to the geomagnetic field is thus attained, but at the expense of complicating the conversion to geographic coordinates. In this case the geographic coordinates of the magnetic poles become

eccentric dipole approximation (1985): $\begin{array}{l}\text{BP } (82.1°\text{N}, 270°\text{E}) \\ \text{AP } (74.8°\text{S}, 119°\text{E})\end{array}$

Agreement of these values with the position of the observed magnetic inclination poles is clearly better. Note again that all magnetic field parameters display slow, but noticeable variations.

An even better approximation to the geomagnetic field can be achieved by using the *invariant magnetic latitude* or the *corrected geomagnetic latitude* (CGL). Rather than going into their definition and derivation, let it suffice to say that these parameters further improve the ordering of geomagnetic field dependent phenomena.

5.3 Charged Particle Motion in the Geomagnetic Field

The motion of charged particles in the geomagnetic field is complicated, even if one approximates that field with only its dipole component. A quantitative description of this motion begins with the force equilibrium condition for a single particle, considering an inertial force $\vec{F}_I$, a velocity dependent magnetic force $\vec{F}_B$ and other, velocity independent, external forces $\vec{F}_j$; see Fig. 5.11. Frictional forces, however, are not considered. We thus assume that interactions with other particles can be neglected, which is often the case. Furthermore, by excluding the reaction effect from particle motion on the existing magnetic field, we acknowledge that our treatment is not self-consistent. Neglecting this feedback is tantamount to the assumption that the currents associated with the charged particle motion and the magnetic fields produced by these currents are relatively small. This condition is also usually fulfilled. We thus look for solutions of the equation of motion

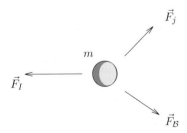

Fig. 5.11. Forces acting on a charged particle. $\vec{F}_I$ = inertial force, $\vec{F}_B$ = velocity dependent magnetic force and $\vec{F}_j$ = velocity independent, external forces.

$$m \frac{\mathrm{d}\vec{v}}{\mathrm{d}t} = \vec{F}_j + q\,(\vec{v} \times \vec{B}) \tag{5.18}$$

Separating this equation into components parallel and perpendicular to the magnetic field, we obtain

$$m \frac{\mathrm{d}\vec{v}_{\parallel}}{\mathrm{d}t} = \vec{F}_{j\parallel} \tag{5.19}$$

$$m \frac{\mathrm{d}\vec{v}_{\perp}}{\mathrm{d}t} = \vec{F}_{j\perp} + q\,(\vec{v}_{\perp} \times \vec{B}) \tag{5.20}$$

The first equation corresponds to the 'normal' equation of motion, valid as well for neutral gas particles, and can be directly integrated if the external forces are time independent

$$\vec{v}_{\parallel}(t) = \vec{v}_{\parallel}(t_0) + \frac{\vec{F}_{j\parallel}}{m}(t - t_0) \tag{5.21}$$

The solution of the second equation proves to be considerably more difficult to derive and is critically dependent on the magnetic field configuration. This is because the magnitude and direction of the magnetic force are themselves functions of the velocity, thereby preventing a direct separation of the variables. In this situation it is helpful to break down the total motion of the particle into its individual components. Because these components evolve at very different time scales, they are nearly independent of each other and can thus be summed linearly to obtain the total motion. For our purposes, it suffices to consider the following special cases:

(1) $\vec{F}_{j\perp} = 0$, $\vec{B}$ uniform $\rightarrow$ gyration

(2) $\vec{F}_{j\perp} = 0$, $\nabla B \parallel \vec{B}$ $\rightarrow$ bouncing

(3) $\vec{F}_{j\perp} = 0$, $\nabla B \perp \vec{B}$ $\rightarrow$ drift

(4) $\vec{F}_{j\perp} \neq 0$, $\vec{B}$ uniform $\rightarrow$ drift

The following discussion of the various types of motion is largely intuitive and perceptual. A more formal derivation of the cited results may be found in the specialized literature.

5.3.1 Gyromotion ($\vec{F}_{j\perp} = 0$, $\vec{B}$ uniform)

Magnetic fields produce an acceleration of charge carriers perpendicular to their direction of motion without changing the magnitude of the velocity. In the absence of other external forces, we have

$$m \frac{\mathrm{d}\vec{v}_{\perp}}{\mathrm{d}t} = q\,\vec{v}_{\perp} \times \vec{B} \tag{5.22}$$

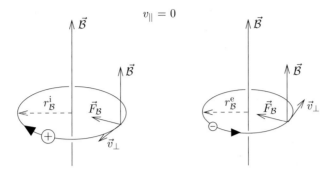

Fig. 5.12. Particle motion in a uniform magnetic field for $v_\parallel = 0$

The transverse acceleration from the magnetic force $F_\mathcal{B}$ on the right side of this equation forces the particle into a spiral orbit with increasingly greater curvature. The inertial force, which is manifested as a centrifugal force F_c that increases for greater curvature, opposes this transverse acceleration. This results in an equilibrium condition for which the magnetic force and the centrifugal force become exactly equal. This state corresponds to constant orbital curvature and thus circular motion. The radius of this circular orbit may be determined by equating these two forces

$$F_\mathcal{B} = |q|\, v_\perp\, \mathcal{B} = F_c = m\, v_\perp^2/r_\mathcal{B}$$

The *gyroradius* or *Larmor radius* is thus given by

$$r_\mathcal{B} = \frac{m\, v_\perp}{|q|\, \mathcal{B}} \tag{5.23}$$

Since the direction of the magnetic force is dependent on the charge polarity, the direction of the circular motion for positively charged particles is opposite to that for negatively charged particles; see Fig. 5.12. For simplicity, we will assume in the following that the positively charged particles are ions and the negatively charged particles are electrons. The centroid of the orbit, for reasons that will become more apparent later, is referred to as the *guiding center*.

The orbital period of the particle is

$$\tau_\mathcal{B} = \frac{2\pi\, r_\mathcal{B}}{v_\perp} = 2\pi\, \frac{m}{|q|\, \mathcal{B}} \tag{5.24}$$

Accordingly, the *gyrofrequency* may be written as

$$\omega_\mathcal{B} = \frac{2\pi}{\tau_\mathcal{B}} = \frac{|q|\, \mathcal{B}}{m} \tag{5.25}$$

The above relations may be derived more formally in a Cartesian coordinate system with the z-axis oriented along the direction of the magnetic field. The x and y components of Eq. (5.22) may then be written as follows

$$m \frac{dv_x}{dt} = q\, v_y\, \mathcal{B}\,, \qquad m \frac{dv_y}{dt} = -q\, v_x\, \mathcal{B} \tag{5.26}$$

Differentiating with respect to time and mutually substituting, we obtain

$$\frac{d^2 v_x}{dt^2} = -\omega_\mathcal{B}^2\, v_x\,, \qquad \frac{d^2 v_y}{dt^2} = -\omega_\mathcal{B}^2\, v_y \tag{5.27}$$

The solutions of these differential equations have the general form

$$v_x = (v_x)_0\, \sin(\omega_\mathcal{B} t + \varphi_x)\,, \qquad v_y = (v_y)_0\, \sin(\omega_\mathcal{B} t + \varphi_y)$$

Inserting these into Eq. (5.26), we obtain the following relations between the constants: $(v_x)_0 = (v_y)_0 = \sqrt{v_x^2 + v_y^2} = v_\perp$ and $\varphi_x = \varphi_y \pm \pi/2$, where the plus (minus) sign holds for electrons (ions). Setting the phase constant $\varphi_y = 0$, one obtains

$$v_x = \pm v_\perp \cos(\omega_\mathcal{B} t)\,, \qquad v_y = v_\perp \sin(\omega_\mathcal{B} t)$$

and thus for the transverse displacement

$$x = \pm r_\mathcal{B} \sin(\omega_\mathcal{B} t)\,, \qquad y = -r_\mathcal{B} \cos(\omega_\mathcal{B} t) \tag{5.28}$$

where we have set the integration constants equal to zero. The plus sign on the x component holds for electrons; the minus sign for ions. These equations thus describe the circular motion illustrated in Fig. 5.12 with the above values of gyroradius, gyroperiod and gyrofrequency.

It is important to note that $\tau_\mathcal{B}$ and $\omega_\mathcal{B}$ do *not* depend on the velocity (energy) of the particles. More energetic particles are faster, but must traverse a longer orbital path $(r_\mathcal{B} \sim v_\perp)$. Also important is that the particles do not gain energy from their forced circular motion in a magnetic field: the magnetic force is always directed perpendicular to the velocity. Finally, it should be noted that the gyromotion of the charge carriers produces an electric current loop and thus a magnetic dipole moment. Considering a reference surface parallel to a magnetic field line, a charge gyrating about this field line will pass through the surface once per gyroperiod; see Fig. 5.13. The current strength associated with this particle motion (i.e. the charge transport through the reference surface per unit time) is thus $\mathcal{I} = |q|/\tau_\mathcal{B}$. Inserting this into Eq. (5.8), one obtains the following expression for the magnetic dipole moment of a gyrating charge

$$\vec{\mathcal{M}}_g = A\, \mathcal{I}\, \hat{r}_{cl} \times \hat{l} = -\frac{|q|\, r_\mathcal{B}^2 \pi}{\tau_\mathcal{B}} \frac{\vec{\mathcal{B}}}{\mathcal{B}} = -\frac{m\, v_\perp^2}{2\, \mathcal{B}^2}\, \vec{\mathcal{B}} = -\frac{E_\perp\, \vec{\mathcal{B}}}{\mathcal{B}^2} \tag{5.29}$$

Every gyrating charge carrier is thus the source of a magnetic field, the strength of which is described by the above given magnetic dipole moment. The far field of this source again has a dipole character and is described by the formulas given in Eqs. (5.6) and (5.7). Note that the dipole moment is proportional to the component of particle energy perpendicular to the magnetic

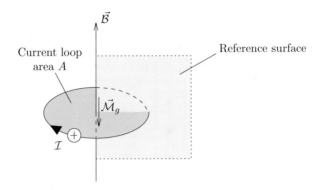

Fig. 5.13. Current and magnetic moment of a gyrating ion

field and inversely proportional to the magnetic field strength. Furthermore, because the magnetic moment is always aligned opposite to the external magnetic field, thereby weakening it, an ensemble of charged particles thus represents a *diamagnetic* medium. Finally, we note that the gyration-induced magnetic moment of a particle remains nearly constant in a 'slowly' varying magnetic field, thus leading to the concept of the *first adiabatic invariant* given by the quantity $J_1 = (4\pi m/|q|)\mathcal{M}_g$. 'Slowly' in this case means that the gyroradius and gyroperiod are small compared with the spatial and temporal scale lengths of the magnetic field variations. We will test this assertion with an example in Section 5.3.2.

Extending to three dimensions, let us now consider the case where the particle also has a velocity component parallel to the magnetic field. This velocity component, oblivious to the magnetic field, is linearly superposed onto the gyromotion. The resulting helical trajectory is sketched in Fig. 5.14. The inclination of this trajectory with respect to the local magnetic field line is denoted the *pitch angle*. The components of velocity parallel and perpendicular to the magnetic field are thus given by

$$v_\| = v \, \cos\alpha \quad , \quad v_\perp = v \, \sin\alpha \tag{5.30}$$

where α is the pitch angle. The field line at the center of the helical motion is denoted here the *guiding (center) field line* and describes the guiding center trajectory of the gyrating particle.

5.3.2 Oscillatory (Bounce) Motion ($\vec{F}_{j\perp} = 0$, $\nabla B \parallel \vec{B}$)

In the following we consider a nonuniform magnetic field that displays an intensity gradient in the direction of the field lines. Such a magnetic field can have convergent (intensity increase) or divergent (intensity decrease) magnetic field lines, whereby the gradient is reflected in the changing density of field lines. This is a direct consequence of the fact that the divergence of a

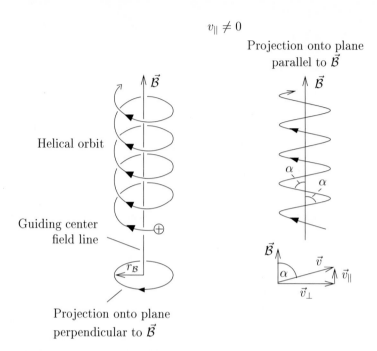

$v_{\parallel} \neq 0$

Projection onto plane
parallel to $\vec{B}$

$\vec{B}$

$\vec{B}$

Helical orbit

α

α

Guiding center
field line

$\vec{B}$

$\vec{v}$

r_B

α

$\vec{v}_{\parallel}$

$\vec{v}_{\perp}$

Projection onto plane
perpendicular to $\vec{B}$

Fig. 5.14. Particle motion in a uniform magnetic field for $v_{\parallel} \neq 0$. The pitch angle α describes the inclination of the helical trajectory to the magnetic field.

magnetic field vanishes. Magnetic field lines, which have no beginning and no end, squeeze together in regions of increasing field strength and spread out in regions of decreasing field strength.

Figure 5.15 shows the gyration orbit of a charged particle in such a magnetic field configuration. The ion clearly passes through regions where the magnetic field has components both parallel and perpendicular to the guiding center field line. This, in turn, leads to magnetic forces that act in directions both perpendicular and parallel to the guiding center field line. While the force perpendicular to the guiding center field line, F_r, is responsible for the gyration of the particle, the force parallel to the guiding center field line, F_z, produces an acceleration in the direction of the diverging field lines. Another way to look at this effect is that the gyrating particle is pushed away from the region of higher magnetic field strength. Using the cylindrical coordinate system of Fig. 5.15, the parallel force may be written as

$$F_z = |q| \, v_{\varphi} \, \mathcal{B}_r$$

The magnitude of $\mathcal{B}_r$ is obtained from the Maxwell equation (A.93), which requires the vanishing divergence of the magnetic field. For cylindrical coordinates and azimuthal symmetry, this means that

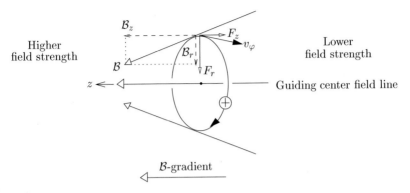

Fig. 5.15. Particle motion in an nonuniform magnetic field with a gradient along the magnetic field lines

$$\text{div}\vec{B} = \frac{1}{r}\,\frac{\partial}{\partial r}\,(r\,B_r) + \frac{\partial B_z}{\partial z} = 0$$

see Eq. (A.28). Multiplication with r and subsequent integration yields

$$\int_0^r \frac{\partial}{\partial r'}\,(r'\,B_r)\,\mathrm{d}r' = -\int_0^r r'\,\frac{\partial B_z}{\partial z}\,\mathrm{d}r'$$

For the case when the gradient $\partial B_z/\partial z$ is independent of r, it follows that

$$B_r(r) = -\frac{r}{2}\,\frac{\mathrm{d}B_z}{\mathrm{d}z} \tag{5.31}$$

At the position of the gyration orbit ($r = r_B$), this becomes

$$B_r(r_B) = -\frac{m\,v_\varphi}{2\,|q|\,B_z}\,\frac{\mathrm{d}B_z}{\mathrm{d}z} \simeq -\frac{m\,v_\varphi}{2\,|q|\,B}\,\frac{\mathrm{d}B}{\mathrm{d}z}$$

where in the second step we have assumed that B_r is only a small perturbation and thus, to a good approximation, B_z corresponds to the total field strength. The field-aligned component of the magnetic force thus assumes the form

$$F_z = -\frac{m\,v_\varphi^2}{2}\,\frac{1}{B}\,\frac{\mathrm{d}B}{\mathrm{d}z}$$

or written in vectorial form

$$\vec{F}_\parallel^{gr} = -\frac{E_\perp}{B}\,\nabla_\parallel B \tag{5.32}$$

The *magnetic gradient force* is therefore proportional to the perpendicular component of the particle energy and to the parallel gradient in the magnetic field strength, but inversely proportional to the magnetic field strength itself.

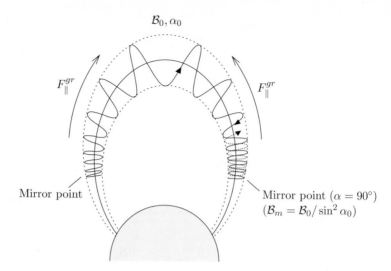

Fig. 5.16. Bounce motion in the Earth's dipole field

The situation that arises upon applying this to the dipole field of the Earth is sketched in Fig. 5.16. A particle moves along a helical path, following its guiding center field line in the direction toward Earth. The increasing field strength – indicated by the convergence of the field lines – leads via the magnetic gradient force to a continual deceleration of the particle until its entire energy parallel to the field line is exhausted and the motion of the guiding center comes to a halt. At this point the pitch angle has reached 90° and the entire kinetic energy of the particle (which is conserved) is concentrated in its gyromotion. The gradient force, which is dependent only on the gyration energy (and not the parallel energy component), continues to act and accelerates the particle upward in the direction toward the apex of the guiding center field line. The particle is thus reflected back into the direction from which it came. The point where the trajectory reversal occurs is appropriately called the *mirror point* and the gradient force is also known as the *mirror force*. The pitch angle decreases as the velocity parallel to the field increases until it reaches its minimum value $\alpha(\varphi = 0) = \alpha_0$ at the apex point. Upon passing through the magnetic equatorial plane, the particle is again decelerated right down to its conjugate mirror point in the opposite magnetic hemisphere. There it is again reflected back in the direction of the field line apex. The particle thus continually bounces back and forth between its two mirror points and is captured in the Earth's magnetic field as if in a magnetic bottle. An approximate expression for the bounce or oscillation period (the exact formula contains an integral that can be only solved numerically) is given by

$$\tau_O = \frac{4\,L\,R_E}{v}\,s_1(\alpha_0) = \sqrt{8m}\,R_E\,s_1(\alpha_0)\,\frac{L}{\sqrt{E}} \tag{5.33}$$

with

$$s_1(\alpha_0) \simeq 1.3 - 0.56 \, \sin \alpha_0 \qquad (5.34)$$

The length of the helical path between mirror points is thus of the order of $2 \, L R_E$. This is only weakly dependent on the particular equatorial pitch angle α_0 ($0.74 \leq s_1(\alpha_0) \leq 1.3$) and thus insensitive to the actual position of the mirror points. The helix is evidently pulled apart like an accordion for lower lying mirror points. If the equatorial pitch angle has the relatively small value $\alpha_0 = 30°$, one obtains the explicit formula

$$\left. \begin{array}{c} \alpha_0 = 30° \\ (s_1(30°) \simeq 1) \end{array} \right\} \quad \left\{ \begin{array}{l} \tau_o^{\mathrm{p}}[\mathrm{s}] \simeq 58 \, L/\sqrt{E[\mathrm{keV}]} \\[2ex] \tau_o^{\mathrm{e}}[\mathrm{s}] \simeq 1.4 \, L/\sqrt{E[\mathrm{keV}]} \end{array} \right. \qquad (5.35)$$

and the oscillation (bounce) period is only 27% shorter for α_0 close to 90°. In any case, the bounce period is directly proportional to the geocentric distance of the guiding center field line at its apex and inversely proportional to the square root of the particle's energy.

In contrast to the bounce period, the *position* of the mirror points is critically dependent on the particular equatorial pitch angle. This may be shown from the following argumentation. A particle cannot gain or lose energy in a static magnetic field – $\vec{F}_B$ always acts perpendicular to the velocity. Under this condition, no accelerating electric fields may exist along the gyration orbit

$$\oint_{\text{gyration orbit}} \vec{\mathcal{E}} \; \mathrm{d}\vec{l} = \mathcal{U}_g = 0$$

The drop in voltage along the gyration orbit $\mathcal{U}_g$, however, is coupled via the Faraday induction law (A.101) to the magnetic flux Φ passing through this gyration loop. This requires that

$$\mathcal{U}_g = -\frac{\mathrm{d}\Phi}{\mathrm{d}t} = -\frac{\mathrm{d}}{\mathrm{d}t} \left(r_B^2 \, \pi \, \mathcal{B} \right) = 0$$

from which it follows that $r_B^2 \, \pi \, \mathcal{B} = \text{constant}$. The area enclosed by the gyromotion is therefore adjusted during the bounce motion in such a way that the magnetic flux passing through the gyration orbit loop remains constant. With $r_B = m \, v_\perp / |q| \, \mathcal{B}$, $v_\perp = v \, \sin \alpha$ and $v = \text{const.}$, this implies that

$$\mathcal{B}/\sin^2 \alpha = \text{const.} \qquad (5.36)$$

The magnetic field strength at the mirror point is thus given by

$$\mathcal{B}_m = \mathcal{B}/\sin^2 \alpha = \mathcal{B}_0/\sin^2 \alpha_0 = \frac{\mathcal{B}_{00}}{L^3 \sin^2 \alpha_0} \qquad (5.37)$$

where $\mathcal{B}_0$ is the magnetic field strength at the apex of the guiding center field line. Knowing the variation of the magnetic field strength along a field line from Eq. (5.17), it is thus possible to determine the magnetic latitude, and therefore the height, of the mirror point. Note that the mirror point height is independent of either the type or energy of the particle. Taking as a specific example a particle with an equatorial pitch angle $\alpha_0 \simeq 5.3°$ at the shell parameter $L \simeq 4$, the mirror point is located at the Earth's surface. This means that particles with equatorial pitch angles smaller then $5.3°$ cannot be preserved on this field line. These particles have pitch angles within the so-called *loss cone*, and they are lost by absorption at the Earth's surface after one bounce. It is a true loss *cone* because the particles are lost without regard for the particular phase of their gyromotion. In reality, the width of the loss cone is somewhat larger, because any particles that dip into the denser upper atmosphere will be deflected by collisions from their bounce trajectory and eventually absorbed.

Finally, it follows from Eq. (5.36) that

$$\frac{\sin^2 \alpha}{\mathcal{B}} \frac{mv^2}{2} = \frac{E_\perp}{\mathcal{B}} = \mathcal{M}_g \sim J_1 = \text{const.}$$

The magnetic moment of the gyromotion is therefore indeed an invariant of the motion. This, of course, is only true as long as the charged particle is not accelerated by external forces, i.e. as long as the motion is 'adiabatic'.

5.3.3 Drift Motion

Gradient Drift $(\vec{F}_{j\perp} = 0, \nabla \mathcal{B} \perp \vec{\mathcal{B}})$. Consider now a nonuniform magnetic field with an intensity gradient *perpendicular* to the field lines. The motion of charged particles in such a magnetic field can be understood with the help of Fig. 5.17, where, for simplicity, the gradient is represented by a discontinuous jump in the magnetic field strength. A particle that traverses this discontinuity will gyrate on a semicircle with a large gyroradius in the region of weak magnetic field, but will transfer to a small semicircular gyration orbit upon entering the region of the strong magnetic field, $r_B \sim 1/\mathcal{B}$. This leads to an effective drift of the particle along the discontinuity, i.e. a drift perpendicular to both the magnetic field and the magnetic field gradient.

The velocity of this *gradient drift* motion, derived formally in Appendix A.12, is given by the expression

$$\vec{u}_D^{gr} = \frac{m \, v_\perp^2}{2q \, \mathcal{B}^3} \, \vec{\mathcal{B}} \times \nabla_\perp \mathcal{B} = \frac{E_\perp}{q \, \mathcal{B}^3} \, \vec{\mathcal{B}} \times \nabla_\perp \mathcal{B} \tag{5.38}$$

This drift velocity corresponds to the time-averaged value of the actual particle velocity $\vec{u}_D^{gr} = \langle \vec{v}_\perp(t) \rangle$. It is obvious that the above expression correctly describes the direction of the drift motion, but its magnitude is also readily understood. In order to demonstrate this, we make use of the approximation $|\nabla_\perp \mathcal{B}| \simeq \mathcal{B}/L_\mathcal{B}$, where $L_\mathcal{B}$ is the scale length over which the magnetic field strength changes significantly. The tenor of this approximation is similar

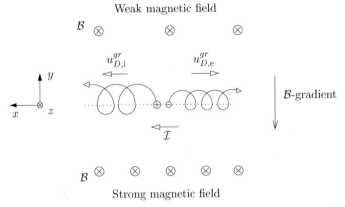

Fig. 5.17. Particle motion in a nonuniform magnetic field. The field gradient is perpendicular to the magnetic field direction. For simplicity, the gradient is taken to be a sudden jump in the magnetic field strength at the dotted line, and the initial motion of the particles is assumed to be perpendicular to this discontinuity.

to that used for the definition of the density scale height or the estimate of ionospheric time constants (see also Appendix A.13.3). The magnitude of the drift velocity may thus be written as follows

$$u_D^{gr} \simeq \frac{1}{2}\frac{r_B}{L_B}v_\perp$$

Since the ratio r_B/L_B is a measure of the field strength differences encountered along the gyration orbit, the dependence of the drift velocity on this quotient is immediately plausible. Also intuitive is the direct proportionality to the gyration velocity.

It follows from Eq. (5.38) that ions and electrons of equal energy drift at the same velocity. The circular orbit segments of the electrons are considerably smaller than those of the ions, but this is compensated by the much greater gyration speed of the electrons. Since ions and electrons drift in opposite directions, a current is created that flows in the direction of the ion drift (see again Fig. 5.17).

The situation in the geomagnetic dipole field is sketched in Fig. 5.18. The largest gradients perpendicular to the geomagnetic field lines occur at the equator and are directed toward the Earth. Accordingly, the drift of positively charged particles is directed to the west; that of negatively charged particles to the east. For an equatorial pitch angle of $\alpha_0 = 90°$ and using Eq. (5.15), the drift velocity is given by

$$\alpha_0 = 90° \atop (\varphi = 0) \qquad \left\{ \begin{array}{l} u_D^{gr} = 3\ L^2\ E/|q|\ B_{00}\ R_E \\[2mm] \text{or, in a readily usable form} \\[2mm] u_D^{gr}[\text{m/s}] \simeq 15.7\ L^2\ E[\text{keV}] \end{array} \right. \qquad (5.39)$$

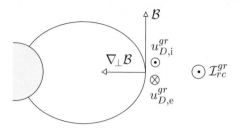

Fig. 5.18. Gradient drift in the geomagnetic dipole field. $\mathcal{I}_{rc}^{gr}$ denotes the ring current associated with the gradient drift motion.

Neutral Sheet Drift. For later application, we consider a case where the magnetic field direction is even reversed at the discontinuity of Fig. 5.17. The particle motion in such a magnetic field configuration is sketched in Fig. 5.19. It is assumed that the magnetic field strength on each side of the discontinuity is the same and that the particles cross the discontinuity surface perpendicularly. The magnetic field strength at the discontinuity itself (more precisely: the magnetic field component perpendicular to the plane of the drawing) must be zero, thus leading to the designation *neutral sheet* for the plane of the discontinuity. As seen, ions and electrons drift with wavelike trajectories along the neutral sheet. Since they are moving in opposite directions, they form the so-called *neutral sheet current.* Neutral sheet currents play an important role in the heliosphere and in the tail of the Earth's magnetosphere.

External Force Drift ($\vec{F}_{j\perp} \neq 0$, $\vec{B}$ uniform). The magnetic field is again assumed to be uniform in the following. We consider the situation when an external force $\vec{F}_{j\perp}$, independent of the charge q, is applied in a direction perpendicular to the magnetic field. Figure 5.20 illustrates the charge carrier

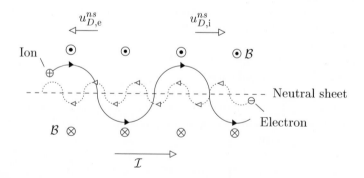

Fig. 5.19. Particle motion along a neutral sheet for perpendicular incidence onto the discontinuity

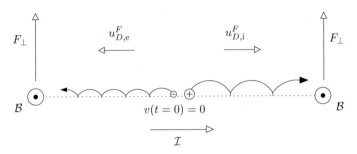

Fig. 5.20. Particle motion in a uniform magnetic field with an external force, independent of charge q

motion in this combination of force fields. The particle, originally at rest, will be accelerated in the direction of the external force immediately after it is applied. As soon as its velocity is nonzero, it will feel the magnetic force and begin to move in a gyration orbit corresponding to its instantaneous velocity. Since the direction of motion steadily changes, the particle will eventually move in a direction opposite to that of the external force. The accompanying deceleration leads to a loss of its newly gained kinetic energy and it comes to rest again at the level of its starting position, but laterally displaced. This motion is continually repeated, resulting in an effective drift of the particle in the direction perpendicular to both the magnetic field and external force. Because of the different sense of gyration, positively and negatively charged particles drift in opposite directions and a current is established in the direction of the ion drift. Note that, for the stationary case, no energy is extracted from the external field because the work expended to accelerate the particle is completely returned during the deceleration phase. Energy is consumed only during the start-up phase in order to accelerate the particle to its effective drift velocity.

An explicit expression for the drift velocity may be derived from the following arguments. As shown in Fig. 5.20, the particles execute a series of identical cycles, implying that the drift velocity must be constant and independent of time. This requires that the external and magnetic forces acting on them, when averaged over time, must exactly compensate each other. Thus we may write

$$\langle \vec{F}_\perp + q\,\vec{v}_\perp \times \vec{\mathcal{B}} \rangle = \vec{F}_\perp + q\,\vec{u}_D^F \times \vec{\mathcal{B}} = 0$$

where the drift velocity again corresponds to the time averaged particle velocity, $\vec{u}_D^F = \langle \vec{v}_\perp(t) \rangle$. Forming the cross product of this vector equation with $\vec{\mathcal{B}}$ yields the relation

$$\vec{F}_\perp \times \vec{\mathcal{B}} + q\,(\vec{u}_D^F \times \vec{\mathcal{B}}) \times \vec{\mathcal{B}} = \vec{F}_\perp \times \vec{\mathcal{B}} - q\,\mathcal{B}^2 \vec{u}_D^F = 0$$

where we have made use of Eq. (A.19). It follows that the velocity of the *external force drift* is

$$\vec{u}_D^F = \frac{\vec{F}_\perp \times \vec{B}}{q \, B^2} \tag{5.40}$$

This drift is thus directly proportional to the magnitude of the external force and inversely proportional to the magnetic field strength.

In order to describe the total motion, we separate the particle velocity into its time dependent and time independent components

$$\vec{v}_\perp(t) = \vec{u}_D^F + \vec{c}_\perp(t)$$

where $\vec{c}_\perp(t)$ plays the role of a random velocity. Inserting this into the equation of motion (5.20) and separating the variable from the constant components, we obtain

$$m \, \frac{d\vec{c}_\perp}{dt} - q \, \vec{c}_\perp \times \vec{B} = \vec{F}_\perp + q \, \vec{u}_D^F \times \vec{B} \; = \; 0$$

It follows that

$$m \, \frac{d\vec{c}_\perp}{dt} = q \, \vec{c}_\perp \times \vec{B}$$

which corresponds exactly to the equation of motion of a charge carrier in a uniform magnetic field without an external force, see Eq. (5.22). The total motion thus consists of a superposition of the gyration motion corresponding to this equation plus the constant external force drift of its guiding center. For a particle starting from rest this motion is a cycloid, i.e. imagine the particle attached to the circumference of a rotating wheel, which rolls in the direction of the external force drift. If the particle has some nonzero starting velocity in the direction of the external force, the curve becomes a more general trochoid (spoke curve). Note the difference between the motion along a helix as shown in Fig. 5.14 and the motion along the cycloidal trajectory described here. While the guiding center moves parallel to a magnetic field line in the first case, it moves perpendicular to the magnetic field in the present case.

Similar to the gradient drift, positive and negative charges move in opposite directions, and – for the same force – at the same speed. Again, the smaller gyroradius of the electrons is just compensated by the higher gyrofrequency.

Ambipolar $\vec{\mathcal{E}} \times \vec{B}$ Drift ($\vec{F}_{j\perp} = q \, \vec{\mathcal{E}}_\perp$, $\vec{B}$ uniform). An important special case of external force drift occurs when the particle is accelerated by an external electric field, $\vec{F}_\perp = q \vec{\mathcal{E}}_\perp$, in which case

$$\vec{u}_D^{\mathcal{E}} = \frac{\vec{\mathcal{E}}_\perp \times \vec{B}}{B^2} \tag{5.41}$$

The charge dependency of the drift motion evidently cancels out in this case. Positive and negative charges move together (ambipolar) with the same velocity in the same direction; see Fig. 5.21. This is often simply called $\vec{\mathcal{E}} \times \vec{B}$ *drift*, although it should not be forgotten that the associated drift velocity is

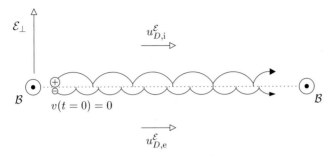

Fig. 5.21. Particle motion in a uniform magnetic field with an electric field imposed perpendicular to the magnetic field lines

inversely proportional to $\mathcal{B}$. It is somewhat paradox that the drift currents disappear upon applying an electric field. A collisionless ensemble of charge carriers in a magnetic field (i.e. a magnetized plasma) is thus an insulator in the direction perpendicular to the magnetic field.

Curvature Drift ($\vec{F}_{j\perp} = \vec{F}_c$, $\vec{\mathcal{B}}$ homogeneous). Centrifugal forces, which arise from the bounce motion of particles along curved magnetic field lines, deliver the most important contribution to force drifting in the Earth's dipole field. The situation in the equatorial plane is sketched in Fig. 5.22: positively charged particles drift to the west; negatively charged particles to the east. Curvature drift and gradient drift thus constructively superpose and reinforce each other. Together, they form a ring current that flows westward around the Earth.

The magnitude of the centrifugal force is needed in order to determine the curvature drift velocity. This may be written in the general form

$$\vec{F}_c = \hat{\rho}_c \, mv_\parallel^2/\rho_c$$

where $\hat{\rho}_c$ is a unit vector in the direction of the radius of curvature. Using the expression (5.14) for ρ_c, we obtain

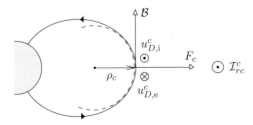

Fig. 5.22. Curvature drift in the Earth's magnetic dipole field. F_c denotes the centrifugal force, $u_{D,i}^c$ and $u_{D,e}^c$ are the curvature drift velocities of the ions and electrons, respectively, and $\mathcal{I}_{rc}^c$ is the ring current associated with the curvature drift.

$$\vec{F}_c = \hat{\rho}_c \, mv_\parallel^2 \, \frac{|\nabla_\perp \mathcal{B}|}{\mathcal{B}} = -mv_\parallel^2 \, \frac{\nabla_\perp \mathcal{B}}{\mathcal{B}} \tag{5.42}$$

The last step accounts for the fact that the radius of curvature vector and the perpendicular gradient in the magnetic field intensity are antiparallel. With $\vec{F}_\perp = \vec{F}_c$ in Eq. (5.40), the velocity of the *curvature drift* is thus given by

$$\vec{u}_D^c = \frac{\vec{F}_c \times \vec{\mathcal{B}}}{q \, \mathcal{B}^2} = \frac{m \, v_\parallel^2}{q \, \mathcal{B}^3} \, \vec{\mathcal{B}} \times \nabla_\perp \mathcal{B} \tag{5.43}$$

In addition to the curvature drift, there exists an oppositely directed gravitational drift. This gravitational drift, however, is small even at thermal energies. This can be demonstrated using the example of a 1 eV proton with an equatorial pitch angle $\alpha_0 = 45°$ on a guiding center field line with the shell parameter $L = 4$. Comparing the two forces near the equator and making use of Eq. (5.16), we obtain

$$F_c = m_\mathrm{p} v_\parallel^2 / \rho_c = 6 \, E_\mathrm{p} \, \cos^2 \alpha_0 / (R_E \, L) \simeq 2 \cdot 10^{-26} \text{ N}$$

$$\gg F_g = m_\mathrm{p} \, g(h=0)/L^2 \simeq 10^{-27} \text{ N}$$

Total Drift. Adding the curvature and gradient drifts together, one obtains an expression for the total drift of a charged particle in the Earth's dipole field

$$\vec{u}_D = \vec{u}_D^{\,gr} + \vec{u}_D^c = \frac{m}{2 \, q \, \mathcal{B}^3} \, (v_\perp^2 + 2v_\parallel^2) \, \vec{\mathcal{B}} \times \nabla_\perp \mathcal{B}$$

$$= \frac{E \, (1 + \cos^2 \alpha)}{q \, \mathcal{B}^3} \, \vec{\mathcal{B}} \times \nabla_\perp \mathcal{B} \tag{5.44}$$

The total drift corresponds to the gradient drift, of course, for the case when the equatorial pitch angle $\alpha_0 = 90°\,(v_\parallel = 0)$. The calculation for $\alpha_0 < 90°$ becomes more complicated because the drift no longer proceeds only in the equatorial plane and one must consider variations of magnetic field and pitch angle along the helical path of the bounce motion. The drift velocity in the vicinity of the lower lying mirror points, however, is considerably smaller than in the equatorial plane. This is not only because the perpendicular field gradient and field line curvature are smaller there, but also because the parallel component of velocity is almost zero near the mirror points. Averaging the drift velocity over the bounce period (which can only be done numerically), one obtains the approximation

$$\langle u_D \rangle \simeq \frac{3 \, L^2 \, E}{|q| \, \mathcal{B}_{00} \, R_E} \, s_2(\alpha_0) \tag{5.45}$$

This corresponds to the gradient drift velocity at the equator derived in Eq. (5.39), but multiplied by the correction function

$$s_2(\alpha_0) = 0.7 + 0.3 \, \sin \alpha_0$$

Evidently, this is only weakly dependent on the actual value of the pitch angle $(0.7 \leq s_2(\alpha_0) \leq 1)$. Ignoring this dependence and using $s_2(\alpha_0) \simeq 1$ to a good approximation, the drift period of a particle is given by

$$\tau_D = 2\pi\, R_E L / \langle u_D \rangle \simeq \frac{2\pi}{3}\, R_E^2\, |q|\, \mathcal{B}_{00}\, \frac{1}{L\,E} \tag{5.46}$$

or, for calculating from a ready-to-use formula

$$\tau_D[\mathrm{h}] \simeq 710/L\,E\,[\mathrm{keV}] \tag{5.47}$$

Evidently, the drift period is inversely proportional to the energy of the particle and to the shell parameter of its guiding center field line.

5.3.4 Composite Charge Carrier Motion

The composite motion of charge carriers in the inner magnetosphere is sketched in Fig. 5.23. This consists of gyration about the local magnetic field line, an oscillatory bouncing motion along that field line and a superposed azimuthal drift around the Earth. Table 5.1 compares the magnitude of the associated time constants for three different combinations of particle energy and geocentric distance (shell parameter L), corresponding to important particle populations in the inner magnetosphere. The time constants are clearly so different, that a linear superposition of the three types of motion is a very good approximation.

Note that the mirror points of the particle trajectories are neatly arranged along a ring of constant height only in the dipole approximation used here. In reality, the heights of the mirror points vary according to the local magnetic field strength. Notorious in this regard is the *South Atlantic anomaly* with its remarkably weak field strength. The mirror points here lie especially deep in the denser atmosphere, leading to enhanced loss of magnetospheric particles. Note further that the above described motions can only take place, of course, if collisions with other particles are sufficiently rare to be neglected.

Table 5.1. Comparison of the gyration, oscillation (bounce) and drift periods for representative particle populations in the inner magnetosphere

Particle type		Protons		Electrons (multiplication factor)
Energy	0.6 eV ($\simeq$ 5000 K)	20 keV	20 MeV	$\sim$
L	3	4	1.3	$\sim$
Period τ_B	0.1 s	0.1 s	5 ms	$5.4 \cdot 10^{-4}$
Period τ_O	2 hours	1 min	0.5 s	$2.3 \cdot 10^{-2}$
Period τ_D	45 years	9 hours	2 min	1

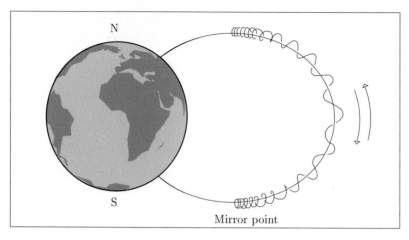

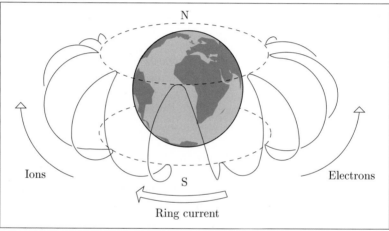

Fig. 5.23. Composite motion of charge carriers in the inner magnetosphere. Upper panel: gyration and bouncing; bottom panel: bouncing and drift

This prerequisite can be tested with knowledge of the collision frequency or, equivalently, the applicable time constant for collision processes.

5.3.5 Coulomb Collisions

Elastic collisions are interactions between particles, by which significant changes in velocity and momentum occur in a short time, but without incurring loss of kinetic energy. The occurrence probability of such events depends on, among other parameters, the collision cross section of the participating particles. The collision cross sections for the previously considered elastic collisions between neutral and charged gas particles, or between two neutral gas particles, were assumed to be independent of particle energy to a very

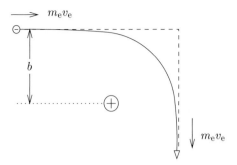

Fig. 5.24. 90° deflection of an electron with velocity v_e in the Coulomb potential of an ion at rest

good approximation. This was because we were dealing with particles in a very limited energy range (ca. 0.1-0.5 eV), and also because the interaction force of these particles varied with the distance taken to a high power (e.g. $1/r^5$). This latter property means that the interacting particles 'notice' each other rather 'suddenly' within a small range of distances. It was thus a good approximation in these cases to introduce an effective distance (a cross section radius) at which the interaction instantaneously sets in. Important for this assumption is that this distance is only weakly dependent on the relative velocity of the colliding particles because the steep increase in the repulsive force prevents faster particles from approaching their collision partners significantly closer than slower particles.

The situation is different for collisions between two charged particles. The interaction here is subject to the Coulomb force, which has a comparatively long-range $1/r^2$ dependence on distance. The interaction thus occurs more gradually for these *Coulomb collisions*, and particles of higher energy must approach their collision partners considerably more closely in order to suffer the same deflection as slower particles. This brings up the issue of just which deflection angle or which change in momentum must be attained before we can designate the interaction as a collision. In order to determine the size of the Coulomb cross section, and especially its dependence on energy, we first consider those interactions for which the deflection angle is 90°. The momentum changes in this case by $\sqrt{2}\,m\,v$, i.e. about 70% of the maximum possible elastic momentum transfer of $2\,m\,v$. As an example, Fig. 5.24 illustrates the situation for a 'collision' between an electron with velocity v_e and an ion at rest. In this case, vector subtraction yields a change in momentum amounting to

$$\Delta I = F\ \Delta t = \sqrt{2}\ m_e\ v_e \tag{5.48}$$

where the force F on the electron due to the electric field of the ion is given by *Coulomb's law*

$$F = \frac{e^2}{4\pi\ \varepsilon_0\ r^2}$$

Approximating the real deflection orbit of the electron (a hyperbola) by a quarter-circle segment of radius b (= impact parameter), the now constant force acting during the 'collision' is given by

$$\langle F \rangle \simeq \frac{e^2}{4\pi\,\varepsilon_0\,b^2}$$

The associated dwell time of the electron in this force field is

$$\Delta t \simeq \frac{2\pi\,b}{4}\,\frac{1}{v_e}$$

Inserting $\langle F \rangle$ and Δt into Eq. (5.48), one obtains the following proportionality

$$b \sim \frac{e^2}{\varepsilon_0\,m_e\,v_e^2}$$

A Coulomb cross section may thus be defined using b, the collision parameter required that the electron undergoes a $90°$ deflection and a momentum change of at least $\sqrt{2}m_e v_e$

$$(\sigma_{e,i}^{Cb})_{90} = \pi\,b^2 \sim \frac{e^4}{\varepsilon_0^2\,m_e^2\,v_e^4} \sim \frac{1}{E_e^2} \tag{5.49}$$

As evident from the derivation, two factors contribute to the strong energy dependence of this cross section:

- The particular deflection of the impacting particle depends on the *relative* change of momentum. If the particle comes in with a large kinetic momentum, and thus a high energy, a correspondingly large additional impulse (i.e. a large force) is needed to deflect the particle significantly from its original trajectory. This requires that the two particles come relatively close to each other, thereby implying a small interaction cross section.

- The impulse suffered by an impacting particle depends not only on the magnitude of the interaction force, but also on the particle's dwell time within this force field. This time becomes shorter as the relative velocity, and therefore the impact energy of the incident particle, increases. While the dwell time is even shorter for smaller impact parameters, this linear decrease is overwhelmed by the quadratic increase in the magnitude of the force.

We will use Eq. (5.49) to guide our approach in deriving a general expression for $90°$ Coulomb collisions in a *thermal* plasma. In this case the electron velocity must be replaced by the relative velocity of the interacting particles, $v_e \rightarrow v_{1,2}$, and the electron mass is replaced by the reduced mass, $m_e \rightarrow m_{1,2}$ (see Section 2.1.1). More precise calculations than those used in our simple estimate above also yield a cross section with a slightly smaller numerical factor

$$(\sigma_{1,2}^{Cb})_{90} = \frac{1}{16\pi^3} \frac{e^4}{\varepsilon_0^2 m_{1,2}^2 v_{1,2}^4} \tag{5.50}$$

Analogous to the 90° case, larger Coulomb cross sections may be derived for smaller deflection angles. Indeed, since collisions with smaller deflections and smaller changes in momentum occur much more often than the 90° variety, their summed effect may be quite large. As such, the *effective* Coulomb cross section is significantly larger than that for 90°collisions. A lower bound on the deflection angle to be considered in such a calculation may be derived from the limited range of the Coulomb potential in a plasma. This comes from a well-known shielding effect for a single charge embedded in an ensemble of charge carriers. The associated shielding distance is denoted the *Debye length* or the *Debye-Hückel radius*

$$l_D = \sqrt{\varepsilon_0 \ k \ T_e/n \ e^2} \tag{5.51}$$

Since charged particles that are further apart than the Debye length do not 'notice' each other any more, this distance serves to limit the magnitude of the Coulomb cross section. At the same time the Debye length defines that scale length, above which a plasma may be considered to be a quasi-neutral gas.

Accounting for all collisions out to the impact parameter corresponding to the Debye length, one would obtain a collision cross section of the order of

$$(\sigma_{1,2}^{Cb})_{scatter} = \pi \, l_D^2$$

Considering the ionospheric F region as an example ($n \simeq 5 \cdot 10^{11}$ m^{-3}, $T_e \simeq 1500$ K), we find that the Debye length is of the order of a few millimeters and the total cross section would be more than 10^{-5} m^2. Such a large cross section, however, is unrealistic and completely inconsistent with the observed diffusion and heat conduction processes in the ionosphere. An interaction cross section compatible with these effects is obtained only if it is based on the *momentum transfer*, the actually important process of each collisional event, rather than on the deflection of the impacting particle. An important difference with momentum transfer is that it decreases much faster with increasing impact parameter than the deflection angle (see, for example, Section 2.3.4). Weighting all possible collisions according to their momentum transfer and considering all impact parameters out to the maximum interaction distance l_D, one obtains the following mean collision cross section, also denoted the *momentum transfer cross section*

$$(\sigma_{1,2}^{Cb})_{momentum} = 4 \ \ln \Lambda \ (\sigma_{1,2}^{Cb})_{90} = \frac{e^4 \ln \Lambda}{4\pi \ \varepsilon_0^2 \ m_{1,2}^2 \ v_{1,2}^4} \tag{5.52}$$

where $\ln \Lambda$ is the so-called *Coulomb logarithm*

$$\ln \Lambda = \ln \left(12\pi \, \varepsilon_0 \, k \, T_e \, l_D/e^2\right) \tag{5.53}$$

The Coulomb logarithm is only weakly dependent on the density and temperature of the given particle population. Again using the ionosphere as an example ($n \simeq 5\cdot10^{11}\,\mathrm{m}^{-3}$, $T_e \simeq 1500\,\mathrm{K}$), one obtains a value $\ln \Lambda \simeq 14$. For the thermal plasma of the inner magnetosphere ($n \simeq 5 \cdot 10^8\,\mathrm{m}^{-3}$, $T_e \simeq 5000\,\mathrm{K}$), this becomes $\ln \Lambda \simeq 19$, and for the solar wind at 1 AU ($n \simeq 6\cdot10^6\,\mathrm{m}^{-3}$, $T_e = 10^5\,\mathrm{K}$), Eq. (5.53) yields $\ln \Lambda \simeq 26$. The mean momentum transfer cross section is therefore 50 to 100 times larger than that for 90° deflections, but still many orders of magnitude smaller than the total deflection cross section $(\sigma_{1,2}^{Cb})_{scatter}$ associated with the Debye length. If the collision partners are charge carrier gases that both conform to a Maxwell velocity distribution, the corresponding mean momentum transfer cross section may be estimated upon inserting Eq. (2.9) into Eq. (5.52), $v_{1,2} \to \bar{c}_{1,2}$. A more precise calculation of the mean value yields the following slightly smaller value

$$\left(\sigma_{1,2}^{Cb}\right)_{momentum}^{Maxwell} = \frac{1}{32\pi} \frac{e^4 \ln \Lambda}{\varepsilon_0^2 (kT_{1,2})^2} \tag{5.54}$$

This expression is valid, as assumed from the beginning, for singly ionized ions. For simplicity, we dispose of all but the most necessary indices in the following. Some important special cases are

$$\sigma_{e,i} \simeq \sigma_{i,e} \simeq \sigma_{e,e} = \frac{1}{32\pi} \frac{e^4 \ln \Lambda}{\varepsilon_0^2 (kT_e)^2}, \qquad \sigma_{i,i} = \frac{1}{32\pi} \frac{e^4 \ln \Lambda}{\varepsilon_0^2 (kT_i)^2} \tag{5.55}$$

For temperatures up to several 1000 K, it may be verified that these collision cross sections are many orders of magnitude larger than those for neutral gas particles. In the ionosphere for an ion temperature of $T_i = 1000$ K, for example, $\sigma_{i,i} \simeq 6 \cdot 10^{-15}\mathrm{m}^2 \gg \sigma_{n,n} \simeq 3 \cdot 10^{-19}\mathrm{m}^2$. The Coulomb cross section for energetic particles in the inner magnetosphere, however, takes on very small values from the dependence $\sigma \sim 1/E^2$, so that collisions occur increasingly less often with rising particle energy.

The collision cross section introduced above may be used to estimate collision frequencies, collision times, mean free paths and momentum transfer collision frequencies. Of particular interest for later applications is the time constant associated with Coulomb collisions. With Eqs. (2.12) and (5.54) one obtains the mean time between collisions in a thermal charge carrier gas as

$$\tau_{1,2} = \frac{1}{\nu_{1,2}} = \frac{1}{\sigma_{1,2}\, n_2\, \sqrt{8k\,T_{1,2}/\pi\, m_{1,2}}}$$

$$= \frac{16\pi^{3/2}\, \varepsilon_0^2\, k^{3/2}}{\sqrt{2}\, e^4} \frac{\sqrt{m_{1,2}}\; T_{1,2}^{3/2}}{\ln \Lambda\, n_2} \tag{5.56}$$

For the special case of proton-proton collisions ($1,2 \to \mathrm{p,p}$) and noting that $16\pi^{3/2}\varepsilon_0^2 k^{3/2}/\sqrt{2}e^4 \simeq 3.85 \cdot 10^{20}$ [SI units], one obtains

$$\tau_{\mathrm{p,p}}[\mathrm{s}] \simeq 10^7 \frac{(T_{\mathrm{p}}[\mathrm{K}])^{3/2}}{\ln \Lambda \; n_{\mathrm{p}}[\mathrm{m}^{-3}]} \simeq 30 \left(\frac{T_{\mathrm{p}}}{T_{\mathrm{e}}}\right)^{3/2} \tau_{\mathrm{e,p}} \tag{5.57}$$

or, for the associated mean free path

$$l_{\mathrm{p,p}}[\mathrm{m}] \simeq 1.6 \cdot 10^9 \frac{(T_{\mathrm{p}}[\mathrm{K}])^2}{\ln \Lambda \; n_{\mathrm{p}}[\mathrm{m}^{-3}]} \tag{5.58}$$

where we have made use of Eq. (2.13). Analogously, the frictional frequency between charge carrier gases, a quantity used in many applications, is given by

$$\nu_{1,2}^* = \frac{4}{3} \frac{m_2}{m_1 + m_2} \frac{1}{\tau_{1,2}} = \frac{1}{\sqrt{72\pi^3}} \frac{e^4}{\varepsilon_0^2 \, k^{3/2}} \sqrt{\frac{m_2}{m_1(m_1 + m_2)}} \frac{\ln \Lambda \; n_2}{T_{1,2}^{3/2}} \tag{5.59}$$

see Eq. (2.55), and for the special case of the frictional frequency between electrons and ions

$$\nu_{\mathrm{e,i}}^* = \frac{1}{\sqrt{72\pi^3}} \frac{e^4}{\varepsilon_0^2 \sqrt{m_{\mathrm{e}}} \, k^{3/2}} \frac{\ln \Lambda \; n}{T_{\mathrm{e}}^{3/2}} \simeq 3.6 \cdot 10^{-6} \frac{\ln \Lambda \; n}{(T_{\mathrm{e}})^{3/2}} \tag{5.60}$$

where the convenient form on the right is valid for SI units.

5.4 Particle Populations in the Inner Magnetosphere

The dipolar configuration of the Earth's magnetic field acts like a huge magnetic bottle, within which many charged particles are confined. These are ordered according to their energy into either the *radiation belt*, the *ring current*, or the *plasmasphere*. The essential properties of these particle populations are summarized in Table 5.2. Some explanatory remarks to this table should be noted:

- The confinement regions of these three particle groups partially overlap
- Only for the collision dominated plasmasphere is it sensible to state a value for the temperature
- Values for the flux and density of bouncing particles will change along a given magnetic field line
- The numerical entries should only be regarded as 'order of magnitude' values

The β^*-parameter listed in Table 5.2 represents a measure for the ratio of kinetic to magnetic energy density

$$\beta^* = \frac{p + p_d}{p_{\mathcal{B}}} \sim \frac{E_{kinetic}^*}{E_{magnetic}^*} \tag{5.61}$$

Table 5.2. Particle populations in the inner magnetosphere

Quantity		Particle Population		
		Radiation Belt	Ring Current	Plasmasphere
Energy	Ions	1 – 100 MeV	1 – 200 keV	< 1 eV ($\simeq$ 5000 K)
	Electrons	50 keV – 10 MeV	< 10 keV	
L shell		$1.2 < L < 2.5$	$3 < L < 6$	$1.2 < L < 5$
Field Line Footpoints		low and middle latitudes	middle and higher latitudes	low and middle latitudes
Particle Density/Flux		H^+ (50 MeV) $< 10^8$ m^{-2}s^{-1}	$\lesssim 10^6$ m^{-3}	$> 10^8$ m^{-3}
Composition		H^+, e^-	H^+, O^+, He^+, e^-	H^+, e^-
Particle Motion		gyration bouncing drift	gyration bouncing drift	gyration corotation
β^*-Parameter		$\ll 1$	< 1	$\ll 1$
Source Region			plasma sheet, ionosphere	ionosphere
Formation Process		CRAND	particle transport & acceleration	charge exchange & transport
Loss Region		upper atmosphere	interplanetary space, upper atmosphere	ionosphere, magnetosphere
Loss Process		deceleration, pitch angle diffusion into loss cone	charge exchange, pitch angle diffusion into loss cone	transport & charge exchange, convection
Significance		radiation damages	magnetic field disturbances	plasma reservoir for ionosphere

where p, p_d and p_B denote the thermodynamic plasma pressure, the dynamic plasma pressure and the magnetic pressure, respectively, from Eqs. (2.23), (2.24) and (6.76). Since p is a measure of the energy density associated with the thermal motion of the plasma particles, p_d a measure of the energy density of the plasma flow and $p_B = \mathcal{B}^2/2\mu_0$ the energy density of the magnetic field, β^* is a quotient that measures the relative importance of kinetic and magnetic energy density. As such, the value of this parameter determines whether the magnetic field must conform to the given particle motion or, vice versa, the particle motion must adapt itself to the given magnetic field structure. The latter is the case in the inner magnetosphere. Moreover, in the absence of a dynamic pressure, $\beta^* \simeq p/p_B$, and this corresponds to the familiar β *parameter* or *plasma β* from fusion physics. The properties of the various particle populations presented in Table 5.2 are described in more detail below.

5.4.1 Radiation Belt

The radiation belt is understood to mean the high energy particle population confined to the inner magnetosphere. The name is taken from the 'radiation' measurement instruments (Geiger-Müller counters), with which it was first detected. An alternative name for this particle population is the *Van Allen belt*.

The story behind the discovery of the radiation belt illustrates the beginning of space research with artificial satellites in a dramatic way. A Geiger counter intended for the measurement of cosmic rays had been placed on board the first American satellite (EXPLORER 1, launched in late January 1958) by the Van Allen research team. In those days the counting rates were recorded in real time when the satellite passed over the ground stations. It was observed that the expected counting rates were recorded at low altitudes (a few 100 km), but surprisingly went to zero at greater heights (> 2000 km). The scientists involved in the experiment found themselves in the embarrassing predicament of having to explain the absence of cosmic rays at larger heights.

The second American satellite (EXPLORER 2, launched in early March 1958) had another Geiger counter on board, but fell shortly after launch into the Atlantic. The third American satellite (EXPLORER 3, launched in late March 1958) was the first spaceflight vehicle to have a tape recorder on board, thereby finally enabling storage of the counting rate measurements during the entire orbit. It was now learned that the counting rate instrumentation had reset to zero due to saturation at large heights and that, in reality, extraordinarily high intensities were measured there. At first it was thought that a new type of radiation had been discovered (as documented by the naming episode mentioned above), and rumors were being floated that outer space was radioactive. The measurements were later interpreted correctly as highly energetic particles stored in the magnetic field of the Earth.

It is remarkable that the second Soviet satellite (SPUTNIK 2, launched in November 1957) had two Geiger counters on board. This satellite just happened to reach its closest approach to the Earth (its perigee) near its Soviet ground stations, i.e. when it was below the radiation belt. Measurements at greater heights had been recorded in Australia, but could not be decoded due to a lack of cooperation. These circumstances prevented the Russian scientists from claiming the first important discovery with the help of an artificial satellite.

Today we know that energetic particles populate the entire storage region of the inner magnetosphere, the position of maximum particle fluxes being dependent on the particle type and energy, as well as the state of the magnetosphere. The radiation belt is understood in the following as the high energy component of this particle population, the lower bound of which is taken here to be roughly 1 MeV for protons and 50 keV for electrons. As an example, Fig. 5.25a shows a cross section through the spatial distribution of the maximum omnidirectional proton flux at energies above 4 and 50 MeV, respectively. The flux maximum lies in the (magnetic) equatorial plane at about $L = 1.8$ ($\simeq 5000$ km height) for 4 MeV particles and shifts downward to roughly 3000 km for 50 MeV particles. The maximum equatorial fluxes reach values of 10^{10} m^{-2}s^{-1} (4 MeV particles) and 10^8 m^{-2}s^{-1} (50 MeV particles). In contrast, maximum flux densities for electrons at energies above 1.6 MeV are observed between $L = 3$ and $L = 4$.

The properties of the highest energy radiation belt particles are relatively stable, implying time invariant source and loss processes. Of course, the mere existence of highly energetic particles deep within the inner magnetosphere is a remarkable phenomenon. The most popular candidate for the main source of the high energy component of the radiation belt particles is the so-called CRAND (= Cosmic Ray Albedo Neutron Decay) process sketched in Fig. 5.26. High energy particles of cosmic origin (generally known under the designation *cosmic rays*, see Section 6.6.1) collide with the nuclei of atmospheric gases in the dense atmosphere. A fraction of the energetic neutrons released by these interactions (a 5 GeV proton produces ca. 7 free neutrons) diffuses in the direction of the magnetosphere (neutron albedo) and decays there in the radiation belt

$$\mathrm{n} \longrightarrow \mathrm{p} + \mathrm{e} + \bar{\nu} \qquad (\bar{\nu} = \text{antineutrino})$$

The electron and proton decay products are caught up in the local magnetic field. Radiation belt particles have also been produced artificially from the explosion of atomic bombs in the atmosphere, e.g. the ARGUS (1958) and STARFISH (1962) experiments.

Once produced, radiation belt particles have a high life expectancy because of their very small interaction cross sections. For example, a 20 MeV proton at 2000 km altitude could remain in the belt for one year, during which it freely gyrates, bounces and drifts in the Earth's magnetic field. The time constants associated with these motions are given in Table 5.1. Note that, in spite of the high drift velocity (around the world in 2 minutes!), the resultant current in this loop is negligible because of the low particle density.

The most efficient loss process for radiation belt particles is removal by collisions with the neutral and ionized gas particles of the upper atmosphere. These continually decelerate the particles, eventually scattering them into the loss cone.

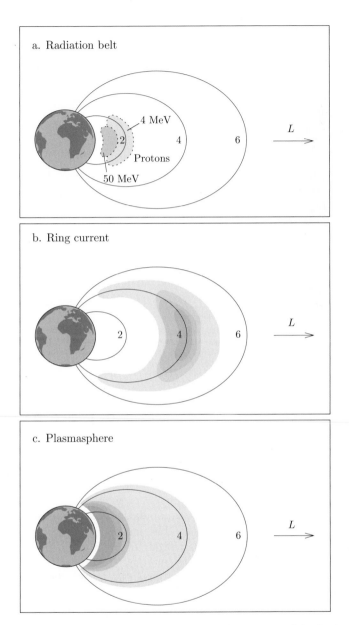

Fig. 5.25. Spatial distribution of the particle populations stored in the inner magnetosphere

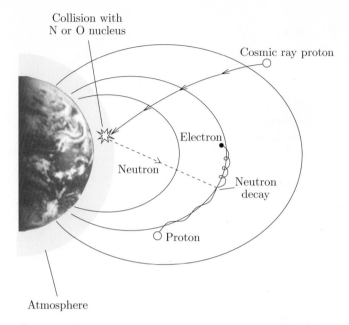

Fig. 5.26. A source of the radiation belt particles (not to scale, adapted from Hess, 1968)

Radiation belt particles are of concern because of the damage they can wreak on satellite electronics and on astronauts. For example, it is suspected that a large number of satellite experiment failures have been caused by radiation belt particles.

5.4.2 Ring Current

The ring current particles represent a population of medium energy that is confined to the inner magnetosphere. The ion energies are typically 1–200 keV, that of the electrons about an order of magnitude lower. The peak particle fluxes and densities at these energies are observed in the middle storage region between $L = 3$ and $L = 6$ (see Fig. 5.25b). The motion of the ring current particles consists of gyration, bounce motion and a compromised azimuthal drift. The drift is compromised because the typical life expectancy of ring current particles is only hours to days and thus of the same order of magnitude as the drift period.

A salient feature of the ring current particles is their great dynamic range. The density can increase during disturbed conditions within a few hours by a factor of 10, in some energy ranges by a factor of 100 above its original value. Simultaneously, the centroid of the ring current shifts toward the Earth. While typical densities during such events can reach a few 10^6 particles per cubic meter, the inner edge of the ring current descends to a distance near $L \simeq 2.5$. This corresponds to field line footpoints at a latitude near $50°$.

Whereas the undisturbed particle population consists mostly of protons ($> 90\%$), a strong increase in the density of atomic oxygen ions (sometimes to more than 50%) occurs during magnetically disturbed conditions. This implies that the ionosphere is one of the sources of the disturbed ring current. These oxygen ions are presumably first accelerated at high latitudes and then transported to the magnetotail plasma sheet (discussed later), before they drift together with particles of interplanetary origin into the ring current region. The exact mechanics of this acceleration and transport process, an area of intensive research, are still insufficiently understood.

Among the major loss processes for ring current particles are charge exchange collisions. The principal charge exchange partners are neutral exospheric hydrogen atoms that lose their electron in the process. For the case of ring current protons, for example, we have

$$\underline{H}^+ + H \rightarrow \underline{H} + H^+ \tag{5.62}$$

where the energetic particle is identified by underlining. The energetic neutral gas particles created by this process are no longer bound by the Earth's magnetic field and usually escape to interplanetary space. A smaller number is injected into the Earth's upper atmosphere. These neutralized particles and their energy are therefore lost to the ring current and are replaced by thermal protons with energies corresponding to those of the exospheric hydrogen atoms from which they originated. Denoting the cross section of the charge exchange reaction by $\sigma_{s,H}^{CE}$, the lifetime of the ring current particles is given by

$$\tau_{s,H}^{CE} = 1/\nu_{s,H}^{CE} = 1/\sigma_{s,H}^{CE} \, n_H \, v_s \tag{5.63}$$

where s stands for the particular ring current ion species and the relative velocity corresponds, to a very good approximation, to the velocity v_s of the energetic ring current particle. As an example, consider a 20 keV proton orbiting at an average distance of $L = 3$. With $\sigma_{H^+,H}^{CE}$ (20 keV) $\simeq 10^{-19}$ m^2 and $n_H(L = 3) \simeq 7 \cdot 10^8$ m^{-3} (see Fig. 2.36), this yields an average lifetime of 2 hours for this particle. Particles of higher energy have a significantly longer lifetime because of their smaller interaction cross section. Ring current particles incur further losses from scattering into the loss cone caused by collisions and electromagnetic waves, and, especially during the initial phase of recovery, from drifting out of the ring current region (see Section 8.1.2).

The observed particle density increase during disturbed conditions leads to a distinct enhancement of the electric current associated with the drift motion. This *ring current* is such a prominent characteristic of the inner magnetosphere that its name has been bestowed upon its causative particle population. We can model the ring current approximately, as sketched in Fig. 5.27, by a current loop in the equatorial plane at a certain distance from

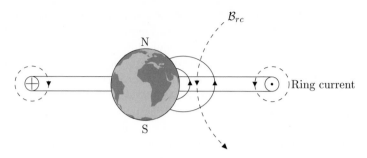

Fig. 5.27. The westward directed magnetospheric ring current as a source of negative magnetic field disturbances at low latitudes

the Earth. Since the magnetic field in the center of a current ring of radius R is given by

$$\mathcal{B} = \mu_0 \mathcal{I} / 2R \tag{5.64}$$

we obtain the following good approximation for the magnetic field disturbance at the Earth's surface

$$\Delta \mathcal{B}_{rc}(R_E) \lesssim \Delta \mathcal{B}_{rc}(r = 0) = -\frac{\mu_0}{2 \, L \, R_E} \frac{\Delta \mathcal{I}_{rc}}{} \tag{5.65}$$

where r is the geocentric distance and $\Delta \mathcal{I}_{rc}$ is the disturbance-induced enhancement of the ring current intensity. Accordingly, the intensity of a ring current at $L = 5$ would need to be increased by 5 MA to produce a magnetic field disturbance of $\Delta \mathcal{B}_{rc} = -100$ nT.

The energy contained in such a *storm ring current* may be estimated as follows. Assuming that all injected ring current particles ΔN have the same energy E and are concentrated in the equatorial plane at a distance $r = L \, R_E$, the drift current enhancement is given by

$$\Delta \mathcal{I}_{rc} = \frac{\Delta N \, |q|}{\tau_D} \simeq \frac{3 \, L}{2\pi \, R_E^2 \, \mathcal{B}_{00}} \Delta E_{rc} \tag{5.66}$$

where we have made use of Eq. (5.46) and $\Delta E_{rc} = E \, \Delta N$ is the increase in ring current energy. Note that we only need to account for the ion component in computing the current because the electron energy is much lower than the ion energy, and $1/\tau_D \sim E$. Inserting Eq. (5.66) into Eq. (5.65), we obtain the following relation between the magnetic field disturbance and the energy enhancement of the ring current

$$\Delta \mathcal{B}_{rc} = -\frac{3 \, \mu_0 \Delta E_{rc}}{4\pi \, R_E^3 \, \mathcal{B}_{00}} = -\frac{3 \Delta E_{rc}}{\mathcal{M}_E} \tag{5.67}$$

Here we have simplified the expression by introducing the dipole moment of the geomagnetic field $\mathcal{M}_E = 4\pi R_E^3 \, \mathcal{B}_{00}/\mu_0$ (see, for example, Fig. 5.6). It follows that a magnetic field disturbance of -100 nT requires an infusion of energy in the amount of $\Delta E_{rc} \simeq 2.6 \cdot 10^{15}$ J.

This estimate, however, neglects the magnetic field disturbance produced by the gyromotion of ring current particles. Recalling Eq. (5.29), the magnetic moment of a gyrating particle is given by

$$\vec{\mathcal{M}}_g = -\frac{E_\perp \, \vec{\mathcal{B}}}{\mathcal{B}^2}$$

The summed magnetic field in the equatorial plane at a given distance r' from ΔN gyrating particles in the same plane is thus given by (see Eq. (5.6) for $\vartheta = 90°$)

$$\Delta \mathcal{B}_g = \frac{\Delta N \, \mu_0 \, \mathcal{M}_g}{4\pi \, r'^3} = \frac{\mu_0 \, \Delta E_{rc}}{4\pi \, r'^3 \, \mathcal{B}(L)}$$

where $\mathcal{B}(L)$ is the magnetic field at the position of the ring current. The disturbance magnetic field at the center of the Earth (note that the magnetic disturbance from the ring current was also calculated for this point), with $r' = L \, R_E$ and $\mathcal{B} = \mathcal{B}_{00}/L^3$, is thus given by

$$\Delta \mathcal{B}_g \simeq \frac{\mu_0 \, \Delta E_{rc}}{4\pi \, R_E^3 \, \mathcal{B}_{00}} = \frac{\Delta E_{rc}}{\mathcal{M}_E}$$

The magnetic field is thus strengthened on the Earth's surface (in contrast to the situation in the ring current region itself). The total disturbance is thus found to be

$$\Delta \mathcal{B} \simeq \Delta \mathcal{B}_{rc} + \Delta \mathcal{B}_g = -2\Delta E_{rc}/\mathcal{M}_E \qquad (5.68)$$

As a result, the energy input required for a magnetic disturbance of $\Delta \mathcal{B} = -100$ nT is raised by 50% to roughly $4 \cdot 10^{15}$ J.

Equation (5.68), or a slightly modified version of it, has become known as the *Dessler-Parker-Sckopke relation*. It represents a remarkably simple relation between a magnetic disturbance observed on Earth and the energy content of the ring current. Even more amazing is that it is also valid for realistic ring current properties, i.e. for realistic density, pitch angle, and energy distributions. This is especially surprising because such distributions are automatically connected with additional curvature drift and gyration-induced ring currents. For example, as graphically illustrated in Fig. 5.28a, the gyromotion about curved field lines leads to currents perpendicular to the plane of curvature. The concentration of the currents associated with the gyration on the inner side of the curved guiding center field line, as well as the simultaneous rarefaction on the outer side, leads to an effective current flow in the direction of the ion motion between neighboring field lines, $\vec{\mathcal{I}}_{gc} \sim -\vec{\mathcal{B}} \times \nabla_\perp \mathcal{B}/\mathcal{B}^2 \sim 1/\rho_c$. This ring current is therefore oppositely directed to the drift current (see Fig. 5.28b).

An additional gyration-induced ring current component is produced by the density gradient in the ring current population. This effect is sketched in

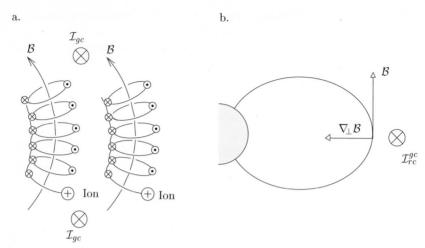

Fig. 5.28. (a) Gyration-induced magnetization currents in magnetic fields with curved field lines and (b) the associated magnetospheric ring current

Fig. 5.29a. The gyration current from particles in the region with a higher density is not fully compensated by the opposing gyration current from particles in the neighboring lower density region. This leads to a net current, perpendicular to both magnetic field and density gradient, given by $\vec{\mathcal{I}}_{gd} \sim \vec{B} \times \nabla n$. In the Earth's dipole field, this current is directed opposite to the drift-associated current on the inner edge of the ring current (positive density gradient). The effect is opposite on the outer edge (negative density gradient) where this current strengthens the drift-induced ring current (see Fig. 5.29b).

The total current is thus composed of the sum of these various drift and gyration-induced components. It is found that the *effective* ring current is directed to the west, in agreement with the disturbance-associated decrease

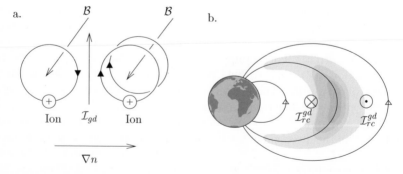

Fig. 5.29. (a) Gyration-induced magnetization currents in plasmas with density gradients and (b) associated magnetospheric ring currents

in magnetic field strength observed on the Earth's surface. Moreover, to a first approximation, this effective ring current corresponds to the magnetization current $\mathcal{I}_{rc}^{gd}$ flowing on the outer edge of the ring current population (see Fig. 5.29b) – certainly a remarkable result.

5.4.3 Plasmasphere

The plasmasphere is a region of relative dense ($n \gtrsim 10^8$ m^{-3}) and cool ($E \lesssim 1$ eV) plasma in the inner magnetosphere. It is nothing more than the continuation of the ionosphere into the magnetosphere; see Fig. 5.25c. The boundary between both regions (designated here as the *plasmasphere base*) is defined as the transition from atomic oxygen to atomic hydrogen as the primary ion constituent. This transition occurs at an altitude between roughly 500 and 2000 km, depending on the particular geophysical conditions, and is characterized by a break in the density profile due to the very different scale heights of the two main constituents, $H_{\mathrm{H^+}} = 16\,H_{\mathrm{O^+}}$; see Fig. 5.30. Due to its composition, the plasmasphere is also referred to as the *protonosphere*. Its outer boundary is a more or less sharp cutoff in the ionization density called the *plasmapause*. The location of this cutoff is variable, depending primarily on the degree of disturbance of the magnetosphere. For average conditions, the plasmapause is delineated by magnetic field lines with apex distances of about $L = 4$–6 and footpoints at magnetic latitudes near 60–65°.

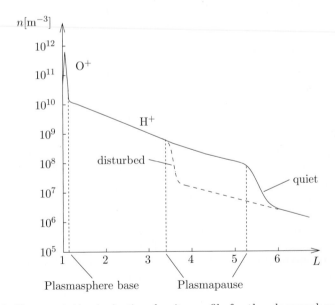

Fig. 5.30. Representative ionization density profile for the plasmasphere at equatorial latitudes

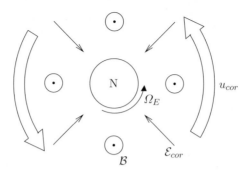

Fig. 5.31. Corotation in the equatorial plane, viewed from the north

We would expect that the motion of plasmaspheric particles is also characterized by gyration, bounce and drift in the geomagnetic field. The corresponding time constants for protons are of the order of $\tau_B^p \simeq 0.1$ s, $\tau_O^p \simeq 2$ hours, $\tau_D^p > 40$ years (see Table 5.1). Comparing these periods with the corresponding Coulomb collision times, it is found that the bounce motion (not to mention the drift motion!), is much too slow to proceed without interruption. According to Eq. (5.57), one obtains a proton collision time of less than a half-hour even in the outermost regions of the plasmasphere ($n \simeq 10^8 \, \mathrm{m}^{-3}$, $T_p \simeq 5000$ K). In other words, even in this tenuously populated region, a proton suffers about four collisions per bounce period, and considerably more if its mirror points lie in the denser plasmasphere. Its motion is thus characterized by a fully established gyration, but a strongly disturbed oscillation between mirror points. Moreover, because of their modest energy, the particles are tightly bound to their guiding center field lines. For example, with $T = 5000$ K, $L = 3$ and $\langle \sin \alpha_0 \rangle \simeq 0.6$, the corresponding equatorial gyroradii are found to be $r_B^p \simeq 60$ m and $r_B^e \simeq 1.5$ m, and these are even considerably smaller near the plasmasphere base.

Surprisingly, observations show that the plasmasphere rotates in its entirety with the Earth. Collisions with neutral gas particles cannot be responsible for this phenomenon because such frictional forces would result in radial drifting. Gradient and curvature drifting can also be quickly rejected. They do not produce the required ambipolar drift, and, as we have seen, the drift speed is simply much too slow. The only remaining explanation is that the corotation represents an ambipolar $\vec{\mathcal{E}} \times \vec{\mathcal{B}}$ drift. For an external, nonrotating observer, it follows that an electric field must exist in the plasmasphere that forces the plasma to corotate with the Earth. Furthermore, this *corotation field* must be directed toward the Earth; see Fig. 5.31. The magnitude of this field is found by equating the corotation speed $u_{cor} = \Omega_E \, r \cos \varphi_g$ to the drift velocity $u_D^{\mathcal{E}} = \mathcal{E}/\mathcal{B}$

$$\mathcal{E}_{cor} = \Omega_E \, r \, \cos \varphi_g \, \mathcal{B} \qquad (5.69)$$

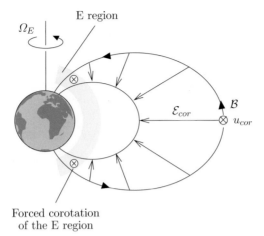

Fig. 5.32. Meridional section through the ionosphere/plasmasphere showing the formation of the corotation field. Note that the length of the vectors does not reflect the strength of the electric field and that the electric fields are everywhere perpendicular to the magnetic field (contrary to this schematized diagram).

where Ω_E is the angular velocity of the Earth's rotation and φ_g is the geographic latitude. Assuming for simplicity that the rotation and dipole axes are aligned, we find for the equatorial plane that

$$\mathcal{E}_{cor}(\varphi_g = 0) \simeq \Omega_E \; R_E \; \mathcal{B}_{00}/L^2 \tag{5.70}$$

or in a ready-to-use form

$$\mathcal{E}_{cor}(\varphi_g = 0)[\mathrm{mV/m}] \simeq 14/L^2 \tag{5.71}$$

This field is produced in the E-region of the ionosphere, more precisely in its *dynamo layer* (see Section 8.1.1). The neutral atmosphere is dense enough at this level to impose its prevailing corotation onto the ionosphere by means of frictional forces. The resulting forced motion of the conducting ionospheric plasma in the geomagnetic field produces an electric dynamo field of magnitude $\vec{\mathcal{E}} = -\vec{u}_{cor} \times \vec{\mathcal{B}}$, which is conveyed along magnetic field lines – as along good conducting copper wires – into the plasmasphere and establishes the corotation; see Fig. 5.32. For an observer rotating with the Earth, of course, the corotation field vanishes. From this vantage point, the plasma is at rest with respect to the magnetic field, both in the E-region as well as in the plasmasphere. Further information on this topic may be found in Sections 6.2.6 and 8.1.1.

It is often assumed that the extent of the plasmasphere is limited by the so-called *electric convection field*. This is understood to be a large-scale electric field that is applied to the magnetosphere by its interaction with the solar wind. To a first approximation it is directed uniformly from the

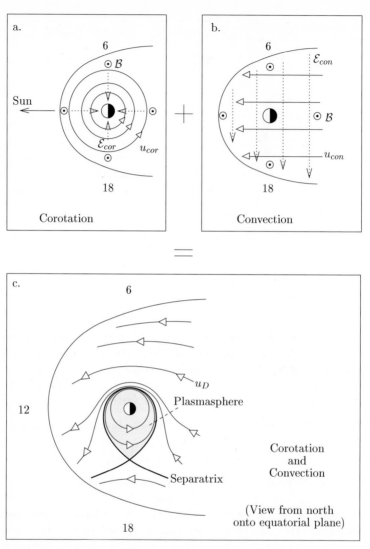

Fig. 5.33. $\vec{\mathcal{E}} \times \vec{B}$ drift of *thermal* plasma in the magnetosphere

dawn to the dusk sector and, depending on the degree of disturbance, has an intensity of $0.2 - 1$ mV/m; see Fig. 5.33b and Section 7.6. This field is superimposed on the Earth's corotation field, which decreases with geocentric distance, and leads to a total drift pattern as sketched in Fig. 5.33c. Note that storage and thus accumulation of plasma is possible only in the region of closed drift trajectories generated by the corotation field (shaded area). Since the convection field dominates outside this region, dense plasma coming from the ionosphere is transported in the direction of the outer magnetosphere. It was thus suggestive to associate the last closed drift trajectory (denoted the separatrix in Fig. 5.33c) with the location of the plasmapause. In the dusk sector (ca. 18 hours LT), the distance to the separatrix is determined by the condition $\mathcal{E}_{cor} = \mathcal{E}_{con}$, as these fields are directly opposed there. With Eq. (5.71) this requires that

$$\text{dusk sector:} \qquad L_{separatrix} \simeq \sqrt{14/\mathcal{E}_{con}[\text{mV/m}]} \qquad (5.72)$$

This is in sufficiently good agreement with the observed location of the plasmapause in this local time sector. Note that today one believes that the extent of the plasmasphere is not only controlled by drift effects, but also by the spatial and temporal variability of plasma replenishment and by plasma instabilities.

The only plausible source for plasmaspheric particles is the ionosphere. This is because local production of hydrogen ions by EUV radiation is rather inefficient due to the extremely long time constant of this process. Taking, for example, a distance $L = 3$, we obtain $\tau_q = n_{H^+}/q_{H^+} = n_{H^+}/J_{H^+}n_H > 50$ d (see Fig. 2.36, Fig. 5.30 and Table 4.2). Compared with this, the transport time constant is quite short. For example, a thermal hydrogen ion needs only about 30 minutes ($\simeq \tau_o/4$), to arrive in the apex region of the field lines at $L = 3$. Even after accounting for collisions, transport effects will dominate. A decisive role in this scenario is played by the charge exchange process $O^+ + H \rightarrow H^+ + O$ near the plasmasphere base. A large fraction of the hydrogen ions produced this way will move upward and populate the plasmasphere.

Losses include the return of plasma to the ionosphere along the magnetic field lines via the reaction $H^+ + O \rightarrow H + O^+$ as well as a 'peeling off' of the outermost layer of the plasmasphere. The former occurs, for example, when the O^+ density at the plasmasphere base is reduced at night or during a negative ionospheric storm (see Section 8.5.1). The second case arises when the convection field increases during a magnetospheric disturbance and the separatrix shifts toward Earth (see Eq. (5.72) and Fig. 5.30).

When determining the density profile of the plasmasphere, it should be noted that transport processes dominate and that the number of Coulomb collisions is generally sufficient to establish a Maxwellian velocity distribution. An aerostatic approach is thus justified in regions near the Earth. Nevertheless, the centrifugal force and magnetic field line tilt should be considered in addition to the pressure gradient force and the height-dependent gravitational field of Earth.

5.5 The Distant Geomagnetic Field

Were the Earth and its magnetic field located in empty space, the dipole structure of the geomagnetic field as sketched in Fig. 5.5 would be preserved

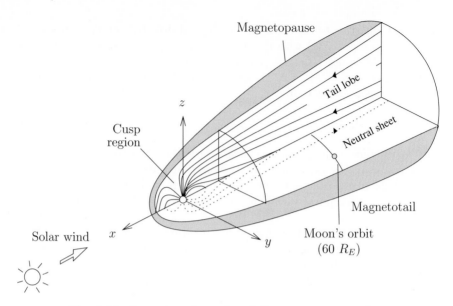

Fig. 5.34. General configuration of the outer magnetosphere

even at large distances. This, as we know today, is not the case. Instead, the Earth is embedded within the 'solar wind', a flow of particles continuously emitted by the Sun. Furthermore, the intrinsic magnetic field of the interplanetary medium interacts with the Earth's dipole field. Both of these, but particularly the solar wind, lead to a fundamental modification of the distant geomagnetic field. The nature of this modification and the current system responsible for it are described in the following.

5.5.1 Configuration and Classification

The most obvious result of the interaction with the interplanetary medium is a confinement of the geomagnetic field to the finite volume called the *magnetosphere*; see Fig. 5.34. The outer boundary of this volume, following our previous nomenclature, is called the *magnetopause*. The magnetosphere assumes an ellipsoidal shape on the sunward side, the geocentric distance to the subsolar point on the magnetopause being about 10 Earth radii (10 R_E $\simeq$ 64 000 km). Fluctuations in this distance of a few Earth radii are observed, depending upon the properties of the interplanetary medium and particularly upon the dynamic pressure of the solar wind. The magnetic field strength is observed to decrease relatively slowly near the magnetopause and still attains values of 40 to 50 nT at the subsolar point. This corresponds to about twice the nominal dipole field strength at this distance. Also noteworthy are the intensity minima in the regions alternately denoted the *cusp* or the *cleft*. These are the separation regions of magnetic field lines that extend to

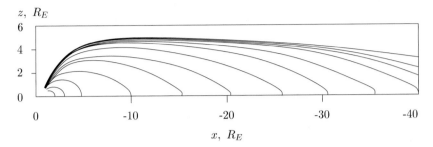

Fig. 5.35. Closed magnetic field lines in the noon-midnight meridian plane of the magnetotail (adapted from Larson and Kaufmann, 1996)

different parts of the magnetosphere, for example, separating field lines that close on the frontside of the magnetosphere from those that extend into space on the Earth's nightside.

The nightside magnetosphere is greatly extended and assumes a cylindrical shape. Because of its similarity with the tail of a comet, this region is referred to as the *magnetotail*. Although the length of the magnetotail is not well established and is probably quite variable, it does extend considerably beyond the Moon's orbit ($\simeq 60\ R_E$). The radius of the magnetotail tends to increase with geocentric distance and reaches values between 25 and 30 R_E at distances near 200 R_E.

A *geocentric solar magnetospheric (GSM)* coordinate system is frequently used to describe the magnetospheric topology. As indicated in Fig. 5.34, the x-axis of this system points to the Sun along the Earth-Sun line. The y-axis is defined by the equation $\hat{y} = \hat{n}_{DP} \times \hat{x}/|\hat{n}_{DP} \times \hat{x}|$, where the unit vector $\hat{n}_{DP}$ points along the geomagnetic dipole axis to the north. As such, the y coordinate measures (perpendicular) distance from the plane defined by the dipole axis and the Earth-Sun line. The z-axis completes the Cartesian coordinate system in the usual way, $\hat{z} = \hat{x} \times \hat{y}$.

As intimated in Fig. 5.34 and more clearly documented in Fig. 5.35, a considerable part of the magnetotail consists of extremely stretched dipole field lines. A polarity reversal in the magnetic field component parallel to the Sun-Earth line takes place across the mid-plane of the magnetotail. The location of this reversal, where $\mathcal{B}_x$ must vanish, is denoted the *neutral sheet*. This surface does contain a weak northward directed component of magnetic field from the extended geodipole, which decreases only very slowly with geocentric distance. Typical values for $\mathcal{B}_z$ are 2 and 0.5 nT at distances of 30 and 200 R_E, respectively. Larger field strengths are measured in the *lobes* of the magnetotail. These are the regions between the neutral sheet and the magnetopause (more precisely, between the magnetotail plasma sheet and the magnetospheric boundary layer, to be discussed later), within which the $\mathcal{B}_x$ component of the magnetic field dominates. The magnetic field strength in

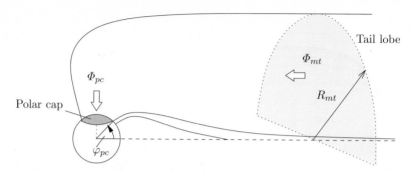

Fig. 5.36. Determining the low-latitude boundary of the polar cap

the lobes decreases even more slowly down the magnetotail, reaching values of $\mathcal{B}_x \simeq 20$ and 8 nT at distances of 30 and $200 \, R_E$, respectively.

The footpoint regions of the tail lobe field lines are of interest for later considerations. These are approximately circular regions around the magnetic poles, a shape that has given them the designation *polar caps*; see Fig. 5.36. The magnetic latitude of their equatorward boundary may be estimated as follows. Ignoring open magnetic field lines extending into interplanetary space (see Section 7.6.4), the magnetic flux of the polar caps, Φ_{pc}, must be equal to the magnetic flux at the beginning of the magnetotail, Φ_{mt}. Denoting the (magnetic) latitude by φ, the longitude by λ, a surface element of the Earth by $dA = R_E^2 \cos\varphi \, d\lambda \, d\varphi$, and the radial component of the geomagnetic field by $\mathcal{B}_r = 2\mathcal{B}_{00} \sin\varphi$, the magnetic flux of a polar cap is given by

$$\Phi_{pc} = \int_{pc} \mathcal{B}_r dA = 2\mathcal{B}_{00} R_E^2 \int_{\lambda=0}^{2\pi} \int_{\varphi_{pc}}^{\pi/2} \sin\varphi \cos\varphi \, d\varphi \, d\lambda$$

$$= 2\pi R_E^2 \mathcal{B}_{00} \cos^2 \varphi_{pc}$$

Conservation of magnetic flux requires that

$$\Phi_{pc} = 2\pi R_E^2 \, \mathcal{B}_{00} \cos^2 \varphi_{pc} \simeq \Phi_{mt} = \mathcal{B}_{mt} \, \pi R_{mt}^2/2$$

or

$$\varphi_{pc} \simeq \arccos \sqrt{\frac{\mathcal{B}_{mt} \, R_{mt}^2}{4 \, \mathcal{B}_{00} R_E^2}} \tag{5.73}$$

Since $\mathcal{B}_{mt} \simeq \mathcal{B}_x$ decreases very slowly with distance down the tail, our choice of just where the magnetotail begins is not especially critical. Taking the values $\mathcal{B}_x(x = -20 \, R_E) \simeq 25$ nT and $R_{mt}(x = -20 \, R_E) \simeq 18 \, R_E$, we calculate a value for the lower polar cap boundary of $\varphi_{pc} \simeq 75°$.

An upper bound for the length of the magnetotail may be calculated from similar reasoning. Assuming that all tail lobe (or, equivalently, all polar cap) magnetic field lines eventually close through the neutral sheet (i.e. we again neglect field line loss to interplanetary space), the polar cap flux must be equal to the magnetic flux through the neutral sheet, $\Phi_{ns} = 2R_{mt}L_{mt}\mathcal{B}_z$. An upper limit on the tail length is thus given by

$$L_{mt} < \frac{\Phi_{pc}}{2R_{mt}\mathcal{B}_z} \tag{5.74}$$

Using a mean tail radius of $R_{mt} \simeq 30R_E$ and a mean magnetic field strength for the normal component of $\mathcal{B}_z \simeq 0.2$ nT one obtains an upper limit on the magnetotail length of $L < 1000\ R_E$.

A heuristic explanation for the observed deformation of the terrestrial dipole field in the outer magnetosphere is presented in the following. By necessity, this must remain qualitative and incomplete as long as the general problem of the interaction between a planetary (or stellar) magnetosphere and an external magnetized plasma flow is not completely solved. Strongly simplified scenarios, however, enable us to understand some of the essential aspects of this interaction.

5.5.2 Dayside Magnetopause Currents

It is known that, in the stationary case, a given magnetic configuration can only be modified by superimposing additional magnetic fields. Moreover, it is clear that such additional fields in space can only be produced by electric currents. It is thus very instructive to identify those electric currents that are responsible for the deformation of the geomagnetic dipole field in the outer magnetosphere. These are almost exclusively *surface currents*, by which one cross section dimension is small with respect to the other two greatly extended dimensions. Figure 5.37 illustrates the topology and magnetic field of such a surface current. The magnetic field strength $\vec{\mathcal{B}}_{sc}$ of a plane, infinitely extended surface current is given by

$$\vec{\mathcal{B}}_{sc} = \frac{\mu_0}{2}\ \vec{\mathcal{I}}^* \times \hat{n} \tag{5.75}$$

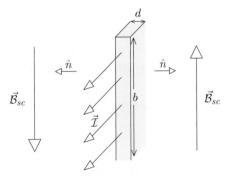

Fig. 5.37. Topology and magnetic field of a surface current (current sheet, $d \ll b$)

where $\vec{\mathcal{I}}^*$ is the surface current density

$$\vec{\mathcal{I}}^* = \vec{\mathcal{I}}/b \tag{5.76}$$

and $\hat{n}$ and $\vec{\mathcal{I}}$ denote the surface normal and the total current strength, respectively. The direction of the magnetic field is immediately obvious if one imagines the surface current to be composed of individual current threads (or long current-carrying wires): the ring-like magnetic field lines of these threads are superposed in such a way that their components normal to the surface cancel each other and only the components parallel to the surface (and perpendicular to the current) remain. This also provides the explanation for the opposing magnetic field directions on opposite sides of the surface current. The magnitude of the field strength may be obtained with the help of Ampère's law

$$\oint_{L(A)} \vec{\mathcal{B}} \, d\vec{l} = \mu_0 \mathcal{I}$$

see Eq. (A.98) with $d/dt \to 0$. Referring to the situation sketched in Fig. 5.37 and letting $d \ll b$, we obtain

$$b \, \mathcal{B}_{sc} + (-b)(-\mathcal{B}_{sc}) = \mu_0 \mathcal{I}$$

or

$$\mathcal{B}_{sc} = \mu_0 \mathcal{I}^*/2$$

in agreement with the relation given above.

The surface current responsible for the limited extent of the dayside magnetosphere is drawn in Fig. 5.38. As indicated in this cut through the noon-midnight meridian plane, the surface current points up out of the paper and

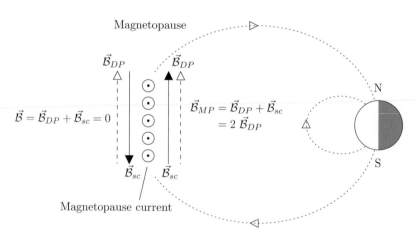

Fig. 5.38. Magnetopause current on the dayside of the magnetosphere (cut through the noon-midnight meridian)

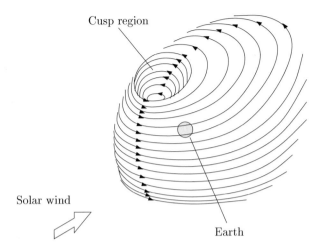

Fig. 5.39. Three-dimensional distribution of magnetopause currents on the dayside magnetopause (adapted from Olsen, 1982)

produces a magnetic field that doubles the original contribution from the dipole field just inside the magnetosphere and – as required – nullifies the magnetic field just outside the surface. According to Eq. (5.75), this *magnetopause current* (also called *Chapman-Ferraro current*) has a current density given by

$$\mathcal{I}_{MP}^* = 2\,\mathcal{B}_{DP}/\mu_0 = \mathcal{B}_{MP}/\mu_0 \qquad (5.77)$$

where $\mathcal{B}_{DP}$ is the unmodified geomagnetic dipole field strength and $\mathcal{B}_{MP}$ is the observed field strength just inside the magnetopause ($= 2\,\mathcal{B}_{DP}$).

The global, three-dimensional distribution of magnetopause currents is determined from the condition that the surface currents must flow in closed loops perpendicular to the magnetopause magnetic field. Figure 5.39 shows this distribution for the dayside magnetopause. Note the funnel-shaped current configuration around the magnetospheric cusp region.

Particle Reflection and Current Formation. Knowing how the magnetopause current system must be arranged to explain the shape of the dayside magnetosphere, we now turn to an explanation of its formation. For this purpose, we consider the interaction of solar wind particles (primarily protons and electrons) with the Earth's magnetic field. Figure 5.40 is a sketch showing such particles incident on the magnetopause field, taken for simplicity as uniform. The view is from the north, looking down onto the equatorial plane, so that the magnetic field lines are directed upward, out of the plane of the paper. Since the flow velocity of the solar wind perpendicular to the magnetopause must vanish, only the thermal velocity component perpendicular to the magnetopause need be considered.

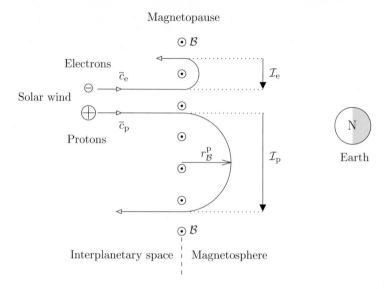

Fig. 5.40. Reflection of normally incident solar wind particles at the magnetopause: viewed from north onto the equator. Arrows show the direction (not the magnitude) of the reflection-induced currents.

The solar wind particles are forced onto gyration orbits as soon as they penetrate into the magnetosphere. This brings them – after completing a half-gyration of radius r_B – right back into interplanetary space, i.e. they are specularly reflected at the magnetopause. This reflection event, however, is accompanied by charge transport that leads to a magnetopause current in the required direction. Protons are deflected to the east – as referenced to the Earth – and electrons are displaced to the west. The net result is an eastward-flowing current.

In order to estimate the magnitude of this magnetopause current, we consider the charge transport through a reference surface oriented perpendicular to both the magnetopause and the equatorial plane; see Fig. 5.41. As indicated by the particle trajectories, only protons that enter the magnetosphere no further away from the reference surface than $2r_B^{\mathrm{p}}$ are actually able to penetrate the surface. It follows that all protons passing through the surface $2r_B^{\mathrm{p}} b$ on the magnetopause will contribute to the current. Assuming again, for simplicity, that the particles conform to a reduced velocity distribution, the number passing through the surface in a time interval Δt will be

$$N = \bar{c}_{\mathrm{p}} \, \Delta t \, 2 \, r_B^{\mathrm{p}} \, b \, n/6$$

where $\bar{c}_{\mathrm{p}}$ and n are the mean thermal velocity and particle density of the solar wind protons, respectively. The current density associated with this charge transport is given by

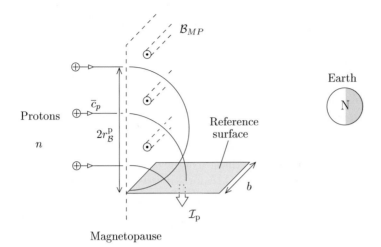

Magnetopause

Fig. 5.41. Estimating the magnitude of the magnetopause current

$$\mathcal{I}_p = \frac{N\,q}{\Delta t} = \frac{\bar{c}_p\,r_B^p\,b\,n\,q}{3}$$

Inserting the relation for the proton gyroradius and expressing the magnetopause current as a surface current $(\mathcal{I}_{MP}^* = \mathcal{I}_{MP}/b)$, we obtain

$$(\mathcal{I}_{MP}^*)_p = \frac{m_p\,n\,\bar{c}_p^2}{3\mathcal{B}_{MP}} = \frac{n\,k\,T_p}{\mathcal{B}_{MP}}$$

where we have made use of the definition for the thermodynamic pressure and the ideal gas law in the second step. Assuming the solar wind plasma to be charge neutral $(n_p = n_e = n)$ and in thermal equilibrium $(T_p = T_e = T)$, and using $\mathcal{B}_{MP} = 2\mathcal{B}_{DP}$, the total current is found to be

$$\mathcal{I}_{MP}^* = (\mathcal{I}_{MP}^*)_p + (\mathcal{I}_{MP}^*)_e = \frac{n\,k\,T}{\mathcal{B}_{DP}} \tag{5.78}$$

This current is indeed sufficient to explain the limited extent of the magnetosphere, as shown by the following estimate of the magnetopause distance based on this current strength. The reflection-induced magnetopause current, according to Eq. (5.78), is inversely proportional to the dipole field strength at the position of the magnetopause. Since $\mathcal{I}_{MP}^* = 2\,\mathcal{B}_{DP}/\mu_0$ from Eq. (5.77), the dipole field strength at the magnetopause must be given by

$$\mathcal{B}_{DP} = \sqrt{\frac{\mu_0\,n\,k\,T}{2}} \tag{5.79}$$

Knowing the radial dependence of the equatorial dipole field strength, $\mathcal{B}_{DP} = \mathcal{B}_{00}/L^3$, it follows that the distance of the magnetopause at the subsolar point may be estimated as

$$L_{MP} = \sqrt[6]{\frac{2\,\mathcal{B}_{00}^2}{\mu_0\,n\,k\,T}} \tag{5.80}$$

Using a solar wind particle density of $n \simeq 25$ cm^{-3} and a temperature T $\simeq 2.2 \cdot 10^6$ K (magnetosheath region for Mach number $M = 8$, see Fig. 6.33), the subsolar magnetopause distance from Eq. (5.80) is $L_{MP} \simeq 11$, in sufficiently good agreement with the observations. The corresponding magnetopause field strength $\mathcal{B}_{MP} = 2\,\mathcal{B}_{DP}$ is slightly greater than 40 nT, again in agreement with measured values. The magnetopause current density of $\mathcal{I}_{MP}^* \simeq 35$ mA/m, considering the size of the dayside magnetopause, produces a remarkable total current (> 4 MA for a north-south extent of 20 R_E). The thickness of the magnetopause layer ($\simeq r_{\mathcal{B}}^{\mathrm{p}}(L_{MP})$) for the above values of T (or $\bar{c}_{\mathrm{p}}$) and $\mathcal{B}_{MP}$ is roughly 60 km, indeed quite small compared to the current sheet width. The magnetopause current is thus a surface current to a very good approximation. It should be mentioned that the actual magnetopause is observed to be considerably thicker, but nonetheless very well approximated by a surface current. An alternative estimate of the magnetopause distance based on pressure balance is presented later in Section 6.4.4.

Charge Neutrality. The reflection scenario sketched in Fig. 5.40 is in need of further explanation because it violates the important principle of charge neutrality. Due to the different penetration depths of electrons and protons ($r_{\mathcal{B}}^{\mathrm{p}} = \sqrt{m_{\mathrm{p}}/m_{\mathrm{e}}}\,r_{\mathcal{B}}^{\mathrm{e}} \simeq 43\,r_{\mathcal{B}}^{\mathrm{e}}$), a negative charge surplus occurs in the thin reflection layer of the electrons and a positive charge surplus is created in

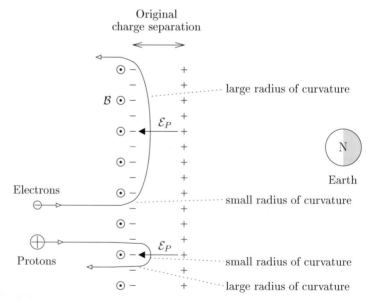

Fig. 5.42. Reflection of solar wind particles at the magnetopause in the presence of an electric polarization field (adapted from Willis, 1971)

the wider reflection region of the protons. This charge separation generates an electric polarization field that, in turn, modifies the particle motion. As indicated in Fig. 5.42, protons are decelerated and electrons are accelerated by an amount such that their penetration depths are nearly equal and the charge separation essentially vanishes. The reduction of the proton gyroradii and the enlargement of the electron gyroradii effectively makes the electrons, with their higher velocities, the main charge carriers of the magnetopause current.

Without firm proof of the existence of such a polarization field, it has yet to be determined whether this latter description of the particle reflection process is better than that offered in the preceeding subsection. Indeed, a population of particles exists just inside the magnetopause (magnetospheric boundary layer), which could nullify the polarization field via oppositely polarized charge separation. Another possibility is that the polarization field triggers a current that flows along magnetic field lines into the ionosphere, thereby eliminating the surplus charge layers. A self-consistent calculation of these effects is difficult and our present understanding of the microscopic processes in the magnetopause layer remains incomplete.

5.5.3 Current System of the Geomagnetic Tail

The magnetotail represents another drastic modification of the Earth's dipole field. It is characterized by extended field lines that are aligned nearly parallel to the tail axis in the antisolar direction and reverse their polarity upon passing through the neutral sheet. Similar to the dayside magnetopause, only surface currents can be held responsible for this type of deformation. Figure 5.43 shows the required current configuration in a cut through the noon-midnight meridian. As indicated, the tail current flows in the neutral sheet in the direction from east to west. The superposition of the associated magnetic field $\vec{B}_{sc}$ and the dipole field $\vec{B}'_{DP}$ leads to an inclination of the total field with respect to the neutral sheet, which decreases with increasing geocentric distance.

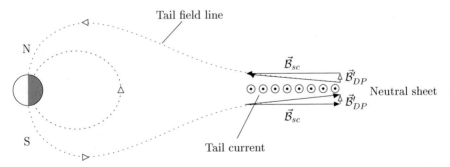

Fig. 5.43. Magnetotail current system (noon-midnight meridian)

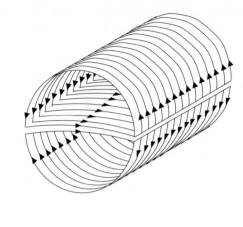

Earth

Fig. 5.44. Three-dimensional distribution of currents in the distant magnetotail (adapted from Olsen, 1982)

The neutral sheet current and the magnetopause current form a closed circuit in the distant magnetotail, the total configuration of which is illustrated in Fig. 5.44. The current distribution corresponds to that of two long solenoids with semicircular cross sections, whose currents join together in the central dividing mid-plane. As required, this configuration produces (a) a natural confinement of the magnetic field to the interior of the magnetotail; (b) magnetic fields that are aligned parallel to the central neutral sheet; and (c) a magnetic polarity reversal across this neutral sheet. The intensity of the magnetotail current can be estimated using the following formula for the magnetic field inside a solenoid

$$\mathcal{B}_{solenoid} = \mu_0 \, n_w \, \mathcal{I} = \mu_0 \, \mathcal{I}^* \tag{5.81}$$

where n_w is the number of windings per unit length, each winding carrying a current $\mathcal{I}$. Taking a representative tail lobe magnetic field of 20 nT at a distance of 30 R_E, we obtain a magnetotail current intensity of $\mathcal{I}^*_{mt}(30 \, R_E) = 2 \, \mathcal{B}_{mt}/\mu_0 \simeq 32$ mA/m $\simeq 200$ kA$/R_E$. The total magnetotail current flowing through a section of the tail 20 R_E in length is thus 4 MA.

Current Formation. The formation of the magnetotail current system is a very interesting, but still poorly understood phenomenon. One might be tempted at first to carry over those results to the magnetotail, that were derived from our treatment of single particle motion in the Earth's dipole field. However, this is only possible at sufficient distance from the neutral sheet. Magnetization currents and drift-associated currents do indeed flow there, the former being dominant, resulting in a total current directed from east to west (dawn to dusk). The situation becomes more complicated in the vicinity of the neutral sheet. Because of the weak magnetic field strength

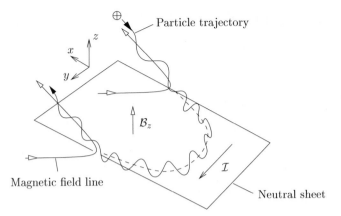

Fig. 5.45. Possible trajectory of a positively charged particle in the neutral sheet of the magnetotail (adapted from Cowley, 1985)

there, the gyroradii become so large that they are no longer small compared with the typical variational scale lengths of the magnetic field. Another factor is the field reversal at the neutral sheet. Both effects lead to such deformation of the particle trajectories that these do not even approximately correspond to modified gyration orbits. As such, the magnetic moment produced by the gyration and the first adiabatic invariant associated with this moment are no longer conserved. A classic example for such *nonadiabatic* particle motion is the neutral sheet drift introduced in Section 5.3.3. An application of this kind of drift to the magnetotail is shown in Fig. 5.45. As indicated, the magnetic force from the $\mathcal{B}_z$ component of the magnetic field in the neutral sheet leads to a continuous change in the drift direction. In spite of the deflection, the electric current associated with this drift motion flows exactly in the direction required of the magnetotail current. Neutral sheet currents thus provide an important contribution to the formation of the magnetotail current system.

Figure 5.45 is only meant to provide a first impression of the possible particle trajectories in the vicinity of the neutral sheet. These trajectories often display chaotic behavior because of their strong dependence on the particular initial conditions. Moreover, an electric convection field of the type sketched in Fig. 5.33b is superposed onto the magnetic field in the tail. This electric field not only modifies the particle orbits, but also accelerates the particles in the direction of the ion drift. Finally, since the magnetic field of the magnetotail is generated by the magnetotail currents, the particle trajectories producing the currents must be mutually consistent with the magnetic field. Self-consistent simulations of this interplay are naturally quite involved and are the subject of intensive efforts.

5.6 Particle Populations in the Outer Magnetosphere

The outer magnetosphere, like its inner magnetospheric counterpart, is also populated by a wide variety of particles. According to where they are found, one distinguishes between their place of residence in the *plasma sheet*, in the *magnetotail lobes*, and in the *magnetospheric boundary layer*. The fundamental properties of these particle populations are summarized in Table 5.3. Note again, that the entries in this table should be considered as nominal values that are subject to distinct spatial and temporal variations. It should also be recognized that the magnetosphere is an enormous object that has not been extensively surveyed by satellite exploration. A particularly spectacular way to obtain information in the magnetotail is reconnaissance from the Moon, which regularly passes through the magnetotail each month at a distance of ca. 60 R_E. The instruments deployed on the Moon's surface as part of the APOLLO-Missions in 1972/73 have provided valuable measurements during these transits. Simultaneous magnetic field measurements were recorded on the moon-orbiting satellite EXPLORER-35.

5.6.1 Magnetotail Plasma Sheet

The *plasma sheet* is a slab-like particle population centered on the mid-plane of the magnetotail. The spatial distribution is sketched in Fig. 5.46. The plasma sheet is threaded by closed, but very extended magnetotail field lines, the footpoints of which are found in a ringlike region around each polar cap called the *polar* or *auroral oval*. The inner edge of the plasma sheet at $|x| \simeq$ 7–10 R_E corresponds to a footpoint latitude of $68° - 72°$.

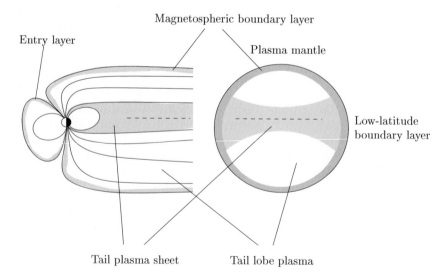

Fig. 5.46. Particle populations in the outer magnetosphere

Table 5.3. Particle populations in the outer magnetosphere. The particle energies and densities for the plasma sheet and lobe regions correspond to typical values at a distance of about 30 R_E (Fairfield, 1987). The magnetic field strengths also correspond to this distance. The values for the magnetospheric boundary layer are from Eastman (1990).

	Particle Population		
Quantity	Magnetotail Plasma Sheet	Magnetotail Lobes	Magnetospheric Boundary Layer
Location	central mid-tail plane	tail lobes	inside of the magnetopause
Magnetic Footpoints	nightside auroral oval	polar cap	cusp region
Energy/ Temp. Ions	$4 \cdot 10^7$ K (5 keV)		$1 - 6$ keV
Electrons	$8 \cdot 10^6$ K (1 keV)	200 eV (< 1 eV $- 2$ keV)	$0.1 - 0.2$ keV
Density	$3 \cdot 10^5$ m^{-3}	$\lesssim 10^4$ m^{-3}	$1 - 20 \cdot 10^6$ m^{-3}
Composition	H$^+$, e$^-$	H$^+$, e$^-$	H$^+$, e$^-$
Particle Motion	variable flow solar/antisolar	gyration; magnetic field-aligned flow	$50 - 250$ km/s antisolar
Mag. Field Strength	$2 - 10$ nT	20 nT	$25 - 40$ nT
β^*-Parameter	$0.1 - 100$	$\ll 1$	$1 - 10$
Source Region	boundary layer; ionosphere	solar wind; ionosphere	solar wind; ionosphere
Formation Process	particle acceleration & transport	direct SW access; evaporation	direct SW access; diffusion
Loss Region	upper atmosphere; interplanetary space	upper atmosphere; interplanetary space	interplanetary space; plasma sheet
Loss Process	pitch angle diffusion, plasmoid ejection	absorption; escape	diffusion and drift
Significance	auroral particle source; particle and energy transfer	polar rain; polar wind	particle and energy transfer

The particle density of the plasma sheet reaches a value of up to 10^6 particles per m^3 at the near-Earth end and decreases slowly with geocentric distance. The main constituents are hydrogen ions and their associated electrons and, during disturbed conditions, a significant admixture of oxygen ions. The ionic plasma displays a relatively isotropic velocity distribution in the central part of the sheet, corresponding to temperatures of a few 10^7 K (or mean particle energies of a few keV) in the near-Earth region. Electron temperatures and energies are significantly lower. The temperatures drop gradually with increasing distance down the tail. Since collisions in the plasma sheet itself are extremely rare, the thermalization of the particles must occur near their mirror points and/or via electromagnetic fields.

Typical flow velocities in the near-Earth region are $\sim$60 km/s, whereby the flow direction can be either earthwards or tailwards. In the distant tail the flow velocities are larger and are directed exclusively away from the Earth. The values for the β^* parameter are typically 0.1–100, the latter being valid for the center of the plasma sheet. The dynamics of the central region is thus controlled by the particle flow and not by the magnetic field configuration.

Both the solar wind via the magnetospheric boundary layer and the polar ionosphere contribute to the plasma sheet particle population. The details of the particle infusion process are not completely understood. Particularly the processes that lead to acceleration and heating of the particles are topics of intensive research. Our knowledge of the loss processes is also only fragmentary. Pitch angle diffusion into the loss cone, stimulated either by plasma waves or by the chaotic motion of the particles in the plasma sheet, certainly plays an important role in the near-Earth region. The resultant precipitating electrons are known to be the most important source of the aurora (see Section 7.4.3). Outflow of particles into interplanetary space on the tailward side (e.g. in the form of *plasmoids*, see Section 8.3.2) is another important loss process.

In addition to its function as a plasma reservoir for the auroral particles, the plasma sheet is significant for the important particle and energy transport processes that occur there. This can be demonstrated with a simple estimate of the rates occurring at the upper and lower boundaries of the plasma sheet. Assuming the following values to be valid in the range of geocentric distances from 10 to 20 R_E: $n = 2 \cdot 10^5$ m^{-3}, $u = 400$ km/s, $E_p = 3$ keV and $A = 0.5$ $R_E \times 30$ R_E, one obtains particle and energy transfer rates of $5 \cdot 10^{25}$ s^{-1} and $2 \cdot 10^{10}$ W, respectively. These are remarkably high values and would be sufficient, for example, to satisfy the particle and energy flux requirements of the aurorae.

5.6.2 Magnetotail Lobe Plasma

Compared with the plasma sheet, the *tail lobe plasma* is a particle population of considerably smaller density and energy. The number densities at the near-Earth end of the tail often lie near the threshold of contemporary particle detectors ($\simeq 10^3$ m^{-3}). The density decrease at the edges of the plasma sheet is correspondingly well pronounced. The lobes gradually become filled with plasma mantle particles in the distant tail, so that the particle density there (but not the energy density) actually exceeds that of the plasma sheet.

It is presently thought that there exists an electron component of the tail lobe population with energies of several 100 eV, which evidently gains direct access from the solar wind to the magnetotail along 'open' magnetic field lines (see also Section 7.6.4). It is conjectured that this electron component is responsible for the *polar rain*, i.e. a moderate intensity precipitation of comparatively low energy particles in the polar cap regions.

Another component is the so-called *polar wind*. This is the suprathermal plasma that continually evaporates from the ionosphere in the polar cap regions. It is the open field line configuration of this region that enables the formation of an ionospheric exosphere. This component is difficult to detect with ordinary particle instruments because of space charge effects.

Finally, particles of higher energy (e.g. $E \simeq 2$ keV) are occasionally observed in the tail lobes, the electrons of which are suspected of being responsible for the *transpolar aurorae*.

5.6.3 Magnetospheric Boundary Layer

The *magnetospheric boundary layer*, located just inside the magnetopause, is the outermost particle population in the Earth's magnetosphere (see Fig. 5.46). One distinguishes here between the polar magnetospheric boundary layer, consisting of the *entry layer*, the *outer cusp region*, and the *plasma mantle*, and the equatorial magnetospheric boundary layer, also called the *low latitude boundary layer, LLBL*). As seen in Fig. 5.46, this latter region is immediately adjacent to the plasma sheet. The thickness of the magnetospheric boundary layer is a few 1000 km at its thinnest position near the subsolar point and can be several Earth radii at the lunar distance.

The properties of the plasma mantle correspond approximately to those of the solar wind in the magnetosheath. Typical particle densities are of the order of 1–$20 \cdot 10^6$ m^{-3}, consisting for the most part of protons with their associated electrons. Only during disturbed conditions does one observe enhancements of charge carriers coming originally from the ionosphere (O^+, He^+). Ion temperatures lie typically in the range 0.5–$2 \cdot 10^7$ K; that of the electrons being a factor of 10 lower. The flow velocity ranges from 100 to 200 km/s at moderate geocentric distances and is directed tailwards. In contrast, stagnant plasma flows ($u \simeq 0$) are often observed in the LLBL.

The most likely source of the plasma mantle particles is the solar wind of the magnetosheath, which can penetrate into the magnetosphere along open magnetic field lines (see Section 7.6.4). Moderate amounts of ionospheric particles may also be found in the cusp regions during disturbed conditions. The origin of the LLBL, however, is not well understood. The significance of the magnetospheric boundary layer lies in its role as a venue of solar wind particle and energy transfer into the magnetosphere, see Section 7.6.4.

5.7 Magnetoplasma Waves in the Magnetosphere

The Earth's magnetosphere is by no means a static object. Indeed, it plays host to a number of fascinating dynamic phenomena, all of which are the subject of intensive research. Among these are fluctuations in the size of the magnetosphere, large-scale plasma circulations, irregular and explosive events during magnetospheric (sub)storms and, finally, a wide spectrum of waves that fills the entire magnetosphere. We are interested here in a special type of these waves. They are characterized by long periods ($\tau \gtrsim 5$ s) and correspondingly low frequencies ($f \lesssim 0.2$ Hz) and belong to the class of *magnetohydrodynamic (MHD) waves*. They are 'magneto' because they occur in magnetically biased plasmas; 'hydrodynamic' because a hydrodynamic approach is often used to describe them quantitatively. It should be remembered, however, that a magnetoplasma is a compressible gas rather than an incompressible fluid. In the following we thus prefer to call them 'magnetoplasma waves' or *ULF (ultra low frequency)* waves in a magnetoplasma.

Such waves can be detected on the ground by their magnetic signature. This is illustrated in the upper part of Fig. 5.47, which shows the horizontal component of the magnetic field at high latitudes recorded during a one-hour interval. The most remarkable feature is the wake-like fluctuation of this component at a period of about 5 minutes. Such fluctuations are designated *geomagnetic pulsations* and are ordered according to their period in classes from *Pc1* to *Pc5* (*Pc* from the French *pulsation continue*), with a range of periods $0.2 \leq \tau \leq 600$ s. We are particularly interested here in the pulsations Pc3 to Pc5 ($10 \leq \tau \leq 600$ s), because these can be attributed to ULF waves in the magnetosphere. The lower part of Fig. 5.47 provides proof that this type of pulsation is indeed the signature of magnetospheric waves. This shows the radial component of the electric field recorded simultaneously with the geomagnetic pulsations on board a satellite at geostationary orbit ($L = 6.6$). The correlation of the two oscillatory data traces is obvious.

Irregular pulsations are also observed (Fr: *pulsation impulsive, Pi*) in addition to the regular pulsations shown in Fig. 5.47. Particularly prominent are the Pi2 pulsations with quasi-periods between 40 and 150 s, which have been associated with magnetospheric substorms.

Magnetoplasma waves play an important role as carriers of energy and information. For example, they represent the only means by which the various plasma regions of the magnetosphere can communicate with each other and adjust themselves to the given environmental circumstances. In particular, they coordinate the temporal variations of the large-scale plasma motions in the ionosphere and magnetosphere. Waves of this type are also frequently observed in other regions of space. In the interplanetary medium, for example,

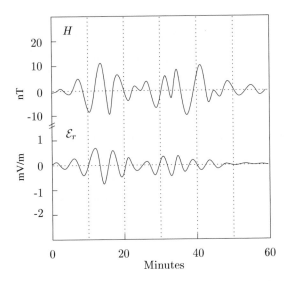

Fig. 5.47. Magnetospheric waves, as observed on the ground as a Pc5 pulsation (H component) and simultaneously in the magnetosphere at an altitude of ca. 36 000 km as a fluctuation of the radial component of the electric field $\mathcal{E}_r$ (adapted from Glaßmeier, 1995)

they play an important role in the heating and acceleration of the solar wind as well as the formation of shock waves – reason enough to treat this wave type in greater detail. This will be postponed to a later chapter, however, after we have fulfilled some necessary prerequisites. Among these is to describe the propagation medium of these waves in more detail. For this purpose we need, in addition to the balance equations valid for a magnetoplasma, the Maxwell equations and a generalized form of Ohm's law (see Section 6.2.7 and the corresponding Appendix A.13). This system of equations is then solved for the special case of a small amplitude harmonic wave propagating in a homogeneous, infinitely extended background plasma (see Section 6.3 and the associated Appendix A.15).

Caution should be exercised, however, when applying these results to the magnetospheric case of interest here. The magnetosphere, after all, is anything but a homogeneous propagation medium. Here, we deal with curved instead of straight magnetic field lines and rather than being constant, the magnetic field strength and plasma densities are dependent upon the geocentric distance (among other parameters). The above cited derivation also only considers plane waves, which, strictly speaking, only occur for infinitely extended excitation or at large distances from compact sources. Neither of these conditions applies for ULF waves in the magnetosphere. Among other places, they are typically excited locally along the flanks of the magnetosphere via the Kelvin-Helmholtz instability (see Appendix A.16). Finally, the derivation is formally valid only for an infinitely extended propagation medium. In contrast, the magnetosphere, with its closed dipolar structure and reflecting boundaries (ionosphere, magnetopause), is a finite volume, whose dimensions are of the same order of magnitude as the wavelengths of interest here. In this case quasi-periodic oscillations can only be excited as standing waves and the magnetosphere will behave like a resonant cavity. The above mentioned pulsations Pc3 to Pc5 at periods between 10 and 600 s are therefore interpreted as eigenmodes of this magnetospheric resonant cavity. In order to describe these eigenmodes quantitatively, locally excited wave propagation must be formulated and derived for an inhomogeneous, spatially bounded magnetoplasma. This task, a topic of recent intense research, has nonetheless proven to be so daunting that it is still incompletely solved.

Exercises

5.1 Calculate the following quantities for your hometown, based on the geocentric dipole approximation:
(a) the magnetic field strength at the Earth's surface;
(b) the inclination and declination angles of the magnetic field;
(c) the shell parameter L and the height of the apex point for the local geomagnetic field line.

5.2 Derive an expression for the path length element ds of a dipole field line. Determine the length of a dipole field line between its footpoint on the Earth and its apex point as a function of the footpoint latitude or shell parameter L.

5.3 The energy density of a magnetic field in vacuum is known to be given by $E^* = B^2/2\mu_0$. Determine the energy content of the geocentric dipole field of the Earth outside of the Earth's surface.

5.4 Confirm, by multiplying Eq. (5.18) with $\vec{v}$, that the kinetic energy of a charged particle moving in a static magnetic field is conserved in the absence of external forces.

5.5 Consider a 50 keV proton with an equatorial pitch angle $\alpha_0 = 15°$ that moves along a dipole field line with shell parameter $L = 3$.
(a) Determine the gyroradius of this particle at the equator and at its mirror point.
(b) What is the geocentric distance (or the height) of its mirror points and what is its oscillation (bounce) period?

(c)   How large must the energy of a proton with pitch angle $\alpha_0 = 90°$ be in order that it always stays at the same point above the Earth (geostationary drift)?

5.6   The dipole moment of a planet is frequently given in units of 'gauss cm^3'. Convert this unit to 'A m^2'. The magnetic dipole moment of Jupiter is $\mathcal{M}_J = 1.6 \cdot 10^{27}$ A m^2.

(a)   Determine the magnetic flux Φ_{pc} through the polar cap of this planet if the (magnetic) polar cap latitude is 80° and the surface radius is $R_J = 71\,000$ km.

(b)   What is the diameter of Jupiter's magnetotail if the magnetic field strength in the tail lobes is $\mathcal{B} = 1$ nT?

References

S. Chapman and J. Bartels, *Geomagnetism I, II*, Clarendon Press, Oxford, 1962

S. Matsushita and W.H. Campbell (eds.), *Physics of Geomagnetic Phenomena I, II*, Academic Press, New York, 1967

W.N. Hess, *The Radiation Belt and Magnetosphere*, Blaisdell Publishing Company, Waltham, 1968

J.A. Jacobs, *Geomagnetic Micropulsations*, Springer-Verlag, Berlin, 1970

J.G. Roederer, *Dynamics of Geomagnetically Trapped Radiation*, Springer-Verlag, Berlin, 1970

A. Nishida, *Geomagnetic Diagnosis of the Magnetosphere*, Springer-Verlag, Berlin, 1978

L.R. Lyons and D.J. Williams, *Quantitative Aspects of Magnetospheric Physics*, Reidel Publishing Company, Dordrecht, 1984

T.Y. Lui (ed.), *Magnetotail Physics*, Johns Hopkins University Press, Baltimore, 1987

A.D.M. Walker, *Plasma Waves in the Magnetosphere*, Springer-Verlag, Berlin, 1993

M. Walt, *Introduction to Geomagnetically Trapped Radiation*, Cambridge University Press, Cambridge, 1994

J.F. Lemaire and K.I. Gringauz, *The Earth's Plasmasphere*, Cambridge University Press, Cambridge, 1998

P.T. Newell and T. Onsager (eds.), *Earth's Low-Latitude Boundary Layer*, American Geophys. Union, Washington, DC, 2002

See also the references in Chapter 1 and the figure and table references in Appendix B.

6. Interplanetary Medium

By 'interplanetary medium' we mean the particles and fields that fill the region between Sun, planets and the interstellar medium. The particles here are mainly those of the solar wind and the most prominent field is the interplanetary magnetic field. The present chapter first discusses the properties of the solar wind and its modeling. This is followed by a description of the interplanetary magnetic field and its associated heliospheric current sheet. In this context a system of equations is introduced which enables us to understand the properties of waves in the interplanetary medium. These play an important role in the formation of shock waves, whereby particular interest is directed to the terrestrial bow shock and the modification of the solar wind caused by it. We then turn our attention to the interface between the solar wind and the interstellar medium, another region where shock waves are an integral component. Finally, we provide an introduction to the various populations of highly energetic particles present in interplanetary space.

6.1 The Solar Wind

Up to the middle of the last century, it was generally assumed that interplanetary space was essentially a vacuum, devoid of any gas whatsoever. By necessity, the presence of a few dust particles was acknowledged because their existence was strongly implied by observations of the zodiacal light. While the occurrence of interplanetary plasma clouds had been discussed even before 1920 as an explanation of the connection between solar flares and geomagnetic storms, these plasma bubbles were considered to be ephemeral phenomena that sporadically passed through interplanetary space during short interludes. The notion that this region might possibly be filled permanently with solar gas was first seriously raised in the 1950s. In order to obtain an initial estimate of the gas density in interplanetary space, we extrapolate the observed coronal density with the help of the barometric law out to large distances from the Sun. We consider an isothermal, charge neutral gas mixture consisting of protons and electrons, for which the following relations hold: $n_p = n_e = n$, $\rho = n\,(m_p + m_e) = n\,m_H$, $T_p = T_e = T = $ const. and $p = p_p + p_e = 2nkT$. As derived in Section 2.4.5, the density profile of such a single gas atmosphere is given by

$$n(r) = n(r_0) \, \exp\left\{ \frac{m_\mathrm{H} \, G \, M_S}{2 \, kT} \left(\frac{1}{r} - \frac{1}{r_0} \right) \right\} \tag{6.1}$$

where r is the heliocentric distance and M_S the solar mass. The factor 2 in the denominator of the exponent comes from the requirement of charge neutrality and implies an electric polarization field that effectively halves the proton mass; see Section 4.4.1. A comparison of this barometric density profile with the empirically determined electron density profile of Eq. (3.7) yields quite good agreement. Accordingly, to a sufficiently good approximation, the corona close to the Sun ($r \lesssim 3R_S$) can be considered to be in aerostatic equilibrium.

Extrapolating this density profile to large distances from the Sun, we may work with a simplified formula

$$n(r \gg r_0) \simeq n(r_0) \, \exp\left\{ -\frac{m_\mathrm{H} \, G \, M_S}{2 \, kT \, r_0} \right\} \tag{6.2}$$

Using $r_0 = 1.5 \, R_S$, $n(r_0) \simeq 10^{13} \, \mathrm{m}^{-3}$ from Eq. (3.7), and $T \simeq 10^6 \, \mathrm{K}$ we obtain $n(r \gg 1.5 \, R_S) \simeq 4 \cdot 10^9 \, \mathrm{m}^{-3}$. Such a large asymptotic value for the gas density is a clear indication that interplanetary space cannot be a vacuum, but must be permanently filled by the solar atmosphere. This conclusion is not invalidated by the fact that the extrapolated values are much too high (indeed, the above value corresponds approximately to that of the near-Earth plasmasphere). Moreover, an asymptotically constant density is physically unrealistic because it leads to an infinitely large atmospheric mass. These deficiencies, however, do not mean that the solar atmosphere must be confined to circumsolar space, but only that the distribution of the solar atmosphere in interplanetary space does not correspond to static equilibrium.

First indications that the solar atmosphere in interplanetary space is indeed a most dynamic phenomenon were obtained from observations of comet tails carried out in the 1940s and 1950s. Comets are known to possess two types of tails: one a diffuse, homogeneous, slightly curved gas/dust tail, the other a strongly structured ion tail that is directed radially outward away from the Sun; see Fig. 6.1. The orientation of the gas/dust tail could be easily explained by the interplay of the solar gravitational attraction and the repulsive solar radiation pressure. The radial orientation of the ion tail, on the other hand, was not immediately understood. Furthermore, the structures observed in the ion tail often exhibited large and strongly variable accelerations. In order to explain both phenomena, Biermann (1951) postulated the existence of a plasma flow that is continually emitted from the Sun with variable flow velocity. The approximate velocity of this plasma flow, which we today refer to as the *solar wind*, was estimated (almost correctly) as 100 km/s. Although Biermann's estimate was partially based on incorrect assumptions, this imperfection had no serious consequences for the subsequent development of the hypothesis. The important point was that the possibility of a dynamic, nonstatic solar atmosphere had been expounded. This notion

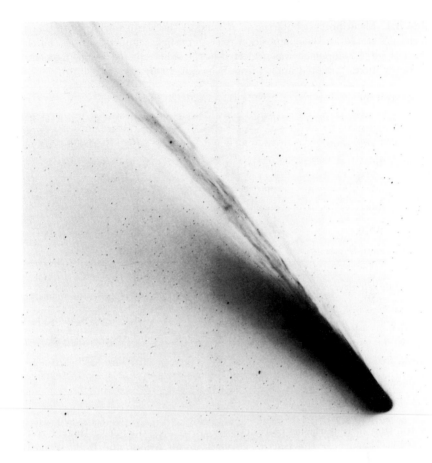

Fig. 6.1. Photometric image (negative) of Comet Mrkos. The straight upper tail is the ion tail (Mt. Wilson/Palomar Observatory).

was eagerly seized by Parker (1958) who was the first to develop a very successful solar wind model. Before introducing this model in its simplest form, it is appropriate to first briefly describe the present state of our knowledge of the solar wind.

6.1.1 Properties of the Solar Wind at the Earth's Orbit

Our present knowledge of the solar wind is based primarily on *in situ* observations recorded on board interplanetary spacecraft. Table 6.1 summarizes the mean properties of the solar wind at the Earth's orbit, i.e. in the ecliptic plane at a heliocentric distance of 1 AU.

Like its coronal origin, the outwardly expanding solar wind consists essentially of protons and electrons with a small admixture of α particles (He^{++}).

Table 6.1. Mean properties of the solar wind at the Earth's orbit. Note that the percentages stated for the composition refer to particle number densities. Furthermore, the He^{++} component was neglected in the values quoted for the momentum and energy fluxes (adapted in part from Schwenn, 1990).

Composition:	$\simeq$ 96% H$^+$, 4% (0–20%) He^{++}, e$^-$		
Density:	$n_p \simeq n_e$	$\simeq$	6 (0.1–100) cm^{-3}
Velocity:	$u_p \simeq u_e = u$	$\simeq$	470 (170–2000) km/s
Proton flux:	$n_p\, u$	$\simeq$	$3 \cdot 10^{12}$ m^{-2}s^{-1}
Momentum flux:	$n_p\, m_H\, u^2$	$\simeq$	$2 \cdot 10^{-9}$ N/m^2
Energy flux:	$n_p\, m_H\, u^3/2$	$\simeq$	0.5 mW/m^2
Temperature:	T	$\simeq$	10^5 (3500–5$\cdot 10^5$) K
Plasma sound velocity:	v_{PS}	$\simeq$	50 km/s
Random velocity:	$\bar{c}_p$	$\simeq$	46 km/s
	$\bar{c}_e$	$\simeq$	$2 \cdot 10^3$ km/s
Particle energy:	E_p	$\simeq$	1.1 keV (flow energy)
	E_e	$\simeq$	13 eV (thermal energy)
Mean free path:	$l_{p,p} \simeq l_{e,e}$	$\simeq$	10^8 km
Coulomb collision time:	$\tau_{p,p} \simeq 30\, \tau_{e,p}$	$>$	20 d

The fraction of α particles can increase up to 20% during disturbed conditions. The density at the Earth's orbit is about 6 ions and 6 electrons per cm^3, whereby considerable fluctuations are observed. For reasons of charge neutrality, the densities adhere to the relation $n_e \simeq n_p + 2n_\alpha$.

The velocity of the solar wind, with a mean value of about 500 km/s, fluctuates between extreme values of ca. 170 km/s and more than 2000 km/s. In this context it has been found useful to distinguish between the slow solar wind ($u < 400$ km/s) and the high speed streams ($u > 600$ km/s), which often occupy sharply delineated adjacent sectors – in particular close to the Sun. This bifurcation is evidently associated with the different coronal origins of the two flows. In any case, the solar wind is a supersonic flow that needs 3 to 4 days to cover the distance from the Sun to the Earth.

In contrast to the density and the velocity, the particle flux is relatively constant and fluctuates by less than a factor of two about its mean. This can be attributed to the fact that the slow wind is mostly associated with high densities and the fast wind with lower densities. Knowing the particle flux, we can make an estimate of the Sun's mass loss rate

$$dM_S/dt \simeq n_p\, u\, m_H\, 4\pi\, (1\ \text{AU})^2 > 10^9\ \text{kg/s}$$

The Sun thus loses more than a million tons of mass each second via the solar wind. With a total mass of $2 \cdot 10^{30}$ kg and a life expectancy of 10^{10} years, however, this mass loss is hardly noticed.

Similar to the particle flux, the momentum flux (or the dynamic pressure) is also observed to be relatively constant. This quantity determines the stagnation pressure exerted by the solar wind on an obstacle such as the terrestrial magnetosphere. It can occasionally display sudden changes, however, for example during the passage of an interplanetary shock.

An essential feature of the solar wind is that its energy density is determined by the flow motion. The thermal energy ($\simeq 2n_{\mathrm{p}}(3\ kT/2)$) is negligibly small by comparison. To a good approximation, the energy flux of the solar wind is thus given by $\phi_{sw}^E(1 \text{ AU}) \simeq n_{\mathrm{p}}\, u\, (m_{\mathrm{H}}\, u^2/2) \simeq 0.5$ mW/m^2, where we have again neglected the helium (α) component. This kinetic energy flux should be compared with the UV radiative energy flux ($\lambda < 175$ nm), which is about thirty times larger.

In order to estimate the total energy loss to the Sun from the solar wind, we must also account for the work performed by the motion of the particles against the solar gravitational attraction. This potential energy is given by

$$E_{pot} \simeq \int_{R_S}^{\infty} m_{\mathrm{H}}\ g_S\ \mathrm{d}r = G\ m_{\mathrm{H}}\ M_S/R_S$$

The flux associated with this potential energy is thus $\phi_{pot}^E \simeq n_{\mathrm{p}}\, u\, E_{pot} \simeq 0.9$ mW/m^2, so that the total energy expenditure of the Sun to the solar wind attains the value $\left(\phi_{sw}^E + \phi_{pot}^E\right) 4\pi(1 \text{ AU})^2 \simeq 4 \cdot 10^{20}$ W. This corresponds to about one millionth of the energy loss due to electromagnetic radiation ($\simeq 4 \cdot 10^{26}$ W).

The solar wind temperature, similar to the density and velocity, is subject to large fluctuations (a factor 100 or more). In fact, citing a single temperature (as in Table 6.1) is a great simplification of the rather complex situation in the real solar wind. Due to the absence of thermal equilibrium, for example, the various components of the solar wind all have different temperatures. Furthermore, the random velocity distributions are anisotropic, displaying different temperatures parallel and perpendicular to the ambient magnetic field. Finally, systematic variations in the temperature are found to be coupled to the solar wind velocity. The proton temperature in the fast solar wind, in particular, is significantly higher than in the slow wind. Whereas the typical slow solar wind ($u \simeq 300$ km/s, $n_{\mathrm{p}} \simeq 8$ cm^{-3}) has $T_{\mathrm{p}} \simeq 0.3 \cdot 10^5$ K, the temperature of the fast solar wind ($u \simeq 700$ km/s, $n_{\mathrm{p}} \simeq 3$ cm^{-3}) is more like $T_{\mathrm{p}} \simeq 2.3 \cdot 10^5$ K. The temperature quoted in Table 6.1 should thus be viewed as a nominal value.

Concerning the gas kinetic quantities, it should be noted that the mean random (thermal) velocity of the protons is only about one tenth of the flow velocity. In contrast, the electron thermal velocity is a factor of four greater than the mean bulk speed. Accordingly, the particle energy of the protons is determined primarily by the flow speed; that of the electrons by their thermal speed. The mean free path of solar wind particles near the Earth's orbit is extremely large and reaches almost an astronomical unit. The time between Coulomb collisions is correspondingly long, amounting to almost three weeks for proton-proton encounters. Considerably more effective moderating interactions occur via electromagnetic fields.

6.1.2 Gas Dynamic Model

Among the essential observations to be reproduced by a solar wind model are

- a strong acceleration of the coronal gas from $u \simeq 0$ near the Sun to supersonic velocities of about 500 km/s at the Earth's orbit
- a decrease of the density from about $2 \cdot 10^{11}$ m^{-3} at the corona ($\simeq 3R_S$) to values of a few 10^6 m^{-3} at the Earth's orbit
- a relatively slow decrease in the temperature from coronal values near $\simeq 10^6$ K to about 10^5 K at 1 AU.

Two very different approaches were taken to construct a theoretical model that reproduces these observations, one using a *gas dynamic* (macroscopic), the other an *exospheric* (microscopic) framework. We describe first the salient features of the gas dynamic model.

In the simplest gas dynamic approach, the solar atmosphere is considered to be a quasi-neutral gas mixture consisting of protons and electrons. Both components are strongly coupled to each other and everywhere have the same particle density, the same velocity and the same temperature, whereby this last parameter is also taken as constant to a first approximation. Furthermore, only the spherically symmetric, stationary case is considered and particle sources and sinks are justifiably neglected. In this case we have

$$
\begin{aligned}
&(1) && n_\mathrm{p} = n_\mathrm{e} = n \\
&(2) && \vec{u}_\mathrm{p} = \vec{u}_\mathrm{e} = \hat{r}u \\
&(3) && T_\mathrm{p} = T_\mathrm{e} = T \simeq \text{const.} \\
&(4) && \partial/\partial\vartheta \;,\; \partial/\partial\lambda \to 0 \\
&(5) && \partial/\partial t \to 0 \\
&(6) && q \;,\; l \to 0
\end{aligned}
$$

The first assumption implicitly contains the requirement of charge neutrality and is guaranteed in practice by the presence of an electric field (the Pannekoek-Rosseland polarization field). The second assumption implicitly requires that no currents flow in the solar wind as it delivers its charge carriers into interplanetary space. To achieve this goal, we need an additional, considerably stronger electric field; see Section 6.1.5. The calculation is simplified by the third assumption without losing essential information. More realistic temperature profiles are considered in Section 6.1.3. The fourth assumption

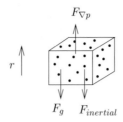

Fig. 6.2. Force equilibrium in the solar wind

implies that the solar wind is spherically symmetric, thereby excluding any variations with the colatitude ϑ or the longitude λ. Time variations are excluded and only the stationary case is treated by reason of the fifth assumption. Finally, the sixth assumption demands that there exist no solar wind particle sources or sinks – a condition which is fulfilled to good approximation.

The calculation of the desired solar wind parameters begins – as was the case for determining the density in the thermosphere or upper ionosphere – with the momentum balance (force equilibrium) equation. We thus consider the sum of all forces acting on a given gas volume of the solar wind. An essential feature of the present situation is that the pressure gradient and gravitational forces are obviously *not* in equilibrium, as this would lead to a static atmosphere. Rather, the outward directed pressure gradient force gains the upper hand and generates a steady acceleration of the given gas volume. This acceleration is opposed by inertial forces that must be included in the momentum balance equation; see Fig. 6.2. Augmenting the fundamental aerostatic equation by an inertial term for the case of dynamical force equilibrium and considering a volume element moving with the solar wind (Lagrange approach), we may write

$$\frac{dp}{dr} = -\rho\, g_S - \rho\, \frac{Du}{Dt} \tag{6.3}$$

Here the convective derivative in the inertial term accounts for acceleration of the gas packet caused by a temporal change of the flow field as well as those by a spatial variation in the flow velocity (e.g. flow constriction, see Section 3.4.3, especially Fig. 3.36). The same approach, of course, is derived directly from Eq. (3.80), if the Coriolis force is absent and the viscosity and frictional forces can be neglected. Since only the stationary, spherically symmetric case is considered, we can make use of Eq. (3.81) and the form of the nabla operator implicitly contained in Eq. (A.29) to obtain

$$\frac{Du}{Dt} = (\vec{u}\,\nabla)u = u\,\frac{du}{dr} \tag{6.4}$$

so that the momentum balance equation (6.3) may be written in the following form

$$\rho \, u \, \frac{du}{dr} = -\frac{dp}{dr} - \rho \, g_S \qquad (6.5)$$

This equation provides a relation between the three variables u, ρ and p. In order to compute one of these quantities, the other two must be eliminated. This is accomplished with the help of an abbreviated density balance equation and the ideal gas law.

With the goal of deriving an equation for the flow field $u(r)$, we first replace the pressure with the particle density n using the ideal gas law. With $p = p_\mathrm{p} + p_\mathrm{e} = 2 \, n \, k \, T$ and $T=$ const., we note that $dp/dr = 2 \, k \, T \, dn/dr$. Inserting this into Eq. (6.5), in which the mass density is converted to the particle density via $\rho = (m_\mathrm{p} + m_\mathrm{e})n \;=\; m_\mathrm{H} \, n$, we obtain

$$u \, \frac{du}{dr} = -\frac{2 \, k \, T}{m_\mathrm{H}} \left(\frac{1}{n} \frac{dn}{dr} \right) - g_S \qquad (6.6)$$

Elimination of the particle density in this equation is accomplished with the density balance equation, which, because of assumptions (5) and (6), assumes the following simple form

$$\mathrm{div}(n \, \vec{u}) = 0 \qquad (6.7)$$

This corresponds to a transport equilibrium relation, similar to those already seen for the cases of the neutral upper atmosphere and the upper ionosphere. In contrast to those two earlier applications, however, we consider here a stationary, not a static case, i.e. we do not make the further assumption that the transport velocity $\vec{u}$ is equal to zero. In fact, here the flow velocity is one of the essential solar wind parameters to be determined. Using the divergence operator in spherical coordinates as given in Eq. (A.30), we obtain for the spherically symmetric case

$$\mathrm{div}(n \, \vec{u}) = \frac{1}{r^2} \frac{\partial}{\partial r} \left(r^2 \, n \, u \right) = 0$$

from which

$$r^2 \, n \, u = \mathrm{const.} \qquad (6.8)$$

This relation states that the number of particles transported per unit time through an arbitrary spherical surface around the Sun ($= n \, u \, 4\pi r^2$) must be constant. This explicitly excludes any increase or decrease of the particle number density between two neighboring spherical shells (which would otherwise correspond to a temporal change). Taking the logarithm of this expression and differentiating with respect to r, we obtain

$$\frac{1}{n} \frac{dn}{dr} = -\frac{1}{u} \frac{du}{dr} - \frac{2}{r}$$

Substituting this into Eq. (6.6) yields the desired equation for u as a function of the heliocentric distance r

$$u \frac{du}{dr} \left(1 - \frac{2\,k\,T}{m_{\mathrm{H}}} \frac{1}{u^2} \right) = \frac{4\,k\,T}{m_{\mathrm{H}}\,r} - \frac{G\,M_S}{r^2} \tag{6.9}$$

This equation can be simplified by introducing the so-called *critical velocity* and the *critical distance*, defined by

$$u_c = \sqrt{\frac{2\,k\,T}{m_{\mathrm{H}}}} \tag{6.10}$$

and

$$r_c = \frac{G\,M_S\,m_{\mathrm{H}}}{4\,k\,T} \tag{6.11}$$

Note that u_c is just the most probable random velocity for a Maxwellian distribution of a hydrogen gas and thus corresponds approximately to the sound speed in this gas; see Eqs. (2.92) and (3.95). Dividing by $-u_c^2$ and substituting for r_c yields

$$u \frac{du}{dr} \left(\frac{1}{u^2} - \frac{1}{u_c^2} \right) = \frac{2\,r_c}{r^2} - \frac{2}{r} \tag{6.12}$$

This equation may be further simplified by temporarily introducing the new variable $x = u^2$. With $dx/dr = 2u\,du/dr$ one obtains the following relation for $x(r)$

$$\frac{dx}{dr} \left(\frac{1}{x} - \frac{1}{u_c^2} \right) = 4 \left(\frac{r_c}{r^2} - \frac{1}{r} \right)$$

Separating the variables and integrating, we obtain

$$\int_{x(r_0)}^{x(r)} \frac{dx'}{x'} - \frac{1}{u_c^2} \int_{x(r_0)}^{x(r)} dx' = 4 \left\{ r_c \int_{r_0}^{r} \frac{dr'}{r'^2} - \int_{r_0}^{r} \frac{dr'}{r'} \right\}$$

where r_0 represents an arbitrary lower boundary. Carrying out the integration and returning to the variable u, we finally arrive at the relation

$$\ln\left(u^2(r)\right) - \frac{u^2(r)}{u_c^2} + 4 \left(\frac{r_c}{r} + \ln r \right) = C \tag{6.13}$$

where the integration constant C, depending only on r_0, comprises all terms not containing variables.

Equation (6.13) describes the variation of the velocity as a function of the heliocentric distance in such a general form that an entire suite of solutions is possible. Additional specifications are needed, e.g. whether or not the solution should display a minimum or a maximum, or should the solution rise or fall monotonically, in order to even narrow down the selection to a certain *class* of solutions; see Fig. 6.3. The classes of solutions (1)–(4) shown in Fig. 6.3 can be understood by examining the differential equation for u prior to its integration. We write Eq. (6.12) in a slightly modified form

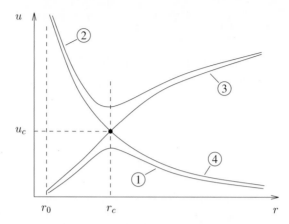

Fig. 6.3. Classes of solutions to Eq. (6.13) for the velocity profile of the solar wind as a function of heliocentric distance. (1) $u < u_c$ for all r; (2) $u > u_c$ for all r; (3) $u(r_c) = u_c$ and $u < u_c$ for $r_0 \leq r < r_c$; (4) $u(r_c) = u_c$ and $u > u_c$ for $r_0 \leq r < r_c$.

$$\frac{1}{u}\frac{du}{dr}\left(1 - \frac{u^2}{u_c^2}\right) = \frac{2}{r}\left(\frac{r_c}{r} - 1\right) \tag{6.14}$$

Consider now the range of solar distances $r_0 \leq r < \infty$, where the upper bound on r_0 is given by the condition $r_0 < r_c$. Since the critical radius $r_c \simeq 6\ R_S$ for typical coronal conditions ($T \simeq 10^6$ K), this restriction on r_0 presents no problem. In this case the right side of the equation is positive for $r_0 \leq r < r_c$, negative for $r > r_c$, and zero for $r = r_c$. Should the latter hold, the left side of the equation must also vanish. This would be the case if $u(r_c)$ were to go to infinity, an unphysical situation that can be discarded; or if $u(r)$ attains an extremum so that du/dr goes to zero at $r = r_c$; or if the expression in the parentheses vanishes, which happens when $u(r_c) = u_c$. The possibility that both the slope and the expression in parentheses go to zero at $r = r_c$ leads to double valued solutions that need not be considered further.

In the case where the velocity profile possesses an extremum at the critical distance, the parenthetical expression on the left side of Eq. (6.14) must be positive or negative over the entire distance range considered. Any change in sign would require an additional null on the right hand side that is not available. It follows that the slope du/dr must be positive when $u < u_c$ (expression in parentheses positive) and $r_0 \leq r < r_c$, because the right side of the equation is positive under the same conditions. By the same token, the slope must be negative for $r > r_c$ because the right side is negative in this case. A maximum must thus occur at $r = r_c$ when $u < u_c$, and this corresponds to the class 1 solution sketched in Fig. 6.3. When u is everywhere greater than u_c, similar argumentation can be used to deduce that the velocity reaches a minimum at r_c (class 2 solutions).

If the expression in parentheses vanishes at the critical point (i.e. $u(r_c) = u_c$), then the slope of the velocity profile du/dr must be either positive ($u <$

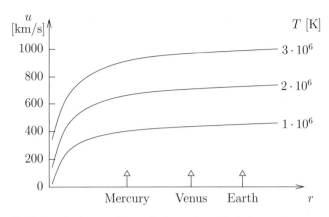

Fig. 6.4. Radial velocity profiles of the solar wind for various temperatures (adapted from Parker, 1963)

u_c for $r_0 \leq r < r_c$) or negative ($u > u_c$ for $r_0 \leq r < r_c$) over the entire range of distances. Otherwise, we would again have a null on the left side at some point other than the critical point, which is not matched by a null on the right hand side. This corresponds to a monotonic velocity profile, either continuously increasing or continuously decreasing, as shown by the solution classes 3 and 4 in Fig. 6.3, respectively.

Comparing these different classes of solutions with the boundary conditions set by the observations (u small in the corona and large in interplanetary space), it is clear that only class 3 solutions are capable of describing the solar wind. This class requires that the velocity reach its critical value u_c at the critical point r_c

$$u(r_c) = u_c \qquad (6.15)$$

Substituting this into Eq. (6.13) enables a determination of the integration constant C

$$C = \ln u_c^2 + 3 + 4 \ln r_c$$

The solar wind equation thus assumes the final form

$$\left(\frac{u}{u_c}\right)^2 - 2 \, \ln\left(\frac{u}{u_c}\right) = 4 \, \ln\left(\frac{r}{r_c}\right) + 4 \, \left(\frac{r_c}{r}\right) - 3 \qquad (6.16)$$

Since the critical radius r_c and critical velocity u_c depend on the temperature of the solar wind, T is a free parameter of Eq. (6.16). Some explicit solutions are shown in Fig. 6.4 for various values of this quantity. The velocities at the Earth's distance predicted by these solutions are clearly compatible with the observations. This is also true for the relatively weak radial dependence of the model velocities at larger heliocentric distances, as confirmed by other observations.

The density profile obtained from the relation (6.8) is

$$n(r) = n(r_1) \left(\frac{r_1}{r}\right)^2 \frac{u(r_1)}{u(r)} \tag{6.17}$$

As originally stipulated, the density goes to zero at large distances. Explicitly, we obtain for the boundary conditions $r_1 = r_c = 6\ R_S$, $n_e(6\ R_S) \simeq 2 \cdot 10^{10}\ \mathrm{m}^{-3}$ from Eq. (3.7), $u(r_1) = u_c(T = 10^6\ \mathrm{K}) \simeq 130\ \mathrm{km/s}$ and $u(1\ \mathrm{AU} \simeq 214\ R_S) \simeq 450\ \mathrm{km/s}$ (Fig. 6.4) a density of $n_e(1\ \mathrm{AU}) \simeq 5\ \mathrm{cm}^{-3}$, again in agreement with the observations.

6.1.3 Temperature Profile

An obvious deficiency of our model is that the values of temperature shown in Fig. 6.4 are compatible with those observed in the corona, but not those observed at the distance of the planets. Furthermore, the solar wind velocity in Eq. (6.16) goes to infinity – even if only very slowly – when $r \to \infty$. Both deficiencies can be attributed to the simplifying assumption of a constant solar wind temperature. Two temperature profiles are therefore considered in the following that display better agreement with the observations and also maintain a finite value for the velocity at large distances.

Adiabatic Expansion. The simplest assumption that leads to a decreasing temperature in the solar wind is that the gas expansion proceeds adiabatically. This assumption requires that no heat exchange take place between an expanding gas packet and its environment. In other words, the work expended by the gas packet during its expansion must be taken entirely from its own internal energy. According to Eq. (2.35), this requires

$$T = T_0 \left(\frac{V}{V_0}\right)^{-2/f} \tag{6.18}$$

The volume of the gas packet will grow in proportion to the square of the distance for the purely radial expansion considered here, $dV = r^2 \sin\vartheta\, d\vartheta\, d\varphi\, dr$, so that Eq. (6.18) may also be written as

$$T(r) = T(r_0) \left(\frac{r}{r_0}\right)^{-4/f} \tag{6.19}$$

Since $f = 3$ for a proton or electron gas, one thus obtains

$$T(r) = T(r_0) \left(\frac{r_0}{r}\right)^{4/3} \tag{6.20}$$

Selecting the boundary condition $T(r_0) = 10^6\ \mathrm{K}$ at $r_0 = 3\ R_S$, this yields a solar wind temperature at the Earth's orbit of $T(1\ \mathrm{AU}) \simeq 3400\ \mathrm{K}$. Such low temperatures are occasionally observed in solar wind sectors with low flow velocity, implying that the expansion in these regions may indeed proceed adiabatically to a first approximation.

Diabatic Expansion. Temperatures observed near 1 AU in sectors of fast solar wind are considerably higher than predicted by the adiabatic model. It follows that the solar wind packets must be further heated in order to at least partially compensate the heat loss caused by radial expansion. Since heat convection can obviously play no role and radiative absorption is negligibly small, it would appear that molecular heat conduction is the only remaining viable heating agent. This mechanism is particularly attractive, because the heat conductivity of an almost collisionless gas is quite high and even the most moderate temperature gradients are capable of generating significant heat fluxes.

In order to determine the solar wind temperature profile in the presence of heat conduction, we consider the energy balance equation for a *stationary* volume element. In the time independent case and neglecting local heat production and loss, this balance equation reduces to the form $d^W = -\mathrm{div}\,\vec{\phi}^W = 0$; see Eqs. (3.51) and (3.52). If the divergence of the heat flux were not zero, our volume would be gradually warmed or cooled, which is incompatible with our assumed time independent state. (On the other hand, the divergence of the heat flux for a nonstationary volume element moving with the solar wind does not vanish and represents the desired heat source for this volume element.) For the spherically symmetric case considered here, the above condition assumes the following form

$$\mathrm{div}\,\vec{\phi}^W = \frac{1}{r^2}\frac{\partial}{\partial r}\left(r^2\,\phi_r^W\right) = 0$$

or

$$r^2\,\phi_r^W = \text{const.} \tag{6.21}$$

where ϕ_r^W denotes the radial component of the heat flux. Considering only molecular heat conduction and referring to Eq. (3.45), this may be written as

$$\phi_r^W = -\kappa\,\frac{\mathrm{d}T}{\mathrm{d}r}$$

where the heat conductivity from Eq. (3.44) is given by

$$\kappa = \xi\,\frac{k^{3/2}\,f\,\sqrt{T}}{\sqrt{m}\,\sigma}$$

In the present case σ is the Coulomb cross section, $\sigma^{Cb} \sim 1/T^2$, so that the temperature dependence of the heat conductivity goes as $\kappa \sim T^{5/2}$ and Eq. (6.21) may thus be written as

$$-r^2\,T^{5/2}\,\frac{\mathrm{d}T}{\mathrm{d}r} = \text{const.}$$

Separating the variables and integrating, we arrive at the solution

$$T^{7/2} = a + \frac{b}{r}$$

where a and b are constants. Requiring that $T(r \to \infty) \to 0$ (a finite temperature would again lead to an infinite solar wind velocity), one finally obtains the radial dependence of the temperature as

$$T(r) = T(r_0) \left(\frac{r_0}{r}\right)^{2/7} \tag{6.22}$$

Taking again the inner boundary condition values $r_0 = 3\,R_S$ and $T(r_0) = 10^6$ K, one calculates a solar wind temperature at 1 AU of $T(1\ \mathrm{AU}) \simeq 3 \cdot 10^5$ K, a value well within the range of observations. Evidently, the extremely small temperature gradient of

$$\frac{\Delta T}{\Delta r} \simeq \frac{3 \cdot 10^5 - 10^6}{1\mathrm{AU}} = -5 \cdot 10^{-6}\,\mathrm{K/m}$$

is adequate to generate a sufficiently large outward-directed heat flux capable of raising the temperature of gas packets at 1 AU from their adiabatic value of 3400 K up to 300 000 K. Clearly, since the thermal motion of the protons is too slow, this heat conduction can only proceed via the electron component. Even at a temperature of 10^6 K, the mean thermal velocity of protons $(\bar{c}_p(10^6\ \mathrm{K}) \simeq 150\ \mathrm{km/s})$ is simply not high enough to overtake a gas packet with a flow velocity of some 500 km/s. In contrast, this is no problem for electrons $(\bar{c}_e(10^6\ \mathrm{K}) \simeq 6000\ \mathrm{km/s})$.

Difficulties with the temperature model presented here arise from the sluggish interaction between particles. The Coulomb collision time at 1 AU is of the order of one day for electron-proton collisions and almost three weeks for proton-proton collisions (see Table 6.1), so that heat exchange via collisions is strongly limited. A direct indicator of an inefficient exchange of heat is the occurrence of anisotropic velocity distributions. As an example, Fig. 6.5 shows a cross sectional cut through a three-dimensional velocity distribution as observed for protons at 1 AU. Note that this is the directionally dependent velocity distribution $g(\vec{c})$, not the distribution of the velocity magnitudes $h(c)$; refer to Section 2.4.3 and particularly Eq. (2.89). A purely Maxwellian distribution would appear in this representation as a series of concentric circles. The actually observed distribution is strongly asymmetric with elliptic contours elongated perpendicular to the magnetic field in the *core* region and distinct elongation along the magnetic field lines in the outer *halo* region of the distribution. It is sensible to characterize these anisotropies with temperatures derived from the distribution function that have different values for the directions parallel and perpendicular to the magnetic field. Whereas the core region of the distribution shown in Fig. 6.5 is anisotropic with $T_\perp > T_\parallel$, the halo region clearly has $T_\perp < T_\parallel$. In general, one observes a mean temperature anisotropy of $T_\parallel/T_\perp \simeq 1.5$ for protons at 1 AU. The corresponding

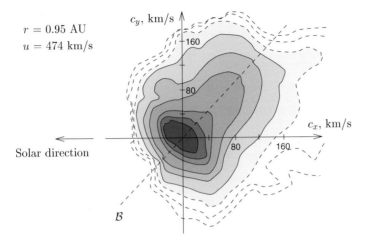

$r = 0.95$ AU
$u = 474$ km/s

c_y, km/s

Solar direction

c_x, km/s

$\vec{B}$

Fig. 6.5. Contour plot of a cross-sectional cut through the measured three-dimensional velocity distribution $g(\vec{c})$ for solar wind protons. Each contour interval shows the number of protons, normalized to a fixed sum total, with thermal velocities between $\vec{c}$ and $\vec{c} + d\vec{c}$ (e.g. between $c_x = 80$ km/s and $c_x = 120$ km/s for $c_y = 0$). The plane of the cross section contains the local magnetic field vector $\vec{B}$. The coordinate system has been chosen such that its origin corresponds to the mean solar wind velocity. The velocity space density increases by a factor of 10 between adjacent contour levels (adapted from Marsch et al., 1982).

electron anisotropy is considerably less pronounced ($T_\parallel/T_\perp \simeq 1.1$). The values of temperature given previously are then obtained as a weighted mean value according to the equation $\langle T \rangle = (T_\parallel + 2T_\perp)/3$.

6.1.4 Extended Gas Dynamic Models

The model discussed above, first introduced by Parker in 1958, represents the simplest gas dynamical approach for a description of the solar wind. An entire suite of considerably more complicated models has been developed in the interim, all of which are based on systems of equations that can only be solved numerically. The improvements provided by these models include

- separate consideration of the proton and electron components
- self-consistent treatment of electromagnetic forces and fields
- incorporation of extended energy balance equations
- explicit retention of temperature anisotropies
- inclusion of wave-particle interactions

and many others. As an example of the results from such refined calculations, Fig. 6.6 shows profiles for the solar wind velocity, density, and both electron and proton temperatures in the inner heliosphere as a function of heliocentric distance. These results should certainly not be considered as final, however,

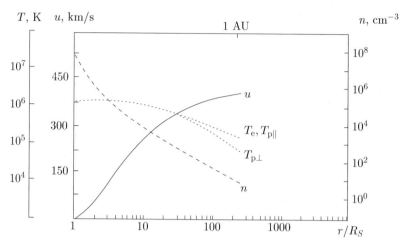

Fig. 6.6. Solar wind velocity, density and temperatures of protons and electrons as a function of heliocentric distance within the inner heliosphere. The profiles have been numerically generated with the help of a complex magnetoplasmadynamic model (adapted from Fichtner and Fahr, 1989).

and many problems remain to be solved. This particularly applies to the modeling of the solar wind acceleration. Apart from this we will return to the topic of self-consistent treatment of electromagnetic forces and fields in Section 6.2.7.

6.1.5 Exospheric Model

As an alternative to the gas dynamical approach, the solar wind can also be described with an exospheric model. The solar wind is interpreted in this case as an evaporative escape flux from the hot solar corona. One proceeds here in a manner similar to that used for the terrestrial exosphere. An exobase is introduced which, to a first approximation, establishes a sharp transition from the collision-dominated to the collisionless regime of the solar atmosphere. The escape flux is subsequently estimated using a Maxwellian velocity distribution, which is still valid below the exobase. Differences from the terrestrial exosphere arise from the environmental conditions, $g_S \gg g_E$, $(T_\infty)_S \gg (T_\infty)_E$, and from the fact that we are dealing with an ionized gas in the case of the solar exosphere. In this sense, the solar case corresponds more to the terrestrial ion-exosphere, which serves as the source of the polar wind in the polar cap regions. In the following we will be content with discussing some important parameters of the solar exosphere such as the exobase height, escape velocity and mean speed of the escape flux.

Exobase Height. As explained in Section 2.4.1, the exobase is defined to be that height where the mean free path of escaping particles is equal to the density scale height of the atmosphere, $l(r_{EB}) = H(r_{EB})$. Both parameters depend on the coronal density profile, taken in the following from the empirically determined formula in Eq. (3.7). For simplicity, we assume that the exobase is located at a coronal height for which the first term in the density formula dominates.

$$n_p(r) = n_e(r) \simeq \frac{10^{14}[\text{m}^{-3}]}{(r/R_S)^6}$$

Inserting this into Eq. (5.58), we obtain the following expression for the mean free path of the charge carriers

$$l_{p,p}(= l_{e,e}) \simeq \frac{1.6 \cdot 10^9 \; T^2}{\ln\Lambda \; n} \simeq 6.7 \cdot 10^{-7} \, T^2 \left(\frac{r}{R_S}\right)^6$$

where in the second step we have used a Coulomb logarithm of $\ln\Lambda \simeq 24$ and expressed all quantities in SI units.

The scale height of the coronal density is estimated in a manner similar to that used for the upper atmosphere in Section 2.3.3. One approximates the density profile at every point by an exponential function of the form $n \sim \exp(-(r - r_0)/H)$ and uses its intrinsic scale height

$$H \simeq \left| \frac{1}{n} \frac{dn}{dr} \right|^{-1} = \frac{r}{6} \tag{6.23}$$

Equating the expressions for $l_{p,p}$ and H, we obtain the following value for the exobase distance

$$r_{EB} \simeq 700 \; T^{-2/5} \; R_S \simeq 2.8 \; R_S \tag{6.24}$$

where we have assumed a coronal temperature of 10^6 K in the second step. This distance does indeed lie within the range of coronal heights for which the assumed density profile is valid.

Escape Velocity. When calculating the escape velocity, it is important to realize that, in addition to the Sun's gravitational acceleration, electric forces must be present to prevent immediate evaporation of the extremely light electron component. We tentatively assume in the following that this electric force is maintained by a Pannekoek-Rosseland polarization field. According to Eq. (4.43) and referred to the coronal situation, this is

$$\mathcal{E}_P = \frac{m_p \; g_S(r)}{2 \; e}$$

where we again assume that the proton and electron temperatures are equal. The escape velocity of the protons is thus calculated from the relation

$$\frac{1}{2} \, m_p \, (v^p_{es})^2 = \int_r^\infty (m_p \, g_S(r') - e \, \mathcal{E}_P(r')) \; dr' = \frac{m_p}{2} \frac{G \, M_S}{r}$$

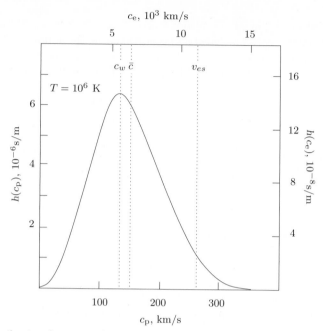

Fig. 6.7. Distribution function of the velocity magnitudes $h(c)$ for coronal protons and electrons. The escape velocity for each component is referred to a heliocentric distance of 2.8 R_S.

and one obtains

$$v_{es}^{\mathrm{p}} = \sqrt{g_S(r)\, r}$$

Evaluating this at the height of the exobase, we obtain an escape velocity of $v_{es}^{\mathrm{p}}(r_{EB}) \simeq 260$ km/s. For the electron component we have

$$\frac{1}{2}\, m_{\mathrm{e}}\, (v_{es}^{\mathrm{e}})^2 = \int_r^\infty (m_{\mathrm{e}}\, g_S(r') + e\, \mathcal{E}_P(r'))\ \mathrm{d}r'$$

and with $(m_{\mathrm{p}}/2)/m_{\mathrm{e}} \gg 1$

$$v_{es}^{\mathrm{e}} = \sqrt{\frac{m_{\mathrm{p}}}{m_{\mathrm{e}}}}\, v_{es}^{\mathrm{p}}$$

The escape velocity of the electrons is thus 43 times larger than that of the protons. However, since their thermal velocity is also larger by this factor, equal amounts of these two components will evaporate. Figure 6.7 illustrates this with the aid of the velocity distribution functions for both gases under coronal conditions. Also apparent in this figure is that a significant fraction of the coronal electrons and protons possesses a sufficiently large thermal velocity to escape from the solar atmosphere.

Escape Flux Velocity. Of particular interest is the mean velocity of the escape flux u_{es} ($\neq v_{es}$), a quantity that corresponds to the solar wind velocity. Here, a superficial examination of the situation sketched in Fig. 6.7

reveals why our present approach is inadequate. As can be seen, the electron component evaporates with a much larger escape flux velocity than the proton component. This leads to a relative motion of these charge carrier gases, i.e. an electric current, that creates a positive charge surplus in the near-Sun region and a negative charge surplus at larger solar distances. The resulting polarization field decelerates the electrons and accelerates the protons in such a way that both attain roughly the same escape flux velocity. It may be concluded that the electric field in the solar wind must be larger than previously assumed. This field is established not only from the requirement of equal proton and electron densities, but also from the requirement for equal escape flux velocities. Note that such an electric field is implicitly contained within the gas dynamic approach by requiring not only equal densities but also equal flow velocities for the two components.

Even when accounting for this considerably stronger electric acceleration field, one obtains escape flux velocities that lie at the lower limit of observed solar wind velocities. Better agreement with the actually observed flow velocities is obtained if one accounts for the energy dependence of the exobase height. According to the approximation represented by Eq. (6.24), the exobase distance is inversely proportional to the mean random velocity

$$r_{EB} \sim T^{-2/5} \sim \left(\overline{c^2}\right)^{-2/5} \simeq \frac{1}{\overline{c}}$$

It follows that particles with higher random velocities can escape from deeper, and thus more densely populated, layers of the corona. This leads to an escape flux with an excessive proportion of particles with higher random velocity and thus a significant increase in the flow velocity.

6.1.6 Large-Scale Solar Wind Structure in the Ecliptic

The solar wind was assumed to be spherically symmetric in the previous sections. This is a blatant oversimplification of the actual situation. Continual changes are observed in the properties of the solar wind over the duration of a solar rotation, whereby the Earth at one time is located in a sector of high solar wind velocity and later in a sector of low solar wind velocity. If one follows these sectors back to their source regions at the Sun, one finds an intimate connection between high speed streams and coronal holes on the one side, and between slow solar wind and regions of closed magnetic configuration (arcs or arcades) on the other; see Fig. 6.8. Every region of the corona in the ecliptic plane (i.e. the orbital plane of the Earth about the Sun, as before sometimes simply denoted 'the ecliptic') evidently emits its own 'flavor' of solar wind. Solar wind particles from each source region differ in their flow velocity, density and temperature from those originating in other source regions.

In order to describe the large-scale structure of this inhomogeneous solar wind, we must know the shape of the different flow sectors as defined by

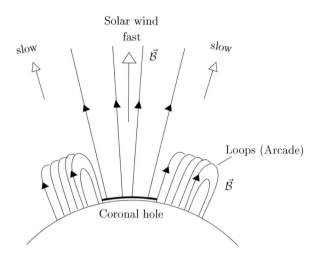

Fig. 6.8. Source regions of the inhomogeneous solar wind in the ecliptic

the associated solar wind *jetlines*. Jetlines connect all outward flowing solar wind particles that have originated from a common, differentially small,
source region. They should not be confused with streamlines, which are always parallel to the velocity vector of the bulk flow. For an observer looking
down onto the ecliptic, the streamlines are purely radial, in compliance with
the radial velocity of the solar wind. Combined with the azimuthal velocity provided by the solar rotation, however, each individual source region on
the Sun produces a jetline that forms an Archimedean spiral. The situation
is similar to the nozzle of a rotating lawn sprinkler that squirts out curved
water jets in the same spiral form. We determine the equation of this spiral
for a solar wind source in the equatorial plane of a Sun-centered heliospheric
coordinate system, for which r is the heliocentric distance and φ and λ are
the heliographic latitude and longitude, respectively; see Fig. 6.9. Since the
solar equator is tilted only by about $7°$ to the ecliptic, the following results
are valid to good approximation for both planes.

Consider a differentially small coronal hole at the solar equator that is
located at longitude $-\lambda$ at the time $t = 0$ and at longitude $-\lambda_{ch}$ at the time
t; see Fig. 6.10. The longitude λ is measured in the mathematically positive
sense (counterclockwise in Fig. 6.10) from the reference longitude $\lambda = 0$. Solar
wind particles emitted at time $t = 0$ at a heliocentric distance r_0 have moved
radially outward a distance $u_{sw} \cdot t$ after a time t, where u_{sw} is approximately
their asymptotic solar wind velocity. Their heliocentric distance at time t is
therefore

$$r(t) = r_0 + u_{sw}\, t$$

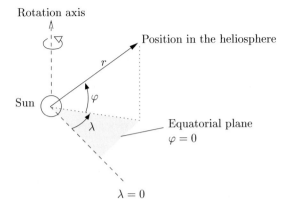

Fig. 6.9. Sun-centered interplanetary or heliospheric coordinate system. Note that this coordinate system is spatially fixed, rather than rotating with the Sun.

Replacing the time with the difference in longitude covered in this time by the solar rotation, $t = (-\lambda + \lambda_{ch})/\Omega_S$, and considering distances that are large compared with the source distance r_0, one obtains the approximation

$$r(\lambda) \simeq -\frac{u_{sw}}{\Omega_S}\,(\lambda - \lambda_{ch}), \qquad \lambda < \lambda_{ch}\,, \; r \gg r_0 \qquad (6.25)$$

This corresponds to an Archimedean spiral (in this connection sometimes called a *Parker spiral*), the curvature of which is determined by the solar wind velocity. Note that the *sidereal* rotation period of the Sun ($\simeq 25$ d) should be used to determine the solar rotation velocity.

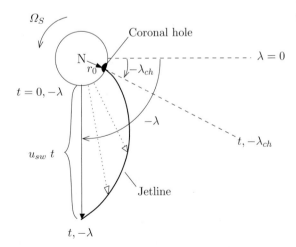

Fig. 6.10. Calculating the jetline equation (view from north onto the heliographic equator)

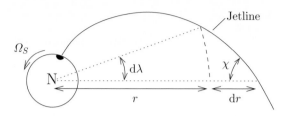

Fig. 6.11. Calculating the jetline angle χ (view from north onto the heliographic equator)

Equation (6.25) is an approximation for two reasons other than the condition $r \gg r_0$. The first is that the solar wind velocity only reaches its asymptotic velocity after a certain distance from the Sun, but we have assumed this (constant) velocity is attained immediately at the source. The second is that we have neglected the – admittedly small – zonal component of the solar wind velocity. The solar wind, after all, rotates with the Sun at least up to the exobase height before its outward motion is decoupled. The magnitude of this zonal velocity component is given by $u_\lambda \simeq \Omega_S\, r_{EB} \simeq 6$ km/s. This means that the relative difference in longitude covered in the time t is somewhat smaller than given above. We neglect this minor effect and also consider in the following only distances large compared with the position of the exobase, $r \gg r_0 \simeq r_{EB} \simeq 3R_S$.

In order to better visualize the form of solar wind jetlines, we determine the angle χ between a given jetline and the radial direction. As shown in Fig. 6.11, this angle may be calculated from

$$\tan\chi \simeq \frac{r\,d\lambda}{dr} \simeq -\frac{\Omega_S}{u_{sw}}\,r, \qquad r \gg r_0 \tag{6.26}$$

where the expression on the right follows from Eq. (6.25). The jetline angle is thus given by

$$|\chi| \simeq \arctan\left(\frac{\Omega_S}{u_{sw}}\,r\right) \simeq \arctan\left(6\cdot10^{-12}\,r[\mathrm{m}]\right) \tag{6.27}$$

The mean jetline angle at the Earth's orbit is 43°; at the Pluto distance ($\simeq 40$ AU) more than 88°.

The large-scale structure of the solar wind can be described with the help of these jetlines. Figure 6.12 illustrates the overall picture with a view from north onto the ecliptic, which approximately corresponds to the heliographic equatorial plane. As indicated, each solar wind sector is characterized by jetlines that form Archimedean spirals with various curvature as determined by the local solar wind velocity. Flow within a given sector is prevented from penetrating into neighboring sectors by the interplanetary magnetic field (treated in the following sections), which is embedded in and carried along by the solar wind. Indeed, contact discontinuities form at the sector boundaries that

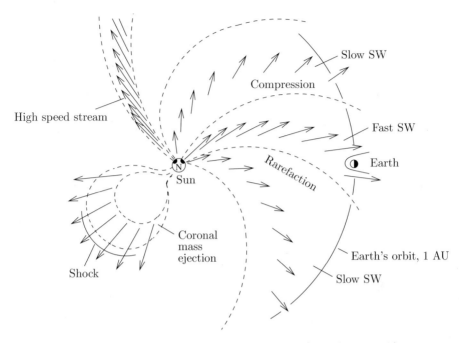

Fig. 6.12. Solar wind structure in the ecliptic (view from north)

separate and tend to isolate the individual sectors. At the same time, jetlines at the boundaries flex and adjust to the different adjacent flow profiles. This leads to the formation of rarefaction and compression regions; slower solar wind not only fails to keep pace with the trailing edge of any faster winds ahead of it, but also gets 'swept up' by the leading edge of faster solar wind flows behind it (see Fig. 6.39 for a more detailed sketch).

In addition to the relatively stable sources of solar wind (equatorial coronal holes have a mean lifetime of about 6 solar rotation periods), there are also ephemeral sources such as coronal mass ejections (see Fig. 6.12). These are considered in more detail in Section 8.6.1.

6.1.7 Solar Wind Outside the Ecliptic Plane

Our previous description of the solar wind is based exclusively on observations in or near the ecliptic and is thus valid essentially for the solar wind at low heliographic latitudes. That the solar wind at higher latitudes might be different was first shown by radio scintillation observations. Those results were impressively confirmed by the first *in situ* measurements outside of the ecliptic plane. Figure 6.13 shows the latitudinal profile of solar wind velocity and density recorded on the ULYSSES spacecraft along its polar orbit around the Sun. The differences observed during this period of low solar activity

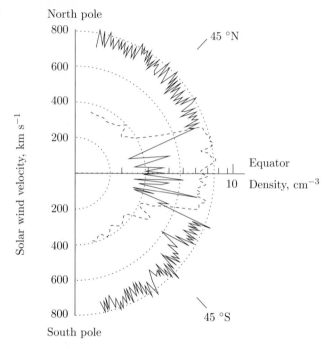

Fig. 6.13. Solar wind velocity (*solid line*) and density (*dashed line*) as a function of heliographic latitude. The north and south poles are aligned with the solar rotation axis and the heliospheric equator is identical to the heliographic equator (see Fig. 6.9). The measurements were recorded on board the ULYSSES spacecraft during a period of low solar activity (1994/95) at solar distances >1 AU (adapted from McComas et al., 2000).

are astounding. The solar wind at mid to high latitudes flows with a very high velocity of about 750–800 km/s and a relatively low density of 3 protons/electrons per cm^3. This compares with the slower and denser solar wind at low latitudes, which is, however, subject to considerable fluctuations. This smaller average velocity and larger density is evidently linked to the primarily closed magnetic field structures that tend to appear at low solar latitudes (see Fig. 6.14b). The strong fluctuations must then reflect the spatial proximity of open and closed magnetic field regions. Comparable fluctuations are observed in the ecliptic and explain the often observed alternation between fast/tenuous and slow/dense solar wind flows near the Earth.

6.2 Interplanetary Magnetic Field

The interplanetary magnetic field is intimately interconnected with the solar wind. We first discuss the pertinent observations and then their modeling.

This is followed by a description of the heliospheric current system responsible for this magnetic field. The section is concluded with some remarks on the interplanetary electric field and the description of the interplanetary medium as a magnetoplasma.

6.2.1 Observations

Coronal images, such as the one shown in Fig. 3.4a, show a distinct structuring of the radiation intensity that can only be caused by the coronal magnetic field. This magnetic field evidently governs the motion and density distribution of the fully ionized coronal gas out to many solar radii. In particular, stronger emission and correspondingly greater particle densities are found in the closed magnetic structures appearing as arcs and rays. The situation here is remotely similar to the concentration of particles in the terrestrial plasmasphere. Open magnetic field structures, on the other hand, are associated with regions of weak emission, e.g. coronal holes. Particles here can escape relatively easily from the corona, which explains the low particle densities and the development of high solar wind flow speeds in these regions. The typical structure of the coronal magnetic field during high and low solar activity, as derived from coronal images, is drawn in Fig. 6.14. As indicated, magnetically closed structures are concentrated at low latitudes on the quiet Sun, but are distributed over the entire solar surface at times of high activity. Another striking feature is the good correspondence between the magnetic field structure shown in Fig. 6.14b and the solar wind distribution of Fig. 6.13, both of which hold for a period of low solar activity (and thus a period of 'ordered conditions'). By the same token, the almost chaotic structures of Figs. 6.14a and 3.4a display a number of similarities. In spite of the great complexity of the coronal magnetic field, it is evident that the magnetic field lines are almost radial beyond roughly $r \simeq 3\ R_S$.

Whereas the structure of the coronal magnetic field can be directly derived from the coronal brightness or density distribution, its field strength is much more difficult to determine. In contrast to the photosphere, magnetic fields of the corona are so weak that they can no longer be determined from the Zeeman effect (i.e. from emission line splitting in a magnetic field). Extrapolation of photospheric fields into the outer solar atmosphere is possible (and regularly practiced), but strictly permissible only under the improbable condition that the corona be free of currents. At the present, the only known method for determining the mean coronal magnetic field is based on radio occultation measurements. This technique exploits the fact that the plane of polarization of a linearly polarized radio signal rotates upon propagating through a magnetized plasma. The magnitude of this *Faraday rotation* is proportional to the strength of the coronal magnetic field. At the exobase distance a mean magnetic field strength of $\mathcal{B}(3R_S) \simeq 10\text{-}30\ \mu\text{T}$ has been determined from these measurements.

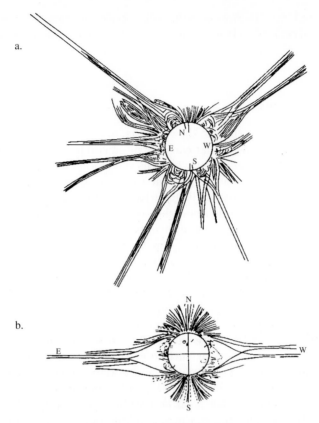

Fig. 6.14. Structure of the coronal magnetic field derived from images of the solar corona in white light: **(a)** solar cycle maximum; **(b)** solar cycle minimum (adapted from Akasofu and Chapman, 1972; Bronshtén, 1960; Vsekhsvjatsky, 1963).

In contrast to the corona, interplanetary space is accessible to direct exploration by spacecraft. Some results of such *in situ* measurements near the Earth are summarized in Table 6.2. The strength of the interplanetary magnetic field at 1 AU is about 3.5 nT, whereby – similar to other properties of the solar wind – considerable fluctuations are observed. The radial and zonal components of the magnetic field, as defined by the coordinate system sketched in Fig. 6.9, are of roughly equal strength. Because of the small inclination of the solar equator with respect to the ecliptic ($\simeq 7°$), these values are also valid to a good approximation for an ecliptic heliocentric coordinate system with the λ-coordinate in the angular direction of the Earth in its motion around the Sun.

One striking feature is that the radial and zonal components change their sign together at more or less regular intervals, the polarity reversal occurring typically twice or four times per solar rotation period. The Earth is thus

Table 6.2. Average properties of the interplanetary magnetic field at the Earth's orbit. The magnetic field components refer to the coordinate system shown in Fig. 6.9.

Magnetic field:	$\mathcal{B}$	$\simeq$	3.5 (0.2–50) nT
	$\mathcal{B}_r$	$\simeq$	$\pm$ 2.6 nT
	$\mathcal{B}_\lambda$	$\simeq$	$\mp$ 2.4 nT
	$\mathcal{B}_\varphi$	$\simeq$	0 ($\pm$ 30) nT
Inclination angle to the radial direction:	χ	$\simeq$	43°
Alfvén velocity:	v_A	$\simeq$	30 km/s
Magnetosonic velocity:	v_{MS}	$\simeq$	60 km/s
β^* parameter:	β^*	$\simeq$	450

embedded sometimes in a magnetic field that is directed away from the Sun (magnetic *positive* sector), and at other times in a field pointing toward the Sun (magnetic *negative* sector); see also Fig. 6.15.

The $\mathcal{B}_\varphi$ component, which is perpendicular to the heliospheric equatorial plane – and thus approximately perpendicular to the ecliptic – displays a mean value of zero, but can assume values of the same order as the other components at times. As explained in Section 7.6, this component plays a key role in the transfer of energy from the solar wind to the magnetosphere.

Knowing the magnetic field strength, a number of additional characteristic quantities can be calculated (see Table 6.2). Among these are the Alfvén and the magnetosonic velocities, to be introduced in the following sections, which represent the propagation and information transfer velocities for all solar wind phenomena. Another important parameter is the ratio of kinetic ($\simeq$ dynamic) to magnetic pressure, β^*, which is very large. This means that the magnetic field configuration must adapt itself to the structure of the solar wind.

The radial dependence of the magnetic field strength and orientation has been measured using interplanetary spacecraft down to a minimum solar distance of approximately 0.3 AU. The average magnetic field was found to be represented by the relations

$$\mathcal{B}_r[\text{nT}] \simeq 2.6/(r[\text{AU}])^2$$
$$\mathcal{B}_\lambda[\text{nT}] \simeq 2.4/r[\text{AU}] \tag{6.28}$$

This may be used to determine the inclination angle of the magnetic field to the radial direction (see Fig. 6.17)

$$|\chi| = \arctan\left|\frac{\mathcal{B}_\lambda}{\mathcal{B}_r}\right| \simeq \arctan\left(6 \cdot 10^{-12}\ r[\text{m}]\right)$$

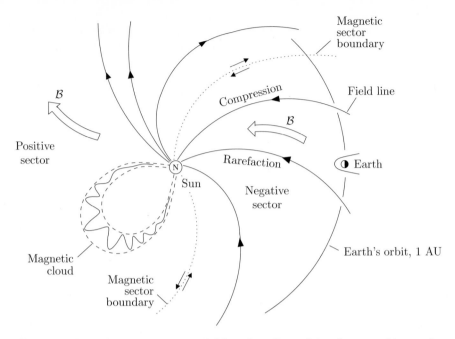

Fig. 6.15. Interplanetary magnetic field in the ecliptic (view from north), see also Fig. 6.12

which corresponds exactly to the jetline angle given in Eq. (6.27). The field lines of the interplanetary magnetic field are evidently oriented parallel to the solar wind jetlines, and are thereby also described by Archimedean spirals. This is sketched schematically in Fig. 6.15, a view onto the north side of the ecliptic plane under solar wind conditions corresponding to those in Fig. 6.12.

6.2.2 Simple Model of the Interplanetary Magnetic Field

In order to understand the field line configuration and the radial dependence of the field strength, we make use of two familiar theorems of ideal magneto-plasmadynamics

- Plasma elements connected at any point in time by a common magnetic field line remain connected by a common field line   (6.29)

- The magnetic flux passing through a closed curve that moves with the plasma flow velocity remains constant   (6.30)

Proofs of these theorems are given in Appendix A.14. The content of the first theorem is graphically illustrated in Fig. 6.16. Although the individual plasma elements move at different speeds in different directions, they remain

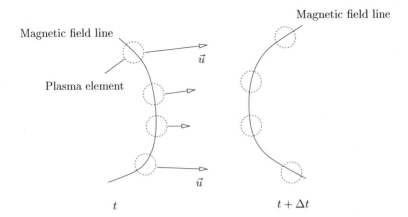

Fig. 6.16. Graphic illustration of theorem (6.29)

magnetically connected via their mutually common field line. Whether or not this is one and the same field line is not stated. Electrodynamic theory tells us nothing about the possible motion of field lines, which, after all, are fictitious quantities. Particularly in those cases where the flow energy of the plasma dominates ($\beta^* \gg 1$, as in the solar wind), the notion has been proven useful that the field lines are 'frozen in' and carried along with the plasma. Following this concept, the field line shown in Fig. 6.16 for times t and $t + \Delta t$ is one and the same and is 'dragged along' by the plasma. If the energy density of the magnetic field dominates ($\beta^* \ll 1$, as in the magnetosphere), however, an often preferred notion is that the plasma elements drift from one common field line to the next one. In this alternative view, the two field lines of Fig. 6.16 are different.

Independent of the preferred interpretation, theorem (6.29) enables an understanding of the interplanetary magnetic field configuration. The magnetic field at its origin in the corona is relatively strong and directed radially outwards. At a heliocentric distance of three solar radii, for example, where $\mathcal{B}(3R_S) \simeq 20 \ \mu T$, $n(3R_S) \simeq 10^{11} \mathrm{m}^{-3}$ and $T < 2 \cdot 10^6$ K, we obtain a value for β^* that is much smaller than unity. The magnetic field therefore determines the motion of the plasma and, in the absence of external forces, the solar wind can only flow along (parallel to) the given magnetic field. This also means that all solar wind packets originating from a differentially small source region must first move outward along a common field line anchored in the source region. It then follows from theorem (6.29) that they must remain connected via a common field line in interplanetary space at some later point in time. Since the line connecting these solar wind packets corresponds to their jetline, the common field line must assume the same form. This explains why the field lines of the interplanetary magnetic field in the ecliptic form Archimedean spirals, the particular curvature of each being determined by the solar wind velocity.

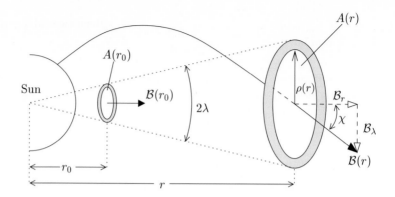

Fig. 6.17. Determining the radial dependence of the magnetic field strength in interplanetary space (view from the north onto the ecliptic)

Knowing the shape of the field lines, the radial dependence of the field strength may be determined. Consider a plasma ring, composed of contiguous solar wind elements, the enclosed area of which is oriented perpendicular to the heliocentric radial direction; see Fig. 6.17. Assuming equal solar wind velocity everywhere, the plasma ring will move radially outward but retain its orientation. According to the second of the above theorems, the magnetic flux passing through the plasma ring must be conserved. With $\Phi = \vec{\mathcal{B}} \, \hat{r} A =$ const., we may write

$$\mathcal{B}(r_0)A(r_0) = \mathcal{B}_r(r)A(r)$$

where the surface area is given by

$$A(r) = \rho^2(r)\pi = (r \tan \lambda)^2 \pi$$

Accordingly, the radial component of the interplanetary magnetic field is given by

$$\mathcal{B}_r(r) = \mathcal{B}(r_0) \left(\frac{r_0}{r}\right)^2 \tag{6.31}$$

Referring to Fig. 6.17 and Eq. (6.26), it also follows that

$$\frac{\mathcal{B}_\lambda(r)}{\mathcal{B}_r(r)} = \tan \chi \simeq -\frac{\Omega_S \, r}{u_{sw}}$$

from which one obtains the azimuthal component of the interplanetary magnetic field in the ecliptic

$$\mathcal{B}_\lambda(r) \simeq -\mathcal{B}(r_0) \frac{\Omega_S \, r_0}{u_{sw}} \left(\frac{r_0}{r}\right), \qquad r \gg r_0 \tag{6.32}$$

Both of these results are in agreement with the observations, summarized by the relations given in Eq. (6.28). A comparison of the radial dependence of $\mathcal{B}_r$

and $\mathcal{B}_\lambda$ also shows that the azimuthal component becomes predominant at large heliocentric distances and the field lines become continually more like concentric circles around the Sun. Already at the orbital distance of Uranus ($\simeq 19$ AU), the field direction deviates from that of a circle by only a few degrees. The radial dependence of the total field strength in the ecliptic is given by

$$\mathcal{B} = \sqrt{\mathcal{B}_r^2 + \mathcal{B}_\lambda^2} \simeq \mathcal{B}(r_0) \sqrt{1 + \left(\frac{\Omega_S \, r}{u_{sw}}\right)^2} \left(\frac{r_0}{r}\right)^2, \qquad r \gg r_0 \qquad (6.33)$$

It is remarkable that, prior to the availability of corresponding measurements, the morphology of the interplanetary magnetic field could be correctly predicted on the basis of the fundamental concepts outlined above (Parker, in the year 1958).

6.2.3 Magnetic Field Structure Outside the Ecliptic

As in the ecliptic, the individual jetlines of the solar wind also delineate the magnetic field lines outside of this plane. We thus consider a solar wind parcel emitted from a differentially small source region at a higher heliographic latitude. As indicated by the dotted lines in Fig. 6.18, all solar wind parcels from this source flow radially outward along the generatrices of a cone. The component of solar wind velocity perpendicular to the solar rotation axis is $u_{r'} = u_{sw} \cos\varphi$, so the steadily increasing distance of the solar wind parcel from this axis may be written as

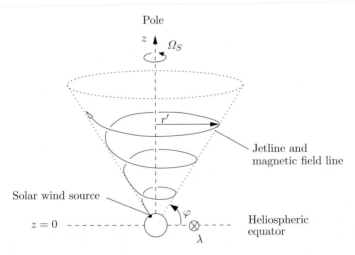

Fig. 6.18. Interplanetary solar wind jetlines and magnetic field lines outside the ecliptic. φ denotes the heliographic latitude of the rotating solar wind source, r' is the distance of the solar wind parcel and/or magnetic field line from the solar rotation axis, and λ is the zonal angle introduced in Fig. 6.10.

$$r'(\lambda) \simeq -\frac{u_{sw}}{\Omega_S} (\lambda - \lambda_{ch}) \cos\varphi, \qquad \lambda < \lambda_{ch}, \; r' \gg r_0$$

where the designations correspond to those used in Eq. (6.25). At the same time, the solar wind parcel moves away from the equatorial plane according to the relation

$$z(\lambda) \simeq -\frac{u_{sw}}{\Omega_S} (\lambda - \lambda_{ch}) \sin\varphi, \qquad \lambda < \lambda_{ch}, \; z \gg r_0$$

In the cylindrical coordinate system used here, these equations describe a spiral with increasing radius and radial inclination, as shown in Fig. 6.18. This jetline simultaneously defines the locus of the interplanetary magnetic field line anchored in the rotating solar wind source. Figure 6.18 thus illustrates not only the shape of the jetlines, but also quite satisfactorily the structure of the interplanetary magnetic field outside the ecliptic.

6.2.4 Heliospheric Current Sheet

As shown above, the two theorems (6.29) and (6.30) represent a formal basis for a quantitative description of the interplanetary magnetic field. We left the issue concerning the origin of these fields, however, as an open question. In search of an answer, we recall the argumentation applied in connection with the deformation of the distant geomagnetic field.

We first note that, at larger distances from an arbitrarily complex magnetized object such as the Sun, the dipole component of the source field will dominate because it decreases most slowly with increasing distance. As we have seen, however, the interplanetary magnetic field is anything but dipolar. In order to explain the observations, we thus need an additional magnetic field of *extrasolar* origin that can be superposed onto the original dipole field. This superposed field can only be produced by electric currents in interplanetary space, the distribution of which is described in the following. Because of our particular interest for the Earth's space environment, we again limit ourselves to the region in and near the ecliptic plane.

The essential element of the interplanetary current system is a current flowing in a huge circumsolar disk, which is denoted the *heliospheric current sheet*. In a meridional cross section (Fig. 6.19a), this current resembles the magnetotail current of the terrestrial magnetosphere. A neutral sheet, across which a reversal of the horizontal magnetic field component occurs, is also embedded in the heliospheric current sheet. Viewed either from above (Fig. 6.19b) or below, the currents flow along logarithmic spirals and thus – as required – perpendicular to the Archimedean spirals of magnetic field lines. The strength of these currents can be determined with the help of Eq. (5.75). Vector multiplication of this relation with the unit surface normal $\hat{n}$ yields

$$\vec{\mathcal{I}}^* = -\frac{2}{\mu_0} \vec{B} \times \hat{n} \qquad (6.34)$$

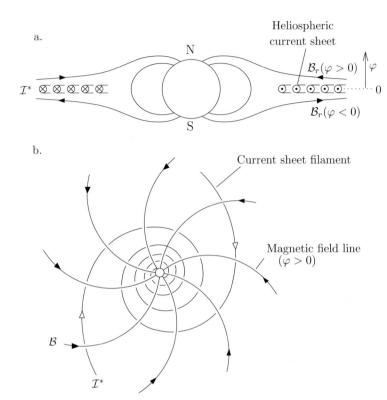

Fig. 6.19. Heliospheric current sheet and associated magnetic field: (**a**) meridional cross section; (**b**) view from north. The phase of the solar cycle is such that a south magnetic pole is located at the northern heliographic pole (Fig. 6.19b adapted from Alfvén, 1981).

For the situation sketched in Fig. 6.19 it may also be verified that $\vec{\mathcal{B}}(\varphi > 0) = -\hat{r}|\mathcal{B}_r| + \hat{\lambda}|\mathcal{B}_\lambda|$. It follows that the two components of the surface current density are given by

$$\mathcal{I}^*_\lambda = -\frac{2\,|\mathcal{B}_r|}{\mu_0} = -\frac{2\,|\mathcal{B}(r_0)|}{\mu_0}\left(\frac{r_0}{r}\right)^2 \tag{6.35}$$

and

$$\mathcal{I}^*_r = -\frac{2\,|\mathcal{B}_\lambda|}{\mu_0} \simeq -\frac{2\Omega_S|\mathcal{B}(r_0)|r_0}{\mu_0\,u_{sw}}\left(\frac{r_0}{r}\right), \qquad r \gg r_0 \tag{6.36}$$

It is remarkable that the radial component of the spiral structure implies a current that continually flows toward the Sun. The charge accumulating from this process must be removed elsewhere. This occurs most simply via line currents that originate over the Sun's poles; see Fig. 6.20. Since the radial component of the surface current density must be balanced by the polar line

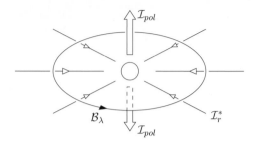

Fig. 6.20. Suggested continuation of the radial component of the heliospheric current sheet

currents, $|\mathcal{I}_r^*| = 2\mathcal{I}_{pol}/2\pi r$, the strength of the polar line current may be expressed as

$$\mathcal{I}_{pol} = 2\pi r |\mathcal{B}_\lambda(r)|/\mu_0$$

Inserting the value $|\mathcal{B}_\lambda(1\ \mathrm{AU})| \simeq 2.4$ nT, one obtains the considerable line current strength of $\mathcal{I}_{pol} \simeq 2 \cdot 10^9$ A.

Theories concerning the origin of the heliospheric current sheet remain speculative. As in the case of the terrestrial magnetotail, magnetization currents and particle drifts are expected to play an important role. The radial component of the current sheet, as once suggested by Alfvén, might be driven by unipolar induction.

The heliospheric current sheet is often located at low heliospheric latitudes (but not necessarily at solar activity maximum) and is associated with a solar wind of low velocity and high density. Because these conditions are similar to those observed in the geomagnetic tail, this kind of solar wind is also sometimes denoted the *heliospheric plasma sheet*.

6.2.5 Sector Structure and $\mathcal{B}_\varphi$ Component

The heliospheric current sheet described in the preceding section provides a plausible explanation for the sector structure of the interplanetary magnetic field observed in the ecliptic. Assuming that the current sheet is inclined with respect to the heliographic equator, the Earth will at one time be located above, but at other times below the current sheet, over the course of a solar rotation; see Fig. 6.21. Accordingly, an observer near the Earth will at one time record a magnetic field directed away from the Sun, but at other times directed toward the Sun. The thickness of this current sheet, as inferred from the duration of the transition between sectors, is estimated to be only a few 1000 km.

In addition to the two large-scale sectors, fluctuations of the field direction have been observed at smaller scales that are attributed to an azimuthal warp in the heliospheric current sheet; see Fig. 6.22. Such warps are likely caused by the nonuniform heliographic distribution of coronal holes. Distortions of

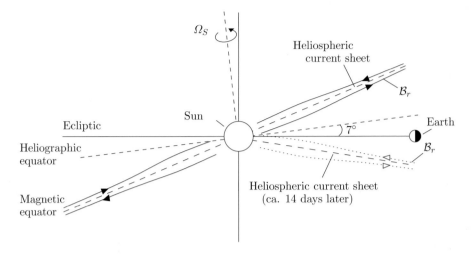

Fig. 6.21. Formation of the two-sector structure of the interplanetary magnetic field

this type rotate more or less rigidly with the Sun and pass over the Earth once per synodic rotation period. An appropriate analogy is to imagine the heliospheric current sheet as the twirling skirt of a ballerina, where the azimuthal warps are folds in the skirt and the rotating Sun is the ballerina.

In addition to the azimuthal warps, radial deformations of the heliospheric current sheet have been observed, that are most likely the result of nonstationary events. Although the meridional component of the interplanetary magnetic field, $\mathcal{B}_\varphi$, has a mean value of zero, radial deformations of the current sheet may lead temporarily to relatively large values of this quantity; see Fig. 6.23.

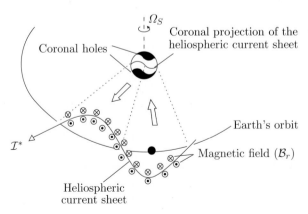

Fig. 6.22. Warp in the azimuthal component of the heliospheric current sheet: formation of additional polarity sectors

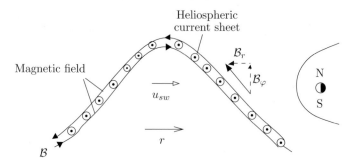

Fig. 6.23. Radial deformation of the heliospheric current sheet resulting first in a positive, then in a negative $\mathcal{B}_\varphi$ component of the interplanetary magnetic field

6.2.6 Interplanetary Electric Field

The structure of the interplanetary magnetic field is taken to be known in the following. Consider the motion of a solar wind particle in this magnetic field, which we now assume to be static and not frozen into the solar wind flow. Near the Sun these particles will flow along and parallel to the magnetic field in the direction of interplanetary space. With increasing solar distance, however, the azimuthal component of the interplanetary magnetic field, $\mathcal{B}_\lambda$, becomes more noticeable. This latter scenario is sketched in Fig. 6.24 for a magnetic field pointing away from the Sun (positive magnetic field sector). Because of the magnetic force associated with the $\mathcal{B}_\lambda$ component, solar wind protons are deflected to the south and solar wind electrons to the north, and are forced into gyration orbits. In order to maintain the actually observed flow of solar wind particles, an electric field is required, directed from south to north in the situation considered here, which exactly compensates the effect of the magnetic force. It is evidently this *interplanetary electric field* that allows solar wind particles to move in a direction perpendicular to the azimuthal component of the interplanetary magnetic field. The intensity of

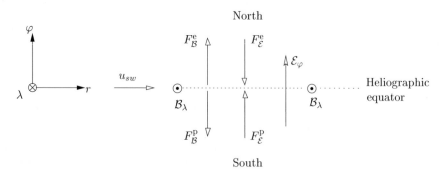

Fig. 6.24. The orientation and magnitude of the interplanetary electric field

this electric field may be obtained from the condition that the magnetic and electric forces acting on the solar wind particles, when averaged over time, must exactly cancel each other, $\langle \vec{F}_\mathcal{B} + \vec{F}_\mathcal{E} \rangle = 0$. Hence we may write

$$\langle \vec{F}_\mathcal{B} \rangle = q \, \vec{u}_{sw} \times \vec{\mathcal{B}} = -\langle \vec{F}_\mathcal{E} \rangle = -q \, \vec{\mathcal{E}}$$

from which it follows that

$$\vec{\mathcal{E}} = -\vec{u}_{sw} \times \vec{\mathcal{B}} \qquad (6.37)$$

This corresponds to the familiar law of electrodynamics, that an electric field given by Eq. (6.37) is induced in a conductor moving laterally through a magnetic field.

Given this electric field, the solar wind motion can also be understood as an $\vec{\mathcal{E}} \times \vec{\mathcal{B}}$ drift. Multiplying Eq. (6.37) with the vector $\vec{\mathcal{B}}$, one obtains just the appropriate drift velocity for this case, $\vec{u}_{sw} = \vec{\mathcal{E}} \times \vec{\mathcal{B}}/\mathcal{B}^2$, see Eq. (5.41). Applying this to the situation drawn in Fig. 6.24, one obtains $|\mathcal{E}_\varphi| = |u_{sw} \mathcal{B}_\lambda|$ or $|u_{sw}| = |\mathcal{E}_\varphi/\mathcal{B}_\lambda|$.

It is important to realize that the interplanetary electric field only exists for an observer at rest in a heliospheric or ecliptic coordinate system. Only a stationary observer notices that the solar wind particles move perpendicular to the azimuthal component of the interplanetary magnetic field, and concludes that the magnetic forces must be compensated by an electric field (or equivalently, the flow moves at the $\vec{\mathcal{E}} \times \vec{\mathcal{B}}$ drift velocity). Moreover, only the observer at rest is able to measure this electric field with the help of electrodes. The situation is different for an observer moving with the solar wind. The solar wind particles are at rest in this system and thereby display no magnetic force effects and also no acceleration from an electric field. The differences in the moving and stationary systems are illustrated in Fig. 6.25. The situation described here reminds us that electric fields are only defined with respect to a given coordinate system, and that the following relations are valid for nonrelativistic coordinate transformations of fields in good conducting plasmas

$$\vec{\mathcal{B}} \simeq \vec{\mathcal{B}}^* \qquad (6.38)$$
$$\vec{\mathcal{E}} \simeq \vec{\mathcal{E}}^* - \vec{u} \times \vec{\mathcal{B}}^* \qquad (6.39)$$

or

$$\vec{\mathcal{B}}^* \simeq \vec{\mathcal{B}} \qquad (6.40)$$
$$\vec{\mathcal{E}}^* \simeq \vec{\mathcal{E}} + \vec{u} \times \vec{\mathcal{B}} \qquad (6.41)$$

The transformation relations are valid as long as the conditions $u \ll c_0$ and $u \, \mathcal{E}/\mathcal{B} \ll c_0^2$ are fulfilled, which is surely the case in the interplanetary medium.

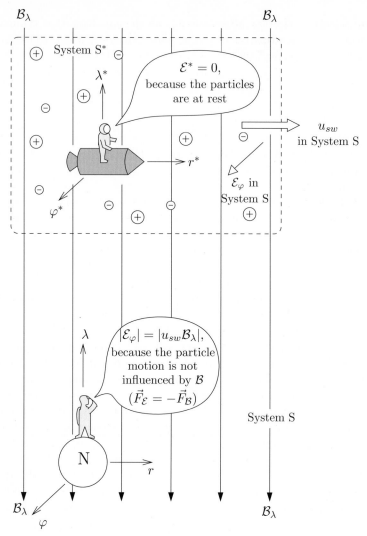

Fig. 6.25. Field transformation in ecliptic coordinates between a system at rest and a system moving with the solar wind (view from the north onto the ecliptic; positive magnetic field sector)

6.2.7 The Interplanetary Medium as a Magnetoplasma

It is actually surprising that the rather simple approach used thus far has been so successful in its description of the interplanetary medium. For example, the solar wind in the gas dynamic model was treated as a neutral gas, although it consists of charge carrier gases that are subject to electric and magnetic forces. Furthermore, the interplanetary magnetic field, the heliospheric currents and the interplanetary electric field were all derived *ex post facto*. We would rather expect that quantities such as the plasma velocity, the electric and magnetic fields and electric currents of magnetically biased, ionized gases are intimately coupled to each other and would generally have to be determined self-consistently. After all, the heliospheric currents created by the slightly different motion of the ions and electrons provide feedback via their magnetic field and associated magnetic force onto the motion of the charge carriers themselves. Recognizing that our previous treatment is in need of reexamination, a self-consistent description of the interplanetary medium is introduced in the following that accounts for the charge carrier properties of the gases.

The basis of this description is provided by the density, momentum and energy balance equations of a plasma. Corresponding to the solar wind, this plasma will be a magnetized, fully ionized gas mixture consisting of ions and their associated electrons. The two gas components strive to attain equal densities everywhere. As we know from Section 4.4, even the slightest deviation from this state produces electric polarization fields that immediately eradicate any charge inequality. We therefore again assume that the ion and electron densities are equal everywhere. In contrast to previous assumptions, however, we now allow for relative motion between the ions and electrons in order to be able to describe heliospheric currents. Our goal will be to derive the balance equations of a magnetized, fully ionized plasma under the conditions

- $n_i \simeq n_e = n$
- $\vec{u}_i \neq \vec{u}_e$

where the indices i and e again denote the ions and electrons, respectively. This derivation proceeds in a manner similar to that used for a neutral gas mixture, namely by adding the balance equations of the participating single gases, in this case the ion and electron gases. The details of this somewhat extensive calculation are presented in Appendix A.13 and will be omitted here. As a result, one obtains the plasma balance equations (6.43) to (6.45) given below. Here, the quantity α is a constant and γ^* is the polytropic index.

Compared with the corresponding balance equations for a neutral gas mixture, these equations contain an additional force term $\vec{j} \times \vec{\mathcal{B}}$, and with it two additional unknowns, the current density $\vec{j}$ and the magnetic field $\vec{\mathcal{B}}$. Accordingly, the five scalar equations (6.43) to (6.45) contain eleven unknowns $\rho, p, u_x, u_y, u_z, j_x, j_y, j_z, \mathcal{B}_x, \mathcal{B}_y$ and $\mathcal{B}_z$. Clearly, six additional equations are required for closure. Since the additional unknowns are electromagnetic quantities, it is intuitive to include the Maxwell equations. While this helps, it is not enough. What we still lack is a generalized form of Ohm's law, which provides a complementary relation between the plasma state parameters and the electromagnetic quantities. The derivation of this additional equation and its radical simplification are also described in Appendix A.13. To a good approximation we obtain

$$\vec{\mathcal{E}} \simeq -\vec{u} \times \vec{\mathcal{B}} \tag{6.42}$$

Evidently, only dynamo (or transformation induced) electric fields are considered in our plasma description, see Eqs. (6.37) and (6.39). This implicitly assumes that the conductivity along the magnetic field lines is so large that only perpendicular electric fields play a role. Equation (6.42) also implies that all plasma motion perpendicular to the magnetic field may be considered to be an $\vec{\mathcal{E}} \times \vec{B}$ drift, or alternatively, that the magnetic fields may be considered to be frozen into the plasma motion. We use this approximation here to eliminate the electric field from the Maxwell equation (A.92) and obtain the following set of equations

$$\frac{\partial \rho}{\partial t} + \nabla(\rho \vec{u}) = 0 \tag{6.43}$$

$$\rho \frac{D\vec{u}}{Dt} = -\nabla p + \rho \vec{g} + \vec{j} \times \vec{B} \tag{6.44}$$

$$p = \alpha \rho^{\gamma^*} \tag{6.45}$$

$$\nabla \times \vec{B} = \mu_0 \vec{j} \tag{6.46}$$

$$\frac{\partial \vec{B}}{\partial t} = \nabla \times (\vec{u} \times \vec{B}) \tag{6.47}$$

where ρ, $\vec{u}$, p and $\vec{j}$ are, respectively, the plasma density, plasma velocity, plasma pressure and current density

$$\rho = \rho_i + \rho_e \tag{6.48}$$

$$\vec{u} = (\rho_i \vec{u}_i + \rho_e \vec{u}_e)/\rho \tag{6.49}$$

$$p = p_i + p_e \tag{6.50}$$

$$\vec{j} = e\,n\,(\vec{u}_i - \vec{u}_e) \tag{6.51}$$

The plasma velocity is obviously the mean velocity of the gas mixture, weighted by the mass of the individual components. This is the ion velocity to a good approximation for a moderate relative velocity between ions and electrons.

The above system of equations corresponds to that of *ideal magnetohydrodynamics (MHD)* – or more appropriately ideal magneto*plasma*dynamics, whereby the term 'ideal' refers, among others, to the neglect of the viscosity in the momentum equation and the radical simplifications leading to Eq. (6.42). As required, eleven scalar unknowns are now matched by eleven scalar equations. It is immediately understandable that this set of partial, nonlinear, coupled differential equations can only be solved numerically with the help of complex algorithms. Our only interest here, however, is to determine whether or not this system of equations is compatible with our previously used approach for a description of the solar wind. For this purpose, we once again consider the originally imposed approximations

$$\vec{u}_{\mathrm{p}} \simeq \vec{u}_{\mathrm{e}} = \vec{u}$$

$$T_{\mathrm{p}} \simeq T_{\mathrm{e}} = T = \mathrm{const.}$$

$$\partial/\partial t \rightarrow 0$$

Clearly, the current density, as well as the magnetic force, vanish in the momentum balance equation at this level of approximation. At the same time, the polytropic law (6.45) may be written in the form valid for an ideal gas and only the time independent terms of the density and momentum balance equations need be retained

$$\nabla(\rho \vec{u}) = 0$$

$$\rho(\vec{u}\nabla)\vec{u} = -\nabla p + \rho \vec{g}_S$$

$$p = n k T$$

Evidently, this system of equations is identical with that used in Section 6.1.2. It follows that, under the above assumptions, the magnetoplasmady-namic system of equations reduces to that used previously in the framework of simple gas dynamics. Differences evolve only upon dropping the very restrictive requirement of equal flow velocity of ions and electrons, thereby allowing relative motion between the charge carrier gases. Only then will currents flow which, in turn, produce magnetic fields and magnetic forces.

One remaining issue is why the current density and the resulting $\vec{j} \times \vec{B}$ force in the momentum equation, are of secondary importance for the solar wind. Recall that our previously used gas dynamics approach, after all, provided a fairly adequate description of the solar wind. Here is the short explanation. Near the Sun where the energy density of the magnetic field dominates ($\beta^* < 1$), the solar wind expands outward along the radially aligned magnetic field and no magnetic forces appear ($\vec{j} \parallel \vec{B}$). In the interplanetary medium, however, the magnetic field is so weak ($\beta^* \gg 1$), that the $\vec{j} \times \vec{B}$ force can be neglected to a good approximation in the momentum balance equation. A simple estimate of magnitudes similar to that given in Appendix A.13.3 shows that the momentum balance equation is indeed dominated at larger distances from the Sun by the inertial term, i.e. by the inertia of the solar wind.

6.3 Magnetoplasma Waves in the Interplanetary Medium

The interplanetary medium is quite variable at a global level, but also very dynamic when considering small scales, usually displaying a wide spectrum of fluctuations in the observed state parameters. This is documented in Fig. 6.26 with an example of solar wind velocity and magnetic field data. Today, it is generally assumed that a large fraction of these fluctuations are caused by

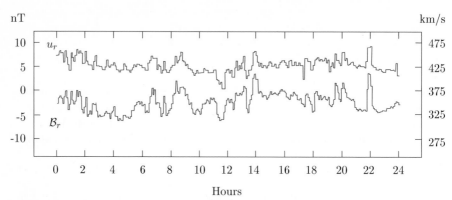

Fig. 6.26. Fluctuations of the radial components of solar wind velocity and interplanetary magnetic field, as recorded on the MARINER 5 spacecraft over a 24-hour period. The good correlation of the oscillations in the two quantities, whereby the magnetic field fluctuations are in phase with velocity fluctuations, implies (for a negative magnetic sector) that the disturbances are caused by Alfvén waves moving outward from the Sun into the outer heliosphere (adapted from Belcher et al., 1969).

waves or their superposition. A key role is played here by magnetohydrodynamic waves (magnetoplasma waves in our nomenclature) as was previously the case for geomagnetic pulsations; see Section 5.7. There we documented the existence of such waves and acknowledged their importance, but only now are we in a position to investigate them more closely. Equipped with the equations of ideal magnetoplasmadynamics introduced in the previous section, we now have a self-consistent description of the wave propagation medium at our disposal. In order to derive analytically the wave solutions of interest here, this system of equations must be further simplified and, most importantly, linearized. This then allows us to determine solutions for the case of plane harmonic waves of small amplitude, emphasizing those cases with a particularly simple constellation between the propagation direction and the magnetic field. One obtains three physically different wave types known as the *plasma acoustic waves*, the *Alfvén waves* and the *magnetosonic waves*. The details of their derivation are summarized in Appendix A.15. Here, we limit ourselves to a presentation of their essential properties.

6.3.1 Plasma Acoustic Waves

As shown in Fig. 6.27, plasma acoustic waves are associated exclusively with perturbations in the plasma state parameters (ρ_1, p_1, u_{1z}). This is not surprising, because the plasma moves (with u_{1z}) only parallel to the magnetic field. It thus feels no magnetic force and, in fact, this wave type can also propagate in a nonmagnetized plasma. Since the current density is also zero,

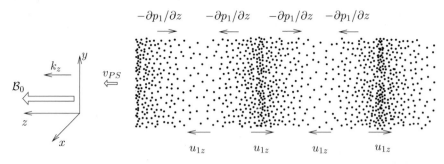

Fig. 6.27. Physics of plasma acoustic waves. This snapshot shows the instantaneous density distribution as well as the locations of maximum pressure gradient force and velocity. Perturbation quantities are indicated by the index 1. The wave propagation proceeds in the direction of the external magnetic field $\mathcal{B}_0$.

it means that the ions and electrons move together with the same velocity, $u_i = u_e = u_{1z}$. The wave is a rhythmic back and forth motion along the magnetic field direction that leads to alternating compression and rarefaction of the plasma and accompanying fluctuations in the pressure. Evidently, this situation corresponds exactly to the one we studied in connection with acoustic waves in a neutral gas (see Section 3.5.2). These waves have therefore been given the designation plasma-*acoustic*. An important characteristic in both cases is that the only restoring force of the wave is provided by the pressure gradient force $-\nabla p$. Accordingly, both wave types are characterized by the same phase velocity v_{ph}

$$v_{ph} = v_{PS} = \sqrt{\frac{\gamma p}{\rho}} \simeq \sqrt{\frac{\gamma k (T_i + T_e)}{m_i}} \qquad (6.52)$$

where v_{PS} is also denoted the *plasma sound speed*. In the solar wind at the Earth's orbit ($T_i \simeq T_e \simeq 10^5 \, \text{K}$), this quantity is typically 50 km/s.

A somewhat more precise expression for v_{PS} is obtained upon accounting for the heat conductivity of the electron gas. For the low frequency fluctuations considered here, the electrons (in contrast to the ions) have enough time to smooth out any compression-induced temperature differences in the electron gas. The changes in their state parameters thus no longer proceed adiabatically, but rather isothermally. It follows from the perfect gas law, $p_e = n k T = \alpha_e \rho_e^{\gamma_e^*}$, that the polytropic index of the electron gas is $\gamma_e^* = 1$. This difference in the adiabatic or polytropic index may be incorporated into Eq. (6.52). Since the energy of a plasma acoustic wave resides primarily in the ionic component, it is also designated as an *ion-acoustic wave* or an *ion-sound wave*.

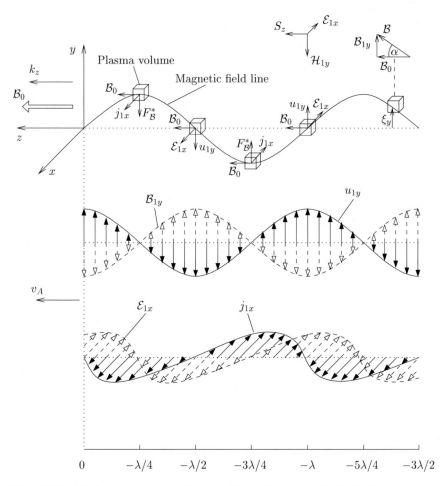

Fig. 6.28. Alfvén wave propagating parallel to the magnetic field. The instantaneous form of the magnetic field line is shown together with the phase and directions of the fluctuations in velocity (u_{1y}), magnetic field ($\mathcal{B}_{1y}$), current density (j_{1x}) and electric field ($\mathcal{E}_{1x}$). For clarity, the relative sizes of the perturbations have been greatly exaggerated. Also included are the restoring force $\vec{F}_{\mathcal{B}}^{*} = \vec{j} \times \vec{\mathcal{B}}$, the electric field resulting from the $\vec{u} \times \vec{\mathcal{B}}$ motion, the direction of the Poynting vector, and the slope of the magnetic field line. Note that $\mathcal{B}_{1y}$ and u_{1y} are in phase for waves propagating antiparallel to the magnetic field direction.

6.3.2 Alfvén Waves

In order to illustrate the physics of this wave type, the coordinated temporal behavior of the various perturbation quantities (index 1) are shown in Fig. 6.28. The diagram shows a snapshot of the conditions along the propagation direction k_z, which is also the direction of the external magnetic field $\mathcal{B}_0$. As indicated, Alfvén waves, in contrast to the plasma acoustic waves, are characterized primarily by fluctuations in their electromagnetic quantities. Perturbations of the magnetic and electric field, as well as the current density, occur together with the transverse velocity u_{1y}. Density and pressure variations, on the other hand, are absent. The form of the magnetic field line drawn in the upper panel may be derived from the known slope of this curve

$$\partial \xi_y / \partial z = \tan \alpha = \mathcal{B}_{1y} / \mathcal{B}_0$$

Here, ξ_y denotes the excursion of the magnetic field line from its mean position (in our case from the z-axis), which is obtained from

$$\xi_y(t, z) = \int (\mathcal{B}_{1y}(t, z) / \mathcal{B}_0) \, dz = \frac{(u_{10})_y}{\omega} \cos(\omega t - k_z z - \pi/2)$$

see Eqs. (A.182) and (A.185).

We first consider an elemental plasma volume at $z = -\lambda/4$ threaded by this magnetic field line. This plasma volume evidently has just reached its maximal excursion and is presently at rest, $u_{1y} = 0$. The magnetic force $\vec{j}_{1x} \times \vec{B}_0$ acts on this plasma volume (the current density $\vec{j}_{1x}$ reaches its maximum value at this position) and accelerates it in the direction of the equilibrium position, i.e. toward the z-axis. The plasma volume thus gains the velocity u_{1y}, which is in the negative y-direction for the case shown here. At the same time the magnetic field line threading this plasma volume 'moves' in the direction of the z-axis with the velocity

$$\partial \xi_y / \partial t = \frac{\partial}{\partial t} \left(\frac{(u_{10})_y}{\omega} \cos(\omega t - k_z z - \pi/2) \right) = u_{1y}$$

The plasma volume and the magnetic field line thus move with the same velocity in the same direction and it would seem that the magnetic field is 'frozen in' to the plasma. This point of view notwithstanding, every motion of a magnetoplasma is coupled with an electric field given by $\vec{\mathcal{E}} \simeq -\vec{u} \times \vec{B}$, see Eq. (6.42), the phase and direction of which is also shown in Fig. 6.28.

The velocity reaches its greatest value as the plasma volume crosses the equilibrium position, i.e. at $z = 0, -\lambda/2$, etc., in our snapshot. The current density here is zero and the plasma is thus free of acceleration. The kinetic momentum associated with this maximum velocity causes the plasma volume to overshoot its equilibrium position. Following this, as the velocity decreases, particle drifts associated with the inertial force again induce currents in the plasma, which decelerate the volume and eventually accelerate it back in the direction of the equilibrium position. We are evidently dealing here with a plasma oscillation for which the magnetic force $\vec{j} \times \vec{B}$ acts as the restoring force.

Formally, this restoring force may also be attributed to the deformation of the magnetic field lines induced by the wave. In order to show this, we use Ampère's law (6.46) to replace the current density with the magnetic field and obtain

$$\vec{j} \times \vec{B} = \frac{1}{\mu_0} \left(\nabla \times \vec{B} \right) \times \vec{B} = -\nabla \left(\frac{B^2}{2\mu_0} \right) + \frac{(\vec{B}\nabla)\vec{B}}{\mu_0}$$

where we have made use of the vector identity (A.35) in the second step. We now write the magnetic field vector in the second term on the right side as a product of a unit vector and the magnetic field magnitude and differentiate this according to the product rule (A.31) to obtain

$$(\vec{B}\nabla)\vec{B} = B(\hat{B}\nabla)(\hat{B}B) = \hat{B}B(\hat{B}\nabla)B + B^2(\hat{B}\nabla)\hat{B}$$
$$= \hat{B}(\hat{B}\nabla)(B^2/2) + B^2(\hat{B}\nabla)\hat{B} = \nabla_\parallel (B^2/2) - \hat{\rho}_c B^2 / \rho_c$$

As indicated, in the final step we have introduced the field-aligned gradient $\nabla_\parallel = \hat{B}(\hat{B}\nabla)$ and, using the relation (A.104), the radius of curvature of a field line. Inserting this into the expression for the magnetic force, the two field-aligned gradients of the quantity $B^2/2\mu_0$ just cancel and we obtain

$$\vec{j} \times \vec{B} = -\nabla_\perp \left(\frac{B^2}{2\mu_0} \right) - \hat{\rho}_c \frac{B^2}{\mu_0 \rho_c} = -\nabla_\perp p_B + (\vec{F}_B^*)_{mt} \qquad (6.53)$$

where, in a purely formal way, we have introduced the *magnetic pressure*

$$p_B = B^2/2\mu_0 \qquad (6.54)$$

The appropriateness of this designation will become clear in Section 6.4.4. We have also introduced a force called the *magnetic tension*

$$(\vec{F}_B^*)_{mt} = -\hat{\rho}_c \frac{B^2}{\mu_0 \rho_c} \qquad (6.55)$$

This force, which evidently acts on a plasma volume threaded by curved magnetic field lines, is proportional to the curvature of the field lines and opposes this curvature. It corresponds to the restoring force that acts upon a laterally displaced piece of string or rope under tension. A field-aligned thread of plasma thus behaves in a manner similar to a rope or string under tension and, like these, can be excited to execute transverse oscillations.

Since neither frozen-in magnetic field nor plasma compression occurs for Alfvén waves, the plasma pressure gradient force and the magnetic pressure gradient force both vanish and the momentum balance equation assumes the general form

$$\rho \partial u/\partial t \simeq (F_B^*)_{mt}$$

This equilibrium underscores the importance of the inertial force and the magnetic tension force for the creation of Alfvén waves.

To make the temporal variation of the various perturbation quantities visible, one should move the sine curve drawn in Fig. 6.28 with the phase velocity of the wave in the direction of the positive z-axis. This phase velocity corresponds to the familiar *Alfvén velocity*

$$v_{ph} = v_A = \sqrt{\frac{B^2}{\mu_0\,\rho}} \tag{6.56}$$

For the interplanetary medium at the Earth's orbit, v_A typically attains values of about 30 km/s (see Table 6.2). Since the Alfvén velocity is independent of the frequency, the wave propagation is nondispersive and the phase velocity corresponds to the group velocity. The energy transport associated with Alfvén waves is thus also nondispersive and proceeds in the direction of the wave. The Poynting vector $\mathcal{S}_z$, defined by the vectors $\mathcal{E}_{1x}$ and $\mathcal{H}_{1y} = \mathcal{B}_{1y}\,/\,\mu_0$, points, as expected, in the positive z-direction. Compared with electromagnetic fields in vacuum, the electric field strength is relatively weak

$$(\mathcal{E}_{1x}\,/\,\mathcal{B}_{1y})_{Alfvén} = v_A \ll (\mathcal{E}_x\,/\,\mathcal{B}_y)_{e.m.} = c_0$$

A characteristic of Alfvén waves essential for the interpretation of observations is that their velocity and magnetic field fluctuations are either correlated or anticorrelated, depending on whether the propagation occurs antiparallel or parallel to the magnetic field; see Eq. (A.185) and Fig. 6.28. From this we may conclude that the in-phase velocity and magnetic field fluctuations shown in Fig. 6.26 were most probably caused by Alfvén waves or superpositions thereof, which propagated antiparallel to the magnetic field.

Alfvén waves can also propagate oblique (but not perpendicular) to the direction of the background magnetic field. In this case the phase velocity decreases according to

$$(v_A)_\vartheta = v_A \cdot \cos\vartheta \tag{6.57}$$

where ϑ is the angle between the propagation and magnetic field directions. The energy transport no longer proceeds in the propagation direction, but rather in the direction of the ambient magnetic field. Moreover, *field-aligned* currents are produced in this case. This turns out to be very important when it comes to establishing or rearranging magnetospheric current systems, because these are almost always associated with field-aligned components (see, for example, Section 8.3.2).

6.3.3 Magnetosonic Waves

As shown in Fig. 6.29, magnetosonic waves, the third type of magnetoplasma waves, display perturbations of the plasma state parameters (ρ_1, p_1, u_{1y}) as well as perturbations of the electromagnetic quantities ($\mathcal{B}_{1z}$, $\mathcal{E}_{1x}$, j_{1x}). The plasma perturbations correspond exactly to those of the plasma acoustic waves, with fluctuations in velocity, density and pressure along the propagation direction and with the pressure gradient force again acting as a restoring force. This is the reason for the designation magneto*sonic* waves. At the same time, however, perturbations also occur in the electromagnetic quantities. The motion of the plasma is directly coupled to an electric field given

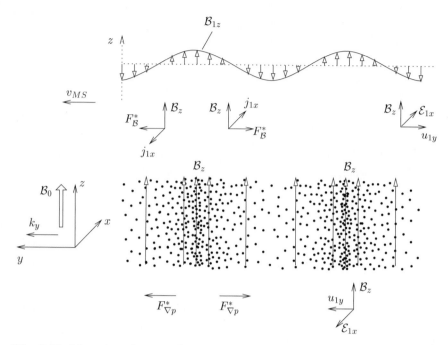

Fig. 6.29. Magnetosonic waves for propagation perpendicular to the ambient magnetic field. The sketch shows a snapshot of the plasma density and magnetic field perturbations (exaggerated in size). Also shown are the restoring forces $\vec{F}_{\mathcal{B}}^* = \vec{j} \times \vec{\mathcal{B}}$ and $\vec{F}_{\nabla p}^* = -\nabla p$, and the electric field resulting from the $\vec{u} \times \vec{\mathcal{B}}$-motion.

by $\mathcal{E}_{1x} = -u_{1y}\mathcal{B}_z$. Magnetic field perturbations are also present, which, according to our assumptions, are aligned with the external magnetic field. In contrast to Alfvén waves, the magnetic field lines are not distorted. Instead, the external field is alternately strengthened and weakened, as indicated by the different field line densities. The magnetic field peaks exactly at the density maxima and is weakest at the density minima. This implies that the magnetic field may again be considered as 'frozen in' to the plasma. Not only does the magnetic field seem to follow the plasma motion, it it also enhanced and diluted with the plasma particle concentration.

The current density induced in the plasma by the wave, in combination with the magnetic field, produces a $\vec{j} \times \vec{\mathcal{B}}$ force, which opposes the increase in magnetic field strength (i.e. the concentration of field lines). Alternatively, this force can also be interpreted as a magnetic pressure gradient force according to Eq. (6.53), which, together with the plasma pressure gradient force, acts as the restoring force. The momentum balance equation thus assumes the general form

$$\rho \frac{\partial u}{\partial t} \simeq -\nabla_\perp (p + p_{\mathcal{B}}) = F_{\nabla p_{total}}^* \tag{6.58}$$

The phase velocity of a magnetosonic wave is the geometric sum of the plasma sound and the Alfvén velocities

$$v_{ph} = v_{MS} = \sqrt{v_{PS}^2 + v_A^2} \qquad (6.59)$$

As with v_{PS} and v_A, v_{MS} is independent of frequency so that the wave propagation proceeds free of dispersion. A typical value in the interplanetary medium at 1 AU is $v_{MS} \simeq 60$ km/s. Magnetosonic waves are also capable of propagating at an arbitrary angle to the ambient magnetic field. In this case, however, they split into two separate modes with different phase velocities. The wave with the higher phase velocity is known as the *fast* magnetosonic wave; its counterpart with lower phase velocity as the *slow* magnetosonic wave. These phase velocities are given by

$$(v_{MS})_\vartheta^2 = \frac{1}{2} \left(v_{PS}^2 + v_A^2 \pm \sqrt{(v_{PS}^2 + v_A^2)^2 - 4 v_{PS}^2 v_A^2 \cos^2 \vartheta} \right) \qquad (6.60)$$

where ϑ again denotes the angle between the propagation direction and the background magnetic field and the two different modes are indicated by the plus and minus signs preceding the square root. Which of the two signs designates the fast mode depends on whether v_{PS} is larger or smaller than v_A. Similarly, in the case of field-aligned propagation ($\vartheta = 0$), the phase velocity changes to either the plasma sound velocity or the Alfvén velocity, whichever one happens to be larger. The smaller of these becomes the phase velocity of the slow mode. It can also be shown that the phase velocity of the Alfvén wave for oblique propagation always lies between those of the slow and fast magnetosonic waves. Alfvén waves are thus also referred to as the *intermediate mode*.

6.4 Modification of the Solar Wind by the Terrestrial Bow Shock

We know from experience that shock waves form in gases in front of supersonic objects (e.g. airplanes, projectiles) or, equivalently, in front of obstacles placed in supersonic flows. These are characterized by discontinuous changes in the state parameters of the gases. Extrapolating this experience to the magnetosphere of the Earth placed in the supersonic flow of the solar wind, one would expect a shock to form in front of this obstacle, and this is indeed the case. This shock is designated as the terrestrial *bow shock* in the following. Figure 6.30 shows its configuration relative to the magnetopause and Fig. 6.31 illustrates the discontinuous changes in the properties of the solar wind and the interplanetary magnetic field at this position.

In full awareness that the bow shock is a rather complicated and not entirely understood phenomenon, we will content ourselves with a very simplified description. In particular, the solar wind is again approximated by

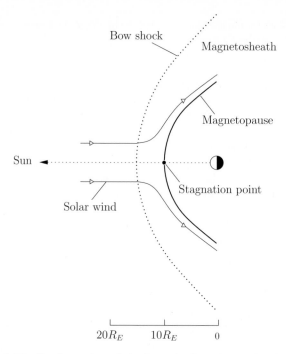

Fig. 6.30. Configuration of the bow shock and magnetosheath region

a quasi-neutral gas flow, which, in the case considered here, encounters a fixed obstacle in the shape of the magnetosphere. Based on this gas dynamical approach, we first discuss the formation of the bow shock and then the jump conditions associated with it. Even under these simplifying conditions, however, a more detailed description of the shape of the bow shock, and particularly the behavior of the solar wind in the *magnetosheath region* between the bow shock and the magnetopause, is only possible with the help of numerical simulations. Some results obtained from these calculations, which assume that the shape of the magnetopause is known, are presented here. This shape may be determined using a pressure balance approach. Finally, we point out some of the many complications that arise when we consider shocks in a magnetoplasma rather than in a neutral gas.

6.4.1 Formation of the Bow Shock

In order to understand the formation of the bow shock, we consider the conditions along the Sun-Earth line. The solar wind velocity must go to zero at the intersection of this line with the magnetopause. At this point, also known as the *stagnation point*, the flow only has a velocity component perpendicular to the magnetopause, but cannot penetrate it. The solar wind velocity must therefore decrease along the Sun-Earth line from some 500

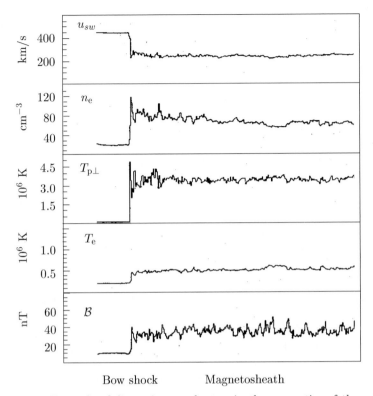

Fig. 6.31. Example of discontinuous changes in the properties of the solar wind and interplanetary magnetic field at the bow shock. u_{sw} is the solar wind velocity, n_e the electron density, $T_{p\perp}$ the proton temperature perpendicular to the magnetic field, T_e the electron temperature and $\mathcal{B}$ the magnetic field strength. $T_{p\perp}$ is assumed to be 10^5 K upstream of the bow shock. The angle between shock normal and interplanetary magnetic field was about $90°$ (quasi-perpendicular bow shock). Measurements are shown for a 27 minute interval, but the flight through the bow shock lasted for only a few tens of seconds (adapted from Sckopke et al., 1990).

km/s in interplanetary space to 0 km/s at the position of the magnetopause. Let us first assume that this deceleration proceeds continuously. Then there must exist some point, the so-called *sonic point*, where the solar wind velocity becomes equal to and then smaller than the speed of sound. We use the speed of sound as a reference, because this is the maximal velocity at which disturbances can propagate in neutral gases. This speed is about 50 km/s for typical solar wind conditions at the Earth's orbit (see Table 6.1). As a result, sound waves that propagate upstream in the solar wind with information about the approaching obstacle magnetosphere can only reach the sonic point. Beyond this point their propagation velocity is smaller than the solar wind flow velocity. This leads to an accumulation of wave energy in the vicinity of this point and leads to the formation of a large-amplitude

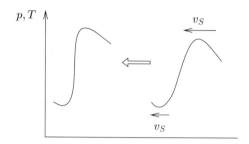

Fig. 6.32. Steepening of a large-amplitude sound wave

disturbance. Sound waves of large amplitude, however, are implicitly associated with nonlinear effects and dissipation of wave energy (see Section 3.5.2). Moreover, the pressure and temperature in the wave crest are significantly larger than in the neighboring wave trough. Since the speed of sound is proportional to the square root of these quantities, however, the sound wave propagates faster at the position of the wave crest than in the adjacent wave trough and the wave front continually steepens; see Fig. 6.32. This process thus leads to formation of a shock wave front, across which the properties of the gas suddenly change. Among the changes is a steplike deceleration of the solar wind from supersonic to subsonic velocities. Much of the original flow energy is transformed via nonlinear effects in the shock front to thermal energy, thereby leading to considerable heating of the gas.

6.4.2 Modification of Solar Wind Properties Across the Bow Shock

Independent of the complicated processes occurring within the shock itself, the mass, momentum and energy of the gas flow must be conserved. As shown in Appendix A.9, this requirement leads to the following discontinuous changes in the state parameters of a gas flow passing through a shock

$$\frac{u_2}{u_1} = \left(\frac{n_2}{n_1}\right)^{-1} = \frac{2 + (\gamma - 1)M_1^2}{(\gamma + 1)M_1^2} \tag{6.61}$$

$$\frac{p_2}{p_1} = \frac{2\gamma M_1^2 - (\gamma - 1)}{\gamma + 1} \tag{6.62}$$

$$\frac{T_2}{T_1} = \frac{(2 + (\gamma - 1)M_1^2)(2\gamma M_1^2 - (\gamma - 1))}{(\gamma + 1)^2 M_1^2} \tag{6.63}$$

and

$$M_2 = \sqrt{\frac{2 + (\gamma - 1)M_1^2}{2\gamma M_1^2 - (\gamma - 1)}} \tag{6.64}$$

These are the familiar *Rankine-Hugoniot equations*. The index 1 refers to upstream, index 2 to downstream, in our case magnetosheath, conditions. As defined previously in Eq. (2.34), γ is the adiabatic exponent ($\gamma = (f+2)/f = 5/3$ for the solar wind) and M is the *Mach number*, defined as the ratio of the flow velocity to the speed of sound

$$M = \frac{u}{v_S} = \frac{u}{\sqrt{\gamma p/\rho}} \tag{6.65}$$

The Mach number is relatively large upstream of the bow shock, $M_1 \simeq 10 \gg 1$. A good approximation is thus $u_2/u_1 = n_1/n_2 \simeq (\gamma - 1)/(\gamma + 1) = 1/4$, which for the density is in good agreement with the observations of Fig. 6.31. For a Mach number of 10, one might be at first surprised that a reduction of the velocity by a factor of four would suffice to produce a subsonic flow. The reason, of course, is that the temperature, and therefore the speed of sound, is considerably larger in the magnetosheath than in the upstream solar wind. For $M_1 \simeq 10$ we obtain $T_2/T_1 \simeq 30$, so that $M_2 < 1$, as required.

6.4.3 Results from Model Calculations

While addressing the issue of how the bow shock is formed, nothing was said about its distance from the magnetopause. Indeed, no simple analytical calculation of this quantity exists. At a minimum, this stand-off distance must be large enough to allow all solar wind passing through the bow shock in the vicinity of the subsolar point to flow around the magnetosphere at subsonic velocities. More quantitative estimates can only be obtained from numerical simulations. Since these also provide information on the conditions in the magnetosheath, they are of particular interest here. The balance equations for density, momentum and energy conservation are solved for a supersonic, quasi-neutral solar wind that encounters a fixed obstacle in the shape of the magnetosphere. Figure 6.33 summarizes some of the results obtained from such calculations. Since the magnetosphere is assumed to be rotationally symmetric, the first three panels can be considered either as equatorial or meridional cross sections.

As indicated, the stand-off distance of the bow shock along the Sun-Earth line is about 1/3 of the geocentric distance of the magnetopause, in agreement with the observations sketched in Fig. 6.30. Furthermore, it may be seen that the simulated bow shock forms at a steeper angle to the solar wind than the magnetopause and acts like a (permeable) shield in front of the magnetosphere. Concerning the conditions in the magnetosheath, Fig. 6.33a shows, with the aid of streamlines, how the solar wind flows around the obstacle magnetosphere. The associated flow velocities are given in Fig. 6.33b. As expected, the velocity at the stagnation point is zero. It is surprising that the flow becomes supersonic again at a distance of only one magnetopause distance from the Sun-Earth line and reaches 80% of its original solar wind

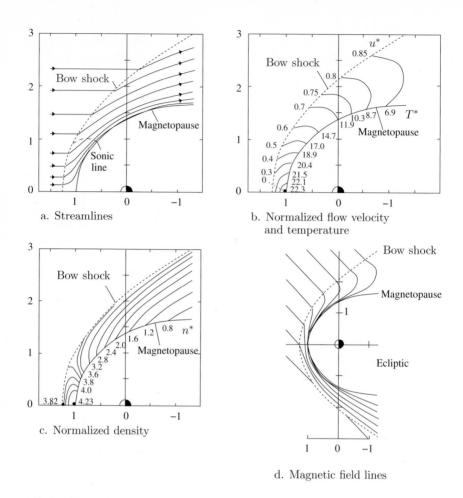

Fig. 6.33. Gas dynamical calculation of the bow shock location and the properties of the solar wind and interplanetary magnetic field in the magnetosheath. The calculations are valid for a Mach number $M = 8$ and an adiabatic exponent $\gamma = 5/3$. Distances are given in units of the magnetopause distance at the subsolar point. Velocity, temperature and density are normalized to their values upstream of the bow shock (adapted from Spreiter et al., 1966).

velocity at two magnetopause distance units. Since the flow velocity and gas temperature are coupled via the Bernoulli equation (see Appendix A.8, Eq. (A.77)), the same contour lines can be used for both parameters. The gas temperature is clearly quite high at the stagnation point and decreases only gradually with increasing flow velocity. Figure 6.33c shows the expected jump in density behind the bow shock.

In order to obtain an impression of the magnetic field structure in the magnetosheath, we again invoke the by now familiar concept of the magnetic field frozen into the solar wind flow. It is assumed in Fig. 6.33d that the interplanetary magnetic field lies in the ecliptic and is inclined at an angle of 45° to the Sun-Earth line in front of the bow shock. As illustrated, the magnetic field lines are draped over the obstacle magnetosphere by the solar wind flow and compacted in the afternoon sector. Such compression effects are actually observed (see Fig. 6.31).

The field line configuration shown in Fig. 6.33d corresponds to a snapshot of the magnetic field as it is convected by the solar wind. It is eventually dragged with the flow around the obstacle magnetosphere, slipping along over the polar regions. Alternatively, one can consider the magnetic field to be fixed in space, in which case it is produced by stationary currents flowing in the bow shock, the magnetosheath and the magnetopause.

6.4.4 Pressure Balance at the Magnetopause

The simulation of solar wind flow around the magnetosphere described above was based on the assumption that the shape of the magnetopause was known. Evidently this shape must be determined beforehand. Here we require that an equilibrium exist at the magnetopause between the external pressure of the solar wind and the internal pressure of the magnetosphere. We determine both of these pressure components, beginning with the particular conditions in the vicinity of the subsolar point.

Solar Wind Pressure. Figure 6.34 sketches the pressure and velocity conditions along the Sun-Earth line. The thermal pressure of the solar wind upstream of the bow shock is denoted p_1. Upon transit through the bow shock, this pressure is enhanced according to Eq. (6.62) by a factor

$$\frac{p_2}{p_1} = \frac{2\gamma M_1^2 - (\gamma - 1)}{\gamma + 1} \tag{6.66}$$

Another increase in the pressure occurs upon approaching the stagnation point. As derived in Appendix A.8, this follows from the *Bernoulli equation* for adiabatic gas flows

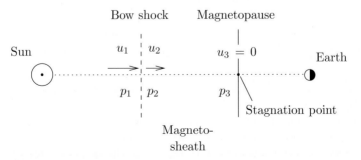

Fig. 6.34. Determining the solar wind pressure at the subsolar magnetopause

$$\frac{u^2}{2} + \frac{\gamma}{\gamma - 1}\frac{p}{\rho} = \text{const.} \tag{6.67}$$

The change in pressure between the bow shock and the magnetopause is thus determined from

$$\frac{\gamma}{\gamma - 1}\frac{p_3}{\rho_3} = \frac{u_2^2}{2} + \frac{\gamma}{\gamma - 1}\frac{p_2}{\rho_2}$$

or

$$\frac{p_3}{p_2} = \frac{\rho_3}{\rho_2}\left(1 + \frac{\gamma - 1}{\gamma}\frac{u_2^2}{2}\frac{\rho_2}{p_2}\right) \tag{6.68}$$

Replacing the density ratio ρ_3/ρ_2 by the pressure ratio corresponding to the adiabatic relation $\rho = \text{const. } p^{1/\gamma}$ and introducing the Mach number $M_2 = u_2/\sqrt{\gamma p_2/\rho_2}$, we obtain

$$\frac{p_3}{p_2} = \left(1 + \frac{\gamma - 1}{2}M_2^2\right)^{\gamma/(\gamma-1)} = \left[1 + \frac{\gamma - 1}{2}\frac{2 + (\gamma - 1)M_1^2}{2\gamma M_1^2 - (\gamma - 1)}\right]^{\gamma/(\gamma-1)} \tag{6.69}$$

where we have made use of Eq. (6.64) in the second step.

Multiplying this relation with Eq. (6.66) and exchanging the pressure p_1 for the velocity u_1 via the expression $p_1 = \rho_1 u_1^2/\gamma M_1^2$, we arrive at the desired relation for the solar wind pressure at the stagnation point

$$p_3 = K\,\rho_1\,u_1^2 \tag{6.70}$$

where the constant K stands for the following expression

$$K = \frac{1}{\gamma M_1^2}\frac{2\gamma M_1^2 - (\gamma - 1)}{\gamma + 1}\left[1 + \frac{\gamma - 1}{2}\frac{2 + (\gamma - 1)M_1^2}{2\gamma M_1^2 - (\gamma - 1)}\right]^{\gamma/(\gamma-1)} \tag{6.71}$$

Using $\gamma = 5/3$ and $M_1 = 10$, one obtains a value $K = 0.884$. The thermal pressure of the solar wind at the stagnation point thus corresponds approximately to the dynamic pressure of the solar wind upstream of the bow shock.

Moving away from the Sun-Earth line, the component of dynamic pressure perpendicular to the magnetopause decreases with $\cos^2\psi$, where ψ is the angle between the Sun-Earth line and the magnetopause normal. The decrease goes as $\cos^2\psi$, because both the momentum flux and the momentum transfer decrease as $\cos\psi$. One may thus write

$$p_3(\psi) = K\rho_1 u_1^2 \cos^2\psi \tag{6.72}$$

Magnetospheric Pressure. Ignoring the relatively minor contribution from thermal pressure in the magnetospheric boundary layer, it is essentially the terrestrial magnetic field which opposes the pressure of the solar wind. As shown in Fig. 6.35, this is done with the aid of the magnetic force acting on the magnetopause current. According to Eq. (5.2), we have

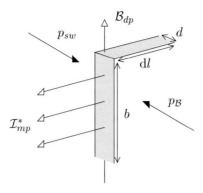

Fig. 6.35. Pressure balance at the magnetopause

$$d\vec{F_B} = dl\ \vec{I} \times \vec{B} = dl\ b\ \vec{I}^*_{mp} \times \vec{B}_{dp}$$

where $\vec{I}^*_{mp}$ is the magnetopause surface current density and $\vec{B}_{dp}$ is the dipole field of the Earth. The internal pressure of the magnetosphere is thus given by

$$p_B = \frac{dF_B}{dl\ b} = I^*_{mp}\ B_{dp} = 2\ B^2_{dp}/\mu_0 = B^2_{mp}/2\mu_0 \qquad (6.73)$$

where we have made use of the relation (5.77). Since the dipole field strength varies as $1/L^3$, this pressure is inversely proportional to the sixth power of the geocentric distance to the magnetopause. The stagnation pressure of the impinging solar wind thus compresses the terrestrial dipole field down to the point where the opposing magnetic pressure of the magnetosphere becomes equal and establishes pressure balance across the current layer at the magnetopause. The situation at the subsolar point is expressed as

$$p_{sw} = K\ m_H\ n\ u^2_{sw} = p_B = I^*_{mp}\ B_{dp} = 2\ B^2_{dp}/\mu_0 \qquad (6.74)$$

For the case when the subsolar point lies in the magnetic equatorial plane where the dipole field strength varies as $B_{dp} = B_{00}/L^3$, it follows that

$$L_{mp} = \sqrt[6]{\frac{2B^2_{00}}{\mu_0\ K\ m_H\ n\ u^2_{sw}}} \qquad (6.75)$$

Inserting typical values for the solar wind ($n \simeq 6 \cdot 10^6 \mathrm{m}^{-3}$, $u_{sw} = 470$ km/s, $M = 10$), we obtain a magnetopause distance of $L_{mp} \simeq 9.5$, i.e. about what is actually observed and had been previously estimated with the help of Eq. (5.80). The essential advantage of the new expression given by Eq. (6.75) is that it depends only on solar wind parameters measured *upstream* of the bow shock. A more accurate calculation of the distance to the magnetopause accounts for the deformation of the dipole field by currents inside the magnetosphere and, particularly, for the curvature of the magnetopause.

It is important to note that the above relation derived for the magnetic force per unit area can be applied quite generally to the interaction between plasmas and magnetic fields. The *magnetic pressure* acting on an ensemble of current carrying particles is always given by

$$p_\mathcal{B} = \frac{\mathcal{B}^2}{2\mu_0} \qquad (6.76)$$

where $\mathcal{B}$ is the total (external and induced) magnetic field; see also Section 6.3.2. This expression is also the one appearing in the definition of the β^* *parameter*, the ratio of kinetic to magnetic pressure; see Eq. (5.61).

Since the dynamic pressure perpendicular to the magnetopause in regions of the distant magnetotail is essentially zero, only the thermal and magnetic pressure of the almost unmodified solar wind are of importance. Ignoring the thermal pressure in the magnetospheric boundary layer, we obtain the following pressure balance condition at the tail magnetopause

$$[nk\,(T_\mathrm{p} + T_\mathrm{e}) + \mathcal{B}^2/2\mu_0]_{sw} \simeq \mathcal{B}^2_{mt}/2\mu_0 \qquad (6.77)$$

where the indices sw and mt stand for solar wind and magnetotail, respectively.

6.4.5 The Bow Shock as a Plasmadynamic Phenomenon

A previously unmentioned (but easily understandable) prerequisite for the formation of a shock is that the Knudsen number, i.e. the ratio of the mean free path to the typical scale of spatial variations, is small with respect to unity. Should this condition not be fulfilled, we are then dealing with a *free molecular flow*, by which, for lack of interaction, the gas particles cannot form a shock and impinge directly onto the obstacle. Well-known examples of this case are satellites, which move through the upper atmosphere of the Earth at supersonic velocities without producing a shock. This would also be the case for the magnetosphere if the interaction of the solar wind particles were limited to Coulomb collisions. As indicated in Table 6.1, the associated mean free paths are almost one AU at the Earth's orbit. In order that phenomena with the thickness of the bow shock can form in the tenuous plasma of the solar wind, it is necessary to invoke much more effective interactions than the extremely rare Coulomb collisions. These interactions must be of electromagnetic nature and can therefore only occur in plasmas. Among these, for example, are the familiar wave-particle interactions, by which the electromagnetic forces of the wave act on the particles in a manner very similar to collisions. Other processes are necessary, however, and the physics of *collisionless shocks*, in fact, is a field of intensive research.

Another fundamental complication comes from the fact that a magnetoplasma supports different wave modes with which a disturbance can propagate. Among these, as we have seen, are the slow, intermediate, and fast modes of magnetoplasma waves; see Section 6.3. Not only are the propagation velocities for these modes different (which necessitates the definition of different Mach numbers), they are also dependent on the direction of the propagation relative to the magnetic field. For example, Fig. 6.31 shows the changes across a bow shock for which the interplanetary magnetic field is oriented very nearly perpendicular to the shock normal (and thus nearly parallel to the bow shock itself). Accordingly, we are dealing here with a so-called *quasi-perpendicular* shock. The Mach number for the fast magnetosonic mode, $M_{MSf(ast)} = u_{sw}/v_{MSf}$, may be utilized to further characterize this bow shock. Supercritical values, i.e. values greater than three, are attained in this case.

It should also be noted that one must account for electromagnetic effects when deriving the Rankine-Hugoniot relations in a magnetoplasma. For example, the momentum balance equation (A.79) must be amended with the magnetic force term $\vec{j} \times \vec{B}$; see Eq. (6.44). This clearly leads to much more complicated jump conditions. The maximum changes expected across a shock, however, are those corresponding to a neutral gas flow. In other words, the velocity and density should change by no more than a factor of four in a magnetoplasma shock.

Even the Rankine-Hugoniot relations modified in this way are still based on a 'single gas' plasma flow, and thus only allow conclusions to be drawn about the plasma as a whole. Observations like those of Fig. 6.31, however, show that the electron and proton components can behave quite differently. In order to account for these differences, one must obviously use a 'two gas' model for shocks that treats the electrons and protons separately.

Finally, we note that the sudden increase in the magnetic field strength documented in Fig. 6.31 implies that currents must be flowing in the bow shock. In view of the relatively small thickness of the shock, these can only be surface currents. Since the magnetic field strength only changes within the magnetosheath, these surface currents must flow upon a closed, solenoidal configuration similar to that shown in Fig. 7.12 for the case of Birkeland currents. The strength of these bow shock currents can be estimated with the help of Eq. (5.81) and Fig. 6.31, $\mathcal{I}_{bs}^* \simeq \Delta \mathcal{B}_{bs}/\mu_0 \simeq 15 \text{ mA/m}$. This corresponds to about one-half of the magnetopause current. There are two reasons why currents are important for the bow shock: first, they decelerate the solar wind via the $\vec{j} \times \vec{B}$ force; second, they lead to Joule heating of the plasma, assuming that plasma instabilities induce an anomalous enhancement of the collision frequency and thus the electrical resistance (*anomalous resistivity*).

The above are only a few examples of the many complications that can arise from a treatment of shock waves in highly rarefied magnetoplasmas. It is all the more amazing that the characteristic parameters of the terrestrial bow shock can be quite accurately estimated from a simple gas dynamical approach.

6.5 Interaction of the Solar Wind with the Interstellar Medium

Shocks also play an important role at the outer boundary of the solar wind, where it eventually encounters the interstellar medium. The general configuration of this fascinating region of interaction is described in the present section. As a short introduction, it is appropriate to review the essential properties of the local interstellar medium.

It is currently thought that the solar system is moving through the outer regions of an extensive, inhomogeneous interstellar gas cloud. Here we are interested in the state of this cloud in the immediate vicinity of our solar system at distances of, say, less than 1000 AU. The cosmic matter and electromagnetic fields of this region are often referred to as the *very local interstellar medium* VLISM. The essential properties of this medium, some of which are subject to large uncertainty, are summarized in Table 6.3.

Assuming standard cosmic abundances, the gas of the local interstellar medium should consist primarily of hydrogen with a modest admixture of helium. A significant fraction of the hydrogen is expected to be ionized and, according to present estimates, the neutral hydrogen density is ca. 0.2 cm^{-3}; the density of the protons and associated electrons about half of this (corresponding to an ionization fraction of 0.3). Relative to the Sun, the interstellar gas flows towards the solar system with a velocity of roughly 25 km/s from a direction close to that of Earth's ecliptic longitude at the time of the June solstice. In analogy to the solar wind, this gas flow has been designated as the *interstellar wind*. Whereas the temperature of the interstellar wind is known relatively well, amounting to about 7000 K, there are considerable uncertainties associated with the intensity and direction of the interstellar magnetic field. As indicated in Table 6.3, we will assume a magnetic field strength of 0.15 nT in the following discussion. The important point is that, for a temperature of 7000 K and a magnetic field strength of 0.15 nT, both the thermal and magnetic energy densities of the interstellar wind are small

Table 6.3. Possible properties of the very local interstellar medium. The discussion in the text is based on the value of the magnetic field strength not in parentheses (adapted from Frisch, 2000).

Composition:	hydrogen ($\simeq 90\%$), helium ($\simeq 10\%$)
Density:	$n_{\mathrm{H}} \simeq 0.2 \text{ cm}^{-3}$
	$n_{\mathrm{p}} = n_{\mathrm{e}} \simeq 0.1 \text{ cm}^{-3}$
Velocity:	$u \simeq 25 \text{ km/s}$
Temperature:	$T \simeq 7000 \text{ K}$
Magnetic field strength:	$B \simeq 0.15 \, (-0.6) \text{ nT}$

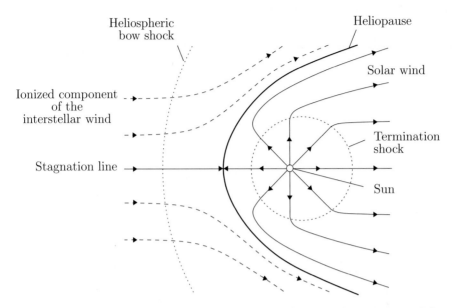

Fig. 6.36. Simulation of the interaction between the *ionized* component of the interstellar wind and the solar wind (adapted from Linde et al., 1998)

compared with that associated with its flow. This also means that the speed of sound v_S and the propagation velocity of the fast magnetosonic mode $v_{MSf} \leq \sqrt{v_{PS}^2 + v_A^2}$ are both smaller than the flow velocity. In other words, just like the solar wind, the flow is supersonic. The Mach number of the interstellar wind, however, is only about $2 - 3$. Another important conclusion drawn from the above energy densities is that the structure of the interstellar magnetic field will adjust itself to the motion of the plasma ($\beta^* \gg 1$), so the magnetic field can again be considered as frozen into the flow.

When investigating the interaction between the interplanetary and interstellar media, one must strictly distinguish between the charge carrier and neutral gas components of the interstellar wind. The neutral gas particles, on the one hand, are more or less unaffected by the solar wind and penetrate deep into the solar system. This is completely different with the ionized component, which is reflected by the interplanetary magnetic field and thus prevented from entering the solar wind dominated region. In this sense, the situation corresponds somewhat to the interaction of the solar wind with the Earth's magnetic field. Conversely, solar wind particles are reflected by the interstellar magnetic field and cannot penetrate into the interstellar wind. In this way, a contact discontinuity is formed between the solar wind and the ionized component of the interstellar wind, which keeps the two immiscible plasma flows separated from each other; see Fig. 6.36. Following the previously used nomenclature, this discontinuity at the outer boundary of the solar wind is called the *heliopause*, and the region of space enclosed by this boundary, dominated by the outward flowing solar wind, is called the *helio-sphere*. As seen in the sketch, the shape of the heliosphere does bear some resemblance to the terrestrial magnetosphere. The solar wind plasma is thus compressed into an ellipsoidal volume on the 'windward' (upstream) side of the heliosphere and a *heliospheric tail* forms on the 'leeward' (downstream) side.

A *heliospheric bow shock* forms in front of the heliosphere that deceler-ates the ionized component of the interstellar wind to subsonic velocities. This component subsequently flows at subsonic velocities in the outer tran-sition region (*heliosheath*) around the obstacle heliosphere. Since the solar wind is also a supersonic flow, it must pass through a shock of its own be-fore it encounters the 'obstacle' heliopause. This shock is commonly denoted the *termination shock* for obvious reasons. After passing through this spheri-cally shaped discontinuity, the solar wind initially flows at subsonic velocities through the inner transition region inside the heliopause, eventually turning its flow direction toward the heliospheric tail.

In order to obtain an estimate of the distances to the flow discontinuities introduced here, we consider the particularly simple conditions along the stagnation line drawn in Fig. 6.36. This is where the ionized component of the interstellar wind meets the solar wind head-on and both flows must decrease their velocity to zero at the position of the heliopause. The equation for pressure equilibrium at this point is correspondingly simple. According to Eqs. (6.17) and (6.70) and using the notation given in Fig. 6.37, we obtain

$$p_{hp}^{is} = K(M_{is})\, m_{\mathrm{H}}\, n_{\mathrm{p}}^{is}\, u_{is}^2 = p_{hp}^{sw} = K(M_{sw})\, m_{\mathrm{H}}\, n_{sw}(r_s)\, u_{sw}^2$$

$$= K(M_{sw})\, m_{\mathrm{H}}\, n_{sw}(r_E)\, (r_E/r_s)^2\, u_{sw}^2 \tag{6.78}$$

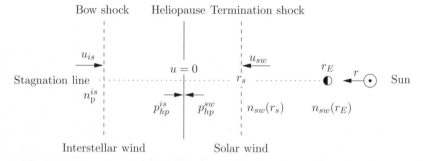

Fig. 6.37. Determining the heliocentric distance of the termination shock r_s along the stagnation line. u_{is} and n_{p}^{is} denote the velocity and proton density of the undis-turbed interstellar wind (upstream of the bow shock); p_{hp}^{is} and p_{hp}^{sw} are, respectively, the (thermodynamic) pressure of the interstellar wind and solar wind at the he-liopause; u_{sw} and $n_{sw}(r_s)$ are the velocity and density of the solar wind upstream of the termination shock; and $n_{sw}(r_E)$ is the solar wind density at the Earth's orbit.

In the second step above, we assume that the solar wind velocity remains essentially constant between the Earth (r_E) and the termination shock (r_s). Neglecting the small difference between the two constants $K(M_{is})$ and $K(M_{sw})$, where M_{is} and M_{sw} are the respective Mach numbers in the interstellar and solar winds, the heliocentric distance to the front side termination shock is found to be given by

$$r_s \simeq r_E \sqrt{\frac{n_{sw}(r_E)\, u_{sw}^2}{n_{\mathrm{p}}^{is}\, u_{is}^2}} \qquad (6.79)$$

Inserting the values for the solar wind density and velocity at the Earth's orbit from Table 6.1 and taking the interstellar parameters n_{p}^{is} and u_{is} from Table 6.3, one obtains a distance to the termination shock of $r_s \simeq 146$ AU. This estimate, however, neglects a number of effects and more recent calculations yield a termination shock distance of roughly 100 AU. This distance will be reached by the two spacecraft VOYAGER 1 (in the year 2006) and VOYAGER 2 (in 2012), both of which are flying outward in the direction near the stagnation line. An *in situ* detection of the termination shock may thus verify the above prediction within the near future.

In contrast to the ionized component, the neutral gas of the interstellar wind, unaffected by magnetic forces, can freely penetrate into the heliosphere. Interstellar neutrals, when passing through the outer transition (heliosheath) region, however, can undergo charge exchange reactions with the decelerated and diverted ionized component of the interstellar wind. This leads to a decrease of the interstellar neutral flux into the heliosphere and a simultaneous build-up of neutrals in the upstream heliosheath (the so-called 'hydrogen wall'). In spite of this culling effect, it is important to realize that the incoming interstellar neutral atoms represent by far the most populous species in the outer heliosphere.

This neutral gas flow only begins to suffer significant losses upon approaching the inner heliosphere, primarily by particle ionization. Responsible for this are the increasingly intense EUV radiation of the Sun and the more frequently occurring charge exchange interactions with solar wind protons. The latter process converts a neutral hydrogen atom into a thermal proton and the participating solar wind proton into an energetic neutral hydrogen atom; compare with the reaction (5.62). Both ionization processes increase quadratically with decreasing heliocentric distance (ϕ^{EUV}, $n_{sw} \sim 1/r^2$). Once ionized, the interstellar gas particles are subject to the local electromagnetic forces and are accelerated by the interplanetary electric field to values of the order of the solar wind velocity. This process thus converts interstellar neutral atoms into so-called *pick-up ions*, which are distinguished from the original solar wind population by their higher thermal energy. The higher thermal energy arises because the pick-up ions are created with a large relative velocity (with respect to the solar wind), which is of the same order of magnitude as the solar wind velocity itself ($u_{sw} \gg u_{is}$). This relative velocity corresponds to a random velocity after the ionization event and is converted at least partially to gyration that is superposed onto the solar wind velocity. Note that the interplanetary magnetic field is nearly azimuthal in the ecliptic and at the distances of interest here and is thus perpendicular to the solar wind velocity. There are many indications that the 'hot' pick-up ions can be accelerated to high energies at the termination shock and thus represent the seed population of the anomalous cosmic rays; see Section 6.6.2.

6.6 Energetic Particles in Interplanetary Space

In addition to the solar wind and the interstellar neutral gas, interplanetary space is populated by a broad spectrum of higher energy particles. 'Higher energy' in this context means that the energy is significantly higher than that of solar wind particles, i.e. for protons with a flow velocity of 500 km/s, this means energies well above 1 keV. The upper bound on the energy spectrum is of the order of 10^{20} eV, which is the highest energy ever detected in an interplanetary particle. This energy is far above that attainable with the help of artificial particle accelerators on Earth. The following discussion of energetic particles is limited to a brief description, ordered according to their different places of origin.

6.6.1 Energetic Particles of Galactic Origin

High energy particles of galactic origin were first observed by Viktor Hess in 1912. Carrying an ionization chamber on board, he ascended in a balloon to a height of 5300 m and recorded a steady increase in the counting rate above 700 m altitude. It was believed at first that a new type of very penetrating electromagnetic (*ultra-*)radiation had been discovered, and this is surely part of the reason why the term *cosmic ray* is still commonly used (and often mis-used for esoteric effects) today. Only later could it be uniquely demonstrated that these 'rays', in reality, were energetic particles of galactic origin. Protons are the most important constituent of these particles, which enter the helio-sphere uniformly from all directions, followed by α-particles and electrons. Nuclei of heavier elements, e.g. carbon, oxygen and iron, are also detected in minute concentrations. The measured energy spectrum for the proton com-ponent, i.e. the particle flux per unit solid angle per unit energy interval as a function of energy, is shown in Fig. 6.38. Apart from the normalization to a unit solid angle, this type of graphic representation resembles that used previously for our description of the solar radiation spectrum.

Knowing the energy spectrum, the energy density associated with cosmic rays can be estimated as about 10^{-13} J m^{-3}. This energy density, at least at the heliocentric distance of the Earth's orbit, is considerably smaller than that of the solar wind ($\simeq 10^{-9}$ J m^{-3}) and also smaller than that of the interplanetary magnetic field ($\simeq 10^{-11}$ J m^{-3}). The latter value means that the cosmic rays, in spite of their high energy, are regulated in their motion

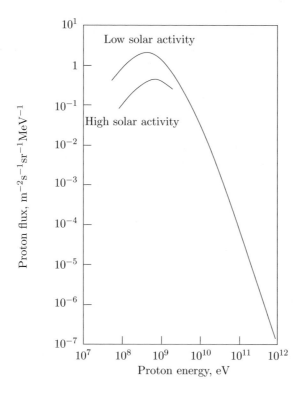

Fig. 6.38. Energy spectrum of cosmic ray protons for times of low and high solar activity. The spectrum is defined as the proton flux per unit energy interval per unit solid angle (adapted from Meyer et al., 1974).

by the structure of the interplanetary magnetic field in the Earth's vicinity. In the ideal case, they move on helical trajectories along the local magnetic field lines and their feedback on the magnetic field structure is negligible. The reason for the meager energy density of cosmic rays, of course, is their very small number density. Deriving a value for the mean cosmic ray energy of 7 GeV from the energy spectrum, the density ratio of cosmic rays to solar wind particles is of the order of 10^{-11}. These few particles with diameters of 10^{-15} m and masses of 10^{-27} kg, however, put up astounding numbers on the energy scale: 10^{20} eV, for example, corresponds to the kinetic energy of a 1 kg mass dropped from shoulder height (1.6 m).

The particle incidence at the Earth displays a remarkable variation with time. At lower energies one observes a distinct anticorrelation with the solar cycle (see again Fig. 6.38). The lower intensity during times of higher solar activity (i.e. during the polarity reversal phase of the solar magnetic dipole field) is partly attributed to the scattering of the cosmic rays on the particularly strong warping of the heliospheric current sheet that occurs at this time. In addition to the solar cycle variation, sporadic reductions in intensity are observed which are caused by scattering on irregularities in interplanetary space following the occurrence of solar mass ejection events (*Forbush effect*).

6.6.2 Energetic Particles of Interplanetary Origin

In addition to the galactic cosmic rays that enter the heliosphere from outside, there are other populations of energetic particles that are produced internally within interplanetary space. Their acceleration occurs exclusively at shocks, the exact processes of which are not entirely understood and are a topic of intensive research. We content ourselves here with a survey of the various particle populations and an identification of the accelerating shocks.

Anomalous Cosmic Rays. Alongside the regular component, an irregular component of cosmic rays is observed during times of low solar activity and the associated favorable conditions for the propagation of energetic particles. This irregular component is distinguished by a different chemical composition, a different charge state, and a differently shaped spectrum from the regular component. For example, a distinct deficit of carbon and iron particles, but a surplus of neon and nitrogen particles, is observed. Furthermore, the particles are only singly ionized, in contrast to the high charge states or complete ionization seen in the constituents of the regular cosmic rays. Finally, these particles are usually observed at the lower end of the normal cosmic ray energy spectrum at energies from 50 to 300 MeV. Because of these peculiarities, this population has become known as the *anomalous cosmic rays*. Since their intensity increases with heliocentric distance, their source must lie somewhere in the distant interplanetary medium. It is presently believed that the anomalous cosmic rays originate from pick-up ions that, because of their enhanced thermal energy, are preferentially accelerated at the termination shock and subsequently scattered back into the inner heliosphere. Here it should be noted, and this holds quite generally, that ions are accelerated much more effectively at shocks than electrons. The unusual composition of the anomalous cosmic rays may be explained by the fact that the neutral gas population of the interstellar wind, which serves as the seed population of the interplanetary pick-up ions, contains neon and nitrogen, but not carbon and iron. Because of their low ionization potential, these latter elements are found only in their ionized form in interstellar space and thus cannot gain entry into the heliosphere.

Corotating Particle Events. By corotating particle events we mean populations of energetic particles that are accelerated at the shocks associated with corotating interaction regions. In order to understand the formation of the shocks responsible for this acceleration, we consider the interaction of two solar wind streams with vastly different flow speeds, as sketched in Fig. 6.39. The interaction is weak near the Sun because the streams flow radially outward and nearly parallel to each other. Only at larger distances

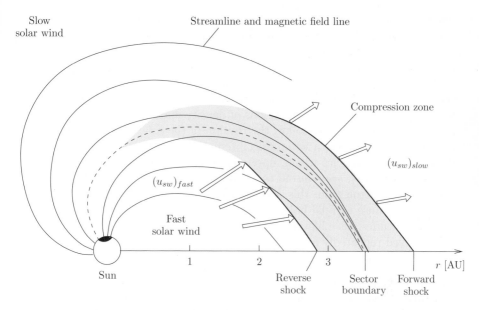

Fig. 6.39. Formation of shocks in corotating interaction regions (view from the north onto the heliospheric equatorial plane)

does the curvature of the jetlines become noticeable so that the two streams begin to flow more in front of and behind each other, rather than side by side. Since the flows are prevented from mixing by the frozen-in magnetic field, the radial velocities must be assimilated at the contact discontinuity located along their common sector boundary. As shown in Fig. 6.39, this proceeds via compression of the plasma and the magnetic field. The slow solar wind is 'swept up' by the trailing fast wind along their common sector boundary, and the fast solar wind is similarly decelerated by the trailing slow wind. The important point is that the two flows move, not only parallel to, but also increasingly perpendicular to the sector boundary within their respective undisturbed regions. At first, because of the small curvature of the jetlines, this perpendicular velocity component is smaller than the largest possible wave propagation speed in the plasma, i.e. smaller than the propagation speed of the fast magnetosonic mode. In this case the two flows can be modified directly by the pressure gradient forces produced at the sector boundary (which accelerates them in the azimuthal direction), so that they flow more nearly parallel to the sector boundary and thus more nearly parallel to each other. At larger distances, however, and usually outside of the Earth's orbit, the curvature of the jetlines becomes so large that the normal component of the relative velocity between the respective solar wind flows and the sector boundary reaches supermagnetosonic values. Relative to the sector boundary, the flow in front is due to the much slower solar wind and the flow in back is from the much faster solar wind, the relative velocities of which are both supermagnetosonic. This leads inevitably to the formation of two shocks, one on each side of the sector boundary, that have been given the (logical) designations *forward* and *reverse shock*.

Since the source regions of the two different solar wind streams rotate with the Sun, the stream-stream interaction region, consisting of the common sector boundary, the compressed regions of density and magnetic field, and the two shocks, is observed to follow the rhythm of the solar rotation. This phenomenon is thus called a *corotating interaction region*. Of interest here are the solar wind particles which, upon acceleration at the associated forward and reverse shocks, are referred to as corotating particle events. The energy gained by a proton from this acceleration process can extend up to about 10 MeV. Because coronal holes, the sources of fast solar wind at low heliographic latitudes, have lifetimes of several months, corotating particle events are observed repeatedly at intervals of one solar rotation (ca. 27 days). This is their singly most definitive characteristic feature.

Storm Particles. Shocks are also observed in front of rapidly expanding solar mass ejections (see Fig. 6.12). The solar wind particles accelerated at these shocks are called *storm particles*. The designation 'storm particles' is appropriate, because solar mass ejections can trigger geospheric storms (see Section 8.6.1) and the two phenomena frequently occur simultaneously. Proton energies of up to 100 MeV can be achieved from this type of acceleration. Since the expansion of solar mass ejections can last for a long time in spite of the high propagation velocities, storm particles are often observed to persist for days, sometimes even weeks.

Bow Shock Particles. The terrestrial bow shock (and planetary bow shocks in general) can accelerate solar wind particles. The energies achieved in this continuously active process, however, are comparatively modest, typically reaching values of a few keV for protons.

6.6.3 Energetic Particles of Solar and Planetary Origin

In addition to the galactic and interplanetary origins, the solar atmosphere and planetary magnetospheres play important roles as sources of energetic particles. For example, it has been known for some time that a wide spectrum of energetic particles is produced during impulsive solar flares (X-ray emission lasting less than one hour). Corresponding to their place of origin, these are designated as *solar energetic particles*. Compared with populations accelerated at shocks, these are overly abundant in electrons, ^{3}He - isotopes and heavy elements in high charge states (e.g. Fe XX, Si XIV). The energies produced are relatively high, reaching values of 1 MeV for electrons and even up to several GeV for protons. Events with such high energies are rare, however, and often not observed for a number of years (these are the so-called

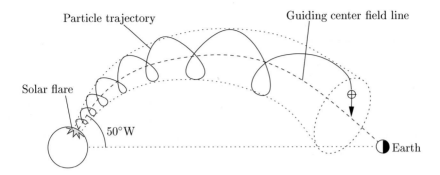

Fig. 6.40. Trajectory of a solar energetic proton in interplanetary space. The particle accelerated during a solar flare reaches the Earth along a spiral interplanetary field line (view from north onto the ecliptic plane).

neutron events, see also Section 8.6.3). A further characteristic of solar energetic particles is that they are accelerated into a relatively narrow cone, moving on helical trajectories along the Archimedean spiral field lines of the interplanetary magnetic field. As a consequence, only events with source regions at heliographic longitudes ca. $50 - 60\ °\mathrm{W}$ of the Earth can be observed at the Earth's orbit; see Fig. 6.40.

The guiding center field line of the particle is drawn as a smooth curve in Fig. 6.40 – surely a great simplification. It is well known, after all, that the interplanetary magnetic field displays considerable fluctuations on all spatial and temporal scales (see, for example, Fig. 6.26). All of these irregularities clearly influence the propagation of energetic particles. In fact, this multifaceted problem is a subject of intensive research.

Besides the Sun, planetary magnetospheres are also sources of energetic particles. The so-called *Jovian electrons* are a classic example. These electrons are accelerated in Jupiter's magnetosphere to energies of up to a few 10 MeV before they escape into interplanetary space. While these particles populate the entire heliosphere, peak intensities at the Earth are only recorded when Jupiter and Earth are magnetically connected by interplanetary field lines, which is the case approximately every 13 months.

Exercises

6.1 Moving at average velocity, how long does the solar wind need to reach the Earth, the outermost planet Pluto, and the termination shock (at 100 AU)? For comparison, how long does a solar photon require to cover such distances?

6.2 What is the minimum time needed for a solar energetic particle (20 MeV proton) to arrive at the heliocentric distances cited in exercise 6.1? Note that such particles move along interplanetary magnetic field lines, the length of which must be calculated.

6.3 How many times has an interplanetary magnetic field line wrapped itself around the Sun before it reaches the distance of the termination shock (100 AU)?

6.4 Compute the magnetopause distance of Mercury's magnetosphere (magnetic dipole moment $\mathcal{M}_{\mathrm{Mercury}} \simeq 5 \cdot 10^{19}$ A m^2). Let the solar wind parameters correspond to average conditions at the location of this planet and let the subsolar point be located in the magnetic equatorial plane.

6.5 During times of extremely disturbed conditions, the Earth's magnetopause can be pushed down to the geostationary orbit distance. How large is the dynamic pressure of the solar wind in this case?

6.6 The Moon has no bow shock, even when it is located outside of the Earth's magnetotail. How can this be explained?

6.7 How large is the β^*-parameter near the Sun ($r = 2R_S$, $\mathcal{B} \simeq 40 \ \mu$T) and at the Earth's orbit?

6.8 Determine the order of magnitude of the individual terms in the momentum balance equation for the stationary solar wind (see also Section A.13.3). Consider both the gas dynamic and the magnetoplasma dynamic forms of this relation. Determine the ratios of the terms for the case when the scale lengths of all quantities appearing in these equations are of comparable magnitude. Compared with the inertial term, can one therefore neglect the $\vec{j} \times \vec{B}$ term at the Earth's orbit?

6.9 Use the wave approach of Eq. (A.189) in the linearized magnetoplasma equations (A.153)–(A.156) to show that magnetosonic waves can exist in a magnetoplasma and that they possess the properties described by Eqs. (A.191) to (A.196).

References

Interplanetary Medium

J.C. Brandt, *Introduction to the Solar Wind*, Freeman and Company, San Francisco, 1970

A.J. Hundhausen, *Coronal Expansion and Solar Wind*, Springer-Verlag, Berlin, 1972

H. Porsche (ed.), *10 Years HELIOS*, BMFT/DLR, Oberpfaffenhofen, 1984

R.G. Stone and B.T. Tsurutani (eds.), *Collisionless Shocks in the Heliosphere: A Tutorial Review*, American Geophysical Union, Washington, 1985

B.T. Tsurutani and R.G. Stone (eds.), *Collisionless Shocks in the Heliosphere: Reviews of Current Research*, American Geophysical Union, Washington, 1985

T.E. Holzer, Interaction between the solar wind and the interstellar medium, *Ann. Rev. Astron. Astrophys.*, *27*, 199, 1989

R. Schwenn and E. Marsch (eds.), *Physics of the Inner Heliosphere 1, 2*, Springer-Verlag, Berlin, 1990/91

K. Scherer, H. Fichtner, and E. Marsch (eds.), *The Outer Heliosphere: Beyond the Planets*, Copernicus Gesellschaft, Katlenburg-Lindau, 2000

Plasma Physics

H. Alfvén and C.-G. Fälthammar, *Cosmical Electrodynamics*, Clarendon Press, Oxford, 1963

H. Alfvén, *Cosmic Plasma*, Reidel Publishing Company, Dordrecht, 1981

F.F. Chen, *Introduction to Plasma Physics and Controlled Fusion*, vol. 1, Plenum Press, New York, 1985

J.A. Bittencourt, *Fundamentals of Plasma Physics*, Pergamon Press, Oxford, 1986

R.O. Dendy, *Plasma Dynamics*, Clarendon Press, Oxford, 1990

P.A. Sturrock, *Plasma Physics*, Cambridge University Press, Cambridge, 1994

R.A. Treumann and W. Baumjohann, *Advanced Space Plasma Physics*, Imperial College Press, 1997

See also the references in Chapters 1 and 5 and the figure and table references in Appendix B.

7. Absorption and Dissipation of Solar Wind Energy

A considerable amount of energy is released as a result of the interaction of the solar wind with the magnetosphere. This provides the basis for a number of important phenomena in the magnetosphere and polar upper atmosphere. Examples of these are electric fields, large-scale plasma motions, currents, aurorae and disturbances of the neutral and ionized upper atmosphere. Since these effects are relatively well documented, especially in the polar upper atmosphere, they will be treated prior to the more speculative details of the energy transfer from solar wind to magnetosphere.

7.1 Topology of the Polar Upper Atmosphere

When describing the observations in the polar upper atmosphere, it is important to distinguish between the following three regions: the *polar cap*, the *polar oval* and *subpolar latitudes*; see Fig. 7.1. As explained earlier, the polar cap is understood to mean a circular area surrounding the magnetic pole with a typical diameter of about 30°. The centroid of this area is displaced by a few degrees toward the nightside from the magnetic pole. The polar oval is an annular region surrounding the polar cap of a few degrees in latitudinal

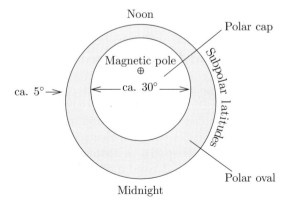

Fig. 7.1. Topology of the polar upper atmosphere. The numerical values correspond to quiet conditions.

width, the narrowest part of which lies in the noon sector. Finally, moving equatorward, one reaches subpolar latitudes, i.e. the region immediately adjacent to the polar oval.

The three regions described above are distinguished by their electric fields, the associated drifts and currents, the intensity of particle precipitation, the degree of upper atmospheric disturbance and, most importantly, by their associated particle populations in the magnetosphere. Whereas the polar cap is connected via magnetic field lines to the magnetotail lobe population, the nightside polar oval maps to the tail plasma sheet and the dayside polar oval to the cusp and magnetospheric boundary layer (see Figs. 5.36 and 5.46). The region of subpolar latitudes, on the other hand, is connected to the trapped particle populations of the inner magnetosphere.

It is important that the definitions of the above regions are referred to magnetic coordinates and the magnetic pole. This means that the polar cap, the polar oval and the subpolar region all rotate together around the geographic pole, keeping their relative structure more or less intact and aligned along the Sun-Earth direction. It may thus happen that an observing station lies within different topological regions at different local times, e.g. in the polar oval at noon, but in the polar cap at midnight. This fact initially caused considerable consternation when attempting to order the observations. Figure 7.2 shows the typical locations of the polar cap and polar oval in the Northern Hemisphere for various universal times.

It should also be stressed that the dimensions of the polar cap and polar oval are subject to strong temporal variations. Whereas both regions contract to a relatively narrow region around the magnetic pole during 'quiet' conditions, the diameter of the polar cap and width of the polar oval both expand during active conditions. This means that, in addition to the variation with universal time, geospheric disturbances also lead to significant changes in the ordering of synoptic measurements. An observing station located at subpolar latitudes during quiet times, for example, may find itself right in the middle of the polar oval during active conditions.

7.2 Electric Fields and Plasma Convection

The electric fields in the polar upper atmosphere may be determined in a number of different ways. Direct measurements are possible with the help of electric field probes carried by orbiting satellites. In this case the potential difference is measured between two electrodes which, because of the microgravity conditions, can be spaced at distances of 200 m and more. Alternatively, electric fields can be derived indirectly from the $\vec{\mathcal{E}} \times \vec{B}$ drift of charge carriers. This requires a measurement of the drift velocity, which can be obtained either from the Doppler shift of reflected radio signals or from the velocity distribution measured with ion mass spectrometers on board satellites. Most spectacular is the method for measuring $\vec{\mathcal{E}} \times \vec{B}$ drift velocities

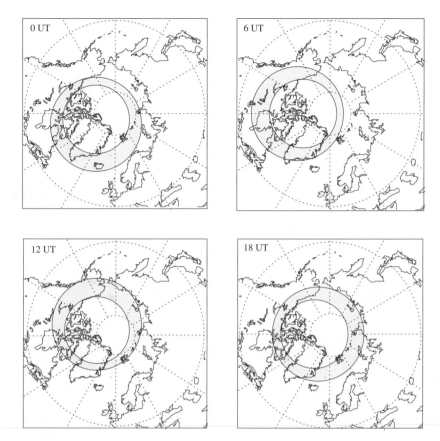

Fig. 7.2. Map showing the locations of the polar cap and polar oval in the Northern Hemisphere at four different universal times (UT). The diameter and thickness of the polar oval correspond to quiet conditions.

from the motion of artificially produced noctilucent ion clouds. Such measurements are only possible in the dawn or dusk sectors, however, when the ionic cloud is in sunlight and the observer is located in the Earth's shadow.

A greatly idealized sketch of the electric field distribution obtained with the help of these different measurement techniques is presented in Fig. 7.3. The field is directed from dawn to dusk within the polar cap, points in the direction of the equator in the polar oval on the dawn side, and is directed toward the pole in the polar oval dusk sector. Typical field strengths during quiet conditions are 10 mV/m in the polar cap and 30 mV/m in the polar oval. This yields a voltage difference across the polar cap along the line from 6 LT to 18 LT of more than 30 kV. This dawn-to-dusk *polar cap potential* is roughly counterbalanced by the sum of the potential drops in the polar oval. The field strength outside these regions at subpolar latitudes is rather moderate (< 5 mV/m) and often near the measurement threshold. Note that

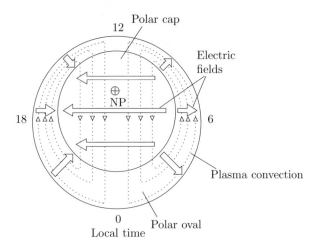

Fig. 7.3. Electric fields and plasma convection in the polar upper atmosphere. NP denotes the north geomagnetic pole. Note that, because only electric fields generated by the solar wind/magnetosphere interaction are of interest here, corotation electric fields are not included in this sketch.

the electric fields are always directed perpendicular to the slightly tilted magnetic field lines in the polar regions. The absence of electric fields parallel to the magnetic field indicates that the conductivity along the field lines is very high so that one may consider them to be conducting wires or equipotential lines.

The transverse electric fields lead to an ambipolar $\vec{\mathcal{E}} \times \vec{B}$ drift of the charge carriers in the F region of the polar ionosphere. The large-scale structure of this drift motion is also shown in Fig. 7.3. The motion is directed antisolar in the polar cap and toward the Sun in the polar oval. Two closed circulation cells, by which the ionospheric plasma drifts clockwise in the dusk sector and counterclockwise in the dawn sector, are created in this way. In order to emphasize the large scale and ambipolar character of this plasma motion, the word 'drift' is usually replaced by 'convection'. It is of note that the streamlines of this *plasma convection*, because they are everywhere oriented perpendicular to the electric field, also correspond to equipotential lines. Furthermore, the convection velocity (and thus the density of streamlines) is directly proportional to the local electric field strength ($u_D = \mathcal{E}/B$). For the field strengths given above, the convection velocity in the polar cap is nearly 200 m/s and about 600 m/s in the polar oval. The 'round trip' for a volume of ionospheric plasma thus lasts many hours, during which it may pass through zones with rather different production and loss rates. This must be accounted for when calculating the ionization density in these regions.

Note that the sketch in Fig. 7.3 is greatly idealized and that details of the electric field distribution and its associated plasma convection are dependent on the properties of the interplanetary magnetic field. One observes distinct asymmetries between the convection cells in the dawn and dusk sectors, for example, which are controlled by the $\mathcal{B}_\lambda$ component of the interplanetary magnetic field (i.e. the $\mathcal{B}_y$-component in a GSM coordinate system). Another common effect is a sometimes moderate, sometimes pronounced rotation of the symmetry axis away from the noon-midnight line. Furthermore, complex field and drift transitions are observed in the noon and midnight sectors. In the midnight sector this has become known as the *Harang discontinuity*. Finally, during disturbed times (and particularly during geospheric storms, see Chapter 8), it often happens that the polar cap potential rapidly increases to values many times larger than those observed during quiet times.

7.3 Ionospheric Conductivity and Currents

In the description of the plasma convection it was tacitly assumed that the drift motion is not modified significantly by collisions of the charge carriers with neutral gas particles. This certainly applies for the F region of the ionosphere. In the lower ionosphere, however, collisional friction plays an important role and changes the motion of the charge carriers in such a way that currents are produced. In this section, the formation of these currents is described and the associated conductivities are calculated. This is followed by an overview of the current distribution in the polar ionosphere.

7.3.1 Modification of Charge Carrier Motion by Collisions with Neutral Gas Particles

The collisionless motion of charge carriers in mutually perpendicular electric and magnetic fields was described already in Section 5.3.3: ions and electrons move on cycloidal trajectories and drift together at the same velocity in the same direction (see Fig. 5.21). Since there is no net transport of charge arising from the ambipolar $\vec{\mathcal{E}} \times \vec{\mathcal{B}}$ drift, there is no current and the conductivity perpendicular to the magnetic field is equal to zero. This situation is maintained in the entire region of the upper ionosphere (and magnetosphere), because collisions with neutral gas particles can be neglected to a good approximation. In the lower ionosphere, however, where the exponentially increasing collision frequency reaches large values, this assumption is no longer valid. The modification to the charge carrier motion resulting from this effect is sketched in Fig. 7.4. Consider a height interval in which the ion-neutral gas collision frequency has reached the same order of magnitude as the gyrofrequency of the ions. On the average, the cycloidal motion of the ion will now be interrupted by a collision once per cycloid loop. Assuming for simplicity that the resulting deflection of the ion occurs uniformly in all directions (isotropic scattering), it will find itself in a state of rest (taken over a statistical average), from which it begins its cycloidal motion anew. Interrupted

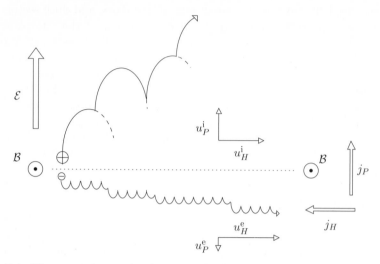

Fig. 7.4. Charge carrier motion in perpendicular electric and magnetic fields for the case when the collision frequency between ions and neutrals corresponds approximately to the ion gyrofrequency. u_k^s denotes the drift velocity of the charge carriers, with s the charge carrier type (i=ions, e=electrons) and k the drift direction (P=Pedersen for parallel/antiparallel to the electric field and H=Hall for perpendicular to the electric field). j_k is the associated current density.

by collisions, the ion thus follows a series of piecewise continuous cycloids, drifting all the while in the direction of the electric field. At the same time, the continual interruptions cause a slowdown of the drift motion in the $\vec{\mathcal{E}} \times \vec{B}$ direction.

In contrast to the ions, the electrons experience only a negligibly small modification to their $\vec{\mathcal{E}} \times \vec{B}$ drift in the height range considered here. This is because the degree of the disturbance depends on the ratio of the collision frequency to the gyrofrequency. Disruptions of the trajectories such as those sketched in Fig. 7.4 only occur when these two frequencies reach nearly equal values. Although for typical collision frequencies in the E region, $\nu_{e,n}/\nu_{i,n} \simeq \sqrt{m_i/m_e}/(4\sqrt{2}) \simeq 40$, see Eq. (2.12), it is also true that $f_B^e/f_B^i = m_i/m_e > 5 \cdot 10^4$, see Eq. (5.25). Therefore, at a height where $\nu_{i,n}(h) \simeq f_B^i$, it must also hold that $\nu_{e,n}(h) \ll f_B^e$ (see also Fig. 7.7). Accordingly, the electron in Fig. 7.4 executes a nearly undisturbed $\vec{\mathcal{E}} \times \vec{B}$ drift.

The main conclusion to be drawn from the scenario depicted in Fig. 7.4 is that the different degree of disturbance imposed onto the motion of ions and electrons leads to a relative drift of these charge carriers. This relative drift, however, corresponds to currents that flow both parallel and perpendicular to the electric field. The former are denoted *Pedersen currents* (current density j_P), and the latter are known as *Hall currents* (current density j_H). Both currents flow perpendicular to the magnetic field, implying a nonzero transverse conductivity of the ionospheric plasma. The magnitude of this conductivity is determined in the following.

7.3.2 Ionospheric Transverse Conductivity ($\vec{\mathcal{E}} \perp \vec{\mathcal{B}}$)

The conductivity of a magnetoplasma, as with any material, is determined by the density and mobility of the available charge carriers. It relates the current density to the electric field via

$$j_k = \sigma_k \, \mathcal{E}_\perp \tag{7.1}$$

where j denotes the current density, σ the conductivity and $\mathcal{E}_\perp$ the electric field applied perpendicular to the magnetic field. The particular component of the current and associated conductivity are designated by the index k, where $k = H$ stands for the Hall component and $k = P$ for the Pedersen component. Evidently, Eq. (7.1) is nothing more than the local form of Ohm's law, $\mathcal{I} = \mathcal{U}/\mathcal{R} = \mathcal{L}\,\mathcal{U}$, where $\mathcal{L}$ is the conductance. For simplicity, let us assume in the following that the lower ionosphere is composed of a single ion species with average properties ($\langle NO^+, O_2^+ \rangle$). Furthermore, let the charge carrier type be denoted by the index s, where $s = i$ stands for the ion component, $s = e$ for the electron component, and $s = n$ for the neutral gas component. According to Eq. (5.4) and with $q_i = e$, $q_e = -e$ and $n_i \simeq n_e = n$, the current density may be written as

$$j_k = \left| \sum_s q_s \, n_s \, \vec{u}_k^s \right| = e \, n \, | \vec{u}_k^i - \vec{u}_k^e |$$

Inserting this into Eq. (7.1), we obtain for the conductivity

$$\sigma_k = e \, n \, | \vec{u}_k^i - \vec{u}_k^e | \, / \mathcal{E}_\perp \tag{7.2}$$

where only the magnitude of this quantity is of interest.

The drift velocities $\vec{u}_k^s$ appearing in this equation can be calculated from the corresponding force equilibrium relations for the charge carrier gases. Only electric, magnetic and frictional forces are considered here, all others being neglected; see Fig. 7.5. Moreover, only collisions with neutral gas particles are taken into account when calculating the frictional forces. This is certainly justified in the lower ionosphere ($n_n \gg n_s$). In this case equilibrium of forces requires that

$$n_s \, q_s \, \vec{\mathcal{E}} + n_s \, q_s \, \vec{u}_s \times \vec{\mathcal{B}} + n_s \, m_s \, \nu_{s,n}^* \, (\vec{u}_n - \vec{u}_s) = 0 \tag{7.3}$$

For a neutral gas atmosphere at rest ($\vec{u}_n = 0$) it follows that

$$q_s \left(\vec{\mathcal{E}} + \vec{u}_s \times \vec{\mathcal{B}} \right) - m_s \, \nu_{s,n}^* \, \vec{u}_s = 0 \tag{7.4}$$

Note that this relation also holds for a single charged particle if only the time independent drift component is of interest. In order to calculate the velocity

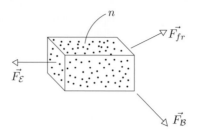

Fig. 7.5. The force equilibrium considered for determination of the charge carrier velocities. $\vec{F}_{\mathcal{E}}$, $\vec{F}_{\mathcal{B}}$ and $\vec{F}_{fr}$ denote the electric, magnetic and frictional forces, respectively, acting on the particles.

components explicitly, we use the coordinate system shown in Fig. 7.6. With $\vec{\mathcal{E}} = \hat{y}\,\mathcal{E}_{\perp}$ and $\vec{\mathcal{B}} = \hat{z}\,\mathcal{B}$ (where $\hat{y}$ and $\hat{z}$ again denote unit vectors in the y and z directions), the vector cross product in the equation of motion (7.4) takes the form

$$\vec{u}_s \times \vec{\mathcal{B}} = \hat{x}\,u_y^s\,\mathcal{B} - \hat{y}\,u_x^s\,\mathcal{B}$$

so that Eq. (7.4) may be resolved into its x and y components

$$q_s\,u_y^s\,\mathcal{B} - m_s\,\nu_{s,n}^*\,u_x^s = 0$$

$$q_s\,\mathcal{E}_{\perp} - q_s\,u_x^s\,\mathcal{B} - m_s\,\nu_{s,n}^*\,u_y^s = 0$$

Solving these equations for u_x^s and u_y^s and introducing the gyrofrequency $\omega_{\mathcal{B}}^s = |q_s|\,\mathcal{B}/m_s$, one obtains

$$u_x^s = \frac{(\omega_{\mathcal{B}}^s)^2}{(\nu_{s,n}^*)^2 + (\omega_{\mathcal{B}}^s)^2}\,\frac{\mathcal{E}_{\perp}}{\mathcal{B}}$$

$$u_y^s = \frac{q_s}{|q_s|}\,\frac{\omega_{\mathcal{B}}^s\,\nu_{s,n}^*}{(\nu_{s,n}^*)^2 + (\omega_{\mathcal{B}}^s)^2}\,\frac{\mathcal{E}_{\perp}}{\mathcal{B}}$$

Inserting the first of these two relations into Eq. (7.2), we obtain the *Hall conductivity* $(u_k^s = u_H^s = u_x^s)$

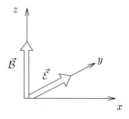

Fig. 7.6. The coordinate system used to calculate the transverse conductivity

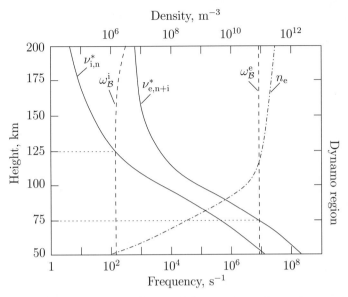

Fig. 7.7. Height profiles of the frictional frequencies and gyrofrequencies for ions and electrons and the electron density in the lower ionosphere. The curves are valid for representative daytime conditions at mid-latitudes. Note that the frictional frequency for electrons takes into account collisions with neutral gas particles as well as Coulomb collisions with ions.

$$\sigma_H = \frac{e\,n}{B} \left| \frac{(\omega_B^i)^2}{(\nu_{i,n}^*)^2 + (\omega_B^i)^2} - \frac{(\omega_B^e)^2}{(\nu_{e,n}^*)^2 + (\omega_B^e)^2} \right|$$

$$= \frac{e\,n}{B} \left\{ \frac{(\omega_B^e)^2}{(\nu_{e,n}^*)^2 + (\omega_B^e)^2} - \frac{(\omega_B^i)^2}{(\nu_{i,n}^*)^2 + (\omega_B^i)^2} \right\} \qquad (7.5)$$

where the last step accounts for the fact that $(\nu_{i,n}^*/\omega_B^i)^2 \gg (\nu_{e,n}^*/\omega_B^e)^2$. The corresponding relation for the *Pedersen conductivity* $(u_k^s = u_P^s = u_y^s)$ is

$$\sigma_P = \frac{e\,n}{B} \left\{ \frac{\nu_{e,n}^* \, \omega_B^e}{(\nu_{e,n}^*)^2 + (\omega_B^e)^2} + \frac{\nu_{i,n}^* \, \omega_B^i}{(\nu_{i,n}^*)^2 + (\omega_B^i)^2} \right\} \qquad (7.6)$$

The variation of these conductivities with height is governed essentially by the height profiles of the ionization density and frictional frequencies; see Fig. 7.7. Below the ionosphere, n is small and $\nu_{s,n}^*$ is large, so both conductivities go quickly to zero there. In the upper ionosphere, $\nu_{s,n}^*$ tends to zero, so the conductivity also vanishes there. Estimating the Hall conductivity with $\nu_{s,n}^* \to 0$ as an example, we note that the first term in the curly brackets goes to $+1$ and the second term goes to -1, so that the expression in the brackets goes to zero.

Both conductivities reach their maxima in the E region. In the case of the Pedersen conductivity, for example, both terms in the curly brackets reach

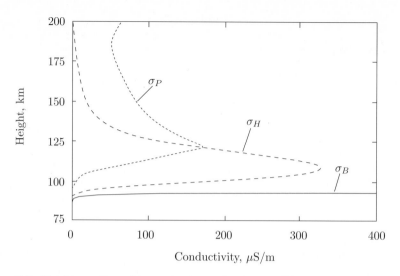

Fig. 7.8. Height profiles of ionospheric conductivities. The curves are representative for noontime conditions in winter at mid-latitudes during moderate solar activity. Note that the conductivities are subject to considerable temporal and spatial variations, particularly at polar latitudes.

their respective maxima at a height where $\nu_{s,n}^*(h) = \omega_\mathcal{B}^s$. This is the case at about 125 km for the ions, but at about 75 km for the electrons. Since the ionization density n is small at 75 km compared with its value at a height of 125 km, the Pedersen conductivity may be written to a good approximation as

$$\sigma_P \simeq \frac{e\, n\, \nu_{i,n}^*\, \omega_\mathcal{B}^i}{\mathcal{B}\,[(\nu_{i,n}^*)^2 + (\omega_\mathcal{B}^i)^2]} \tag{7.7}$$

and this expression attains its maximum at a height slightly above 125 km. Note that the condition for maximum conductivity, $\nu_{i,n}^* = \omega_\mathcal{B}^i$, corresponds closely to the situation $\nu_{i,n} \simeq f_\mathcal{B}^i$ illustrated in Fig. 7.4. Representative height profiles for the Hall and Pedersen conductivities are shown in Fig. 7.8. Evidently, the transverse conductivity - and here specifically the Hall conductivity - is confined to a rather narrow range of heights designated the *dynamo layer*. Just like with a dynamo, electric fields and currents are induced when this conducting layer moves across the terrestrial magnetic field; see Section 8.1.1.

7.3.3 Parallel Conductivity ($\vec{\mathcal{E}} \parallel \vec{\mathcal{B}}$)

The conductivity *along* the magnetic field lines is very different from the transverse conductivity. The general approach for determining this *parallel* or *Birkeland conductivity*, however, is the same. We start again with the local form of Ohm's law

$$j_B = \sigma_B \, \mathcal{E}_\parallel \qquad (7.8)$$

and obtain

$$\sigma_B = e \, n \, |\vec{u}_\parallel^i - \vec{u}_\parallel^e| \, / \mathcal{E}_\parallel \qquad (7.9)$$

The values for the parallel velocities may again be derived from the force equilibrium relations for the charge carrier gases, this time omitting the magnetic force

$$u_\parallel^s = \frac{q_s}{m_s \, \nu_{s,n}^*} \, \mathcal{E}_\parallel$$

This leads to the following expression for the Birkeland conductivity

$$\sigma_B = e^2 \, n \left(\frac{1}{m_i \, \nu_{i,n}^*} + \frac{1}{m_e \, \nu_{e,n}^*} \right) \simeq \frac{e^2 \, n}{m_e \, \nu_{e,n}^*} \qquad (7.10)$$

where in the second step we have accounted for the fact that $m_i \, \nu_{i,n}^* \gg m_e \, \nu_{e,n}^*$. The height profile of this conductivity again reflects the density variations of the charge carriers and their mobility. Since n is small and $\nu_{e,n}^*$ is large below the ionosphere, $\sigma_B \to 0$. Moving upward into the ionosphere, however, where n increases and $\nu_{e,n}^*$ decreases, σ_B very rapidly grows to quite large values (see Fig. 7.8). For a finite ionization density and a vanishing frictional frequency, in fact, it might even seem that the conductivity would go to infinity. This is not the case, however, because the mobility of the electrons in the upper ionosphere ($h \gtrsim 200$ km), and even more so in the magnetosphere, is not so much constrained by collisions with neutral gas particles as by collisions with ions. Accordingly, the frictional frequency in Eq. (7.10) must be replaced by that for electron-ion collisions

$$\sigma_B(h \gtrsim 200\text{km}) \simeq \frac{e^2 \, n}{m_e \, \nu_{e,i}^*} \qquad (7.11)$$

Here, we recall that the electron-ion frictional frequency is proportional to the plasma density n; see Eq. (5.60). As a result, the density cancels out of the expression in Eq. (7.11) and the Birkeland conductivity approaches a constant value. Whereas smaller densities do reduce the number of carriers available for charge transport, they simultaneously provide for greater mobility of those charge carriers. We thus obtain the ready-to-use formula

$$h \gtrsim 200 \text{ km}: \qquad \sigma_B[\text{S/m}] \simeq 8 \cdot 10^{-3} (T_e[\text{K}])^{3/2} / \ln \Lambda \qquad (7.12)$$

Combining the Pedersen, Hall and Birkeland components together into a total current density, one obtains the following form for Ohm's law

$$\vec{j} = \sigma_B \vec{\mathcal{E}}_\parallel + \sigma_P \vec{\mathcal{E}}_\perp + \sigma_H \vec{\mathcal{B}} \times \vec{\mathcal{E}}_\perp / \mathcal{B} \qquad (7.13)$$

The remaining question is whether this equation is consistent with the generalized Ohm's law (A.133) derived in Appendix A.13.2. Comparing the starting relations (A.115) and (A.116) with Eq. (7.3), we see that the approach used here is based on a different form of the momentum balance equation. Since stationary conditions were assumed for the calculation of the ionospheric conductivities, the inertial term was omitted. Furthermore, a simple estimate of magnitudes, such as that performed in Appendix A.13.3, shows that the inertial force, the pressure gradient force and the gravitational force all play negligible roles in the present case. Taking the example of the pressure gradient force, we have $\nabla p_s \simeq n_s k T_s / H_s \ll n_s e \mathcal{E}$. This inequality is fulfilled to a very good approximation, even in the vertical direction where the scale lengths are relatively small ($H_s \leq 100$ km), as long as the applied electric field is significantly stronger than the Pannekoek-Rosseland field strength ($< 1\mu$V/m). On the other hand, a determination of the ionospheric conductivity must certainly account for the friction imposed on the charge carriers by the neutral gas particles, which, at the height of the dynamo layer, outnumber the ions and electrons by a factor of about 10^6. This also implies that the frictional force between the charge carriers themselves, in spite of the relatively large Coulomb cross section, can be neglected. Applying these modifications, the momentum balance equations (A.115) and (A.116) reduce to the force equilibrium relation (7.3). It should also be noted that a further simplification was made for the derivation of the ionospheric conductivity presented here, which was specifically avoided in Appendix A.13. This concerns the neglect of the feedback of the currents on the geomagnetic field. Use of this simplification means that the calculation is not self-consistent. In the present case this approximation is justified, however, because the magnetic perturbation fields caused by the currents are small ($\lesssim 2\,\mu$T), when compared with the Earth's magnetic field $\gtrsim 30\,\mu$T. Finally, it should be recalled that Eq. (7.13) is valid only for a neutral gas at rest or in a coordinate system moving with the neutral gas. This is important when specifying the electric fields to be considered.

7.3.4 Ionospheric Currents

The height profiles of the ionospheric currents correspond to those of the ionospheric conductivities. The Hall and Pedersen currents thus attain their greatest intensities in the dynamo layer of the E region. Field-aligned Birkeland currents, which play a key role in closing the circuit of the Pedersen currents, are also present. Figure 7.9 illustrates this current configuration, using the dawn polar oval as an example.

The associated horizontal distribution of ionospheric currents is presented in Fig. 7.10. The sketch corresponds to the greatly idealized situation shown in Fig. 7.3. Following the direction of the electric field, Pedersen currents flow from dawn to dusk in the polar cap and radially in the polar oval. They are part of a circuit closed via Birkeland currents flowing at the boundaries of the polar oval, designated as *region 1 currents* on the poleward side and *region 2 currents* on the equatorward side. Obviously, the region 1 currents have to be stronger than the region 2 currents. The Hall current distribution corresponds

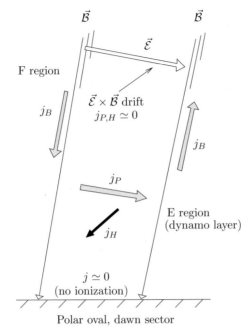

Polar oval, dawn sector

Fig. 7.9. Height distribution of ionospheric currents. j_B, j_P and j_H denote the Birkeland, Pedersen and Hall current densities, respectively.

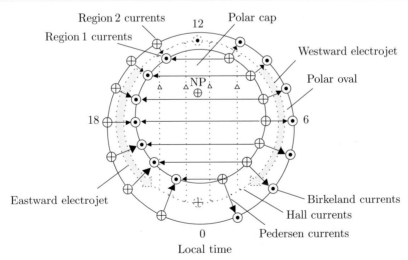

Fig. 7.10. Horizontal distribution of electric currents in the polar ionosphere. NP denotes the north geomagnetic pole. The situation corresponds to that shown in Fig. 7.3.

to that of the plasma convection, except that, because the ions drift slower than the electrons (see Fig. 7.4), the current flow is oppositely directed to the drift motion. The Hall current thus flows in the sunward direction within the polar cap and antisunward within the polar oval. This circuit, for the most part, is closed within the ionosphere, but also via Birkeland currents concentrated in the noon and midnight sectors.

The intensity of the currents depends on the strength of the electric field on the one hand, and the magnitude of the ionospheric conductivity on the other. As mentioned before, this latter quantity does display strong spatial and temporal variations. The conductivity of the polar cap region, for example, is strongly suppressed in winter due to absence of solar irradiation. The current densities are correspondingly low. The conductivity in the polar oval is less sensitive to seasonal variations because the ionization density there is controlled primarily by energetic electron precipitation – the particles producing the aurora. The combination of high ionospheric conductivity and strong electric fields in the polar oval leads to very intense currents, which, in the case of the Hall component, are designated the *polar* or *auroral electrojets*. These current jets flow westward in the dawn sector and eastward in the dusk sector. Depending on the degree of disturbance, their integrated intensity amounts to 0.1-1 MA. Of comparable magnitude are the summed Pedersen and Birkeland currents ($\simeq$1-3 MA in each half of the polar oval).

7.3.5 Magnetic Field Effects

The presence of the above described current system is revealed indirectly by its associated magnetic fields. Successful detection of the currents from the ground is especially favorable for the polar electrojets. These resemble a current sheet for an observer located below them, the associated magnetic field of which is known to be $\mathcal{B}_{PEJ} \simeq \mu_0 \, \mathcal{I}^*_{PEJ}/2$, where $\mathcal{I}^*_{PEJ}$ corresponds to the height integrated current density of the polar electrojet

$$\mathcal{I}^*_{PEJ} = \int_{dynamo\ layer} j_H \, dh \qquad (7.14)$$

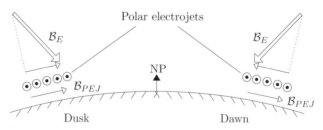

Fig. 7.11. Disturbance of the geomagnetic field ($\mathcal{B}_E$) caused by the magnetic fields of the polar electrojets ($\mathcal{B}_{PEJ}$)

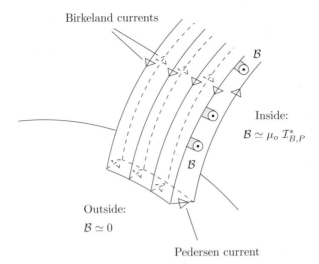

Birkeland currents

Inside:
$$\mathcal{B} \simeq \mu_o \, \mathcal{I}_{B,P}^*$$

Outside:
$$\mathcal{B} \simeq 0$$

Pedersen current

Fig. 7.12. Solenoidal flow configuration of the Birkeland-Pedersen-Birkeland currents within the (dawn) polar oval and its associated magnetic disturbance

As seen in Fig. 7.11, the eastward electrojet in the dusk sector strengthens the horizontal component of the geomagnetic field ('positive disturbance'), and the westward electrojet in the dawn sector weakens the geomagnetic field ('negative disturbance').

Detection of the Pedersen currents is more difficult. Practically, it can only be done with the help of satellite measurements. The closed system consisting of Birkeland-Pedersen-Birkeland currents, because of its great extent in geographic longitude, takes on the characteristics of a solenoid; see Fig. 7.12. The magnetic field of a solenoid, however, is concentrated on the inside of the coil – external leakage fields have much smaller intensity. Accordingly, a ground-based observer is very poorly positioned to measure the magnetic field of these currents (and thereby determine the current intensity itself). A polar-orbiting satellite, on the other hand, will record a distinct magnetic signature upon flying through this current system.

7.4 Aurorae

The aurorae, or polar lights, are surely among the most spectacular apparitions that nature has to offer. They are a literal sensation to the observer and have captured the fantasy of humankind since ancient times. Depending on the intensity and appearance, the polar lights have been considered to be either an impressive visual display or a foreboding portent of the heavens and one still finds many historical reports on these 'great signs of wonder'. In contrast to inhabitants of the lower and middle latitudes, the aurorae are a permanent fixture in the life and mythology of Norse and Eskimo cultures. The Inuits, for example, reportedly considered the polar lights to be spirits or souls of their deceased ancestors, with which one could conduct a conversation. Although the polar lights surely represent the longest known phenomenon of space research, they are in many ways still an enigmatic apparition today. We describe first the morphology of the aurorae and then consider their underlying physics.

7.4.1 Morphology

From a purely phenomenological point of view, the polar lights are airglow emission from the polar upper atmosphere caused by the incidence of energetic particles. The forms of this emission, however, can be extraordinarily diverse and variable. A brief summary of the salient features and properties of the aurorae are presented in Table 7.1.

Table 7.1. Essential characteristics of the aurorae

Forms:	discrete, e.g. arcs, bands, rays, patches	
	diffuse	
Height:	$\gtrsim 100$ km	
Orientation:	vertical:	along magnetic field lines
	horizontal:	primarily east-west direction
Dimensions:	north-south:	discrete aurorae from ca. 100 m (rayed bands) to a few 1000 m (arcs); diffuse aurorae from 100 to 1000 km
	east-west:	a few 100 to a few 1000 km
	vertical:	a few 10 km (patches) up to a few 100 km (rays)
Colors:	$\mathrm{O}\ \begin{cases} 557.7 \text{ nm (yellow-green)} \\ 630.0 \text{ nm (red)} \\ 636.4 \text{ nm (red)} \end{cases}$	
	N_2^+ 391.4 – 470 nm (blue-violet)	
	N_2 650 – 680 nm (dark red)	
Intensity:	up to a few 100 kR	
Dynamics:	occasional strong variations in form, intensity and color within seconds; horizontal velocities of up to a few 10 km/s	
Global distribution:	polar oval (aurora borealis and aurora australis)	

a.

b.

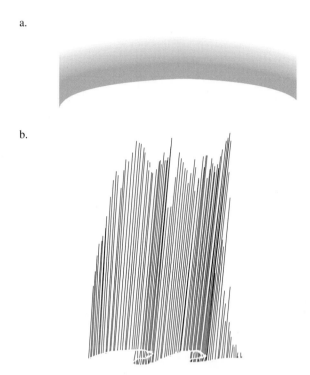

Fig. 7.13. Examples of discrete auroral forms: (**a**) homogenous arc; (**b**) rayed band, also called drapery or curtain (adapted from Akasofu, 1979)

A fundamental morphological difference exists between *discrete* and *diffuse* aurorae. The former are spatially confined apparitions with a more or less clearly recognizable shape. Typical examples are the *auroral arcs* or the *rayed bands* that resemble neatly folded curtains; see Fig. 7.13. In contrast, diffuse aurorae, as the name implies, are luminous features of areal extent that are vaguely confined and almost structureless.

The altitude of the polar lights was determined for the first time by means of triangulation at the beginning of the twentieth century. It was found that the lower edge of the luminosity was located at a height of about 100 km. The upward extent varies, depending on whether the emission is areal or rayed, between a few tens and a few hundreds of kilometers. For aurorae with great vertical extent, the alignment of luminosity with the Earth's magnetic field lines may be clearly recognized. Viewed from below, the distortion from this perspective makes the 'glowing field lines' appear to come together at a distant point, creating a wreath-like luminous feature known as the *auroral corona*.

Aurorae reach their greatest extent in the zonal direction, sometimes spreading out over hundreds or even thousands of kilometers. Their lati-

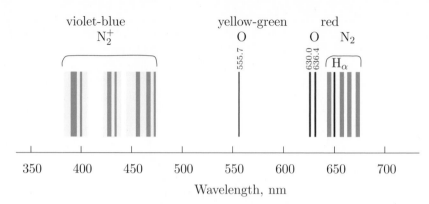

Fig. 7.14. Auroral spectrum in the visible range (adapted from Akasofu, 1979)

tudinal width is comparatively modest and amounts to only a few hundred meters for rayed bands.

Unlike sunlight, the polar lights have no continuum spectrum. Indeed, its color is determined by only a few discrete lines and bands; see Fig. 7.14. Dominant are the yellow-green line of atomic oxygen at 557.7 nm, the red line of the same constituent at 630.0 and 636.4 nm, the blue-violet band system of singly ionized molecular nitrogen and the dark red bands of neutral molecular nitrogen. Depending on the relative excitation of these transitions, the aurorae may be yellow-green, red, blue-violet, or take on a milky white appearance if all colors are uniformly blended together. Infrared, ultraviolet and X-ray auroral emissions are produced in addition to visible light. Particularly intense, for example, are the lines of atomic oxygen at 130.4 and 135.6 nm. This UV-emission, of course, can only be observed with special detectors from space.

Although it has often been claimed that aurorae are also audible, scientifically acceptable evidence for this phenomenon is still missing. Infrasonic waves in the frequency range 0.05 – 0.5 Hz (audible sound $\gtrsim$ 20 Hz), however, have been well documented and appear to be associated with the occasionally observed supersonic motion of aurorae.

By nature, the brightness of the polar lights is quite variable, ranging from subvisual to brightly glowing radiance. As with all airglow phenomena, the intensity is measured in units of rayleighs (see Section 3.3.8). It suffices here to note that 1 kR corresponds approximately to the brightness of the Milky Way, 100 kR to the luminosity of cumulus clouds illuminated by the full moon. The maximal intensity of several 100 kR cited in Table 7.1 is thus a remarkably bright glow.

The dynamics of the aurora is strongly dependent on the degree of disturbance. Quiet conditions are thus characterized by weakly luminous, quasi-stationary auroral forms. In contrast, a veritable fireworks of changing intensities, colors and shapes can occur during active conditions, with horizontal and vertical motions (the latter especially in *flaming* aurorae) playing an important role. Special mention should be made here of the *pulsating* aurorae characterized by quasi-periodic fluctuations in intensity.

The occurrence of polar lights is essentially confined to the northern and southern polar ovals – referred to in this connection as the *auroral ovals*. Transpolar aurorae in the center of the polar cap are observed only during very quiet conditions; during very disturbed conditions aurorae can extend down to mid-latitudes. For example, numerous publications described the spectacular auroral activity observed in Europe, the USA and Japan during the great geospheric storms in March and October 1989, in April and July 2000, and in October/November 2003. It is important to realize that polar lights are often *conjugate* phenomena, by which similar forms are observed simultaneously in both the northern and southern polar auroral ovals. This implies that these aurorae are excited near the footpoints of closed field lines.

7.4.2 Dissipation of Auroral Particle Energy

Today we know that the aurorae are caused by the incidence of energetic particles onto the upper atmosphere. The particles are mostly electrons in the energy range from a few 100 to a few 10 000 eV, but ions are also observed – here especially responsible for the proton aurorae, i.e. Doppler-shifted emission of the red $H\alpha$ line of hydrogen. The physics is similar to the process occurring in a television set: accelerated electrons also impinge here on an absorber and stimulate it to emit light. The upper atmosphere, however, is a vastly more tenuous absorption medium than the commonplace TV picture screen. Accordingly, the electrons are decelerated slowly by a series of elastic and inelastic collisions. Similar to the extinction of solar UV-radiation, some processes associated with the primary collision events are

- scattering (elastic collisions)
- collisional ionization
- collisional dissociation
- collisional excitation

and various combinations of the above. The energy of the impinging auroral particles is thus gradually transferred to the upper atmosphere via these collisions and the resulting energy deposition profile is sketched in Fig. 7.15. Among other factors, the height of maximum energy deposition depends on the collision cross sections, which are themselves functions of the energy of the incident electrons. The pitch angle of the helical trajectory of the incident particles also plays an important role. Roughly speaking, the energy of 0.1 keV electrons is absorbed above 200 km, 1 keV electrons at about 130 km, and 10 keV electrons at about 100 km altitude.

The redistribution of the absorbed energy is complex and characterized by numerous secondary processes such as

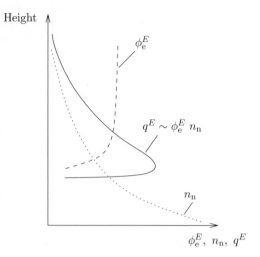

Fig. 7.15. Energy deposition profile in the upper atmosphere resulting from precipitating auroral electrons. ϕ_e^E, n_n and q^E denote the energy flux of the auroral electrons, the density of the neutral atmosphere and the energy deposition rate per unit volume, respectively. The maximum energy deposition rate evidently occurs at that height where the energy flux of the incident electrons is still large enough, and the density of the atmosphere has become large enough to effectively absorb the incident energy.

– secondary ionization
– secondary dissociation
– secondary scattering
– charge exchange
– dissociation exchange
– excitation exchange
– dissociative recombination
– radiative recombination
– collisional quenching

and various combinations of the above. In the case of collisional ionization, for example, we can directly apply the block diagram of Fig. 3.20 and the list of chemical reactions given in Table 4.3. Obviously, a quantitative analysis of these processes is only possible with the help of extensive model calculations. We content ourselves here by summarizing some of the more interesting results of these numerical simulations.

Conversion of the Deposited Energy. Contrary to initial presumptions, only a very small part ($\lesssim 1\%$) of the energy provided by the precipitating electrons is converted to radiation. The largest fraction (ca. 50%) is expended as heat, another important part (ca. 30%) is converted to potential chemical energy, and the rest is scattered back into the magnetosphere.

When estimating the heating rate, it is noted that 35 eV of primary energy, on the average, must be expended for production of an ion-electron pair below about 150 km altitude. Since photochemical equilibrium holds in this height range, the ionization production rate is proportional to the square of the ionization density, $q^I = \alpha\, n^2$, see Eq. (4.27). The heating rate is thus given by

$$q^W = 0.5\ (35\ \text{eV})\ \alpha\, n^2$$

where we have assumed a heating efficiency of $\eta^W \simeq 0.5$. Letting the ionization density produced by auroral electrons be $2 \cdot 10^{11}$ m^{-3}, the heating rate calculated from the above equation becomes $q^W \simeq 10^{-8}$ W/m^3. This heating rate is not especially impressive compared with the heating of the upper atmosphere by solar UV-radiation (see Fig. 3.22). Nevertheless, it should be remembered that this energy source is also active during the night and, particularly, during the long polar winter.

Excitation and De-excitation Mechanisms. Another surprising result of numerical simulations is the fact that only a relatively small part of the observed light emission comes from direct collisional excitation of atmospheric gases. Among the directly excited emissions is the blue-violet line of singly ionized molecular nitrogen at 391.4 nm

$$\text{N}_2 + \text{e}_p \longrightarrow \text{N}_2^+ (B^2 \textstyle\sum_u^+) + \text{e}_p + \text{e}_s$$
$$\downarrow$$
$$\text{N}_2^+ (X^2 \textstyle\sum_g^+) + \text{photon}(391.4\ \text{nm})$$

where e_p denotes a primary electron, e_s a secondary electron and the expressions in parentheses indicate the different excitation states of the molecular ion. It is estimated that one out of every 25 ionizations of an N$_2$ molecule leads to the emission of a 391.4 nm photon. This means that about $25 \cdot 35$ eV $= 875$ eV of primary energy is required to produce one 3.2 eV photon (at $\lambda = 391.4$ nm), corresponding to an excitation efficiency of less than 0.4%.

More important contributions to the excitation are delivered by chemical reactions. The excitation for the atomic oxygen line at 557.7 nm, for example, is suggested to proceed via the following scheme

$$\text{N}^+ + \text{O}_2 \longrightarrow \text{NO}^+ + \text{O}(^1S)$$

$$\text{O}(^1S) \ \xrightarrow{\ 1\ \text{s}\ }\ \text{O}(^1D) + \text{photon}(557.7\ \text{nm})$$

Here, $\text{O}(^1S)$ denotes an excited oxygen atom in the metastable ^1S state which, after a typical lifetime of about 1 s, de-excites to the ^1D state, thereby emitting a photon at a wavelength of 557.7 nm (see also Fig. 3.30). The small probability of this 'forbidden' transition originally led to some difficulties in identifying the line. In fact, even as late as the 1920s, a special gas, the

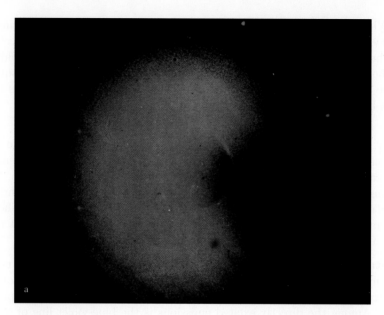

Fig. 7.16. Emission phenomena of the upper atmosphere. (**a**) The geocorona in the light of Lyman-α radiation (121.6 nm; false color image). The hydrogen envelope of the Earth is irradiated by the Sun (from the left). Note the brightening of the night shadow due to multiply scattered photons. Faint auroral emissions are visible at high latitudes. The apparent banded structure is an artifact of the imaging process (Carruthers et al., 1976). (**b**) Aurora observed from the ground. The red coloring comes from the 630.0/636.4 nm emission of atomic oxygen, the milky-white coloring from the blending of this emission with the yellow-green line of atomic oxygen at 557.7 nm (N. Rosing, Grafrath). (**c**) Auroral rayed bands recorded from the flight deck of the Space Shuttle DISCOVERY at a height of about 350 km. The Earth's cloud deck in moonlight may be seen at the bottom of the image. About 100 km higher and parallel to the horizon, clearly visible even through the green aurora, one can see the thin, tan-colored emission layer of the nightglow (image courtesy of NASA). (**d**) Auroral oval imaged in the wavelength range 123-155 nm. The dominant emission in this image, recorded from the DE1 satellite at an altitude of about 20 000 km, is that of atomic oxygen at 130.4 nm. The faint green contours, superimposed afterwards on the image, represent coastlines. On the left side of the image one sees the intense dayglow of the sunlit upper atmosphere, an emission consisting predominantly of the atomic oxygen line at 130.4 nm, excited primarily (but certainly not exclusively) by photoelectrons (L.A. Frank, University of Iowa). (**e**) Northern and southern auroral ovals seen from the side. Similar to the previous plate (d), this UV image was also recorded from the DE1 satellite at an altitude of about 20 000 km. This time, however, a filter (117–165 nm) was used that also passes Lyman-α radiation at 121.6 nm. The geocorona is thus visible (red false-color) as a diffuse background glow (L.A. Frank, University of Iowa; see also Frank and Craven, 1988).

Fig. 7.16. Emission phenomena of the upper atmosphere (continuation)

so-called geocoronium, was held responsible for this emission. Note that the parent ion of the above reaction chain, N^+, is produced not so much by direct collisional ionization (N is only a trace gas in the upper atmosphere!) as by dissociative collisional ionization and by secondary chemical reactions. It is also important that a significant amount of energy is released during the course of the above reactions.

The $^1D \rightarrow {}^3P$ transition, which is responsible for the red oxygen lines at 630.0 and 636.4 nm, has an even smaller probability than the $^1S \rightarrow {}^1D$ transition; see again Fig. 3.30 and Eq. (3.66). The long mean lifetime of this state (110 s) means that the oxygen atom, at altitudes below about 200 km, does not lose its excitation energy so much by photon emission, but rather by collisional quenching, in which case the transition energy is converted into heat. This is one of the reasons why aurorae dominated by the red oxygen emission tend to occur at greater heights.

Finally, it should be pointed out that auroral X-ray emission is not produced by collisional excitation (and also clearly not by chemical reactions), but by the sudden, collisionally-induced deceleration of energetic primary electrons. In other words, we are dealing here with bremsstrahlung radiation.

Height Dependence of the Emission. Model calculations show that the variation of the aurorae with height depends critically on the particular emission line under consideration, the energy and pitch angle of the incident electrons, and the state of the absorbing upper atmosphere. Different excitation and de-excitation mechanisms, for example, are responsible for the fact that the 630.0/636.4 nm line of atomic oxygen is emitted at greater heights than the 557.7 nm line. It is also understandable that particles with higher energies and smaller pitch angles will penetrate deeper into the upper atmosphere. Accordingly, the level of maximum energy dissipation and auroral emission will shift in this case to lower heights. As an example, Fig. 7.17 shows the emission height profile of the 557.7 nm line of atomic oxygen for two different energies of the incident electrons. The emission layer drops down by about 70 km when the electron energy increases from 0.8 keV to 10 keV. A similar drop in the height of the emission layer is not seen for the 630 nm line. This is because of the strong increase in collisional quenching of the excited ^1D state at lower altitudes.

7.4.3 Origin of the Auroral Particles

Contrary to many of the popularized accounts of the phenomenon, polar lights are *not* produced by solar wind particles. Solar wind electrons have neither access to the nightside polar oval nor do they possess adequate energy to generate the aurorae. In searching for better alternatives, we note that the night sector of the polar oval is magnetically connected with the magnetotail plasma sheet and the dayside polar oval with the magnetospheric boundary layer. These two plasma reservoirs are clearly responsible for the supply of

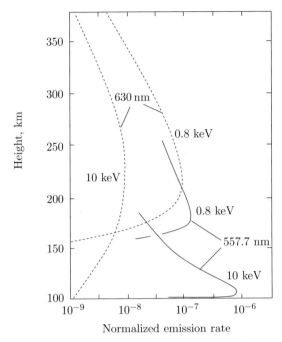

Fig. 7.17. Height profile of auroral emission for various emission lines and energies of the incident electrons. Electrons are considered which start at an altitude of 600 km and impinge onto an upper atmosphere with a thermopause temperature of 1000 K. Their pitch angle distribution is isotropic and limited to the Earth-facing hemisphere. In other words, all pitch angles are equally probable and only particles moving toward the Earth are considered. A Gaussian energy distribution is assumed for the precipitating particles, the mean energy of which is 0.8 eV in one case and 10 keV in another case. The emission rate is normalized to the intensity of the incident electron flux (adapted from Banks et al., 1974).

auroral particles. An open issue is how these particles are precipitated into the upper atmosphere and, if necessary, accelerated to the observed energies.

Diffuse Aurorae. It is generally assumed today that convection and subsequent pitch angle diffusion of plasma sheet particles are responsible for producing the diffuse aurora in the night sector. The envisioned scenario is sketched in Fig. 7.18. A dawn-to-dusk electric convection field in the central plane of the magnetotail leads to an $\vec{\mathcal{E}} \times \vec{\mathcal{B}}$ drift of the plasma sheet particles in the direction of (and around) the Earth (see also Section 7.6.6). While these particles drift in the magnetotail region connected by magnetic field lines to the auroral oval, they scatter stochastically, primarily by electromagnetic plasma waves, into the local loss cone. These are the plasma sheet electrons (and to a lesser extent the plasma sheet ions) which plunge down the field lines and are made responsible for producing the diffuse aurora. We will not

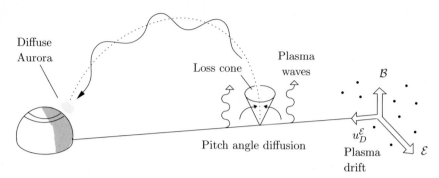

Fig. 7.18. Process responsible for diffuse aurorae. The magnetic field $\mathcal{B}$ corresponds here to the $\mathcal{B}_z$ component of the teardrop-shaped extended dipole field in the plane of the neutral sheet, and $\mathcal{E}$ is the $\mathcal{E}_y$ component of the convection field across the magnetotail.

attempt to go into the many complicated aspects of the wave-particle interaction involved in this process. It suffices to realize that the electric fields of the waves randomly accelerate some of the particles in the direction along the magnetic field line. The pitch angle of these particles is thus decreased and, if the original pitch angle was at the edge of the loss cone, it may now lie inside the loss cone. This process is known as *pitch angle diffusion* into the loss cone.

Discrete Aurorae. Measurements show that discrete aurorae are produced by higher energy electrons ($E_e \gtrsim 1$ keV). In this case we thus require additional acceleration of the electrons populating the magnetotail plasma sheet ($E_e \lesssim 1$ keV) and the magnetospheric boundary layer ($E_e \lesssim 200$ eV). Just how this is accomplished is still an open issue. One widespread concept assumes that the acceleration proceeds via magnetic field-aligned electric fields, a possible scenario of which is shown schematically in Fig. 7.19.

This scenario postulates an unspecified generator in the magnetotail plasma sheet or in the magnetospheric boundary layer that produces Birkeland currents along the magnetic field lines threading the polar auroral oval. These currents are closed by Pedersen currents in the ionosphere, but under the assumption that the current circuit is strictly localized and exists independently of the large-scale region 1/ region 2 current system described in Section 7.3.4. Let us first assume that the field lines are good conductors. According to Eq. (7.12) and using $\ln \Lambda \simeq 18$, the Birkeland conductivity in the magnetosphere is given by

$$\sigma_B[\text{S/m}] \simeq 4 \cdot 10^{-4} (T_e[\text{K}])^{3/2} \tag{7.15}$$

A very high conductivity of $\sigma_B > 30$ S/m is obtained here even for a relatively low electron temperature of 2000 K. This conductivity is much too high to allow the build-up of significant electric fields. Assuming that an electron flux

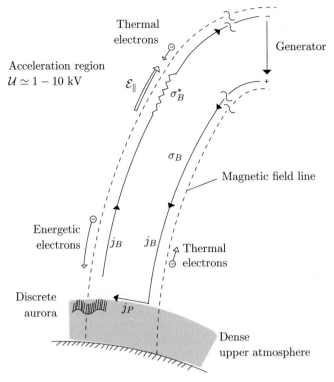

Fig. 7.19. Process possibly responsible for discrete aurorae

of a few 10^{13} particles/m^{2}s is necessary to produce a discrete aurora, such a flux of incident electrons corresponds to an upward directed current of a few μA/m^2, $j = e\,\phi_{\mathrm{e}}$. With $\mathcal{E}_\parallel = j_B/\sigma_B$ and taking a generously measured field line length of 15 R_E, this yields a voltage drop of less than 10 V, much too small to explain the acceleration of auroral electrons!

The expression for the conductivity given in Eq. (7.15) does not account for the fact that the mobility of the electrons in the geomagnetic dipole field is constrained not only by collisions, but also by magnetic gradient (mirror) forces. In the absence of other external forces, electrons moving toward the Earth can only penetrate down to the height of their mirror points, see Section 5.3.2. In order to establish a sufficiently intense, upward-directed current carried by the electrons, the mirror points must be lowered for those electrons with pitch angles outside the loss cone. According to Eq. (5.37), this requires a decrease in their pitch angle, i.e. an increase in their velocity component parallel to the field (see Fig. 5.14). An additional field-aligned electric field is evidently necessary to compensate the magnetic gradient forces and to accelerate the electrons beyond their nominal mirror point in the direction toward the Earth. The mirror force is thus clearly responsible for a reduction in the effective parallel conductivity for upward-directed currents carried by precipitating electrons. Pertinent calculations yield the following simple relation between the Birkeland current density in the ionosphere and the required voltage difference along the connecting dipole field line

$$j_B \simeq K \, \mathcal{U}_{\parallel} \tag{7.16}$$

The proportionality factor in the above equation represents a type of conductivity that attains the following value for a Maxwell distribution of magnetospheric electrons

$$K = e^2 \, n / \sqrt{2\pi \, m_e \, k \, T_e} \tag{7.17}$$

If the current carrying electrons originate in the magnetotail plasma sheet, where typical values of the free parameters in Eq. (7.17) are $n \simeq 3 \cdot 10^5 \mathrm{m}^{-3}$ and $T_e = 8 \cdot 10^6$ K, a current density of 3 $\mu\mathrm{A/m}^2$ can be driven with a potential difference of $\mathcal{U}_{\parallel} = 3$ kV. This would be sufficient to produce the higher energy electrons responsible for discrete aurorae.

The above described reduction of the Birkeland conductivity by magnetic mirror forces is not the only way to build up field-aligned potential differences and accelerate auroral electrons. Another possibility assumes that, above a certain intensity, the upward-directed currents excite plasma instabilities that lead to the formation of localized potential differences at greater heights (ca. 6000 – 12 000 km). Quite often mentioned in this connection is a particular type of standing plasma wave, a so-called *double layer*. Still other types of plasma instabilities can lead to enhanced momentum exchange between electrons and ions, thereby increasing the effective frictional frequency. In this case, using the form given in Eq. (7.11), we obtain a strongly reduced parallel conductivity given by

$$\sigma_B^* = \frac{e^2 \, n}{m_e \, (\nu_{e,i}^*)_{eff}} \tag{7.18}$$

The situation with σ_B^* is indicated in Fig. 7.19. Finally, the possibility exists that the auroral electrons are not accelerated by quasi-static potential differences, but rather by wave fields. Particularly well suited for this purpose would seem to be a special type of magnetoplasma waves, the lower hybrid variety, which may be excited at the upper and lower edges of the plasma sheet.

Evidence for the acceleration region hypothesized in Fig. 7.19 comes from the fact that it is a source of radio waves. This radio emission is so intense, that it would probably be the first detectable terrestrial radiation when traveling toward the solar system from the depths of interstellar space. The frequency of this radiation lies close to the local gyrofrequency of the electrons

$$f_{\mathrm{radio}} \simeq f_B^e \simeq \frac{e \, \mathcal{B}_{00}}{\pi \, m_e} \left(\frac{R_E}{r}\right)^3 \simeq 1.7 \cdot 10^6 \left(\frac{R_E}{r}\right)^3 \, [\mathrm{Hz}] \tag{7.19}$$

At a height of 6000 km ($r \simeq 2R_E$), for example, the resulting frequency is about 200 kHz ($\lambda \simeq 1.5$ km). Corresponding to the wavelength range, this emission is referred to as *auroral kilometric radiation, AKR*. How this emission is generated – just like the issue with the auroral particle acceleration – is still another incompletely solved and thus challenging problem of space research.

7.5 Neutral Atmospheric Effects

Ionospheric convection and currents, as well as auroral particle precipitation, have important consequences for the dynamics and energetics of the polar upper atmosphere. The neutral gas is accelerated and heated via collisions with the drifting charge carriers. Further heat input is provided by incident auroral particles. The resulting changes in wind velocity, temperature and density are an important feature of the polar upper atmosphere.

7.5.1 Drift-induced Winds

Collision-induced friction associated with the ionospheric convection leads to a continuous acceleration of the neutral atmospheric gases in the direction of the drift motion. This creates a horizontal wind circulation that corresponds to the convection pattern shown in Fig. 7.3 and is superimposed on the tidal wind circulation. Since collisions are relatively rare in the F region, it takes a certain amount of time before the wind velocity adjusts itself to the convection velocity. The associated time constant may be calculated as follows. Consider a thermospheric gas packet (mass density ρ_n, spatially uniform velocity u_n) that is accelerated by ion frictional forces. If only the inertial force opposes this acceleration, the equilibrium case requires that

$$\rho_n \frac{\partial u_n}{\partial t} \simeq \rho_n \, \nu^*_{n,i} \, (u_i - u_n)$$

Assuming that the convection velocity u_i is constant ($\partial u_i / \partial t = 0$), and introducing the velocity difference $\Delta u = u_i - u_n$, we have

$$- \frac{\partial (\Delta u)}{\partial t} \simeq \nu^*_{n,i} \, \Delta u$$

This equation may then be integrated to obtain

$$\Delta u(t) = \Delta u(t = 0) \, e^{-t/\tau_W} \tag{7.20}$$

It follows that the wind velocity, starting from zero, would attain a value of about $2/3$ of the convection velocity in the time $\tau_W = 1/\nu^*_{n,i}$. As an example, using Eq. (2.56) and assuming F region conditions ($n_i \simeq n_{O^+}$, $n_n \simeq n_O$, $T_{O^+} \simeq T_O \simeq 1000$ K, $\sigma_{O,O^+} \simeq 8 \cdot 10^{-19} \mathrm{m}^2$, see also Section 3.4.4), one derives a time constant of $\tau_W [\mathrm{s}] \simeq 10^{15}/n_{O^+} [\mathrm{m}^{-3}]$. Taking a typical ionization density of $5 \cdot 10^{11}$ m^{-3}, one obtains an acceleration time of about thirty minutes. Clearly, the adjustment of the thermospheric motion to the ionospheric convection in the dayside F region is distinctly delayed. It is retarded even further on the nightside because of the reduced ionization density there.

7.5.2 Heating

The friction between drifting charge carriers and neutral gas particles not only leads to acceleration, but also to heating of the thermospheric gases. Because of the deflections suffered during the collisions, neutral gas particles are accelerated not just in the direction of the drift, but also perpendicular to it. This creates an energy enhancement of the random motion and corresponds to an input of heat. The same process in reverse heats the charge carrier gases. In both cases energy is extracted from the ordered drift motion and transferred to the gases in the form of frictional heat. If the drift motion is to be maintained in spite of these losses, this must be done at the expense of the electric field energy responsible for the convection. A quantitative description of these processes is possible with the help of the energy balance equations of the interacting gases. In this exercise it is important that we also account for the heat exchange between the gas components. It turns out that the effective heating rate of the neutral gas is proportional to the square of the relative velocity between the ions and the neutral gas particles. This is understandable when we consider that the kinetic energy is proportional to the square of the velocity, that the friction vanishes when the relative velocity goes to zero, and that electrons, because of their small mass, are very ineffective in the transfer of energy.

Here we consider a different approach for a determination of the effective heating rate, recalling from electrodynamics that currents lead to *Joule heating* of the conductors in which they flow. This may be expressed as

$$q_J = \vec{j}\,\vec{\mathcal{E}} = \sigma_P\,\mathcal{E}^2 \qquad (7.21)$$

where q_J is the heat increment per unit volume and time caused by the current flow. This relation clearly corresponds to the local form of the power dissipation in an electric circuit, $P = \mathcal{I}\,\mathcal{U} = \mathcal{U}^2/\mathcal{R} = \mathcal{L}\,\mathcal{U}^2$, where $\mathcal{L}$ is again the conductance. Note that only the Pedersen currents, but not the Hall currents, contribute to the heating. The latter currents, of course, flow perpendicular to the electric field.

For a conductor at rest, $\vec{\mathcal{E}}$ is the externally applied electric field (e.g. via a voltage difference). If the conductor is moving in a magnetic field, one must also account for the induced dynamo electric field $\vec{\mathcal{E}}_{dyn} = -\vec{u}\times\vec{B}$ (see Section 6.2.6). This happens to be the case under consideration here: the dynamo region of the upper atmosphere is a conductor moving in the magnetic field of the Earth. The motion of this conductor proceeds at the plasma velocity of this region, $u = (\rho_i u_i + \rho_e u_e + \rho_n u_n)/(\rho_i + \rho_e + \rho_n) \simeq u_n$, where Eq. (6.49) has been extended to include the neutral gas component, and we obtain

$$q_J = \sigma_P\,(\vec{\mathcal{E}}_{con} - \vec{\mathcal{E}}_{dyn})^2 \simeq \sigma_P\,(\vec{\mathcal{E}}_{con} + \vec{u}_n \times \vec{B})^2 \qquad (7.22)$$

Here, $\vec{\mathcal{E}}_{con}$ denotes the electric convection field and the minus sign in the first set of parentheses accounts for the fact that the induced field weakens the (accelerating) convection field. Alternatively, Eq. (7.22) can also be understood as follows. An observer at rest with the neutral gas (system index *) sees that the charge carriers move with an $\vec{\mathcal{E}} \times \vec{\mathcal{B}}$ drift velocity of magnitude $\vec{u}^* = \vec{u}_i - \vec{u}_n$. Responsible for this motion, from the observer's point of view, is the electric field $\vec{\mathcal{E}}^* = -\vec{u}^* \times \vec{\mathcal{B}} = -\vec{u}_i \times \vec{\mathcal{B}} + \vec{u}_n \times \vec{\mathcal{B}} = \vec{\mathcal{E}}_{con} + \vec{u}_n \times \vec{\mathcal{B}}$, where the quantities $\vec{u}_i$, $\vec{u}_n$ and $\vec{\mathcal{E}}_{con}$ are referred to a coordinate system at rest with the Earth.

If the neutral gas velocity reaches the convection velocity, one obtains

$$q_J = \sigma_P \left(\vec{\mathcal{E}}_{con} + (\vec{\mathcal{E}}_{con} \times \vec{\mathcal{B}}) \times \vec{\mathcal{B}}/\mathcal{B}^2 \right)^2 = 0$$

and this is understandable, because friction is no longer active when the flow velocities of the neutral gas particles and the charge carriers are equal.

Usually, the neutral gas velocity at heights within the dynamo region is relatively small ($\lesssim 150$ m/s). Therefore, the induced field can be neglected here in comparison with the prevailing convection field. In particular, a good approximation within the range of the polar oval is

$$q_J(\text{polar oval}) \simeq \sigma_P \, \mathcal{E}_{con}^2 \tag{7.23}$$

Inserting typical values for the dynamo layer at an altitude of 130 km ($\sigma_P \simeq 10^{-4}$ S/m, $\mathcal{E}_{con} \simeq 30$ mV/m), one obtains a heating rate of $q_J(130$ km$) \simeq 10^{-7}$ W/m^3. This is many times larger than the heating rate from solar UV radiation for this range of heights (compare with Fig. 3.22). Pedersen currents are thus an important source of heat for the polar upper atmosphere. Additional heat input comes from the energy dissipation of incident auroral particles.

7.5.3 Composition Disturbances

An immediate consequence of the intense heating by currents and particle precipitation is a distinct increase in temperature in the region of the polar oval, leading to an expansion of the atmospheric gases. This expansion proceeds preferentially upward because this is the direction of minimum backpressure. As the growing high pressure zone reaches greater heights, it becomes capable of driving horizontal winds that distribute the excess heat over a wider area. In this way, wind circulation patterns such as those indicated in Fig. 7.20 are established.

The vertical winds associated with this circulation have important consequences for the composition of the upper atmosphere. It is observed, for example, that the relative abundance of the lighter gases (e.g. atomic oxygen and helium) decreases and, simultaneously, the relative abundance of heavier gases (e.g. molecular nitrogen and argon) increases. Figure 7.20 provides an explanation for this striking effect. Consider the mid-thermospheric region at heights of about $110 - 170$ km. The primary gas, molecular nitrogen, can

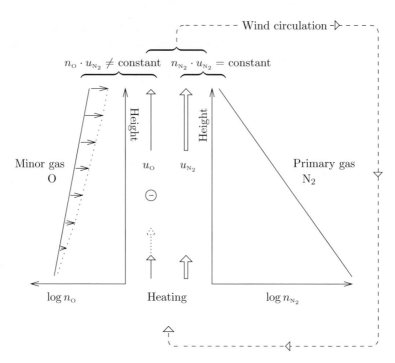

Fig. 7.20. Formation of wind circulation cells and density decrease of a light minor gas (here atomic oxygen) from a wind-induced disturbance of the equilibrium distribution

flow upward there with the vertical velocity u_{N_2} without significantly disturbing its diffusive equilibrium. This requires, to a first approximation, that the particle flux $n_{N_2} u_{N_2}$ is constant and independent of height. Because of the exponentially decreasing density, the velocity must increase in a way consistent with the scale height of the primary gas. The strong collisional friction in the middle thermosphere forces the atomic oxygen, a minor gas, to move upward at essentially the same velocity as the primary gas. In this case, however, the particle flux $n_O u_{N_2}$ is no longer constant. Instead, more oxygen atoms are taken away from the top than resupplied from the bottom, creating a deficit of this component. The dotted arrow indicates how large the vertical velocity of the oxygen atoms would have to be in order to maintain continuity of the particle flux corresponding to the scale height of this gas.

The scenario for the disturbance sketched in Fig. 7.20 may be described formally as follows. If $n_{N_2} u_{N_2} = $ constant, then

$$u_{N_2} \sim \frac{1}{n_{N_2}} \sim e^{(z - z_0)/H_{N_2}}$$

where we have made the rough approximation of neglecting the height dependence of the temperature. Invoking the continuity equation for the minor
gas s and setting $u_s \simeq u_{N_2}$ we obtain

$$\frac{\partial n_s}{\partial t} \simeq - \frac{\partial}{\partial z}(n_s \, u_{N_2}) \sim - \frac{\partial}{\partial z}\left\{ e^{-(z-z_0)/H_s} \, e^{(z-z_0)/H_{N_2}} \right\}$$

or

$$\frac{\partial n_s}{\partial t} \sim \frac{1}{H_{N_2,s}} \, e^{-(z-z_0)/H_{N_2,s}} \tag{7.24}$$

where the effective scale height $H_{N_2,s}$ can assume both positive and negative
values

$$H_{N_2,s} = \frac{H_{N_2} \, H_s}{H_{N_2} - H_s} = \frac{H_{N_2}}{\mathcal{M}_s/\mathcal{M}_{N_2} - 1} \tag{7.25}$$

Accordingly, one will observe either an increase or a decrease of the particular
minor gas. For lighter gases ($\mathcal{M}_s < \mathcal{M}_{N_2}$), we have $\partial n_s/\partial t < 0$ and the
density of gas s decreases; for heavier gases ($\mathcal{M}_s > \mathcal{M}_{N_2}$), however, $\partial n_s/\partial t > 0$ and the minor gas density increases. Composition disturbances of this type
are a characteristic feature of the upper atmosphere at polar and sometimes
even mid-latitudes; see Section 8.4.1 and here particularly Fig. 8.15.

7.6 Energy Transfer from Solar Wind to Magnetosphere

Maintaining the various phenomena observed in the polar upper atmosphere
requires a substantial amount of energy. Assuming that this energy flows
into the upper atmosphere via the region 1 Birkeland currents (we justify this
assertion later), and letting $\mathcal{U}_{pc}$ denote the polar cap potential, then the power
dissipated in both polar regions is $P \simeq 2\,\mathcal{U}_{pc}\,\mathcal{I}_B(\text{region } 1) \simeq 2 \cdot 30 \text{ kV} \cdot 1 \text{ MA} \simeq 10^{11}$ W, and this must somehow be extracted from the solar wind. An initial
estimate demonstrates that this is possible in principle. According to Table
6.1, the energy flux at the Earth's orbit is about 0.5 mW/m^2. For a total
magnetospheric cross section of $(15 \, R_E)^2 \pi$, one calculates the available power
to be more than 10^{13} W. Absorption of only a small fraction of this power
would thus suffice to supply the polar upper atmosphere with the required
energy. Just how this happens in detail is a hotly debated problem and a topic
of intensive research. Firmly established is that it must be the flow energy
of the solar wind which is eventually tapped and that a large fraction of the
diverted kinetic energy is initially converted into electrical energy. A dynamo
configuration is introduced in the following that is capable of explaining the
essential aspects of the energy absorption and conversion processes.

7.6.1 Solar Wind Dynamo

In a dynamo, kinetic energy is converted to electrical energy by the motion
of a conductor in a magnetic field. In the case of the solar wind dynamo, it

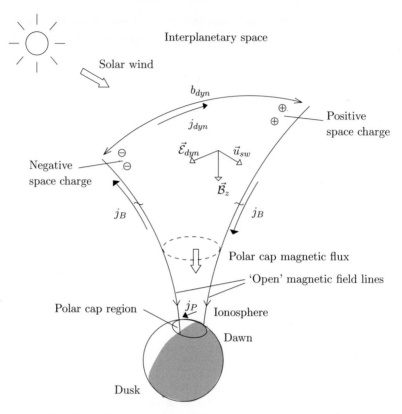

Fig. 7.21. Operational principle of the solar wind dynamo

is the conducting plasma of the solar wind that flows across magnetic field lines connected to the Earth. In order to better understand the operational principle of this dynamo, let us first imagine that the magnetic field lines from the polar cap do not connect to the magnetotail lobes, but rather stick right out into interplanetary space. In other words we are dealing with an *open magnetosphere*, for which the polar cap field lines have one footpoint on the Earth, but otherwise extend outward into the interplanetary medium. Figure 7.21 illustrates the *modus operandi* envisioned for this scenario. The solar wind, a good conductor, flows from the dayside across the open magnetic field of the polar cap region and induces an electric dynamo field given by

$$\vec{\mathcal{E}}_{dyn} = -\vec{u}_{sw} \times \vec{B}_z \tag{7.26}$$

see Eq. (6.37). The situation is analogous to that of a magnetohydrodynamic (MHD) generator, by which an artificially created plasma flows perpendicular to a magnetic field, producing electric fields and currents. On a microscopic scale, the solar wind protons, upon entering the open polar cap magnetic field region, are deflected by the magnetic force toward the dawn side. Similarly, the electrons are deflected toward the dusk side. This leads to charge separation and creation of a polarization electric field directed from dawn to dusk. The charge separation and build-up of excess charge continues until the electric force exerted by the polarization field exactly compensates the magnetic force

$$n \, q_s \, \vec{\mathcal{E}}_P + n \, q_s \, \vec{u}_{sw} \times \vec{\mathcal{B}}_z = 0$$

from which it follows that

$$\vec{\mathcal{E}}_P = -\vec{u}_{sw} \times \vec{\mathcal{B}}_z = \vec{\mathcal{E}}_{dyn}$$

Upon establishing this polarization field, the particles continue their journey – slightly deflected to the dawn or dusk sides. It may be easily verified that the deflection necessary for compensating the magnetic force is extremely small (centimeters) compared with the maximum possible deflection given by $r_B^i + r_B^e$ (> 1000 km).

The essential point of the situation outlined here, is that the polarization field built up by this process is transferred to the polar ionosphere along the highly conducting magnetic field lines. One should thus observe an electric field directed from dawn-to-dusk in the polar cap region and, as we have seen, this is indeed the case. Using the previously given polar cap potential of about 30 kV, an assumed magnetic field strength of $\mathcal{B}_z \simeq -2$ nT, and a typical solar wind velocity of 500 km/s, the required width of the dynamo region is estimated to be $b_{dyn} = \mathcal{U}_{pc}/(u_{sw} \, \mathcal{B}_z) \simeq 5 \, R_E$, a dimension that is certainly realistic.

At the same time, the conducting ionosphere in the dynamo layer represents a load for the solar wind dynamo. A current flows there and, because of the Joule heating it produces, consumes energy that must be extracted from the solar wind. This occurs via a reduction of the solar wind flow velocity. In general, any closure current flowing within the region of the solar wind dynamo leads to a deceleration of the plasma. According to Eq. (5.3), the magnetic force $\vec{F}_B^* = \vec{j}_{dyn} \times \vec{\mathcal{B}}_z$ generated by this current is directed oppositely to the solar wind velocity. Vice versa, it is easy to understand that any induced reduction in the solar wind velocity automatically leads to a current flow in the required direction. In the equilibrium case, the decelerating force is exactly compensated by the inertial force and the force balance equation becomes

$$\vec{F}_B^* = \vec{j}_{dyn} \times \vec{\mathcal{B}}_z = \vec{F}_{inertial}^* = \rho_{sw} \, \frac{\partial \vec{u}_{sw}}{\partial t}$$

Vector multiplication from the left with $\vec{\mathcal{B}}_z$ yields the following expression for the intrinsic dynamo current density

$$\vec{j}_{dyn} = \frac{\rho_{sw}}{\mathcal{B}_z^2} \, \vec{\mathcal{B}}_z \times \frac{\partial \vec{u}_{sw}}{\partial t} \tag{7.27}$$

Since $\partial \vec{u}_{sw}/\partial t$ is negative for a deceleration, the dynamo current flow is directed from the dusk to the dawn sector. Accordingly, the solar wind protons drift slowly toward the dawn side, the electrons toward the dusk side. The electric polarization field is evidently no longer capable of completely compensating the magnetic force. This is understandable, because the flow of current through the ionosphere leads to a decrease of the polarization space charge, which must be continually built up again by the lateral deflection of solar wind particles. Note that the Birkeland currents j_B that close the current circuit correspond to the region 1 currents of Fig. 7.10.

In order to get a feeling for the power capacity of our solar wind dynamo, let us assume that its exposed cross sectional area is of the order of b_{dyn}^2 and that the solar wind loses 5% of its velocity during transit through the dynamo. This means that

$$P_{dyn} = b_{dyn}^2 \, \frac{1}{2} \, n \, m_{\mathrm{H}} \, u_{sw}^3 \, [1 - (0.95)^3] \simeq 10^{11} \text{ W}$$

This is clearly adequate to satisfy the energy requirement of a polar region.

7.6.2 Open Magnetosphere

The solar wind dynamo described above requires the existence of 'open' magnetic field lines that extend into interplanetary space. Figure 7.22 presents a strongly simplified scenario that suggests how such field lines might be formed. An interplanetary magnetic field consisting of a negative B_z component (GSM coordinates, see Fig. 5.34) is considered to be superimposed onto the dipole field of the Earth, which, for the sake of simplicity, is assumed not to be distorted by the solar wind. The topology of the resulting field configuration may be broken down into three different subregions: The first is the region of interplanetary magnetic field lines, for which no footpoints are located on the Earth; the second is the region of open field lines, for which

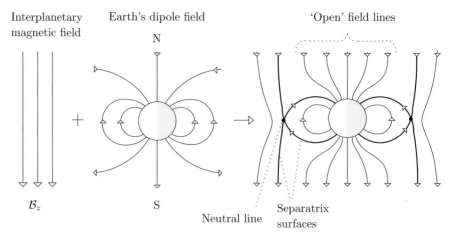

Fig. 7.22. Formation of 'open' magnetic field lines

one footpoint is on the Earth and the other in interplanetary space; and the third is the region of closed magnetic field lines, for which both footpoints are anchored at the Earth. These regions are separated by surfaces (called separatrices) that intersect in an x-shaped neutral line where the magnetic field strength must be equal to zero.

It may be easily verified that the corresponding superposition with a positive B_z component leads to a *closed magnetosphere*, for which all magnetic field lines associated with the Earth have both footpoints on the ground. It follows that the effectivity of the solar wind dynamo considered in the previous section should depend on the direction of the B_z component of the interplanetary magnetic field. In particular, energy transfer from solar wind to magnetosphere should be expected only when the B_z component is negative, because only this enables the formation of open magnetic field lines. This remarkable prediction of our model, in fact, has been essentially confirmed by observations, lending some credibility to the arguments on which it is based.

The topology of the resulting total field becomes significantly more complicated if, in addition to the B_z component, the dominant horizontal components B_x and B_y are superimposed onto the Earth's dipole field. The fundamental connection between a negative B_z component of the interplanetary magnetic field and open field lines, however, remains intact.

7.6.3 Plasma Convection in the Open Magnetosphere

What is the large-scale plasma motion to be expected in an open magnetosphere? In any case, the electric field across the open magnetic field lines created by the solar wind dynamo should produce an antisolar $\vec{\mathcal{E}} \times \vec{B}$ drift of solar wind and magnetospheric plasma across the polar cap. A prerequisite for this common drift motion, however, is that solar wind and magnetospheric plasma first get attached to open magnetic field lines. How this might be accomplished is sketched in Fig. 7.23. Plasma, strung out along the negative B_z component of the interplanetary magnetic field, moves at the local solar wind velocity in the direction of the open magnetosphere. Here, it comes into contact with plasma that is attached to closed dipole field lines. At the position of the neutral line, however, where the magnetic field strength is zero, magnetic field lines are not defined. As a result, the attachment of the plasma to a given common magnetic field line, normally required by theorem (6.29), does not hold at this particular location. Solar wind plasma threads attached to interplanetary magnetic field lines and magnetospheric plasma threads attached to dipole field lines can thus be separated and mutually combined at the location of the neutral line. This process is commonly referred to as *reconnection*. Once connected by a common open magnetic field line, part of the interplanetary plasma and part of the magnetospheric plasma now drift together across the polar cap region. Upon arriving at the neutral line on the nightside, the process of separating and joining the plasma threads is

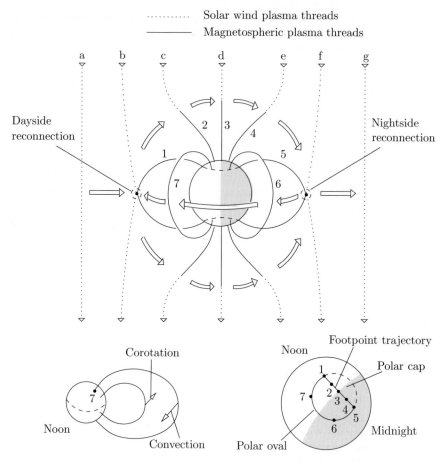

Fig. 7.23. Plasma convection in the open magnetosphere (Dungey model)

repeated (this time truly deserving of the designation *re*-connection), again creating purely interplanetary and purely magnetospheric plasma threads. While the interplanetary plasma continues on with the solar wind in the direction of the outer heliosphere, the magnetospheric plasma, bound by closed dipole field lines, drifts back in the direction of the dayside sector. It is immediately clear that a return transport is necessary. Without it, in fact, the entire dayside magnetospheric plasma would be transported away and lost as a result of its drift across the polar cap into the night sector. This return transport takes place in the outer magnetosphere, as indicated by the small sketch inserted at lower left in Fig. 7.23. This also shows the corotational motion of the plasmasphere in the inner magnetosphere, which, as we have seen, actually moves in the opposite direction to the return transport in the afternoon - dusk sector (see also Fig. 5.33). When the plasma arrives back in the dayside sector, it can again be reconnected with interplanetary plasma.

This continuing process defines the large-scale plasma circulation in an open magnetosphere, denoted by the arrows in Fig. 7.23. As demonstrated by the small sketch inserted at the lower right, this is compatible with the plasma convection observed in the polar ionosphere, which, of course, describes the footpoint motion of the plasma threads (compare also with Fig. 7.3).

Quite useful here – as applied earlier with the solar wind – is the intuitive notion that the magnetic field is 'frozen' into the plasma. In this concept we imagine the plasma threads and the magnetic field lines to move together and reconnection takes place not only between different plasma populations, but also between magnetic field lines. This is also referred to as *field line merging*. As such, the magnetic flux frozen into the plasma is also transported from the front side of the magnetosphere into the magnetotail and, after reconnection there, carried again back to the front side of the magnetosphere. Solar wind plasma with an embedded negative B_z component can thus lead to temporary 'erosion' of the dayside magnetosphere and to an increase of the magnetic flux on the nightside.

7.6.4 Open Magnetosphere with Tail

A mendable defect of the open magnetosphere considered so far is the absence of a tail. Incorporating a magnetotail into the model requires the vast majority of the open magnetic field lines to first extend into the tail lobes before entering interplanetary space at more or less large distances from Earth; see Fig. 7.24. This also means that the nightside neutral line is carried far down the tail. An upper bound on its distance may be derived from the following arguments. Assuming a polar cap drift velocity of 200 m/s (which corresponds roughly to an electric field strength of 10 mV/m and a polar cap potential of 30 kV), plasma attached to the footpoint of an open magnetic field line needs about 5 hours to traverse the polar cap of typically 30° in diameter. During this time, plasma attached to the interplanetary end of this field line, moving at a solar wind velocity of 500 km/s, will have traveled a distance of 1400 R_E, which should correspond to the maximum possible distance of the nightside neutral line. In reality, the neutral line is located much closer to the Earth, the estimated mean distance being in the range from 100 to 200 R_E. One factor leading to this shorter distance is the solar wind velocity, which is considerably smaller in the magnetosheath than in interplanetary space. Furthermore, the polar cap potential is larger than 30 kV during average conditions, leading to a further reduction in the distance. It is important to realize in this context that the magnetotail does not terminate at the neutral line. Instead, it extends far beyond and tail-like structures have still been detected at distances of more than 1000 R_E. Note that open magnetic field lines and their reconnection produce only a subtle, but nonetheless quite important, modification of the magnetotail.

The open magnetic field lines extending far into the magnetotail provide an explanation for the formation of the plasma mantle. The solar wind plasma

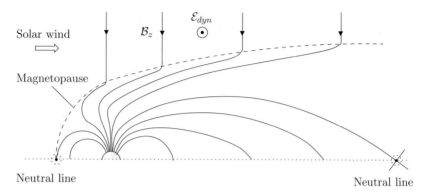

Fig. 7.24. Configuration of 'open' magnetic field lines in a magnetosphere with magnetotail

is, of course, free to penetrate into the magnetosphere along open magnetic field lines and move in the direction of the polar caps. The particles inside the magnetotail lobe regions must propagate with their thermal velocity in a direction opposite to their mean velocity, i.e. opposite to the reduced solar wind velocity of the magnetosheath region. Protons, for which the thermal velocity is much smaller than the mean velocity, cannot penetrate very far and reach the polar caps only in extreme cases. They penetrate far enough, however, that a plasma layer consisting mostly of solar wind particles, the plasma mantle, is formed on the inner edge of the magnetopause.

The behavior of solar wind electrons is different. Owing to their small mass, their random velocity is significantly larger than the solar wind velocity. As a result, these particles easily penetrate down to the polar caps. This is particularly true for the superthermal electrons in the beam shaped outer region of the velocity distribution. These so-called *strahl electrons* with energies of a few 100 eV are recorded in the polar caps as *polar rain*.

Energetic particles from solar flares are also afforded access to the polar caps, provided the polar cap in question is magnetically connected by open field lines to the flare region. Since this can only happen with one of the two polar caps (the other end of the severed interplanetary magnetic field line must point in the direction of the outer heliosphere), one should observe a distinct asymmetry in the incident flux rates, and this is indeed the case.

7.6.5 Reconnection

As plausible as our previous description of the solar wind dynamo and its associated large-scale plasma convection may appear at first, serious flaws arise upon closer examination. This is especially true for our conception about the formation of open magnetic field lines as illustrated in Fig. 7.22. Contrary to this superposition scenario, it was assumed earlier that the interplanetary magnetic field is frozen into the solar wind and follows its motion. Solar wind particles, however, are reflected from the magnetic field of the Earth and thus cannot penetrate into the magnetosphere. It would only seem logical that this should also hold for the frozen-in magnetic field. Indeed, we assumed earlier that the interplanetary magnetic field, apart from a certain compressional effect in the magnetosheath, was carried around the magnetospheric obstacle; see Section 6.4.3. Observations generally support these earlier assumptions. Hence, the fraction of solar wind plasma that penetrates into the magnetosphere must be comparatively small. Moreover, the magnetosphere really does consist primarily of closed field lines and direct evidence of open magnetic field lines remains sparse. Finally, by a wide margin, most of the solar wind energy flows around the magnetosphere and only a small fraction of the available energy is absorbed. Nevertheless, without this comparatively modest energy absorption there would be no electric fields or currents in the polar upper atmosphere, no plasma convection and no aurorae. Therefore, even if only on a limited scale, the possibility must exist to circumvent the rigid separation between the magnetoplasma of the magnetosphere and the magnetoplasma of interplanetary space. This is exactly what happens in the phenomenon we have called reconnection. Although the details of this process are not completely understood, an x-type magnetic neutral line must form during reconnection that enables penetration of solar wind plasma and its associated magnetic field into the magnetosphere – thereby leading to the formation of a partially open magnetosphere.

Probing deeper into the physics of this phenomenon, it is clear that the theorem (6.29) must somehow be rendered invalid, because it explicitly forbids reconnection of interplanetary and magnetospheric plasmas and their embedded field lines. Since this theorem is based essentially on the relation (6.42), one of the fundamentals of ideal magnetoplamadynamics, it is clear that this must also lose its validity at the location of reconnection. Evidently, those radical simplifications we made during its derivation from the generalized Ohm's law are no longer justified. Particularly suspicious here is our neglect of the second term on the right side of this equation, $\vec{j}/\sigma_B$. The circumstances under which this term becomes important follow from a simple comparison of terms. The order of magnitude (designated by OM) of the ratio of the first to the second term in Eq. (A.137) ($= 1/R_2$ in this relation) is given by

$$R_m = \mathrm{OM}\left(\frac{\vec{u} \times \vec{B}}{\vec{j}/\sigma_B}\right) \simeq \mu_0\, \sigma_B\, u\, L_{\mathcal{B}} \tag{7.28}$$

where, as in Section A.13.3, we have approximated the curl of the magnetic field by $\mathrm{OM}\,(\nabla \times \vec{B}) \simeq \mathcal{B}/L_{\mathcal{B}}$; see Eq. (A.136). The ratio R_m has become known as the *magnetic Reynolds number*. Clearly, the term $\vec{j}/\sigma_B$ can only be neglected as long as $R_m \gg 1$. Conversely, the previously used fundamental relation of magnetoplasmadynamics, Eq. (6.42), loses its validity if the parallel conductivity σ_B and/or the scale length $L_{\mathcal{B}}$ take on small values. Both of these possibilities may exist – at least locally. We recall from our discussion on the polar lights, for example, that the anomalous increase of the frictional frequency can greatly reduce the parallel conductivity. Furthermore, we know that the width of the surface current, which represents a measure of the scale length L_B in regions of transition from one magnetized plasma population to another, can assume relatively small values. In principle, the possibility thus exists that Eq. (6.42), and therefore also theorem (6.29), can lose their validity at the position of this boundary layer, thereby enabling the onset of reconnection. Alternatively, other terms neglected in the transition from Eq. (A.133) to Eq. (A.134) may become important.

Various models have been developed to describe the reconnection process quantitatively. These assume, for example, the continued validity of ideal magnetoplasmadynamics at sufficiently large distances from the actual reconnection region. Accordingly, the surroundings, not the reconnection region itself, may be described using the equations of ideal plasma dynamics. The situation corresponds to the one we encountered earlier in the quantitative treatment of shock waves: independent of the processes active in the shock wave itself, we can infer the changes it induces in the state parameters by integrating the gas equations at some distance from the shock wave. Calculations performed with these kinds of models have shown, for example, that particles should be very effectively accelerated and heated by the reconnection process. Microscopic processes in the reconnection region itself have also been investigated. This work simulates the motion of electrons and ions in field line configurations with an x-type neutral line. By their nature, such simulations are quite complex and subject to limitations in performance, but essentially confirm the results of magnetoplasmadynamic calculations.

In spite of these successes, many questions remain. For example, it is still unclear whether reconnection is a quasi-stationary or nonstationary process and, if nonstationary, what is its associated time scale. The spatial scale of the reconnection process is also unknown. Clarification of these issues is one of the greatest challenges of space research, not only because reconnection is important for the terrestrial magnetosphere, but also because of its great significance for other magnetized plasmas in space and astrophysics.

7.6.6 Origin of Birkeland Currents

An issue remaining from our description of the ionospheric Birkeland currents is the question of their continuation in the magnetosphere. A possible configuration of the entire current system, compatible with an open magnetosphere, is introduced in this section.

Region 1 Currents. Extending the region 1 currents of the dayside polar cap boundary along their associated magnetic field lines, most of them are found to continue into interplanetary space. It is thus plausible to associate them with the solar wind dynamo. Birkeland currents of the same strength

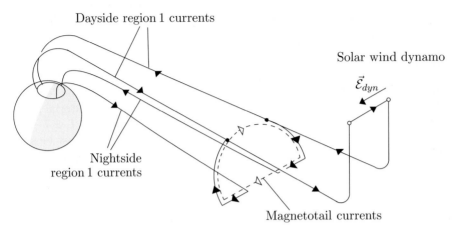

Fig. 7.25. Possible current circuit configuration of the region 1 currents (adapted from Stern, 1983)

and direction also flow, however, at the nightside polar cap boundary, and their extensions along the associated field lines end primarily in the neutral sheet of the magnetotail. Assuming that these currents are also generated by the solar wind dynamo, one requires a current system connecting open and closed magnetic field lines that might look something like the configuration sketched in Fig. 7.25. In this model, part of the dynamo current is redirected through the magnetopause to the neutral sheet, from which it then flows into the nightside ionosphere.

The current configuration of Fig. 7.25 is also capable of explaining how the electric field of the solar wind dynamo is transferred to the closed magnetic field lines of the central plane of the magnetotail. As stated earlier, a dawn-to-dusk electric convection field is measured there, the strength of which depends on the efficiency of the solar wind dynamo. This field points in the direction of the current flow so that energy is dissipated. The magnitude of the current flowing at $x_{GSM} = -30\,R_E$ through a $20\,R_E$ long piece of the magnetotail was estimated in Section 5.5.3 to be roughly 4 MA. Assuming that the voltage drop across the magnetotail corresponds approximately to that across the polar cap (for a polar cap potential of 30 kV and a tail width of $40\,R_E$, this corresponds to a convection field strength of 0.1 mV/m), the power dissipated in the above mentioned piece of the magnetotail would be 10^{11} W, and considerably more when considering the entire magnetotail. A dissipation rate of this order of magnitude clearly plays an important role in the energy budget of the magnetotail.

The above described extrapolation of the region 1 currents also raises the question as to the exact location of the solar wind dynamo that generates them. According to Fig. 7.21, currents in the inner dynamo region should flow in a direction opposite to that of the electric field ($\vec{j} \cdot \vec{\mathcal{E}} < 0$). This is the case along the upper and lower boundaries of the magnetotail, where the magnetopause currents are oppositely directed to the dynamo field $\vec{\mathcal{E}}_{dyn}$ (see Fig. 7.25). It has therefore been suggested that this region is at least part of the solar wind dynamo.

Region 2 Currents. The situation is different for the case of the region 2 currents. Here, the associated magnetic field lines end in the equatorial plane of the dawn and dusk magnetosphere at a geocentric distance of $L \simeq 7 - 10$. Lacking a suitable dynamo process in this region, the currents must be supplied from excess charges. The origin of these excess charges may be understood from the following.

Consider the drift trajectories of plasma sheet particles that emerge from the magnetotail and enter the region of influence of the terrestrial dipole field. For the sake of simplicity, we are only interested here in the conditions within the central plane of the magnetotail (which merges into the equatorial plane of the dipole field), and also neglect any day-night asymmetries. In order to simulate the field configuration in the central magnetotail plane, we superimpose a weak, northward directed, homogeneous magnetic field onto the geomagnetic dipole field. This additional uniform field is meant to describe the B_z component of the dipole field lines that have been stretched out and pulled back into the magnetotail. The electric convection field applied to the magnetosphere is simulated by an additional homogeneous electric field, directed from dawn to dusk, which is superimposed onto the magnetic fields. Figure 7.26 shows these magnetic and electric fields and, specifically, the drift orbits of plasma sheet ions resulting from this combination of fields. Plasma sheet particles far from the Earth will execute a relatively undisturbed $\vec{\mathcal{E}} \times \vec{B}$ drift in the direction of the geomagnetic dipole. Upon approaching the dipole field, the gradient drift becomes increasingly important and the charge carriers are deflected in the transverse direction. This follows immediately from the ratio of the respective drift velocities

$$\left(\frac{u_D^{gr}}{u_D^{\mathcal{E}}} \right)_{dipole} \sim \frac{E_\perp(L)}{L} \tag{7.29}$$

where $E_\perp(L)$ is the energy of the drifting charge carriers and L is the shell parameter of the nearly undisturbed dipole field; see Eqs. (5.39) and (5.41). Since the gradient drift proceeds at first in the direction of the electrical force, the energy of the particles increases with decreasing L, so that both numerator and denominator of the above ratio contribute to the increasing importance of the gradient drift. As soon as this becomes dominant, the charge carriers are guided around the dipole field as indicated for the ions in Fig. 7.26.

Depending on their energy, the particles can only drift to within a certain distance of the dipole, defined by a trajectory of closest approach. The region within this trajectory is a forbidden zone that is devoid of drifting particles.

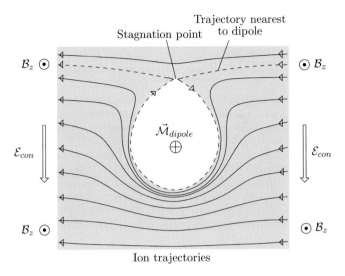

Fig. 7.26. Drift of energetic ions in the equatorial plane of a dipole field, upon which uniform electric and magnetic fields are superimposed in the indicated directions. The view is from the north onto the equatorial plane (adapted from Alfvén and Fälthammar, 1963).

The superposition of the trajectories of closest approach for plasma sheet ions and electrons is a crucial point for the present discussion. As illustrated in Fig. 7.27, there exist regions in the dawn and dusk sectors of the dipole field which are accessible only to electrons or ions, respectively. These regions of positive and negative excess charge are designated as the *Alfvén layers*, or in their entirety as the *shielding layer*.

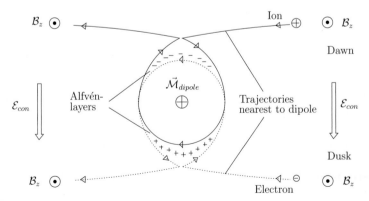

Fig. 7.27. Formation of the Alfvén layers. Ions and electrons have equal energies for the situation sketched here (adapted from Schield et al., 1969).

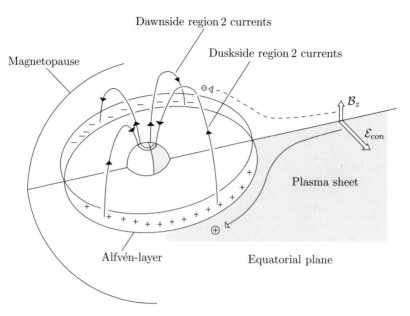

Fig. 7.28. Possible current configuration of the region 2 currents

Applying this scenario to the terrestrial magnetosphere, one obtains a picture similar to that sketched in Fig. 7.28. The convection field directed from dawn to dusk leads to a build up of positive excess charge in the dusk sector and negative excess charge in the dawn sector. This charge separation is the source of an electric field that is oppositely directed to the convection field and keeps it excluded from the inner magnetosphere (hence the designation 'shielding layer'). More interesting in our context is the continual destruction of these excess charges by means of field-aligned currents. These currents are directed toward the ionosphere on the dusk side and away from the ionosphere on the dawn side, corresponding exactly to the directions required for the region 2 currents sketched in Fig. 7.10. Alfvén layers thus represent a plausible explanation for the formation of these currents.

It is important to realize that the currents are carried essentially by the thermal plasma, not so much by the more energetic Alfvén layer particles, which must first be scattered into the loss cone before they can be precipitated. These precipitated particles, however, are the ones made responsible for the formation of diffuse aurorae (see Fig. 7.18). Another important point is that, because of their relative motion, the ions and electrons drifting past the Earth maintain a current, which is an essential component of the magnetospheric ring current. In fact, it is this drift motion, carried primarily by the more energetic ions, that is made responsible for a significant dawn-dusk asymmetry of the ring current intensity (see Section 8.1.2).

Note also that the closest approach to Earth of the drift trajectories serves to define the inner edge of the magnetotail plasma sheet. In order to derive an estimate for this distance, we consider the stagnation points of the trajectories, i.e. the points for which the x_{GSM} components of the $\vec{\mathcal{E}} \times \vec{B}$ and the gradient drift are equal, but of opposite sign (see Fig. 7.26). Recalling Eqs. (5.39) and (5.41), we obtain

$$L_{stagnation} = \frac{3E_{\perp}}{|q| \, R_E \, \mathcal{E}_{con}} \tag{7.30}$$

where the original energy of the ions can be inserted for $E_{\perp}$ to a first approximation. Equation (7.30) indicates that the stagnation point, and thus the inner edge of the plasma sheet, shifts closer to the Earth for increasingly strong convection fields. For a plasma sheet ion energy of $E_{\perp} = 5$ keV and a convection field strength of $\mathcal{E}_{con} = 0.2$ mV/m, for example, one obtains a stagnation point distance of $L_{stagnation} \simeq 12$. The trajectory passing through this stagnation point is the one that comes closest to Earth for the case considered here and its distance is much less than $L \simeq 12$ in the midnight sector (see Fig. 7.27). This is quite compatible with the plasma sheet distance of about $7 - 10$ R_E actually observed in this local time sector. Finally, it should be remembered that the gradient drift velocity is directly proportional to the energy of the particles. Hence, thermal particles play no role in the drift motion depicted in Fig. 7.26.

A quantitative treatment of all these processes for more realistic conditions becomes quite complicated and requires complex numerical models. These account for (1) the day/night asymmetry of the terrestrial magnetosphere; (2) the different density and energy distributions of the drifting charge carriers; (3) the electric corotation field; (4) the feedback of the Alfvén layer field on the convection field; (5) the modification of the ionospheric conductivity due to those energetic particles of the drift population that scatter into the loss cone and precipitate into the upper atmosphere; (6) the feedback of this increase in conductivity on the formation of the Alfvén layer; and many other complications. Preliminary results from such model calculations are encouraging and support the above described drift and current scenario. Less success has been achieved in the attempts to develop a self-consistent treatment of the region 1 and region 2 currents.

7.6.7 Low-Latitude Boundary Layer Dynamo

The internal magnetospheric dynamo of the low-latitude boundary layer (LLBL) may play a certain role alongside the solar wind dynamo. The configuration of this dynamo is sketched in Fig. 7.29. Particles of the LLBL population, moving with velocities of 50 – 200 km/s in the antisolar direction, flow across the φ-component of the dipole field. This produces an electric dynamo field which, for a magnetic field strength of $B_{\varphi} = 30$ nT, a plasma velocity of $u_{BL} = 100$ km/s and an LLBL thickness of 2000 km, yields a potential difference of 6 kV. Since the field lines of the magnetopause

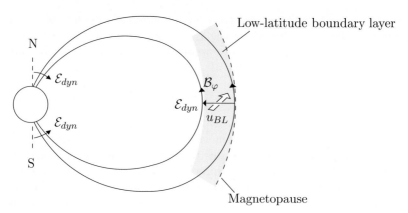

Fig. 7.29. Operational principle of the low-latitude boundary layer dynamo (meridional cut through the dusk sector)

region come together in the noon sector of the polar cap, one should observe twice this voltage there.

Before the different plasma populations of the magnetosphere were discovered it was assumed that this entire region was filled with thermal plasma. It was also conjectured that the motion of this plasma across the Earth's magnetic field was responsible for the observed dynamo effect. The idea was

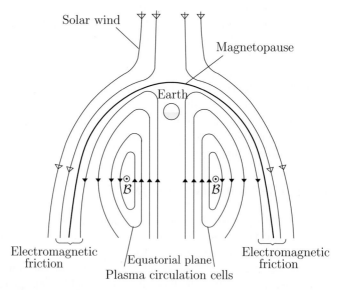

Fig. 7.30. Circulation of thermal plasma in a closed magnetosphere induced by electromagnetic 'friction' from the solar wind, as viewed from north onto the equatorial plane (Axford-Hines model)

that the solar wind stimulates this plasma to circulatory motion via a viscous interaction as shown in Fig. 7.30. This description, it should be noted, does not account for the modification of the flow from the electric corotation field. Furthermore, gradient drift of the particles is unimportant because of their low energy. The resulting flow pattern resembles that of a fluid that is set into motion by momentum transfer at the opposite sides of a container (in our case the flanks of the magnetotail). Combined with the return flow of the fluid in the central region of the container (in our case the central region of the magnetosphere and its associated magnetotail), this situation results in the formation of two circulation cells as seen in Fig. 7.30. These circulation cells, projected via their electric dynamo field along the magnetic field lines, are reflected in the polar upper atmosphere where they correspond to the convection cells observed there. A remaining open question concerned the nature of the momentum transfer at the magnetospheric flanks. Collisional friction could be excluded with certainty because the densities there were much too low. Instead, it would have to be an electromagnetic coupling that transferred momentum via plasma waves and/or instabilities. Today it is assumed that the convection dynamo associated with the thermal plasma circulation plays only a minor role. The electromagnetic interaction between solar wind and LLBL is thus considered to be too weak. A conclusive evaluation of this dynamo, however, is still pending.

Exercises

7.1 Current strengths, energy dissipation rates and magnetic field perturbations are to be estimated in the following for the moderately disturbed polar upper atmosphere. To simplify the calculation, we approximate the two semicircular segments of the dawn and dusk polar ovals as two rectangular polar ovals, each with a width of 4° in latitude and a length along the noon-midnight direction of 30°. Their separation distance along the dusk-dawn direction is also assumed to be 30°. The polar cap thus becomes a square (30° × 30°), bounded by the rectangular polar ovals in the dawn and dusk sectors. Let the Pedersen conductivity in the polar cap region be $(\sigma_P)_{pc} \simeq 25~\mu S/m$ and five times larger in the polar ovals. Consider these conductivities to be constant over the height interval of the dynamo layer ($\simeq 50$ km).

(a) Calculate the current strengths of the region 1 and region 2 currents for a polar cap potential of 50 kV, neglecting the resistance of the magnetospheric part of the region 2 current circuit. How large is the energy dissipation rate in one polar region?

(b)   How large is the current strength of the auroral electrojet, if the Hall conductivity is twice as large as the Pedersen conductivity? Estimate the magnitude of the magnetic field perturbation that would be observed on the ground underneath the auroral electrojet. How large is the magnetic field perturbation that would be recorded by a satellite traversing the region between the region 1 and region 2 currents?

7.2   That aurorae are indeed caused by precipitating energetic particles was first verified in the middle of the previous century from Doppler shifted hydrogen emission lines. Let a proton in the loss cone move along its guiding center field line in the direction of the upper atmosphere. Upon penetrating into the denser gas layers, it is neutralized by charge exchange and stimulated to emit an Hα photon ($\lambda = 656.3$ nm). At what wavelength will this emission be detected on the ground if the energy of the hydrogen atom is 5 keV and it is moving directly toward the observer at the instant of emission?

7.3   Calculate the height profile of the electron temperature in the nighttime subpolar ionosphere. Similar to the discussion in Section 6.1.3, assume stationary conditions and neglect the small heating and cooling rates. Contrary to that situation, however, we are now dealing with a static, planar plasma layer, for which a constant input of heat $\phi_z^W(h_p)$ from the plasmasphere is observed at its upper boundary $h_{p(lasmasphere)}$. The temperature at the lower boundary of the region of interest, h_0, is assumed to be known.

7.4   The magnitude of the solar wind/magnetosphere energy transfer rate is frequently estimated using the so-called ε parameter

$$\varepsilon = C \, u \, \mathcal{B}^2 \, l_0^2 \, \sin^4(\vartheta/2)$$

where C is a constant, u is the solar wind velocity, $\mathcal{B}$ the strength of the interplanetary magnetic field, and l_0 a characteristic length. The angle ϑ is determined from the relation $\tan \vartheta = \mathcal{B}_y/\mathcal{B}_z$ in a GSM coordinate system. Show that the factor $u\mathcal{B}^2$ is proportional to the Poynting vector of the solar wind, whereby only the component of the interplanetary magnetic field perpendicular to the solar wind velocity is considered. The product $u\mathcal{B}^2 \, l_0^2$ then describes a power and the additional factor $\sin^4(\vartheta/2)$ determines the fraction of this power transferred to the magnetosphere (maximum when the interplanetary magnetic field points in the negative z-direction, $\vartheta = 180°$).

References

A. Omholt, *The Optical Aurora*, Springer-Verlag, Berlin, 1971

A.V. Jones, *Aurora*, Reidel Publishing Company, Dordrecht, 1974

H. Volland, *Atmospheric Electrodynamics*, Springer-Verlag, Berlin, 1984

S.W.H. Cowley, Magnetic reconnection, in *Solar System Magnetic Fields* (E.R. Priest, ed.), 121, Reidel Publishing Company, Dordrecht, 1986

Y. Kamide and W. Baumjohann, *Magnetosphere-Ionosphere Coupling*, Springer-Verlag, Berlin, 1993

A. Brekke, *Physics of the Upper Polar Atmosphere*, John Wiley & Sons, Chichester, 1997

D. Bryant, *Electron Acceleration in the Aurora and Beyond*, Institute of Physics Publishing, Bristol and Philadelphia, 1999

S. Ohtani, R. Fujii, M. Hesse, and R. Lysak (eds.), *Magnetospheric Current Systems*, Geophys. Monograph No. 118, American Geophys. Union, Washington, 2000

G. Paschmann, S. Haaland, and R. Treumann (eds.), Auroral Plasma Physics, *Space Sci. Rev., 103, No. 1-4*, 2002

P.E. Sandholt, H.C. Carlson and A. Egeland, *Dayside and Polar Cap Aurora*, Kluwer Academic Publishers, Dordrecht, 2002

R.D. Hunsucker and J.K. Hargreaves, *The High-Latitude Ionosphere and its Effects on Radio Propagation*, Cambridge University Press, Cambridge, 2003

Popularizations: the Aurorae

R.H. Eather, *Majestic Lights*, American Geophysical Union, Washington, 1980

A. Brekke and A. Egeland, *The Northern Light*, Springer-Verlag, Berlin, 1983

See also the references in the previous chapters as well as the figure and table references in Appendix B.

8. Geospheric Storms

The term *geospheric storm* is used here to designate an event of strongly enhanced dissipation of solar wind energy in the near-Earth space environment. Such activities typically last for 1 to 3 days and are characterized by energy dissipation rates of up to 10^{12} W. This is many times greater than the usual energy transfer rate from solar wind to magnetosphere. As implied by the label 'geospheric storm', all regions of the space environment are affected and the symptoms of the disturbance syndrome are correspondingly varied

$$
\text{Geospheric storm} \begin{cases} \text{Magnetic storm} \\ \text{Auroral storm} \\ \text{Magnetospheric storm} \\ \text{Thermospheric storm} \\ \text{Ionospheric storm} \\ \quad \vdots \end{cases}
$$

The earliest discoveries of storm phenomena, of course, were those available to ground-based observation. Disturbances in the Earth's magnetic field, which were detected in the 18th century, are a prominent example. The long tradition and the global coverage of these measurements are reasons why magnetic field disturbances have been used to characterize the intensity and temporal behavior of geospheric storms up to the present day. This chapter thus begins with a description of the nature of these magnetic field perturbations and their quantitative representation by magnetic indices. We follow this with a discussion of the accompanying intensity enhancements of the aurorae and then attempt to trace these two disturbance phenomena back to magnetospheric activities. A large fraction of the solar wind energy is dissipated in the upper atmosphere and the resulting perturbations in the thermosphere and ionosphere are discussed in subsequent sections. Since geospheric storms represent an important link in the chain of solar-terrestrial relations, it is also appropriate to address their interplanetary and solar origins. Solar flares are also discussed in this context. In closing, we point out some technological consequences of geospheric storms.

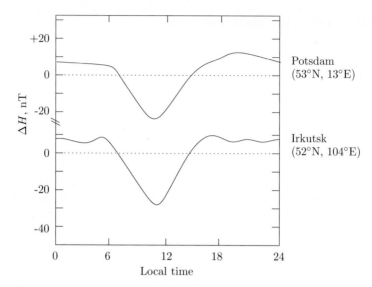

Fig. 8.1. Diurnal deviation of the horizontal intensity of the geomagnetic field from its mean value, as measured at mid-latitudes during quiet conditions (adapted from A. Schmidt in Chapman and Bartels, 1951, see figure references)

8.1 Magnetic Storms

It is well known that the magnetic field observed at the Earth's surface is by no means constant, but rather subject to variations on all time scales. Of primary interest here are fluctuations with periods from a few tens of minutes up to several days, whereby one should be careful to distinguish between regular and irregular variations, the latter being denoted as *magnetic activity*. We first investigate the origin of the regular magnetic field fluctuations. Following this, we describe the distinct signatures of magnetic activity at low, middle and high latitudes and their quantitative characterization by magnetic indices.

8.1.1 Regular Variations

During times of low solar wind energy dissipation, also designated as *quiet* or *undisturbed*, variations of the magnetic field observed on the Earth's surface are characterized by regular, small-amplitude fluctuations, which repeat themselves in similar fashion from day to day. Figure 8.1 illustrates this phenomenon with data from two mid-latitude stations. Gauss, in fact, considered upper atmospheric currents (known as *solar quiet* or simply *Sq currents* in the contemporary jargon) to be a possible cause of these magnetic fluctuations and Stewart (1883) held tidal winds responsible for the formation of these currents. Support for this explanation comes from the ordering of these variations with local rather than universal time (see Fig. 8.1).

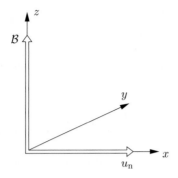

Fig. 8.2. Coordinate system used for the derivation of wind-induced charge carrier drifts

That thermospheric winds are indeed capable of producing currents in the ionosphere may be understood with the help of the momentum balance equations for electrons and ions. For simplicity, we consider stationary conditions in an infinitely extended ionosphere, homogeneous in the horizontal direction, which features a purely vertical magnetic field and a homogeneous, purely horizontal, neutral atmospheric wind. Neglecting viscosity effects, we need only account for magnetic field and frictional forces in the horizontal components of the momentum balance equations

$$n_s q_s \vec{u}_s \times \vec{B} + n_s m_s \nu^*_{s,\mathrm{n}}(\vec{u}_\mathrm{n} - \vec{u}_s) = 0 \qquad (8.1)$$

The index s again stands for the ion or electron component ($s = \mathrm{i}, \mathrm{e}$) and the index 'n' for the neutral gas component. In contrast to the momentum balance equation (7.4) used for the derivation of the ionospheric conductivity, the externally applied electric field, but not the neutral gas velocity, is set equal to zero here. Using the coordinate system given in Fig. 8.2, we may separate Eq. (8.1) into its x and y components

$$q_s u^s_y B + m_s \nu^*_{s,\mathrm{n}}(u_\mathrm{n} - u^s_x) = 0 \qquad (8.2)$$
$$-q_s u^s_x B - m_s \nu^*_{s,\mathrm{n}} u^s_y = 0 \qquad (8.3)$$

It follows that the flow velocities of the charge carriers are

$$u^s_x = \frac{m_s \nu^*_{s,\mathrm{n}} u_\mathrm{n}}{q_s^2 B^2 / m_s \nu^*_{s,\mathrm{n}} + m_s \nu^*_{s,\mathrm{n}}} = \frac{(\nu^*_{s,\mathrm{n}})^2}{(\nu^*_{s,\mathrm{n}})^2 + (\omega^s_B)^2} u_\mathrm{n} \qquad (8.4)$$

$$u^s_y = -\frac{q_s}{|q_s|} \frac{\omega^s_B}{\nu^*_{s,\mathrm{n}}} u^s_x = -\frac{q_s}{|q_s|} \frac{\nu^*_{s,\mathrm{n}} \omega^s_B}{(\nu^*_{s,\mathrm{n}})^2 + (\omega^s_B)^2} u_\mathrm{n} \qquad (8.5)$$

Referring to the height profiles of $\nu^*_{s,\mathrm{n}}$ and ω^s_B shown in Fig. 7.7, one may distinguish between three regions:
Lower heights ($h \lesssim 50$ km). Since $\nu^*_{s,\mathrm{n}} \gg \omega^s_B$, we obtain $u^s_x \simeq u_\mathrm{n}$ and $|u^s_y| \ll$

$|u_x^s|$. The ions and electrons evidently collide so frequently with the neutral gas particles that they are unable to execute a complete gyration orbit. This leads to a situation where the charge carriers move together with the same velocity as the neutral gas particles. Of course, no current flows as a result of this forced 'ambipolar' drift.

Larger heights ($h \gtrsim 175$ km). Since $\nu_{s,n}^* \ll \omega_B^s$, we obtain $|u_x^s| \ll |u_y^s| \simeq (\nu_{s,n}^*/\omega_B^s)u_n \to 0$. The charge carriers at larger heights – in accordance with the frictional force drift – move in a direction perpendicular to both the wind and the magnetic field, the magnitude of the drift velocity decreasing with increasing height. Since the ions and electrons move in opposite directions, an electric current is produced, the intensity of which, however, decreases exponentially with increasing altitude.

Central dynamo layer ($h \simeq 100$ km). When $\nu_{i,n}^* \gg \omega_B^i$ and $\nu_{e,n}^* \ll \omega_B^e$, the x component of the charge carrier drift may be written as

$$u_x^i \simeq u_n$$
$$u_x^e \simeq (\nu_{e,n}^*/\omega_B^e)^2 u_n \ll u_x^i$$

and the y component becomes

$$u_y^i \simeq -(\omega_B^i/\nu_{i,n}^*)u_n \ll u_x^i$$
$$u_y^e \simeq (\nu_{e,n}^*/\omega_B^e)u_n \ll u_x^i$$

In this case the ions, because of the strong collisional friction they experience, move essentially with the neutral gas particles. The electrons, however, apart from a rather slow frictional force drift, hardly move at all. Accordingly, a current is produced of the order of

$$j_x \simeq e\, n\, u_n$$

We can make a quick estimate whether or not this current is sufficiently strong to explain the observed magnetic field variations. The typical fluctuation amplitude of thermospheric tidal winds at mid-latitudes at the height of the dynamo layer is of the order of 75 m/s. Assuming a charge carrier density of 10^{11} m^{-3}, such a wind would produce a current density fluctuation with an amplitude of about 1.2 μA/m^2. Assuming that this current density flows in a dynamo layer with a thickness of 50 km, one obtains a surface current density of $\mathcal{I}^* = j \cdot \Delta h \simeq 60$ mA/m. This then leads to magnetic field variations of the order of $H = \mu_0 \mathcal{I}^*/2 \simeq 40$ nT, in basic agreement with the observations shown in Fig. 8.1.

Alternatively, the generation of wind-induced currents can also be understood as a dynamo effect. The weakly ionized upper atmosphere is then considered to be a conducting plasma that moves perpendicular to the magnetic field with the plasma velocity $\vec{u} = (\rho_i \vec{u}_i + \rho_e \vec{u}_e + \rho_n \vec{u}_n)/(\rho_i + \rho_e + \rho_n) \simeq \vec{u}_n$ (corresponding to Eq. (6.49) with an additional neutral gas component). According to Eq. (6.42), this induces an electric dynamo field given by

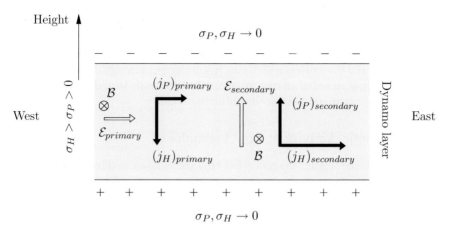

Fig. 8.3. Formation of the equatorial electrojet. The field and current directions correspond to daytime conditions.

$\vec{\mathcal{E}}_{dyn} \simeq -\vec{u}_n \times \vec{\mathcal{B}}$. Combining this with the conductivities presented in Section 7.3.2, one calculates electric currents that are in agreement with the current densities derived above.

Particularly intense tidal currents are observed along the magnetic inclination equator. The so-called *equatorial electrojet*, flowing there in a band of a few hundred kilometers width, owes its high intensity to the particular configuration of electric and magnetic fields in the equatorial dynamo layer; see Fig. 8.3. The electric field $\mathcal{E}_{primary}$, generated by the global dynamo at low latitudes, is directed from west to east during the daytime. Combined with the northward-directed geomagnetic field $\mathcal{B}$, this produces an eastward-directed Pedersen current $(j_P)_{primary}$ and a downward-directed Hall current $(j_H)_{primary}$. This Hall current leads to an accumulation of charge, negative at the upper edge and positive at the lower edge of the dynamo layer, which results in the formation of an upward-directed electric polarization field $\mathcal{E}_{secondary}$. This field continues to increase in strength until its own Pedersen current exactly compensates the existing Hall current, $(j_P)_{secondary} = (j_H)_{primary}$. At the same time it produces its own Hall current, $(j_H)_{secondary}$, which flows in the same direction as the primary Pedersen current. The total current along the equator may thus be written as

$$j_{equator} = (j_P)_{primary} + (j_H)_{secondary} = \sigma_P\,\mathcal{E}_{primary} + \sigma_H\,\mathcal{E}_{secondary}$$

$$= \sigma_P\,\mathcal{E}_{primary} + \frac{\sigma_H^2}{\sigma_P}\,\mathcal{E}_{primary} = \sigma_C\,\mathcal{E}_{primary} \qquad (8.6)$$

where we have accounted for the requirement $(j_H)_{primary} = \sigma_H\,\mathcal{E}_{primary} = (j_P)_{secondary} = \sigma_P\,\mathcal{E}_{secondary}$ in the intermediate step and, in the final expression, have introduced the *Cowling conductivity*

$$\sigma_C = \sigma_P + \sigma_H^2/\sigma_P \qquad (8.7)$$

Since the ratio of Hall to Pedersen conductivity is nearly 4 at the height of interest here (ca. 110 km), the Cowling conductivity is well more than ten times as large as the Pedersen conductivity. This explains the great intensity of the equatorial electrojet, which can reach current strengths of 100 kA and more. Since it is directed eastward at daytime, it leads to a distinct increase of the surface magnetic field strength at the equator.

8.1.2 Magnetic Activity at Low Latitudes

In addition to the regular magnetic field fluctuations caused by tidal currents, irregular deviations are observed that can attain considerable amplitudes. Here we address their characteristics at low latitudes. Figure 8.4 shows an example of disturbances recorded in the horizontal component of the magnetic field at four magnetic observatories during a geospheric storm. The observatory locations are shown in Fig. 8.5. Evidently, the observed variations are rather complex and Gauss once justifiably called them 'mysterious hieroglyphics of nature' waiting to be decoded. In spite of considerable effort, the decoding project is still 'work in progress' up to the present day.

In order to capture the essential characteristics of such disturbances and, at the same time, establish a global measure for their intensity, the data of Fig. 8.4 are processed according to the following scheme. First, all regular fluctuations, such as the diurnal variation, are eliminated. Hourly mean values, sequentially centered at the half-hour, are then calculated. The hourly means from the four stations are then combined into a global mean value, accounting for the relative location of each station with respect to the magnetic equator. The result of this procedure is the *Dst index* (*D*isturbance *st*orm), which is plotted in the bottom panel of Fig. 8.4.

The main feature of the disturbance described by the Dst index is the characteristic depression in the horizontal intensity, which, in the example shown here, reaches a value of -150 nT. Disturbances of such intensity, following a suggestion by Humboldt, are denoted *magnetic storms*, whereby the word 'storm' connotes the degree of disturbance rather than the phenomenon itself. Whereas storms of the intensity shown in Fig. 8.4 are nothing extraordinary, more severe storms with field depressions of, say, -300 nT are sometimes not observed for years (or even decades) and are thus significant geophysical events. Figure 8.4 also shows that one distinguishes between the main and recovery phases of a storm. The former phase is defined by the decrease in magnetic field strength; the latter by its gradual return to undisturbed conditions.

The field depression during the main phase is attributed to an enhancement of the magnetospheric ring current, a scenario that has already been described in Fig. 5.27. We recall that a magnetic disturbance of $\Delta H \simeq \Delta \mathcal{B}_{rc} = -150$ nT corresponds to an effective ring current of many million amperes and an energy content of a few 10^{15} J. It is generally agreed that the enhancement of the ring current is produced by the injection of energetic particles into the

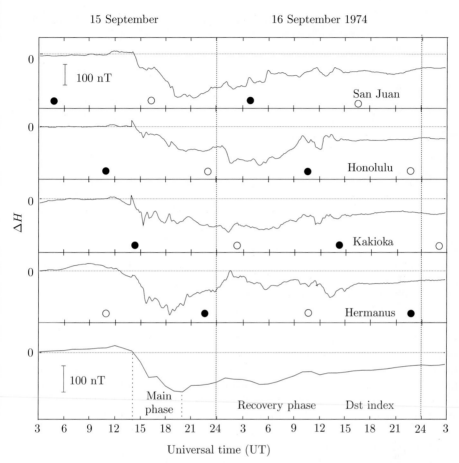

Fig. 8.4. Example of a magnetic storm. Deviations of the horizontal component of the magnetic field from its quiet time reference value (denoted by the dotted line at '0'), measured at four different low-latitude stations during disturbed conditions, are plotted as a function of universal time. See Fig. 5.9 for the definition of H and refer to Fig. 8.5 for the locations of the contributing observatories. The intensity scale valid for all measurements is given in the upper panel (San Juan data); open and solid circles denote the times of local (magnetic) noon and midnight, respectively. The bottom curve shows the Dst index derived from these measurements (adapted from *Solar Geophysical Data*, 1975).

inner magnetosphere. There are different opinions, however, just how this particle injection is triggered and sustained. One such scenario is explained in more detail later (Section 8.3).

Having established the source of the magnetic field depression, it is clear that the magnetic field relaxation during the recovery phase is attributed to a decay of the ring current. One of the main loss processes for the ring

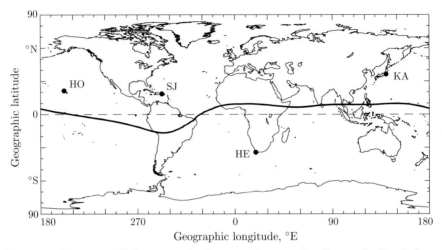

Fig. 8.5. Locations of the magnetic observatories contributing to the Dst index. The abbreviations: HE is Hermanus, KA is Kakioka, HO is Honolulu and SJ is San Juan. The thick solid line indicates the location of the magnetic inclination equator. Note that, in order to minimize the influence of the equatorial electrojet, the stations are located off the magnetic equator.

current ions is charge exchange with exospheric neutral hydrogen atoms, see Eq. (5.62). Moreover, because of their greater interaction cross section, oxygen ions are expected to have shorter lifetimes than hydrogen ions. This can lead to a distinct two-step recovery phase, beginning with a rapid, continuing with a slower, relaxation rate. An alternative explanation argues that the rapid initial recovery is caused by drift-induced removal of ring current particles from the inner magnetosphere.

Originally, an 'initial phase' was defined in addition to the main and recovery phases. This was characterized by a moderate, but sudden increase in the magnetic field strength. Today, since this effect is observed in only about half of the cases, it is no longer considered to be an integral component of a magnetic storm. The former designation *sudden storm commencement* (SSC), indicating the sudden rise in magnetic field strength at the beginning of the initial phase, has since been replaced by the term *sudden impulse* (SI). Sudden impulses, however, can also occur during the main phase of a storm or even outside the interval of a storm period. They are attributed to a compression of the magnetosphere by an interplanetary shock wave. We recall, according to Eq. (6.74), that the magnetopause surface current density is proportional to the ram pressure of the solar wind flow. Should this pressure be increased by an interplanetary shock that hits the magnetosphere (or by a more gradual increase in the solar wind dynamic pressure), the magnetopause currents, as well as their associated magnetic field, are enhanced. Figure 5.38 indicates that this leads to an increase of the geomagnetic field at the Earth's surface.

When using the Dst index, it is important to remember that this describes only the zonally averaged disturbance, and not its local time dependence. As a result, it does not account for the obvious asymmetry between the stronger disturbance in the afternoon/dusk sector and the weaker disturbance in the dawn sector documented in Fig. 8.4. It is not yet completely clear just how these differences arise. Partial ring currents, associated with the ion drift paths in the afternoon/dusk sector, and also field-aligned currents, have variously been deemed responsible for this phenomenon.

8.1.3 Magnetic Activity at High Latitudes

Figure 8.6 shows that magnetic activity at high latitudes clearly differs from that at low latitudes. The deviations of the horizontal component of the magnetic field from its quiet time reference value, observed at various high-latitude locations, are plotted for the same geospheric storm considered in Fig. 8.4. The locations of the contributing observatories are indicated in Fig. 8.7. It may be seen that the disturbances are considerably more intense than at low latitudes and deviations of more than 1500 nT are not uncommon. It may also be noticed that both positive and negative disturbances are observed. Deviations in the local afternoon sector tend to be positive and those in the night and morning sectors are mostly negative. As a rule, negative disturbances are considerably larger than their positive counterparts. The large temporal fluctuations, however, are the most striking and, at the same time, the most confusing feature. Each station seems to display its own individual disturbance signature.

The great intensity, as well as the spatial variability, indicate that the source of these disturbances should be found close to the Earth and, indeed, ionospheric currents are held responsible for the observed deviations. The fact that positive disturbances occur mostly in the afternoon sector and negative disturbances in the morning sector is evidence that the currents in question, at least partially, are the eastward-directed and westward-directed polar electrojets, respectively (see Fig. 7.11). Since an increase in the energy transfer from solar wind to magnetosphere leads to enhanced polar electric fields and currents, this interpretation is intrinsically plausible. Part of the observed temporal variability may therefore be attributed to fluctuations in this energy transfer rate.

An attempt to predict all geomagnetic activity in this way, however, runs into serious difficulties. This is especially true for the intense, impulse-like depressions of the magnetic field that are observed preferentially in the night sector. These types of disturbances are designated *magnetic substorms*, implicitly indicating that many of them are observed during a magnetic storm. Particularly noteworthy examples for this phenomenon were recorded on 15 September at about 14:00 UT at the stations Fort Churchill and College, on 16 September at about 01:30 UT at the stations Leirvogur, Fort Churchill and Dixon Island, and on the same day at about 13:00 UT at the

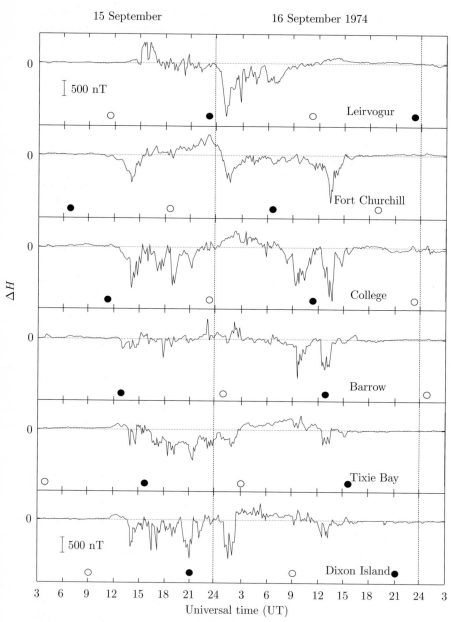

Fig. 8.6. Example of magnetic activity at high latitudes during a magnetic storm. Deviations of the horizontal component of the magnetic field from its quiet time reference value (denoted by the dotted line at '0'), measured at six high-latitude stations during the geospheric storm of 15/16 September 1974, are plotted as a function of universal time. Open and solid circles denote the times of local (magnetic) noon and midnight, respectively (adapted from *Solar Geophysical Data*, 1975).

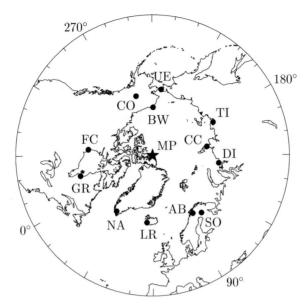

Fig. 8.7. Locations of the magnetic observatories contributing to the auroral electrojet indices (1974). Abbreviations: NA = Narsarsuaq, LR = Leirvogur, AB = Abisko, SO = Sodankyla, DI = Dixon Island, CC = Cape Chelyuskin, TI = Tixie Bay, UE = Cape Uelen, BW = Barrow, CO = College, FC = Fort Churchill, and GR = Great Whale River. The position of the magnetic pole (MP) is marked by a star and the magnetic longitude is denoted around the perimeter of the circle.

stations Fort Churchill, College and Barrow. This type of disturbance is also attributed to ionospheric currents, this time, however, to a special electrojet that flows in the night sector of the polar oval, the general location of which is sketched in Fig. 8.12.

The difficulty of predicting the occurrence of magnetic substorms is usually taken as an indication that they are a spontaneous phenomenon, rather than being driven by an external agent. In general, they are attributed to magnetospheric instabilities, correspondingly denoted *magnetospheric substorms.* We will examine this phenomenon more closely in Section 8.3. It is important for magnetospheric substorms (and for their related magnetic substorms) that they are preceded by a period of enhanced solar wind-magnetosphere energy transfer. This time interval represents the *growth phase* of the substorm. It may be recognized by a moderate increase in the eastward and westward electrojets, as well as by their associated magnetic field disturbances. For reasons that will be apparent later, the sudden decrease in magnetic field strength during a substorm is denoted the *expansion phase* and its return to the original field strength is called the *recovery phase.*

Special magnetic indices have been introduced in order to describe the magnetic activity at polar latitudes. In contrast to the methods used for

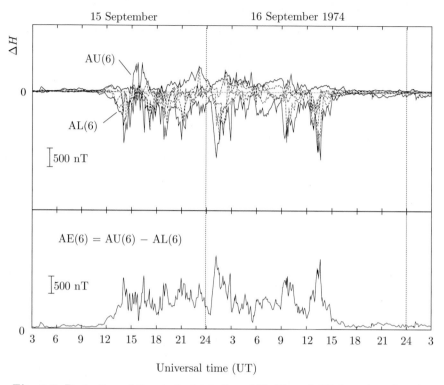

Fig. 8.8. Derivation of the electrojet indices AU, AL and AE, based on data from six (6) stations

the Dst index, these cannot be calculated from directly averaging the data. If this were done, positive and negative disturbances would partially cancel each other and the characteristic variability of the disturbance would be obscured. This is avoided by the following procedure. As with the derivation of the Dst index, only the disturbance of the horizontal component of the magnetic field is considered. All regular variations, particularly the diurnal variation, are first eliminated. Following this, the deviations of these 'purified' H components from their nominal values at all contributing stations, ΔH, are plotted in superposition as a function of universal time, see Fig. 8.8, the temporal resolution being typically 1 to 2.5 minutes. As a final step, the upper and lower envelopes of these superposed curves are determined. The value of the upper envelope then defines the *upper auroral electrojet index AU* and the lower envelope does the same for the *lower auroral electrojet index AL*. Whereas AU describes essentially the intensity of the eastward-directed electrojet, AL characterizes the combined effect of the westward-directed and the substorm associated electrojets. The information contained in the values of AU and AL can be further combined into one index, formed by subtracting the two indices. This difference yields the *auroral electrojet index AE*, which is plotted in the lower panel of Fig. 8.8. It may be seen that this index reflects the high temporal variability of the magnetic activity at high latitudes quite well. In practice, the electrojet indices are based on measurements from not just 6, but rather 10 to 12 stations. The locations of these additional stations are also given in Fig. 8.7. Evidently, the illustrated ring of stations is designed to capture the local time dependent intensity enhancements of the polar electrojets, a goal that usually, but certainly not always, meets with success.

8.1.4 Magnetic Activity at Mid-latitudes

The magnetic activity observed at mid-latitudes is caused by the magnetospheric ring current as well as the outlying effects of the polar electrojets. Additional disturbances are caused by field-aligned currents. In view of these different, often hard to distinguish, contributions, one is usually content with a rather crude characterization of this activity. The *Kp index* (from the German *Kennzahl, planetar*), introduced by Bartels in 1949, is the most frequently used coding scheme for this purpose. The derivation of this characteristic number is somewhat complicated and will not be further described here. It suffices to know that the Kp index is based primarily on data from magnetic observatories at middle and high northern latitudes; that its values are generated with a time resolution of 3 hours; that it represents a quasi-logarithmic measure of the disturbance range; and that it takes on values between 0 (very quiet) and 9 (very disturbed). The great advantages of this index are that it may be quickly obtained and that it represents a sufficiently good measure for the general level of magnetic activity (and thus indirectly for the intensity of the solar wind energy dissipation rate). Disadvantages are the coarse temporal resolution and its lack of association with a specific current system, thereby impeding its physical interpretation.

When interested in calculating sums and averages, one should use the linear *ap index* instead of the quasi-logarithmic Kp index. Conversion between the two indices may be performed with the help of Table 8.1. An example of such an averaging procedure is the *Ap index*, which is the mean value of

Table 8.1. Conversion of the Kp and ap indices. The subscripts on the Kp values provide a finer gradation for the degree of the disturbance.

Kp =	0_0	0_+	1_-	1_0	1_+	2_-	2_0	2_+	3_-	3_0	3_+	4_-	4_0	4_+
ap =	0	2	3	4	5	6	7	9	12	15	18	22	27	32
Kp =	5_-	5_0	5_+	6_-	6_0	6_+	7_-	7_0	7_+	8_-	8_0	8_+	9_-	9_0
ap =	39	48	56	67	80	94	111	132	154	179	207	236	300	400

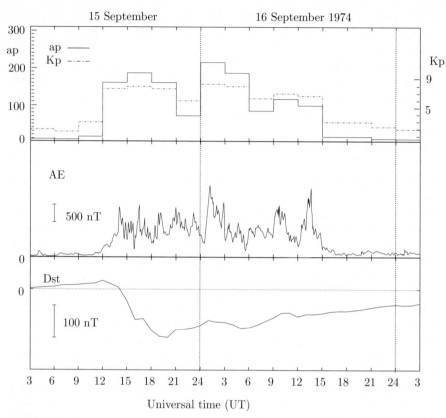

Fig. 8.9. The magnetic storm of 15 and 16 September 1974 as characterized by the ap, Kp, AE and the Dst indices

ap for one (universal time) day. Figure 8.9 shows the time profiles of the Kp and ap indices, together with the AE and Dst indices, during the previously considered storm of 15/16 September 1974.

8.2 Auroral Substorms

As implied by the name, auroral substorms are intimately connected with magnetic substorms. The phenomenon, as seen from the ground, may be described as follows. We are observing an auroral arc located at the equatorward edge of various discrete auroral forms. Suddenly, exploding in a dazzle of activity, rays, folds and seams are created in a brilliant lightshow until the arc finally breaks up into isolated pieces. At about the same time, the intense glow propagates toward the pole at high velocity (many 100 m/s). This *expansion* takes place simultaneously with the sharp depression in the magnetic field strength during the associated magnetic substorm, lending its name to this phase of the disturbance. In addition to the poleward motion, expansion occurs in the east and west directions. Since the western edge of this expansion can occasionally take the shape of a breaking wave, it has become known as a *westward traveling surge*.

While very detailed observations are possible from the ground, they fail to convey the global picture. This is obtained with the help of satellite surveys from very great heights. Discrete aurorae, which are often only a few hundred meters wide, can no longer be resolved in such images. The large-scale luminous apparitions, however, can also be strongly structured and display a broad and often confusing variety of forms. In order to keep the description as simple as possible, we consider in Fig. 8.10 only the rough structure of an isolated, moderately intense, substorm. The view is downward from above the north polar region and, for better orientation, the direction to the Sun, the position of the geomagnetic north pole and the invariant magnetic latitudes at 65°and 75°are indicated. That region, for which the intensity of the auroral oxygen emission line at 557.7 nm exceeds 4 kR, is drawn in black.

As seen in the first picture of the sequence, the substorm is first noticed as a strongly localized brightening of the auroral oval in the night sector. This brightening spreads rapidly, first in longitude and then in latitude, gaining intensity along the way until, at 14:42 and 14:54 UT (4th and 5th picture in the sequence), it reaches its maximum intensity of almost 50 kR. The illumination gradually becomes weaker during the subsequent recovery phase, but still occupies an increasingly larger area that eventually fills the entire midnight sector of the auroral oval.

Figure 8.10 can (and should) only provide a first impression of the global auroral activity. A large number of additional emission features are observed during larger or repeating substorms, both in the night as well as in the day sector. Among these are filamentary structures that extend into the region of the polar caps. The classification and explanation of this complex auroral distribution is the subject of intensive research and we will not dwell further on it here. Independent of these complexities, one should note that the particle precipitation responsible for the aurora, by means of its collisional ionization, enhances the charge carrier density in the dynamo layer of the polar ionosphere. As a result, the black areas shown in Fig. 8.10 are also regions of enhanced conductivity, through which, specifically, the substorm-associated electrojet flows.

8.3 Magnetospheric Substorms

The substorm activity observed at polar latitudes is intimately coupled to notable changes in the magnetosphere. Satellite measurements recorded in the near-Earth magnetotail during the growth phase of a magnetic substorm, for example, show the following modifications

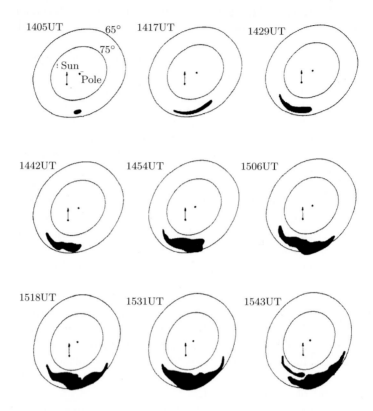

Fig. 8.10. Example of an auroral substorm as seen from a satellite perspective. The sketch format is described in the text (adapted from Craven and Frank, 1985).

– enhanced magnetic field strength in the magnetotail lobes
– narrowing of the plasma sheet thickness
– increased field line stretching
– displacement of the plasma and current sheet closer to the Earth, and
– greater inclination of the tail magnetopause to the solar wind

The ensuing expansion phase is then characterized by a sudden return to a less disturbed state, the transition to more dipole-like field lines being designated as *dipolarization*, see Fig. 8.11. Among the phenomena observed during this phase are the following

– strong electric and magnetic field fluctuations
– injection of energetic particles into the inner magnetosphere (especially into the ring current)
– plasma sheet heating

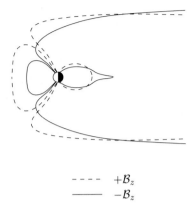

$$---- \quad +\mathcal{B}_z$$
$$\rule{3em}{0.4pt} \quad -\mathcal{B}_z$$

Fig. 8.11. Reconfiguration of the magnetosphere during a substorm. The solid lines describe the shape of the magnetosphere during the growth phase of a substorm (interplanetary $\mathcal{B}_z$ component negative); the dashed line shows the shape after the expansion phase ($\mathcal{B}_z \geq 0$) (adapted from McPherron et al., 1973).

– impulsive high velocity flows in the plasma sheet (bursty bulk flows), and
– tailward wandering field and particle disturbances (plasmoids)

Finally, the recovery phase is characterized by an abatement of activity and a gradual return to the 'ground state' of the magnetosphere. Our comprehension of these phenomena, including our attempts to assimilate them into a global description together with the activities in the polar upper atmosphere, is still inadequate. Indeed, magnetospheric substorms belong to the most challenging topics of magnetospheric physics. We content ourselves here with plausible explanations for selected aspects of this phenomenon, at the same time cautioning the reader that these are of inherently speculative nature.

8.3.1 Growth Phase

The energy required for a substorm must first be accumulated and stored before it can be suddenly released. This occurs during the growth phase of a substorm, which is initiated by the arrival of a southward-directed and thus negative $\mathcal{B}_z$ component of the interplanetary magnetic field at the magnetosphere. This leads via reconnection to the formation of open magnetic field lines and thus to an activation of the solar wind dynamo. The resulting enhanced voltage, applied across the magnetotail, leads to an intensification of the magnetotail currents (see Fig. 7.25). The magnetotail, because of its solenoidal shape, has a high inductance so that the current strength cannot rise instantaneously, but must proceed slowly with a time constant $\tau = \mathcal{L}_{tail}/\mathcal{R}_{tail} \simeq 1$ h ($\mathcal{L}_{tail}$ and $\mathcal{R}_{tail}$ are the inductance and resistance of the magnetotail current loop). Correspondingly, the magnetic field strength in the magnetotail lobes also increases slowly. This gradual accumulation of magnetic field energy corresponds to the growth phase of a substorm. As the field strength builds up, the magnetic pressure in the magnetotail lobes also rises and leads to the observed compression of the plasma sheet $p_{plasma\ sheet}(\simeq nk(T_i + T_e)) \simeq (p_B)_{mt} (\simeq B_{mt}^2/2\mu_0)$. Moreover, the increase in the magnetotail current causes the magnetic field lines to stretch in a manner similar to that sketched in Fig. 5.43. This effect is particularly impressive in the near-Earth zone $(6 - 10R_E)$. This may be attributed to the fact that, associated with the growing current strength, the $\vec{j} \times \vec{B}$ force also increases and points toward the Earth for the northward-directed magnetic field in the central plane of the magnetotail. The tail current and the entire magnetotail thus shift closer to the Earth. This acceleration in the direction of Earth acts to oppose the enhanced ram pressure of the solar wind. As sketched in Fig. 8.11, the tail magnetopause assumes a steeper inclination to the solar wind direction during the growth phase. This also compensates for the increased pressure in the magnetotail lobes. Finally, the higher voltage of the solar wind dynamo leads to an enhancement of the electric convection field, enabling the inner edge of the plasma sheet to shift closer to Earth, see Eq. (7.30).

Concerning the polar region, activation of the solar wind dynamo leads to a gradual intensification of the entire polar current system, particularly the eastward and westward electrojets. The magnetic field disturbances arising from these enhanced currents are a most characteristic feature of the growth phase of a substorm. At the same time, the build-up of the magnetotail lobe field causes an enlargement of the polar cap diameter. The polar oval and its embedded auroral arcs thus shift to lower latitudes in agreement with the observations.

8.3.2 Expansion Phase

While the underlying physics of the growth phase is considered to be roughly understood, this is not the case for the expansion phase. It is generally agreed that it is the excess magnetic energy stored in the magnetotail lobes during the growth phase that is suddenly released during the expansion phase. In any case, this requires a reduction of the tail currents that maintain this extra magnetic energy. There are different opinions, however, about just how this reduction takes place. One school assumes that it results from a partial disruption of the currents in the central plane of the near-Earth magnetotail, see Fig. 8.12. As shown, this disruption induces an electric field of the order of

$$\mathcal{E}_{ind} \sim \frac{d\Phi_{tail}}{dt} = -\mathcal{L}_{tail}\frac{dI_{tail}}{dt}$$

where $\mathcal{L}$ again denotes the inductance. The situation corresponds to the more familiar one when an induction or transformer current circuit is suddenly broken. The electric field induced in the disrupted circuit can be so strong that it is capable of maintaining the current flow, at least temporarily, by breakdown

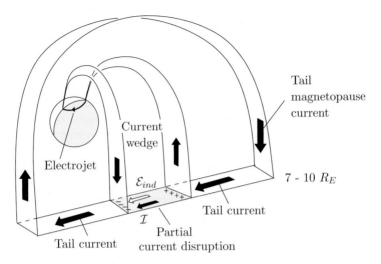

Fig. 8.12. Formation and structure of the current wedge associated with a substorm. $\mathcal{I}$ denotes the remaining tail current in the region of partial disruption (7-10 R_E down the tail). $\mathcal{E}_{ind}$ is the electric field induced there, which generates the electrojet in the polar ionosphere responsible for the substorm (adapted from McPherron et al., 1973; Weimer, 1994).

arcing. A less spectacular alternative is present in the magnetotail. In this case the induced electric field is also projected along the highly conducting magnetic field lines into the polar ionosphere. This allows the current, once disrupted in the magnetotail, to divert its flow along the field lines through the ionosphere and thus at least partially maintain its original strength. As shown in Fig. 8.12, this diversion causes intense currents in the midnight sector of the polar ionosphere and it is this additional electrojet that is made responsible for the generation of magnetic substorms. Because of its tapered structure along the field lines, this diversion current loop is denoted the *current wedge*.

As is the case for any other current circuit, formation of the current wedge requires a certain amount of time, proceeding in this case via Alfvén waves. For typical magnetospheric propagation velocities of 1000–2000 km/s, such a wave needs about one minute for the round trip tail-ionosphere-tail. These upward and downward propagating Alfvén waves produce Pi2 pulsations at periods of 40 to 150 seconds which are an essential feature that marks the beginning of the substorm expansion phase. Furthermore, one may speculate whether the upward directed Birkeland component of the current wedge is (at least partially) maintained by electrons extracted from the plasma sheet and accelerated downward along the field lines. This would explain the formation of particularly intense aurorae on the west side of the auroral protrusion, i.e. within the region of the westward traveling surge.

Another consequence of the electric induction field is the energization of particles. A sudden increase in the flux of energetic particles is observed on satellites in geostationary orbit ($L = 6.6$), for example, at the start of the expansion phase. This increase occurs without evidence of dispersion, i.e. no propagation associated time delay depending on the energy of the particles is observed. Local acceleration of the particles by electric fields is a possible explanation of this observation. Furthermore, the electric induction field is held responsible for the injection of energetic particles into the ring current region. Drift trajectories of plasma sheet particles are shifted unusually close to the Earth because of the intensity of this field. Should the electric field decrease again, these particles are then stranded in the inner magnetosphere (indicated by the unshaded region in Fig. 7.26), where only drift motion around the Earth is possible. Such erratic convection is evidently necessary in order to transfer plasma sheet particles onto ring current trajectories.

There are different opinions as to the cause of the critical partial disruption of the tail current. One school assumes that the plasma sheet is capable of maintaining only a limited current density, above which plasma instabilities set in and suppress the current flow. A second school supports the notion that the observed current reduction is controlled externally, for example by a sudden decline or complete disappearance of the southward component of the interplanetary magnetic field. And still another school holds reconnection responsible for the current decrease. According to this concept, sketched in Fig. 8.13, the exaggerated stretching of the field lines, combined with enhanced pressure in the magnetotail lobes, leads to the formation of an x-type neutral line near the Earth (*Near-Earth Neutral Line, NENL*), whereby near the Earth in this context means a down-tail distance of about 20–30 R_E. Plasma threads, together with their frozen-in magnetic fields, drift toward the neutral sheet and reconnect at this newly formed x-line, leading to a decline in magnetic energy, heating of the plasma sheet and formation of earthward directed *bursty bulk flows*, which reach the inner edge of the tail in a matter of minutes. According to Eq. (7.27), because of their sudden deceleration by the Earth's dipole field, eastward directed electric currents are produced. These are intensified by magnetization currents that flow on the inner edge of the decelerated and compressed plasma volume, as illustrated in Fig. 5.29, an example showing the same effect at the inner edge of the ring current. Both currents are directed such that they lead to a weakening of the tail current and thus to a dipolarization of the near-Earth magnetic field. At the same time, via a current wedge similar to that sketched in Fig. 8.12, they sustain the electrojet in the polar ionosphere associated with the substorm. An attractive aspect of this somewhat more complicated disturbance scenario is that it can immediately explain the creation of so-called *plasmoids* (see again Fig. 8.13). These are cylindrically shaped plasma and magnetic field perturbations that move down the magnetotail with high velocity. Typi-

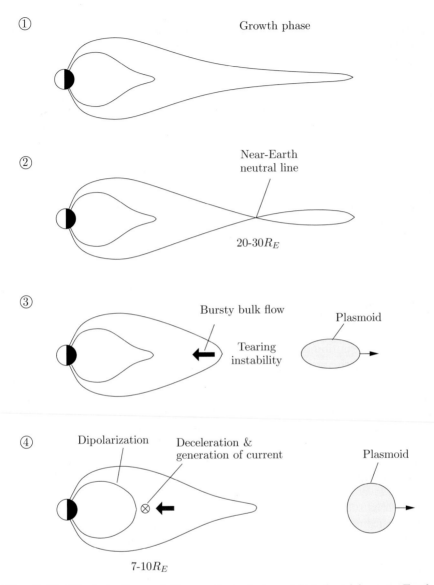

Fig. 8.13. Dipolarization and the creation of plasmoids induced by near-Earth reconnection. Note that the indicated disturbances take place in the inner regions of the magnetotail and, apart from the compression effects of the escaping plasmoid, the outer tail lobe region remains in its original form (adapted from Hones, 1977; Shiokawa et al., 1998).

cal dimensions of these far from the Earth ($\simeq 100\ R_E$) are about $5\ R_E$ for the cylinder radius and about $40\ R_E$ for the cylinder length in the y direction. Such a plasmoid, traveling at a velocity of 700 km/s, transports an appreciable amount of energy, 10^{14}–10^{15} J, which the magnetotail effectively loses to interplanetary space.

Independent of the disturbance scenario considered, we are naturally interested to know if the amount of energy released by this process is sufficient to maintain the observed energy dissipation in the polar upper atmosphere. This is indeed the case, as demonstrated by the following estimate, based on the current disruption scenario sketched in Fig. 8.12. In order to explain the field line stretching observed at the geosynchronous orbit distance during the growth phase, it is necessary to boost the tail current in the near-Earth region $(7 - 10\ R_E)$ by about 100 mA/m. According to Eq. (5.81), this corresponds to an increase in the tail lobe field strength of $\Delta \mathcal{B}_{mt} = \mu_0(\mathcal{I}_{mt}^*/2) \simeq 63$ nT. Moreover, in order to maintain a substorm associated electrojet of 0.5 MA in both polar regions, this excess tail current must be disrupted and rerouted over a length of about $2\ R_E$. The excess magnetic energy stored in this segment of the magnetotail, however, amounts to $(\Delta \mathcal{B}_{mt}^2/2\mu_0)(R_{mt}^2\pi/2)\ 2R_E \simeq 3 \cdot 10^{14}$ J, where we have assumed a tail radius of $R_{mt} \simeq 15\ R_E$. For our specified power requirement of 10^{11} W, this energy is certainly sufficient to drive the expansion phase of a polar substorm (duration $\lesssim 0.5$ h). The current disruption must be extended to regions further down the tail in the case of longer lasting activity. Such tailward expanding disruption or reduction of the current has indeed been observed.

In view of the speculative nature of the various substorm scenarios, it might seem surprising that *in situ* measurements have contributed so little to clarify the situation. Remember here that the magnetosphere is a huge volume of space, the various regions of which have only been surveyed sporadically by individual satellites on a point-by-point basis. The probability that a given satellite would be fortuitously located at the right place at the right time is thus relatively low. Furthermore, since measurements from a single satellite cannot discriminate between spatial and temporal variations, they are severely limited in their capability to investigate dynamical phenomena such as substorms. In response to this dilemma, more recent endeavors (like, for example, the CLUSTER mission) aim to resolve this ambiguity with the help of a fleet of spatially separated satellites flying in formation.

8.4 Thermospheric Storms

A significant part of the solar wind energy absorbed during a geospheric storm is dissipated by electric currents and particle precipitation in the polar upper atmosphere. The resultant heating can be so intense that it produces not only local, but even global disturbances of the thermosphere. Following our previous naming convention, we denote such an event as a *thermospheric storm*. Due to the observed high wind velocities, this designation does come close to our concept of a storm, but then all other state parameters such as temperature, mass density and composition are affected as well. Of the many phenomena associated with a thermospheric storm, we address two specific aspects here, composition disturbances at mid-latitudes and density disturbances at equatorial latitudes. These two very striking effects are also intimately connected with ionospheric perturbations.

8.4.1 Composition Disturbances at Mid-latitudes

The type of composition perturbation occurring during a thermospheric storm is, in principle, the same as that observed at polar latitudes during less disturbed conditions: heavier gases display an increase in density, lighter gases a decrease. These density changes are considerably stronger during a storm, however, and also not confined to the polar heating zone. This expansion effect is shown schematically in Fig. 8.14. A satellite with a neutral gas analyzer aboard, moving northward along the indicated polar orbit, for example, would enter the zone of disturbed neutral gas composition already at mid-latitudes. This is documented with actual data in Fig. 8.15, where *relative* changes in density are shown as a function of magnetic latitude (lower panel). The plotted curves, $R(n)$, are the density values measured along the satellite trajectory during the storm, divided by the density values obtained during an undisturbed orbit. $R(n) = 1$ thus means no change with respect to quiet conditions and $R(n) = 10$ indicates a tenfold increase in density. The magnetic activity during the storm and reference orbits, as reflected by the Kp index, is plotted in the upper panel. The measurements clearly show that substantial density disturbances can occur at an altitude of 280 km. The rise in the argon density reaches a factor of 80 and that of molecular nitrogen a

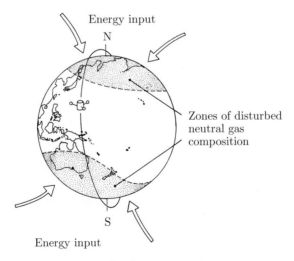

Fig. 8.14. Energy injection and formation of two zones of disturbed neutral gas composition during a thermospheric storm

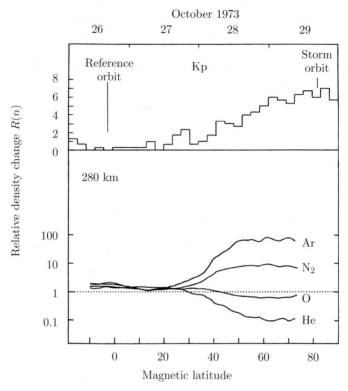

Fig. 8.15. Latitudinal dependence of relative density changes as observed during a thermospheric storm. The form of the data presentation is described in the text. The density measurements refer to an altitude of 280 km and 9 hours local time.

factor of 10. At the same time, the atomic oxygen density drops to nearly one half, the helium density to even one tenth, of the respective quiet-time values. Also remarkable is that the zone of disturbed composition extends down to a latitude of 30°.

While the reasons for the observed density perturbations were already elucidated in Section 7.5.3, their extension down to mid-latitudes is still in need of an explanation. One assumes today that the density disturbances do indeed originate at polar latitudes, but are subsequently carried to mid-latitudes by strong winds. The details of this perturbation transport are still the subject of intensive investigation. It is well established, however, that the propagation occurs in the night sector and this is immediately understandable. The ion drift in the polar caps leads to a strong acceleration of the airmass toward the night sector (see Fig. 7.3 and Section 7.5.1). Moreover, two additional high pressure areas form in the polar heating zones during a storm, inducing winds which are superposed onto the normal diurnal wind circulation; see Fig. 8.16. In the night sector, this amplifies the already es-

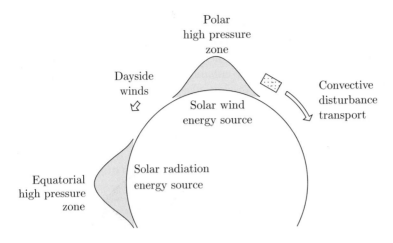

Fig. 8.16. Storm induced changes in the large-scale wind circulation and convective transport of composition perturbations to mid-latitudes

tablished equatorward atmospheric circulation, generating storm winds with velocities of 1000–2000 km/h. These are the storm winds held responsible for the transport of the composition perturbations to mid-latitudes, whereby particular significance is attributed to the somewhat slower winds in the lower thermosphere. At lower heights, because of the much larger diffusion time constants there, composition perturbations persist much longer than in the upper thermosphere (compare with Section 2.3.6).

In contrast to the night sector, perturbation transport on the dayside is prevented by the predominantly poleward directed winds. Even if the polar high pressure is strong enough to reverse the direction of the nominally poleward flowing airmass (as implied in Fig. 8.16), the velocity of the resultant flow is simply too slow to produce significant expansion of the perturbation. The question then arises of why the composition disturbances are also observed on the dayside (the data of Fig. 8.15, for example, apply to 9 hours local time). A plausible explanation is that, since the mid-latitude thermosphere essentially rotates with the Earth, any disturbance produced in the night sector will eventually be carried into the dayside by the corotation.

8.4.2 Density Disturbances at Low Latitudes

The density perturbations observed during magnetic activity are not restricted to high and mid-latitudes. They also extend all the way down to equatorial regions, where they, however, assume a different form. This is documented in Fig. 8.17, again using satellite data. These show the temporal changes in the argon, molecular nitrogen, atomic oxygen and helium densities, measured during a magnetic storm, at 290 km altitude for latitudes between 5 and 10°S. In response to the magnetic activity shown in the upper panel, a fluctuating, but systematic, rise in the density is observed for *all* the gases. Evidently, other disturbance mechanisms are active here than at higher latitudes. Whereas the larger density increase of the heavier gases indicates a rise in the temperature, the density enhancement of the lighter gases can be more easily explained as a compression effect. The temperature increase of $\Delta T_\infty \lesssim 160$ K consistent with the nitrogen measurements, for example, would produce only a four percent increase in the helium density.

Regardless of the immediate cause of the density increase, it is worth asking just why the observed disturbances can happen so fast and so far removed from the polar heating zone. Indeed, the first sign of a density increase

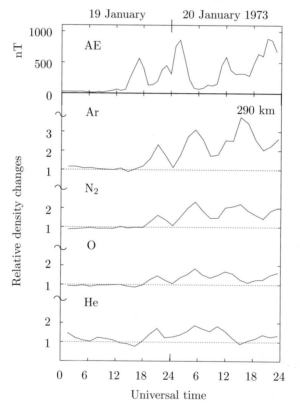

Fig. 8.17. Thermospheric storm effects at low latitudes. The upper panel shows the development of polar magnetic activity during a two-day interval. The response of the upper atmosphere to this activity at low latitudes ($5°-10°$S) is displayed in the lower panels, again with plots of the relative density changes. The reference density values were measured on the day before the magnetic storm. All densities refer to an altitude of 290 km and a local time of 01:00 hours (adapted from Prölss, 1997; for simplicity we mention here that Figs. 8.18 and 8.19 were also taken from this source and Figs. 8.20 – 8.22 are from Prölss, 1995).

is observed less than four hours after the beginning of the magnetic activity. One possible explanation is that the density increase is produced by a so-called *traveling atmospheric disturbance* (TAD). This is understood to be an impulse-like disturbance arising from a superposition of atmospheric gravity waves that propagates at high velocity (500 – 1000 m/s) from polar to equatorial regions. Figure 8.18 illustrates the essential features of this disturbance scenario.

During a substorm – here indicated by an increase in the AL index – the polar upper atmosphere is subjected to a sudden injection of energy and heating. The resulting expansion of the gases leads to excitation of a broad spectrum of atmospheric gravity waves that propagate outward from their source region over the entire Earth. When these arrive at mid-latitudes, the higher frequency waves are already strongly damped and the low-frequency (long wavelength) components juxtapose to form an impulse-like disturbance that propagates equatorward at high velocity. Model calculations show that in-

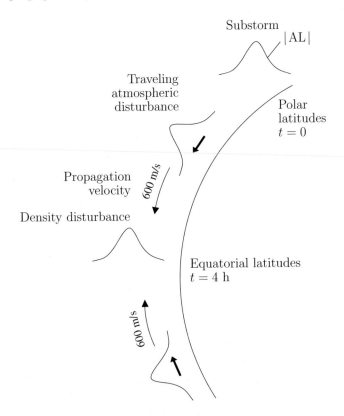

Fig. 8.18. Explanation of equatorial density perturbations during magnetic (sub)storm activity. The black arrows embedded in the traveling atmospheric disturbances characterize the associated wind disturbances.

creases in the density and temperature, but also equatorward directed winds, should be associated with such a disturbance. The typical duration of a TAD passing overhead should be of the order of one to two hours.

Starting at a latitude of 70° with a mean propagation velocity of 600 m/s, the traveling disturbance reaches the equator in less than 4 hours. There it encounters the TAD coming from the opposite polar region. The superposition of these two disturbances, especially the clash of the equatorward directed disturbance winds, as one might easily imagine, leads to both compression and heating of the gases, in agreement with the density changes shown in Fig. 8.17.

8.5 Ionospheric Storms

Ionospheric disturbances during magnetic activity were first documented in 1929 by Hafstad and Tuve. Indications for this phenomenon had been noted earlier, however, as radio engineers observed distinct modifications of their ionospheric transmission paths during magnetic storms. Meanwhile hundreds of publications on this topic have appeared, reflecting the complexity, but also the significance, of this phenomenon. All state parameters are affected over the entire range of the ionosphere. Most prominent and certainly best studied are perturbations in the F region, however, particularly those near the density maximum. For simplicity, the following discussion is devoted exclusively to these types of disturbances.

Early observations at mid-latitudes showed that both an anomalous density increase as well as an anomalous density decrease can occur over the course of a magnetic storm. The former manifestation has been classified in the literature as a *positive*, the latter as a *negative ionospheric storm*. The effects of these two types of storms on the ionospheric density profile are shown in Fig. 8.19. Apparently, the density disturbances can be rather severe and deviations in the peak electron density by more than 100% are not uncommon.

The starting point for an explanation of these disturbances is the density balance equation (4.23)

$$\frac{\partial n_s}{\partial t} = q_s - l_s + d_s$$

In principle, all three of the terms on the right side of this equation could cause the observed density changes and, in fact, all three, including various combinations thereof, have been held responsible for the storm effects. In the following we describe a disturbance scenario based on an intimate coupling between thermospheric and ionospheric storms. In this scenario negative ionospheric storms at mid-latitudes are due to changes in the neutral gas composition and positive storms are predominantly caused by thermospheric winds.

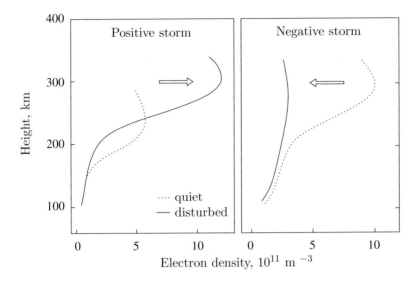

Fig. 8.19. Changes in electron density profiles as observed during a positive (left) and negative (right) ionospheric storm. The dotted profiles serve as the respective quiet-time references.

8.5.1 Negative Ionospheric Storms

It is immediately clear that disturbances of the neutral gas composition, like those documented in Fig. 8.15, should have important consequences for the ionosphere. After all, the decrease in the atomic oxygen density must produce a decrease in the production of the oxygen ions and the increase in the molecular nitrogen density must produce an increase in the loss rate for this ion species (see Section 4.2). These two neutral gas density changes thus combine to decrease the ionization density in the F region. Accordingly, all ionosonde stations located underneath the zone of disturbed composition should observe negative ionospheric storm effects, and this is indeed the case. The middle panel of Fig. 8.20 once again shows the latitudinal profiles of the O and N_2 density perturbations previously documented in Fig. 8.15, but now referred to a constant pressure, rather than a constant geodetic height (whereby the indicated pressure level of 8 μPa corresponds approximately to an altitude of 300 km). The reason for this change in coordinates is that ionospheric layers are not tied to a specific height, but rather to the isobars of the thermosphere, and therefore move up and down in dependence on the temporal and spatial variations of the pressure levels. It may be seen that the density decrease of atomic oxygen is considerably more pronounced in the pressure coordinate system. The density increase of the molecular nitrogen, on the other hand, is strongly diminished.

The locations of two ionosonde stations, situated beneath the satellite trajectory, are also indicated in the latitudinal profile and measurements from

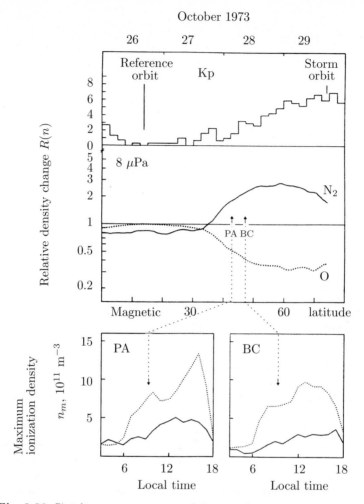

Fig. 8.20. Simultaneous occurrence of thermospheric composition disturbances and negative ionospheric storms. The upper panels correspond to Fig. 8.15, except that the relative density changes of the gases N₂ and O are now referred to a constant pressure level of 8 μPa. The lower part of the figure shows the diurnal variations of the peak electron density measured at the two ionosonde stations Pt. Arguello (PA) and Boulder (BC) on the reference day (26 October, dotted line) and on the day of the storm (29 October, solid line). The latitudinal position of the stations relative to the composition disturbance zone and the time of the satellite measurement relative to the diurnal variation are each indicated by arrows.

these stations are given in the lower panels of Fig. 8.20. These show the diurnal variation of the peak electron density, n_m, as observed during an undisturbed reference day and during the storm. Obviously, a severe depression in the maximum electron density is recorded on the day of the storm.

A simple estimate shows that the neutral gas density changes are indeed sufficient to explain the observed ionospheric storm effects. Recalling Eq. (4.65), the maximum ionization density is approximately given by

$$n_m \simeq \frac{J_O \, n_O(h_m)}{k^{CE}_{O^+,N_2} \, n_{N_2}(h_m) + k^{CE}_{O^+,O_2} \, n_{O_2}(h_m)} \tag{8.8}$$

Considering only relative density changes, $R(n_s) = (n_s)_{\text{disturbed}}/(n_s)_{\text{reference}}$, and assuming that the relative density change of molecular oxygen is about equal to that of molecular nitrogen (after all, the two gases have nearly the same molecular weight), we may then use $R(n_{O_2}) \simeq R(n_{N_2})$ to obtain

$$R(n_m) = \frac{R(n_O)}{R(n_{N_2})} \frac{k^{CE}_{O^+,N_2} + k^{CE}_{O^+,O_2} (n_{O_2}/n_{N_2})_{\text{reference}}}{k^{CE}_{O^+,N_2} + k^{CE}_{O^+,O_2} (n_{O_2}/n_{N_2})_{\text{reference}} R(n_{O_2})/R(n_{N_2})}$$

$$\simeq \frac{R(n_O)}{R(n_{N_2})} \tag{8.9}$$

where all neutral gas density values refer to the location of the ionization density peak. Noting that Fig. 8.20 indicates $R(n_O) \simeq 0.5$ and $R(n_{N_2}) \simeq 2$ at heights near the ionization maximum, Eq. (8.9) predicts that the ionization density should be reduced to about one fourth of its original value, in sufficiently good agreement with the actually observed density decrease.

8.5.2 Positive Ionospheric Storms

An important characteristic of traveling atmospheric disturbances, a key element for explaining positive ionospheric storms, is that they exhibit equatorward directed winds. The velocity of these winds, typically 50–200 m/s, should not be confused with the much larger propagation velocity of the disturbance, see Fig. 8.21. We are interested here in the frictional force exerted by these neutral winds on the charge carriers of the ionosphere. Because charge carriers are constrained in their motion by the geomagnetic field, it is useful to decompose this frictional force into its components parallel and perpendicular to the magnetic field. According to Eq. (5.40), the perpendicular component of the frictional force produces a horizontal drift which, assuming to a first approximation that the ionosphere is uniform and infinitely extended in the horizontal direction, has no consequences for the vertical density distribution. On the other hand, the charge carriers are free to move along the magnetic field, and the field-aligned component of the frictional force will drive the ionization along the inclined field lines to greater heights. The expected reaction to a traveling atmospheric disturbance should thus be an elevation of the ionosphere, as indicated in Fig. 8.21 for the location of the layer peak (Δh_m). Lifting the ionization maximum relative to the neutral gas atmosphere, however, leads to an increase in the peak ionization density (Δn_m). This is because molecular nitrogen and molecular oxygen, which are

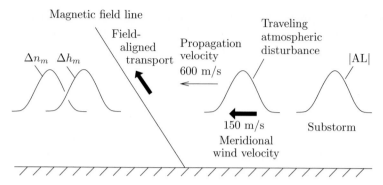

Fig. 8.21. A short-duration positive ionospheric storm caused by a traveling atmospheric disturbance. Substorm activity is again indicated by the AL index, Δh_m is the change in the height of the maximum ionization density, and Δn_m is the subsequent change in maximum ionization density.

responsible for the loss of ionization, decrease much faster with height (i.e. they have smaller scale heights) than atomic oxygen, the species that governs the ion production rate. The net result of elevating the ionization layer is thus an effective increase in the ionization density, similar to the situation present in the lower F region as explained in Section 4.3.3.

Measurements consistent with the above described scenario are shown in Fig. 8.22. The upper panel of this figure shows the AL index, which clearly indicates the occurrence of an isolated substorm. In response to this injection of energy in the polar regions, a distinct upward shift of the ionization layer peak is observed at mid-latitudes after a time delay of about 2 hours (Δh_m, middle panel). This upward shift is clearly responsible for the sharp increase in the maximum ionization density, as documented in the middle panel as a difference (Δn_m), and in the bottom panel as a comparison between the diurnal profiles of n_m for quiet and storm conditions. The time delay of the density increase following the upward shift of the layer peak (middle panel) has a straightforward explanation. Recalling Eq. (4.57), the expression for the ionospheric production time constant, and using $n \simeq 7 \cdot 10^{11} \mathrm{m}^{-3}$ and $q_{O^+} \simeq J_O n_O(h_m) \simeq 18 \cdot 10^7 \, \mathrm{s}^{-1}\mathrm{m}^{-3}$, we obtain $\tau_q \simeq 1$ hour.

That the observed upward shift of the layer peak is indeed sufficient to explain the observed density increase may be shown by the following estimate. Using again Eq. (8.8) for calculating the undisturbed peak ionization density, the disturbed peak density for a layer shifted upwards by Δh_m is given by

$$(n_m)_{disturbed} \simeq \frac{J_O \, n_O(h_m) \, e^{-\Delta h_m/H_O}}{k^{CE}_{O^+,N_2} \, n_{N_2}(h_m) \, e^{-\Delta h_m/H_{N_2}} + k^{CE}_{O^+,O_2} \, n_{O_2}(h_m) \, e^{-\Delta h_m/H_{O_2}}}$$

$$(8.10)$$

where it is assumed that isothermal conditions prevail in the region of the density maximum. Neglecting the relatively small difference in the scale heights

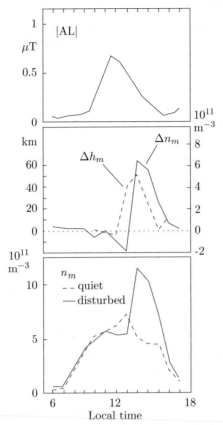

Fig. 8.22. Example of a short-duration positive ionospheric storm. Reacting to an isolated substorm (AL index, upper panel), an ionosonde station at mid-latitudes records first an impulse-like upward shift of the ionization layer peak (Δh_m) and then a sharp increase in the maximum electron density (Δn_m, middle panel). The bottom panel compares the diurnal variations of the maximum ionization density for a quiet reference day and the storm day.

of N_2 and O_2 and introducing a combined scale height H_{N_2,O_2}, Eq. (8.10) takes the following form

$$(n_m)_{disturbed} \simeq \frac{J_O \, n_O(h_m)}{k^{CE}_{O^+,N_2} \, n_{N_2}(h_m) + k^{CE}_{O^+,O_2} \, n_{O_2}(h_m)} \, e^{-\Delta h_m (1/H_O - 1/H_{N_2,O_2})}$$

The relative density change produced by the upward shift of the peak layer is thus given by

$$R(n_m) \simeq e^{-\Delta h_m (1/H_O - 1/H_{N_2,O_2})} = e^{\Delta h_m / H_{ql}} \tag{8.11}$$

where we have again made use of the scale height H_{ql} defined in Eq. (4.31). Noting that $\mathcal{M}_{N_2,O_2} \simeq 30$ and $\mathcal{M}_O = 16$, H_{ql} corresponds approximately to the scale height of atomic oxygen. As shown in Fig. 8.22, $\Delta h_m \simeq 50$ km and is thus of the same order of magnitude as H_{ql}. As a consequence, one would expect a density increase by a factor of about 2 to 3, and this is indeed observed.

In addition to the short duration events (as considered in Fig. 8.22), longer lasting positive storms are observed. These could be produced, for example, by a series of rapidly successive traveling atmospheric disturbances or a modification of the global wind circulation. Concerning the latter possibility, recall (as sketched in Fig. 8.16) that the poleward directed winds on the dayside can be diminished or even reversed. In both cases this will lead to a more enduring upward shift of the layer peak and thus to persistent positive storm effects.

The above discussion may have left the impression that ionospheric storm effects are completely understood. This is by no means the case. There are still a number of open issues concerning both the morphology as well as the physics of this phenomenon. Electric fields, for example, were completely omitted from consideration in our explanations of the storm effects. Continued research in this field is thus clearly necessary and rewarding, and hereby specifically encouraged.

8.6 The Sun as the Origin of Geospheric Storms

The immediate reason for the occurrence of geospheric storms is a strongly enhanced transfer of energy from solar wind to magnetosphere. Conditions for enhanced energy transfer are particularly favorable, in turn, when the solar wind velocity is high and the $\mathcal{B}_z$ component of the interplanetary magnetic field assumes significantly negative values (see Section 7.6). Since the increase in solar wind velocity is subject to a relatively narrow bound, however, the energy transfer is essentially controlled by the $\mathcal{B}_z$ component of the interplanetary magnetic field. This is verified by Fig. 8.23, which displays plots of the mean variation of the solar wind velocity and the $\mathcal{B}_z$ component of the interplanetary magnetic field during the course of a magnetic storm. In contrast to the relatively weak increase in solar wind velocity ($< 30\%$), the $\mathcal{B}_z$ component displays an unambiguous, strong correlation with the depression in the Dst index. Since the energy contained in the interplanetary magnetic field is comparatively small, the $\mathcal{B}_z$ component can only function as a type of control valve that regulates the solar wind-magnetosphere energy transfer. The initial question regarding the source of geospheric storms is thus reduced to finding the source of significant negative $\mathcal{B}_z$ components in the interplanetary magnetic field. Among the various formation mechanisms that have been discussed are radial warping of the heliospheric current sheet (compare with Fig. 6.23, where, to a first approximation, the $\mathcal{B}_\varphi$ component corresponds to the $\mathcal{B}_z$ component considered here), amplification of weak negative $\mathcal{B}_z$ fields by shock waves, and large-amplitude Alfvén waves.

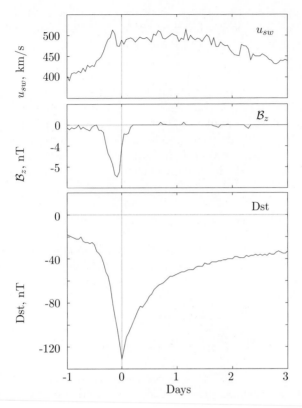

Fig. 8.23. Mean behavior of the interplanetary medium during magnetic storms. The quantities plotted from top to bottom are the solar wind velocity u_{sw}, the $\mathcal{B}_z$ component of the interplanetary magnetic field (GSM coordinate system), and the degree of magnetic activity measured by the Dst index. The curves are average values (more precisely: *median* values) of 121 strong storms with Dst-Minima between -100 and -200 nT. The individual curves were superposed using as a common reference the time of the respective minimum in the Dst index.

In the following, two disturbance scenarios are described that account not only for the presence of a negative $\mathcal{B}_z$ component in interplanetary space, but also for those events on the Sun responsible for its formation. These are the solar mass ejections with their associated magnetic clouds and the coronal hole/coronal streamer interfaces with their corotating interaction regions. Within the framework provided by these examples of solar terrestrial relations, it is also appropriate to address the topic of solar flares and their consequences.

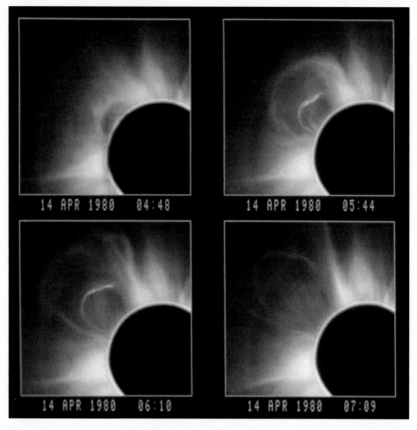

Fig. 8.24. Solar mass ejection as observed in visible (white) light. The images were recorded with a coronagraph on board the SMM satellite. The apparent radius of the dark occulting disk, which is needed to block out the bright glare of the photosphere, is 1.4 solar radii (adapted from Hundhausen, 1988).

8.6.1 Solar Mass Ejections and Magnetic Clouds

In the early 1970s, as continuous observations of the solar corona above the disturbing atmosphere were first made possible with the help of coronagraphs on Earth-orbiting satellites, solar investigators soon became witness to a highly remarkable event. Figure 8.24 presents a visual record of the observed phenomenon in a sequence of four images. Consider first the loop-like structure in the upper left of the first image that bounds a dark (and thus low density) region of the inner corona. Less than an hour later (second image), this loop has moved to the outer edge of the visible corona; simultaneously, a second loop has appeared in the field of view closer to the Sun. About a half-hour later (third image), the two loops have moved outward from the Sun so far that the outer loop is now only faintly visible. Finally, after an additional hour (fourth image), even the inner loop has all but vanished into

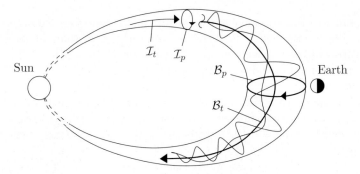

Fig. 8.25. Flux rope model of a magnetic cloud (adapted from Lepping et al., 1990)

interplanetary space and only vestiges of the outer loop can still be recognized close to the Sun. Since the dynamics of these emission features reflect the motion of the associated gases, Fig. 8.24 clearly documents the expulsion of an enormous amount of gaseous solar material into interplanetary space. Such an event is denoted a *coronal* or *solar mass ejection* (CME/SME), and the mass of the ejected gas typically lies in the range $10^{12} - 10^{13}$ kg. Meanwhile, a great number of such mass ejections have been investigated. Our interest in these studies concentrates on the mean ejection velocity of roughly 500 km/s (with a large scatter between 50 and 1800 km/s) and the mean kinetic energy with a range from 10^{23} to 10^{25} J.

There are various ideas about how such mass ejections arise. We discuss a model here for which the outer (inner) visible loop in Fig. 8.24 is associated with the outer (inner) edge of a *magnetic flux rope*. This is understood to be a magnetic field configuration possessing both a toroidal component $\mathcal{B}_t$ (i.e. parallel to the loop) and a poloidal component $\mathcal{B}_p$ (i.e. perpendicular to the loop, circular); see Fig. 8.25. The field is purely toroidal along the central axis of the rope and $\mathcal{B}_t$ reaches its largest value there. At the outer edge of the rope, the toroidal field vanishes and the poloidal component $\mathcal{B}_p$ attains its largest value. Between these extremes, $\mathcal{B}_t$ and $\mathcal{B}_p$ superpose to form helical field lines that look like the strands of a rope. Such a field configuration is generated by electric currents that also possess toroidal and poloidal components.

With a proper apportionment of the composite currents, flux ropes of this type are in equilibrium with their coronal surroundings. If, however, a large amount of poloidal magnetic energy (e.g. 10^{25} J) is generated by a sudden enhancement of the toroidal current strength $\mathcal{I}_t$, the apex region will be accelerated outward by internal magnetic forces ($\sim \mathcal{I}_t \, \mathcal{B}_p$) to the velocities observed in solar mass ejections. Following this outward motion with simulation calculations all the way to the Earth, one obtains a configuration similar to that sketched in Fig. 8.25.

Such a flux rope can assume any arbitrary inclination to the ecliptic plane. If its axis lies in the ecliptic, a characteristic rotation of the magnetic field would be observed on a spacecraft located in front of the magnetosphere as the flux rope passes by. Depending on the orientation of the poloidal field, the $\mathcal{B}_z$ component would assume first positive (negative) and then negative (positive) values, see upper curve of Fig. 8.26. If the axis is oriented perpendicular to the ecliptic, however, the instruments on the spacecraft would record an increase and then a decrease of either the negative or positive $\mathcal{B}_z$ component (Fig. 8.26, lower curve).

Variations of this type have indeed been observed during the passage of so-called *magnetic clouds*. These are understood to be interplanetary disturbances of moderate extent (diameter $\simeq 0.2$ AU at the Earth's orbit), which are characterized by enhanced magnetic field strength, low temperature, low density, low β parameter and by the above specified magnetic field variations. Most important in this context is that magnetic clouds can be associated with strong negative $\mathcal{B}_z$ components, as emphasized by the shaded parts of the curves in Fig. 8.26. It is thus not surprising that an intimate correlation has been observed between the encounter of a magnetic cloud with the Earth and the onset of a geospheric storm. On the other hand, it has not been unambiguously demonstrated that magnetic clouds are really outward accelerated flux ropes. Indeed, solar mass ejections frequently assume a more

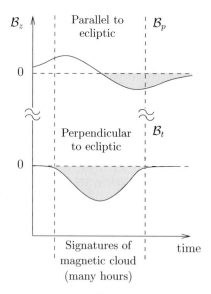

Fig. 8.26. Variation of the $\mathcal{B}_z$ component of the interplanetary magnetic field during the passage of a magnetic cloud. The variations correspond to the polarities of the fields and currents shown in Fig. 8.25. Note that $\mathcal{B}_p$, in contrast to $\mathcal{I}_t$, $\mathcal{I}_p$ and $\mathcal{B}_t$, is not restricted to the interior of a flux rope.

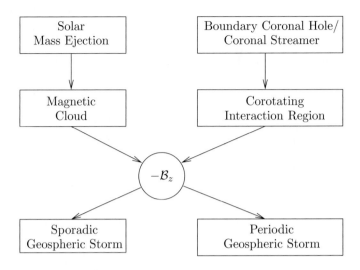

Fig. 8.27. Solar and interplanetary origin of geospheric storms

bubble-like shape. In any case, as reflected in the left branch of the causality diagram in Fig. 8.27, a close statistical correlation does exist between the occurrence of mass ejections and magnetic clouds.

8.6.2 Corotating Interaction Regions

In addition to the geospheric storms induced sporadically by magnetic clouds, there are also examples that exhibit a tendency to repeat. In particular, geospheric storms with a 27-day periodicity are commonly observed during the phase of declining solar activity. Their source is clearly active for quite some time and follows the solar rotation. Bartels once designated such sources as *M regions* (from *m*agnetically effective), but their true nature was obscure for many decades. Today we identity M regions as the boundaries between the coronal holes and the coronal streamers. While coronal holes are known to be source regions of fast solar wind flows, coronal streamers (and especially the streamer belt at equatorial latitudes) are sources of slow solar wind flows (see Fig. 6.8). The different velocities of these flows in interplanetary space lead to compression of the solar wind plasma on the forward edge of the high-speed flow and thus to a compression of the frozen-in magnetic field (see, for example, Fig. 6.39). Should this ambient magnetic field already possess a negative $\mathcal{B}_z$ component, it can be amplified to the point where a geospheric storm is triggered as the compression region passes the Earth. These *corotating interaction regions*, as shown in the right branch of the diagram in Fig. 8.27, are thus held responsible for the periodically repeating geospheric storms.

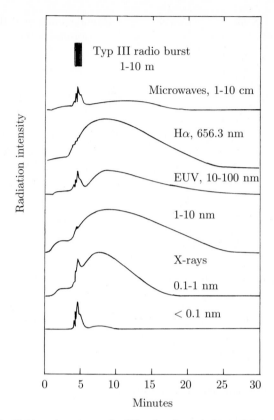

Fig. 8.28. Radiative signature of a 2B solar flare (adapted from Kane, 1974)

8.6.3 Solar Flares

Another phenomenon that, for many years, had been held responsible for causing geospheric storms is the *solar flare*. To use a very generalized definition, this is a strongly localized ($\lesssim 0.1\ R_S$) and short-duration ($\lesssim 1$ h) release of a great amount of energy (up to about 10^{25} J) in the solar atmosphere. Serious doubts concerning any such causality role exist today. That we are nevertheless interested in solar flares is not just our fascination for this phenomenon, but also because solar flares are an essential component of solar-terrestrial relations that cause subtle, but distinctly measurable, disturbances in the upper atmosphere.

Knowledge of a solar flare reaches us first in the form of electromagnetic radiation released by the event. The affected regions suddenly brighten and become detectable in various wavelength ranges. Figure 8.28 shows the typical temporal signature of an eruptive flare as observed in the X-ray, EUV, Hα, and radio wavelength regions. In rare cases, the brightening can even be detected in 'white light', i.e. in the visible range, indicating that the photosphere, in addition to the chromosphere and corona, is involved in the event.

Solar flares are most easily observed in the light of the Hα and CaII lines, or at radio wavelengths, since these emissions can all be received directly at the Earth's surface. Accordingly, the customary classification of the flare strength is based on these observations. A frequently used scaling accounts for both the extent (areal importance classes: S(ubflare), 1–4) and the maximum brightness (classes: F(aint), N(ormal), B(right)) of the Hα emission. A flare reaching the category 3B, for example, is a rather large, but correspondingly rare, eruptive event. The occurrence frequency of solar flares generally follows the rhythm of the solar cycle, whereby about 5 larger events (areal importance ≥ 2) occur every year during lower solar activity and 70 such events are recorded yearly during times of higher solar activity. It is the increase in the EUV and X-ray radiation that is particularly important for the upper atmosphere. For large solar flares, this can amount to $10 \ \mathrm{mW/m^2}$ in the EUV range (which is roughly twice the usual intensity) and up to $1 \ \mathrm{mW/m^2}$ in the soft X-ray range at $\lambda = 1$–10 nm (which is an increase by a factor of 10^4).

Besides the electromagnetic emission, energetic particles (primarily protons and electrons) are also released during solar flares. One distinguishes here between *solar high energy particles* with proton energies above about 0.5 GeV and a lower energy component, called simply *solar energetic particles*, for which the typical energies range from 1 to 100 MeV. Flares with solar energetic particles are relatively rare, however, and those with accompanying solar high energy particles are often not observed for years or even decades.

Much of the physics of solar flares remains inadequately understood. Among the open issues is the unknown source of the energy that gets released during the flare. A process most often cited in this context is that of spontaneous reconnection, by which (as explained earlier) magnetic energy is converted into the kinetic energy of the particles. One of the possible scenarios accounting for flare radiation is sketched in Fig. 8.29.

Consider a loop-like magnetic flux tube from the Sun that extends out into the corona. The energy released by the reconnection process leads to a strong acceleration of charged particles at the apex of this flux tube. The particles are then injected on their helical trajectories into the chromosphere, the electrons emitting synchrotron radiation at radio wavelengths along the way. This explains the impulsive rise in the intensity of this emission (see Fig. 8.28). As the electrons strike the denser gases of the chromosphere, they are decelerated by collisions and produce emission by bremsstrahlung. This lies in the EUV and X-ray range (and particularly in the hard X-ray range), explaining the rapid increase in the short wavelength emission. At the same time, precipitating protons at energies of several 10 MeV can stimulate nuclear emission in the γ range or, at higher energies, release neutrons and positrons.

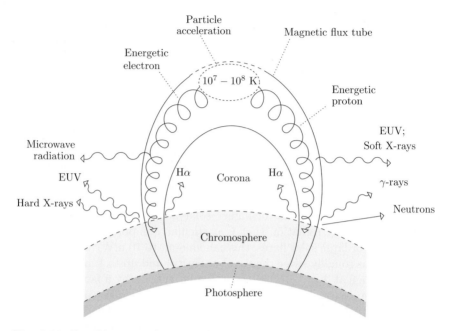

Fig. 8.29. Possible source location of the various solar flare emission components (adapted from Gurman, in Walker, 1988)

The thermalization of the incident energetic particles in the chromosphere leads to strong ionization and heating of the resident gases. This heating is so intense that the outer layer of the chromosphere evaporates as a plasma into the coronal extension of the flux tube. Meanwhile, the hot plasma of the apex region created by the acceleration of the particles simultaneously heats the adjacent lower-lying regions so that eventually the entire flux tube is filled with a completely ionized hot gas at a temperature of 10 to 100 million K. It is the thermal bremsstrahlung of this plasma that is held responsible for the longer lasting increase of the EUV and soft X-ray radiation following the impulsive phase of the flare emission.

Turning to an explanation of the Hα radiation, it may be attributed to direct excitation of chromospheric hydrogen from the thermalization of incident energetic particles on the one hand, and to emission stimulated from the recombination of collisionally ionized hydrogen on the other hand. It should be noted that the hot plasma in the coronal magnetic flux tube has an energy of several keV and remains in continuous contact with the chromosphere.

The energy released by the flare accelerates particles not only downward in the direction of the denser solar atmosphere, but also into interplanetary space along magnetic field lines that have been opened by the reconnection process. These particle beams, which often display a pulsating behavior, stimulate the background plasma of the corona to undergo oscillations at the local

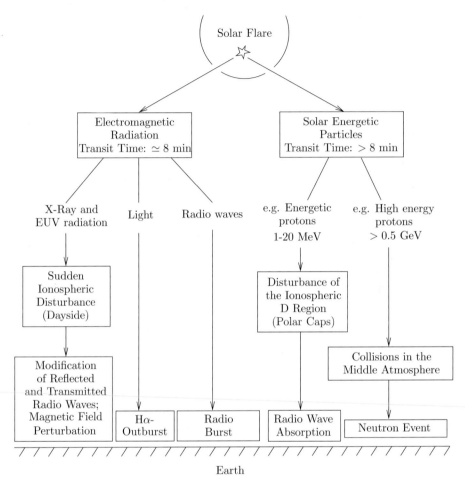

Fig. 8.30. Solar flare effects observed at the Earth

plasma frequency that is recorded at the Earth in the form of Type III radio emission.

The disturbances induced at the Earth by solar flares are modest, consisting essentially of the additional ionization produced by the EUV and X-ray emission, but also by energetic particles; see Fig. 8.30. The rapid increase in the ionization density (*sudden ionospheric disturbance*, SID) leads to an abrupt change in the reflection conditions for radio waves that are observed as sudden shifts in the received frequency and phase. Moreover, the density increase in the D region produced by harder X-ray emission leads to short-term attenuation of the radio waves (*Mögel-Dellinger effect*). Finally, the conductivity of the ionosphere is increased, triggering a sudden intensification of the ionospheric currents and their associated magnetic field disturbances (*Crochet effect*). While solar radiation effects are restricted to the dayside, energetic protons and electrons only have access to the polar upper atmosphere. On the other hand, high energy protons behave like cosmic rays and their gyroradii are so large that they can penetrate easily into the magnetosphere, even at mid-latitudes.

8.7 Disturbance Effects on Technological Systems

Geospheric storms and solar flares are of great interest not only from a scientific point of view. They are also keenly monitored for their technological impacts, especially their propensity for disrupting operational systems and utilities. The damages attributed to these phenomena are estimated to average more than 10 million dollars per year.

Magnetic Storms. As shown in Fig. 8.6, intense fluctuations of the magnetic field strength are observed in the vicinity of the polar electrojets. The fluctuations can become so large during magnetic storms that they induce significant transient voltages and currents in the Earth and in extended conductors such as high voltage transmission cables and oil pipelines. In the more benign cases, this can trip safety relays and temporarily interrupt electric circuits. In severe cases, the additional current can shift the operational range of high voltage transformers into saturation, and the resulting overload can lead to a meltdown of the transformer windings. This happened, for example, during the great storm of March 1989, when the entire electric power system broke down in the Canadian province Quebec (6 million customers) for 9 hours. Over longer time scales, the accumulated effect of even smaller storms contributes to the premature aging of materials in transformers and oil pipelines.

Magnetic storms are of less consequence, but nevertheless costly for magnetic prospecting. This technique utilizes high precision airborne observations of the geomagnetic field in search of anomalies that may provide evidence of valuable mineral deposits. Obviously, magnetic storms can render such observations worthless.

Thermospheric Storms. As evident in Fig. 8.17, a considerable enhancement of the neutral gas density (in extreme cases of up to several 100%) can occur during a thermospheric storm. Because of the resulting increase in air drag, this leads to a stronger and more erratic deceleration of low-orbiting satellites, thereby necessitating a recalculation of their orbital elements (ephemerides). Furthermore, the life expectancy is shortened by the enhanced drag and the satellite prematurely reenters the denser atmosphere. Cases have been documented whereby satellites have gone completely out of control as a result of the strong increase in air density.

Ionospheric Storms. Because of the decreasing attenuation toward shorter wavelengths, the preferred transmission frequency for ionospheric radio communications is set near the maximum plasma frequency of the F region. If this maximum reflection frequency is significantly lowered because of a negative ionospheric storm, the transmitted radio signal is lost. Simply lowering the transmission frequency is often impossible, because the frequency bands in this radio range are completely allocated.

Global navigation satellite systems, like the *Global Positioning System* (GPS), are also affected by ionospheric storms. While passing through the dispersive ionosphere, the GPS signals are slightly delayed. For single-frequency transmissions this results in ranging errors of about 50 m that need to be corrected. The actual magnitude of this error depends on the total electron content along the radio propagation path. The total electron content of the ionosphere, however, is a highly variable quantity during disturbed conditions, thereby rendering such corrections difficult. To make things worse, storms frequently cause rapid fluctuations of the amplitude and phase of the GPS signals (*scintillations*), which may prevent a position-fixing altogether.

Energetic Particles. Energetic particles, whether they are accelerated in a solar flare, at interplanetary shocks, or within the magnetosphere, can inflict considerable damages. For example, communication electronics on board a satellite could be so badly damaged by incident high energy particles that spurious commands are issued and the satellite goes out of control. In the less severe incidents, one observes anomalous switching events and premature aging of electronic components. It is also known that high energy particles can destroy biological components such as cells or chromosomes. If the APOLLO astronauts had been surprised on the moon by a strong particle storm, for example, they would have had little chance of survival in their relatively thin-layered spacesuits. Finally, the precipitation of energetic particles onto the polar upper atmosphere leads to a strong increase in the ionization density of the D region (see Fig. 8.30). This results in stronger absorption of radio waves, which, in the more severe cases, can disrupt all transpolar radiocommunications (*polar blackout*). While the effects of solar flares are usually of short duration, energetic particles accelerated on interplanetary shocks are responsible for longer-lasting disturbances.

In view of these dangers and potential damages, we are all naturally very interested in obtaining the soonest possible knowledge about solar flares and impending geospheric storms. In other words, we want a reliable prediction of *space weather*. Only limited success has been achieved in this endeavor up to the present day.

Exercises

8.1 Wind-induced charge carrier drifts and currents are calculated in Section 8.1.1. Verify these results using the electric dynamo field induced by winds in the upper atmosphere and the conductivities given in Section 7.3.2.

8.2 A discontinuous jump in the Dst index (sudden impulse, SI) is observed following the arrival of an interplanetary shock at the magnetosphere. This increase may be estimated quantitatively as follows

$$\Delta \text{Dst} = K \left(\sqrt{p_{sw2}} - \sqrt{p_{sw1}} \right)$$

where K is a constant ($\simeq 7 \cdot 10^5$ $[\gamma/(\text{J/m}^3)^{1/2}]$) and $p_{sw1,2}$ is the dynamic pressure of the solar wind in front of and behind the shock, respectively. Justify this expression. As an example, consider a shock for which the particle density increases from 6 to 15 cm^{-3} and the solar wind velocity rises from 450 to 500 km/s. What will be the order of magnitude of the Dst increase?

8.3 For a quantitative description of the energy transfer into the magnetospheric ring currrent during a geospheric storm, one frequently uses the following formula

$$P_{rc}[\text{W}] \simeq -4 \cdot 10^{13} \left(\frac{\partial \text{Dst}}{\partial t} + \frac{\text{Dst}}{\tau_{rc}} \right)$$

where P_{rc} is the energy transfer rate, Dst is the Dst index in γ and τ_{rc} is the time constant for energy loss from the ring current in hours. Justify this expression using the relations (4.56) and (5.68).

8.4 A large fraction of the solar wind energy transferred to the magnetosphere during a geospheric storm is dissipated by currents in the polar upper atmosphere. In order to estimate this Joule heating rate, it is assumed to be proportional to the AE index

$$P_J \sim \text{AE}$$

Justify this assumption. Recall for this purpose that the particle precipitation in the polar upper atmosphere observed during geomagnetic activity – as well as the enhancement in conductivity caused by this particle precipitation – can be considered to be proportional to the AE index. Furthermore, to a first approximation, assume that the ratio of the Pedersen conductivity to the Hall conductivity remains constant.

8.5 Determine the current space weather situation and obtain a forecast as reported on the internet (key words: *space weather forecast*).

References

A.P. Mitra, *Ionospheric Effects of Solar Flares*, Reidel Publishing Company, Dordrecht, 1974

S.-I. Akasofu, *Physics of Magnetospheric Substorms*, Reidel Publishing Company, Dordrecht, 1977

J.T. Gosling, The solar flare myth, *J. Geophys. Res.*, *98*, 18 937, 1993

R.D. Elphinstone, J.S. Murphree, and L.L. Cogger, What is a global auroral substorm, *Rev. Geophys.*, *34*, 169, 1996

G.W. Prölss, Magnetic storm associated perturbations of the upper atmosphere, in *Magnetic Storms* (B.T. Tsurutani, W.D. Gonzalez, Y. Kamide, J.K. Arballo, eds.), AGU Monograph No. 98, 227, Washington, 1997

B.T. Tsurutani, W.D. Gonzalez, Y. Kamide, and J.K. Arballo (eds.), *Magnetic Storms*, American Geophysical Union, Washington, 1997

A. Ieda, S. Machida, T. Mukai, Y. Saito, T. Yamamoto, A. Nishida, T. Terasawa, and S. Kokubun, Statistical analysis of the plasmoid evolution with GEOTAIL observations, *J. Geophys. Res.*, *103*, 4453, 1998

S. Kokubun and Y. Kamide (eds.), *Substorms-4*, Kluwer Academic Publishers, Dordrecht, 1998

A. Hanslmeier, *The Sun and Space Weather*, Kluwer Academic Publishers, Dordrecht, 2002

See also the references in the previous chapters and the figure references in Appendix B.

Appendix A
Formulas, Tables and Derivations

A.1 Selected Mathematical Formulas

Differential and Integral Formulas

(1) Derivative of a definite integral

$$\frac{d}{dx}\int_a^x f(t)dt = -\frac{d}{dx}\int_x^a f(t)dt = f(x) \tag{A.1}$$

(2) Definite integral of a derivative

$$\int_a^b \left(\frac{d}{dx}f(x)\right)dx = f(x)\Big|_a^b = f(b) - f(a) \tag{A.2}$$

(3) Derivative of a product

$$\frac{d}{dx}(f(x)g(x)) = g(x)\frac{df(x)}{dx} + f(x)\frac{dg(x)}{dx} \tag{A.3}$$

(4) Integration by parts

$$\int_a^b u\,v'dx = u\,v\Big|_a^b - \int_a^b v\,u'dx \tag{A.4}$$

where the prime superscript denotes the derivative with respect to x, e.g. $v' = dv/dx$

(5) Indefinite integrals

$$\int x\,\sqrt{a^2 - x^2}\,dx = -(a^2 - x^2)^{3/2}/3 \tag{A.5}$$

$$\int \sin^2(cx)dx = \frac{x}{2} - \frac{\sin(cx)\cos(cx)}{2c} = \frac{x}{2} - \frac{\sin(2cx)}{4c} \tag{A.6}$$

$$\int \cos^3 x\,dx = \sin x - \frac{\sin^3 x}{3} \tag{A.7}$$

$$\int x\,e^{ax}\,dx = e^{ax}\left[\frac{x}{a} - \frac{1}{a^2}\right] \tag{A.8}$$

(6) Definite integrals

$$\int_0^\infty e^{-ax^2}\, dx = \frac{\sqrt{\pi}}{2a} \qquad \text{for } a > 0 \qquad (A.9)$$

$$\frac{2}{\sqrt{\pi}} \int_0^x e^{-t^2}\, dt = \mathrm{erf}(x)$$

$$\simeq 1 - \frac{e^{-x^2}}{x\sqrt{\pi}}\left[1 - \frac{1}{2x^2} + \frac{1\cdot 3}{2^2 x^4} - \frac{1\cdot 3\cdot 5}{2^3 x^6} + \cdots\right] (A.10)$$

$$\int_0^\infty x^n e^{-ax^2}\, dx = \begin{cases} \frac{1}{2}a^{-(n+1)/2}((n-1)/2)! & \text{for } n = \text{odd} \\[2mm] \frac{1}{2}a^{-(n+1)/2}((n-1)/2\cdots 3/2\cdot 1/2)\sqrt{\pi} & \text{for } n = \text{even} \end{cases}$$
$$(A.11)$$

(7) Taylor series

$$f(x_0 + h) = f(x_0) + hf'(x_0) + \frac{h^2}{2!}f''(x_0) + \frac{h^3}{3!}f'''(x_0) + \cdots \quad (A.12)$$

The prime again denotes the derivative with respect to x. If h, on a macroscopic scale, is a small quantity Δx, it is sufficient to use a linear extrapolation of the function, for which

$$f(x_0 + \Delta x) \simeq f(x_0) + \Delta x \frac{df}{dx}\bigg|_{x_0} \qquad (A.13)$$

(8) Exponential series

$$e^{\pm x} = 1 \pm x/1! + x^2/2! \pm x^3/3! + \cdots \qquad (A.14)$$

Vector Algebra

In the following A and a are arbitrary scalar quantities, $\vec{A}, \vec{B}$ and $\vec{C}$ are arbitrary vectors, and a hat '^' denotes a unit vector.

(1) Vector product in Cartesian coordinates ($\rightarrow$ scalar)

$$\vec{A}\vec{B} = \vec{B}\vec{A} = A_x B_x + A_y B_y + A_z B_z \qquad (A.15)$$

(2) Cross product in Cartesian coordinates ($\rightarrow$ vector)

$$\vec{A} \times \vec{B} = -\vec{B} \times \vec{A} = \begin{vmatrix} \hat{x} & \hat{y} & \hat{z} \\ A_x & A_y & A_z \\ B_x & B_y & B_z \end{vmatrix} = \hat{x}(A_y B_z - A_z B_y)$$

$$+ \hat{y}(A_z B_x - A_x B_z) + \hat{z}(A_x B_y - A_y B_x) \qquad (A.16)$$

(3) Dyadic product in Cartesian coordinates ($\rightarrow$ tensor)

$$[\vec{A}\vec{B}] = \begin{bmatrix} A_x B_x & A_x B_y & A_x B_z \\ A_y B_x & A_y B_y & A_y B_z \\ A_z B_x & A_z B_y & A_z B_z \end{bmatrix} \tag{A.17}$$

(4) Vector identities

$$\vec{A}(\vec{B} \times \vec{C}) = \vec{B}(\vec{C} \times \vec{A}) = \vec{C}(\vec{A} \times \vec{B}) \tag{A.18}$$

$$\vec{A} \times (\vec{B} \times \vec{C}) = (\vec{C} \times \vec{B}) \times \vec{A} = (\vec{A}\vec{C})\vec{B} - (\vec{A}\vec{B})\vec{C} \tag{A.19}$$

Nabla - Differential Operators

(1) Cartesian Coordinates (x, y, z)

$$\nabla = \hat{x}\frac{\partial}{\partial x} + \hat{y}\frac{\partial}{\partial y} + \hat{z}\frac{\partial}{\partial z} \tag{A.20}$$

Note that the *nabla* (or *del*) operator represents a vector quantity *by definition*, even if this is not indicated by a vector arrow.

$$\nabla A = \text{grad } A = \hat{x}\frac{\partial A}{\partial x} + \hat{y}\frac{\partial A}{\partial y} + \hat{z}\frac{\partial A}{\partial z} \tag{A.21}$$

$$\nabla\vec{A} = \text{div } \vec{A} = \frac{\partial A_x}{\partial x} + \frac{\partial A_y}{\partial y} + \frac{\partial A_z}{\partial z} \tag{A.22}$$

$$\nabla \times \vec{A} = \text{curl } \vec{A} = \hat{x}\left(\frac{\partial A_z}{\partial y} - \frac{\partial A_y}{\partial z}\right) + \hat{y}\left(\frac{\partial A_x}{\partial z} - \frac{\partial A_z}{\partial x}\right)$$
$$+ \hat{z}\left(\frac{\partial A_y}{\partial x} - \frac{\partial A_x}{\partial y}\right) \tag{A.23}$$

As indicated, the nabla operator applied to a scalar quantity yields its gradient, to a vector quantity its divergence and via cross product to a vector quantity its curl.

$$\nabla^2 A = \underline{\Delta} A = \frac{\partial^2 A}{\partial x^2} + \frac{\partial^2 A}{\partial y^2} + \frac{\partial^2 A}{\partial z^2} \tag{A.24}$$

where $\underline{\Delta}$ denotes the Laplace operator (or Laplacian)

$$\underline{\Delta} = \frac{\partial^2}{\partial x^2} + \frac{\partial^2}{\partial y^2} + \frac{\partial^2}{\partial z^2} \tag{A.25}$$

$$\nabla^2 \vec{A} = \underline{\Delta}\vec{A} = \hat{x}\underline{\Delta}A_x + \hat{y}\underline{\Delta}A_y + \hat{z}\underline{\Delta}A_z$$

$$= \hat{x}\left(\frac{\partial^2 A_x}{\partial x^2} + \frac{\partial^2 A_x}{\partial y^2} + \frac{\partial^2 A_x}{\partial z^2}\right) + \hat{y}\left(\frac{\partial^2 A_y}{\partial x^2} + \frac{\partial^2 A_y}{\partial y^2} + \frac{\partial^2 A_y}{\partial z^2}\right)$$

$$+ \hat{z}\left(\frac{\partial^2 A_z}{\partial x^2} + \frac{\partial^2 A_z}{\partial y^2} + \frac{\partial^2 A_z}{\partial z^2}\right) \tag{A.26}$$

(2) Cylindrical coordinates (r, φ, z)

$$\nabla A = \hat{r}\frac{\partial A}{\partial r} + \hat{\varphi}\frac{1}{r}\frac{\partial A}{\partial \varphi} + \hat{z}\frac{\partial A}{\partial z} \tag{A.27}$$

$$\nabla\vec{A} = \frac{1}{r}\frac{\partial(rA_r)}{\partial r} + \frac{1}{r}\frac{\partial A_\varphi}{\partial \varphi} + \frac{\partial A_z}{\partial z} \tag{A.28}$$

(3) Spherical coordinates (r, ϑ, λ), see, for example, Fig. 5.3

$$\nabla A = \hat{r}\frac{\partial A}{\partial r} + \hat{\vartheta}\frac{1}{r}\frac{\partial A}{\partial \vartheta} + \hat{\lambda}\frac{1}{r\sin\vartheta}\frac{\partial A}{\partial \lambda} \tag{A.29}$$

$$\nabla\vec{A} = \frac{1}{r^2}\frac{\partial(r^2 A_r)}{\partial r} + \frac{1}{r\sin\vartheta}\frac{\partial(\sin\vartheta A_\vartheta)}{\partial \vartheta} + \frac{1}{r\sin\vartheta}\frac{\partial A_\lambda}{\partial \lambda} \tag{A.30}$$

Vector Algebra with the Nabla Operator

(1) Identities

$$\nabla(a\vec{A}) = \vec{A}\nabla a + a\nabla\vec{A} \tag{A.31}$$

$$\nabla \times (a\vec{A}) = a\nabla \times \vec{A} + \nabla a \times \vec{A} \tag{A.32}$$

$$\nabla \times (\vec{A} \times \vec{B}) = (\nabla\vec{B})\vec{A} - (\nabla\vec{A})\vec{B} = \vec{A}(\nabla\vec{B}) - \vec{B}(\nabla\vec{A})$$
$$+ (\vec{B}\nabla)\vec{A} - (\vec{A}\nabla)\vec{B} \tag{A.33}$$

$$\vec{A} \times (\nabla \times \vec{B}) = (\nabla\vec{B})\vec{A} - (\vec{A}\nabla)\vec{B}$$
$$= \nabla(\vec{A}\vec{B}) - (\vec{A}\nabla)\vec{B} - (\vec{B}\nabla)\vec{A} - \vec{B} \times (\nabla \times \vec{A}) \tag{A.34}$$

$$(\nabla \times \vec{A}) \times \vec{A} = -\nabla A^2/2 + (\vec{A}\nabla)\vec{A} \tag{A.35}$$

$$\nabla \times (\nabla \times \vec{A}) = \nabla(\nabla\vec{A}) - \underline{\Delta}\vec{A} \tag{A.36}$$

$$\nabla \times (\nabla A) = 0 \tag{A.37}$$

$$\nabla(\nabla \times \vec{A}) = 0 \tag{A.38}$$

(2) Vector quantity $(\vec{B}\nabla)\vec{A}$ in Cartesian coordinates

$$(\vec{B}\nabla)\vec{A} = \hat{x}\left(B_x\frac{\partial A_x}{\partial x} + B_y\frac{\partial A_x}{\partial y} + B_z\frac{\partial A_x}{\partial z}\right)$$
$$+ \hat{y}\left(B_x\frac{\partial A_y}{\partial x} + B_y\frac{\partial A_y}{\partial y} + B_z\frac{\partial A_y}{\partial z}\right)$$
$$+ \hat{z}\left(B_x\frac{\partial A_z}{\partial x} + B_y\frac{\partial A_z}{\partial y} + B_z\frac{\partial A_z}{\partial z}\right) \tag{A.39}$$

(3) Divergence of a Tensor

$$\nabla[T] = [\partial/\partial x \ \partial/\partial y \ \partial/\partial z] \begin{bmatrix} T_{xx} & T_{xy} & T_{xz} \\ T_{yx} & T_{yy} & T_{yz} \\ T_{zx} & T_{zy} & T_{zz} \end{bmatrix}$$

$$= \hat{x}\left(\frac{\partial T_{xx}}{\partial x} + \frac{\partial T_{yx}}{\partial y} + \frac{\partial T_{zx}}{\partial z}\right) + \hat{y}\left(\frac{\partial T_{xy}}{\partial x} + \frac{\partial T_{yy}}{\partial y} + \frac{\partial T_{zy}}{\partial z}\right)$$
$$+ \hat{z}\left(\frac{\partial T_{xz}}{\partial x} + \frac{\partial T_{yz}}{\partial y} + \frac{\partial T_{zz}}{\partial z}\right) \tag{A.40}$$

Integral Theorems

(1) Gauss's law

If V denotes a volume, dV' a differentially small element of this volume, $A(V)$ its surface, $d\vec{A}' = \hat{n}dA'$ a differentially small element of this surface ($\hat{n}$ = surface normal, positive when directed out of the volume) and $\vec{B}$ is an arbitrary vector field, then

$$\int_V \nabla\vec{B}\,dV' = \oint_{A(V)} \vec{B}\,d\vec{A}' \tag{A.41}$$

The flux of a vector quantity through a closed surface is equal to the integral of the divergence of this vector quantity over the volume enclosed by the surface.

(2) Stokes's theorem

If A denotes a surface, $d\vec{A}' = \hat{n}dA'$ a directed, differentially small element of this surface ($\hat{n}$ = surface normal), $L(A)$ the (closed) boundary curve of this surface, $d\vec{l}$ a directed, differentially small element of this curve, and $\vec{B}$ an arbitrary vector field, then

$$\int_A (\nabla \times \vec{B})d\vec{A}' = \oint_{L(A)} \vec{B}\,d\vec{l} \tag{A.42}$$

The line integral of a vector quantity along a closed curve is equal to the surface integral of the normal component of the curl of this vector quantity over the surface enclosed by the curve.

A.2 Physical Parameters of the Earth

Mean radius R_E	6371 km
Polar radius	6357 km
Equatorial radius	6378 km
Mass M_E	$5.974 \cdot 10^{24}$ kg
Mean density	5.515 g/cm^3
Normalized gravitation GM_E	$398.6 \cdot 10^{12}$ m^3 s^{-2}
Earth's gravitational acceleration (sea level)	$9.780(1 + 0.0053 \sin^2 \varphi)$ [m/s^2] φ = geographic latitude [deg]
Ellipsoid coefficient of Earth's gravitational acceleration	$C = 5.267 \cdot 10^{25}$ m^{5}s^{-2}
Escape velocity (sea level)	11.19 km/s
Sidereal rotation period	86 164 s (23^h 56^m 4^s)
Angular velocity Ω_E	$7.292 \cdot 10^{-5}$ rad/s
Centrifugal acceleration at the equator (sea level)	$33.92 \cdot 10^{-3}$ m/s^2
Obliquity of the rotation axis to the ecliptic	$23° \ 26' \ 20''$
Orbital period about the Sun	$31.557 \cdot 10^6$ s = 365.2422 d $= 365^\text{d} \ 5^\text{h} \ 48^\text{m} \ 42.5^\text{s}$
Mean orbital velocity	29.78 km/s
Magnetic Moment $\mathcal{M}_E$	$7.7 \cdot 10^{22}$ A m^2
Age	$4.5 \cdot 10^9$ years

Adapted from Lang, 1992

A.3 Planetary Data

Planet	Mean Distance from Sun [AU]	Orbital eccentricity	Relative radius	Relative mass	Orbital period [years]	Rotation period	Inclination to ecliptic [deg]
Mercury	0.39	0.206	0.38	0.055	0.24	59 d	7.0
Venus	0.72	0.007	0.95	0.82	0.62	243 d (r)	3.4
Earth	1	0.017	1	1	1	24 h	0.0
Mars	1.5	0.093	0.53	0.11	1.9	25 h	1.9
Jupiter	5.2	0.048	11.2	318	12	10 h	1.3
Saturn	9.6	0.056	9.4	95	29	11 h	2.5
Uranus	19.2	0.046	4.0	15	84	17 h (r)	0.8
Neptune	30.1	0.009	3.9	17	164	16 h	1.8
Pluto	39.5	0.249	0.18	0.002	248	6.4 d	17.1

1 AU $\simeq 150 \cdot 10^6$ km, $R_E \simeq 6371$ km, $M_E \simeq 6 \cdot 10^{24}$ kg;
r = retrograde. Adapted from Lang, 1992

A.4 Model Atmosphere

Representative height distribution of thermospheric quantities for a thermopause temperature of 1000 K. Note that H_n denotes the density scale height (not the pressure scale height), and that the units refer to the respective physical quantity, not to its logarithm (adapted from Jacchia, 1977).

Height km	T K	log[N₂] m⁻³	log[O₂] m⁻³	log[O] m⁻³	log[Ar] m⁻³	log[He] m⁻³	log[H] m⁻³	log[n] m⁻³	log[p] Pa	$\mathcal{M}$	H_n km	ρ kg/m³	log ρ kg/m³
90	188.0	19.746	19.170	17.390	17.824	14.573		19.854	−.732	28.91	5.63	3.43E−06	−5.465
92	188.1	19.592	19.009	17.547	17.669	14.418		19.700	−.886	28.85	5.59	2.40E−06	−5.620
94	188.5	19.437	18.843	17.646	17.514	14.263		19.545	−1.040	28.76	5.55	1.67E−06	−5.776
96	189.4	19.281	18.673	17.686	17.359	14.108		19.390	−1.193	28.65	5.53	1.17E−06	−5.933
98	191.0	19.126	18.499	17.687	17.204	13.953		19.235	−1.344	28.52	5.54	8.13E−07	−6.090
100	193.7	18.971	18.323	17.665	17.049	13.798		19.081	−1.492	28.36	5.62	5.67E−07	−6.247
102	197.9	18.820	18.149	17.600	16.837	13.772		18.928	−1.635	28.21	5.62	3.97E−07	−6.401
104	204.3	18.668	17.972	17.543	16.626	13.744		18.777	−1.773	28.02	5.63	2.78E−07	−6.556
106	213.4	18.515	17.792	17.486	16.417	13.713		18.626	−1.905	27.80	5.65	1.95E−07	−6.710
108	225.8	18.364	17.607	17.426	16.212	13.679		18.477	−2.029	27.55	5.72	1.37E−07	−6.862
110	241.7	18.216	17.419	17.360	16.013	13.644		18.332	−2.145	27.28	5.88	9.72E−08	−7.012
115	293.2	17.871	16.982	17.172	15.557	13.555		17.996	−2.397	26.64	6.86	4.39E−08	−7.358
120	350.5	17.579	16.637	16.985	15.172	13.476		17.716	−2.599	26.15	8.26	2.26E−08	−7.646
125	409.8	17.328	16.356	16.816	14.844	13.408		17.480	−2.768	25.73	9.65	1.29E−08	−7.890
130	469.6	17.112	16.117	16.667	14.561	13.349		17.277	−2.911	25.36	11.22	7.97E−09	−8.098
135	526.9	16.924	15.909	16.538	14.314	13.298		17.103	−3.035	25.01	12.91	5.26E−09	−8.279
140	580.0	16.758	15.725	16.425	14.095	13.254		16.951	−3.146	24.68	14.69	3.66E−09	−8.436
145	627.7	16.610	15.561	16.326	13.899	13.217		16.817	−3.245	24.36	16.50	2.66E−09	−8.576
150	669.8	16.476	15.412	16.238	13.720	13.184	11.756	16.698	−3.336	24.06	18.29	1.99E−09	−8.701
155	706.6	16.353	15.276	16.158	13.555	13.156	11.697	16.590	−3.421	23.76	20.02	1.53E−09	−8.814
160	738.5	16.240	15.148	16.085	13.401	13.130	11.646	16.491	−3.501	23.48	21.67	1.21E−09	−8.918
170	790.4	16.032	14.916	15.953	13.118	13.087	11.563	16.314	−3.648	22.93	24.69	7.84E−10	−9.106
180	829.9	15.843	14.703	15.836	12.857	13.050	11.498	16.157	−3.784	22.40	27.34	5.34E−10	−9.273
190	860.4	15.667	14.504	15.729	12.613	13.017	11.446	16.015	−3.911	21.89	29.69	3.76E−10	−9.425
200	884.4	15.501	14.315	15.629	12.381	12.987	11.392	15.883	−4.030	21.40	31.81	2.72E−10	−9.566
210	903.5	15.341	14.134	15.534	12.157	12.960	11.357	15.760	−4.144	20.94	33.71	2.00E−10	−9.699
220	918.8	15.186	13.959	15.442	11.939	12.935	11.327	15.644	−4.253	20.50	35.49	1.50E−10	−9.824
230	931.3	15.036	13.787	15.354	11.727	12.910	11.302	15.533	−4.358	20.08	37.16	1.14E−10	−9.944
240	941.5	14.888	13.620	15.268	11.519	12.887	11.281	15.427	−4.459	19.69	38.74	8.74E−11	−10.058
250	949.9	14.744	13.455	15.183	11.315	12.864	11.262	15.325	−4.557	19.32	40.24	6.79E−11	−10.168

Representative height distribution of thermospheric quantities for a thermopause temperature of 1000 K (first continuation).

Height km	T K	$\log[N_2]$ m^{-3}	$\log[O_2]$ m^{-3}	$\log[O]$ m^{-3}	$\log[Ar]$ m^{-3}	$\log[He]$ m^{-3}	$\log[H]$ m^{-3}	$\log[n]$ m^{-3}	$\log[p]$ Pa	$\mathcal{M}$	H_n km	ρ kg/m^3	$\log\rho$ kg/m^3
260	956.8	14.601	13.293	15.101	11.113	12.843	11.247	15.227	-4.652	18.96	41.68	5.32E-11	-10.274
270	962.6	14.461	13.133	15.019	10.914	12.821	11.233	15.132	-4.744	18.66	43.06	4.20E-11	-10.377
280	967.5	14.322	12.975	14.939	10.717	12.800	11.220	15.040	-4.835	18.37	44.37	3.34E-11	-10.476
290	971.6	14.185	12.818	14.860	10.522	12.780	11.209	14.949	-4.923	18.10	45.63	2.68E-11	-10.573
300	975.1	14.049	12.663	14.781	10.328	12.760	11.199	14.862	-5.009	17.85	46.83	2.16E-11	-10.667
310	978.0	13.914	12.509	14.704	10.136	12.740	11.190	14.776	-5.094	17.62	47.98	1.75E-11	-10.758
320	980.6	13.779	12.356	14.627	9.945	12.720	11.182	14.691	-5.177	17.41	49.07	1.42E-11	-10.848
330	982.7	13.646	12.204	14.550	9.756	12.701	11.174	14.609	-5.259	17.22	50.10	1.16E-11	-10.935
340	984.6	13.514	12.053	14.474	9.567	12.681	11.166	14.527	-5.339	17.04	51.08	9.53E-12	-11.021
350	986.2	13.382	11.902	14.399	9.380	12.662	11.160	14.447	-5.419	16.87	52.00	7.85E-12	-11.105
360	987.6	13.251	11.753	14.323	9.193	12.643	11.153	14.368	-5.497	16.72	52.88	6.48E-12	-11.188
370	988.9	13.121	11.604	14.249	9.007	12.624	11.147	14.291	-5.574	16.58	53.71	5.37E-12	-11.270
380	990.0	12.991	11.456	14.174	8.822	12.605	11.141	14.214	-5.650	16.44	54.48	4.47E-12	-11.350
390	990.9	12.862	11.308	14.100	8.638	12.587	11.135	14.138	-5.726	16.31	55.22	3.72E-12	-11.429
400	991.7	12.733	11.161	14.027	8.455	12.568	11.129	14.064	-5.800	16.18	55.92	3.11E-12	-11.507
420	993.1	12.477	10.869	13.880	8.091	12.531	11.118	13.917	-5.946	15.93	57.20	2.18E-12	-11.661
440	994.2	12.223	10.579	13.735	7.729	12.495	11.108	13.773	-6.089	15.68	58.37	1.54E-12	-11.811
460	995.1	11.971	10.291	13.591	7.369	12.459	11.098	13.633	-6.229	15.42	59.46	1.10E-12	-11.99
480	995.8	11.721	10.005	13.448	7.013	12.423	11.089	13.496	-6.365	15.13	60.50	7.88E-13	-12.103
500	996.4	11.472	9.722	13.306	6.658	12.387	11.079	13.363	-6.498	14.81	61.52	5.68E-13	-12.24
520	996.9	11.225	9.439	13.165	6.306	12.352	11.070	13.234	-6.627	14.44	62.56	4.11E-13	-12.386
540	997.3	10.980	9.159	13.025		12.317	11.061	13.110	-6.752	14.02	63.65	3.00E-13	-12.523
560	997.6	10.736	8.881	12.885		12.282	11.052	12.989	-6.872	13.54	64.82	2.19E-13	-12.659
580	997.9	10.494	8.604	12.747		12.247	11.043	12.874	-6.986	13.00	66.13	1.62E-13	-12.791
600	998.2	10.253	8.329	12.609		12.213	11.034	12.765	-7.096	12.40	67.63	1.20E-13	-12.921
620	998.4	10.013	8.055	12.472		12.178	11.025	12.662	-7.199	11.74	69.35	8.95E-14	-13.048
640	998.5	9.775	7.783	12.336		12.144	11.017	12.565	-7.295	11.04	71.38	6.74E-14	-13.172
660	998.7	9.539	7.513	12.201		12.110	11.008	12.475	-7.386	10.32	73.77	5.11E-14	-13.291
680	998.8	9.303	7.244	12.067		12.077	11.000	12.391	-7.469	9.59	76.63	3.92E-14	-13.407
700	998.9	9.070	6.977	11.933		12.043	10.991	12.314	-7.546	8.86	80.02	3.03E-14	-13.518
720	999.0	8.837	6.712	11.800		12.010	10.983	12.244	-7.617	8.17	84.07	2.38E-14	-13.624
740	999.1	8.606	6.448	11.668		11.977	10.974	12.179	-7.682	7.53	88.87	1.89E-14	-13.724

Representative height distribution of thermospheric quantities for a thermopause temperature of 1000 K (second continuation).

Height km	T K	$\log[N_2]$ m^{-3}	$\log[O_2]$ m^{-3}	$\log[O]$ m^{-3}	$\log[Ar]$ m^{-3}	$\log[He]$ m^{-3}	$\log[H]$ m^{-3}	$\log[n]$ m^{-3}	$\log[p]$ Pa	M	H_n km	ρ kg/m^3	$\log\rho$ kg/m^3
760	999.2	8.376	6.185	11.537		11.944	10.966	12.119	−7.741	6.94	94.52	1.52E−14	−13.819
780	999.3	8.148		11.407		11.911	10.958	12.065	−7.795	6.41	101.12	1.24E−14	−13.908
800	999.3	7.920		11.277		11.879	10.950	12.015	−7.845	5.94	108.75	1.02E−14	−13.991
820	999.4	7.694		11.148		11.847	10.941	11.969	−7.891	5.53	117.43	8.55E−15	−14.068
840	999.4	7.470		11.019		11.815	10.933	11.926	−7.934	5.19	127.16	7.26E−15	−14.139
860	999.5	7.246		10.892		11.783	10.925	11.886	−7.975	4.89	137.89	6.24E−15	−14.205
880	999.5	7.024		10.765		11.751	10.917	11.848	−8.012	4.64	149.50	5.43E−15	−14.265
900	999.6	6.803		10.639		11.719	10.909	11.812	−8.048	4.43	161.80	4.77E−15	−14.321
920	999.6	6.583		10.513		11.688	10.901	11.778	−8.082	4.26	174.56	4.24E−15	−14.373
940	999.6	6.365		10.388		11.657	10.894	11.745	−8.115	4.11	187.53	3.80E−15	−14.421
960	999.6	6.148		10.264		11.626	10.886	11.714	−8.146	3.98	200.42	3.42E−15	−14.466
980	999.7			10.141		11.595	10.878	11.684	−8.176	3.88	213.01	3.11E−15	−14.508
1000	999.7			10.018		11.564	10.870	11.654	−8.206	3.79	225.07	2.84E−15	−14.547
1050	999.7			9.714		11.488	10.851	11.584	−8.276	3.61	252.00	2.30E−15	−14.638
1100	999.8			9.414		11.413	10.832	11.518	−8.342	3.48	273.85	1.90E−15	−14.721
1150	999.8			9.118		11.339	10.813	11.454	−8.406	3.37	291.09	1.59E−15	−14.797
1200	999.8			8.826		11.266	10.795	11.394	−8.466	3.28	304.82	1.35E−15	−14.870
1250	999.9			8.538		11.194	10.777	11.335	−8.525	3.19	316.29	1.15E−15	−14.940
1300	999.9			8.254		11.123	10.759	11.279	−8.581	3.11	326.39	9.82E−16	−15.008
1350	999.9			7.973		11.052	10.741	11.225	−8.635	3.03	335.74	8.45E−16	−15.073
1400	999.9			7.696		10.983	10.724	11.174	−8.686	2.94	344.79	7.29E−16	−15.137
1450	999.9			7.423		10.915	10.706	11.124	−8.736	2.86	353.92	6.32E−16	−15.199
1500	999.9			7.153		10.847	10.689	11.077	−8.783	2.78	363.29	5.50E−16	−15.260
1600	999.9			6.623		10.715	10.656	10.987	−8.873	2.61	383.21	4.20E−16	−15.376
1700	999.9			6.106		10.585	10.623	10.906	−8.954	2.44	405.55	3.26E−16	−15.487
1800	1000.0					10.459	10.592	10.832	−9.028	2.28	430.68	2.57E−16	−15.590
1900	1000.0					10.336	10.561	10.764	−9.096	2.13	459.34	2.05E−16	−15.688
2000	1000.0					10.216	10.530	10.702	−9.158	1.99	491.85	1.66E−16	−15.780
2100	1000.0					10.099	10.501	10.646	−9.214	1.86	528.47	1.37E−16	−15.865
2200	1000.0					9.984	10.472	10.594	−9.266	1.74	569.81	1.14E−16	−15.944
2300	1000.0					9.873	10.444	10.547	−9.313	1.64	615.98	9.61E−17	−16.017
2400	1000.0					9.763	10.416	10.504	−9.356	1.55	667.00	8.22E−17	−16.085
2500	1000.0					9.656	10.390	10.463	−9.397	1.48	723.19	7.12E−17	−16.148

A.5 Diffusion Equation for Gases

Consider a secondary gas in a primary gas (or gas mixture), which is nonuniformly distributed due to some disturbance and has thus formed density gradients. As discussed in Section 2.3.5, the thermal motion of the secondary gas particles leads to a gradual elimination of the density differences and this moderating process is designated as diffusion. In some situations – in particular when describing the vertical density distribution in an atmosphere – it has been shown useful to extend the meaning of this concept and to understand diffusion generally as the motion of a secondary gas through a main gas under the influence of *all* forces acting on the secondary gas. As a result, diffusion then corresponds to the total motion of the secondary gas relative to the primary gas, not only to the particle transport due to density gradients. This generalization leads to a diffusion flux that can be determined directly from the momentum balance equation of the secondary gas. Denoting the secondary gas with the index 1 and the main gas (or the remaining gas mixture) with index 2, we may use Eq. (3.80) and Eq. (A.60) to obtain

$$\rho_1 \frac{D\vec{u}_1}{Dt} = -\nabla p_1 + \eta_{1,2}\underline{\Delta}\vec{u}_1 + \rho_1\vec{g} + \rho_1\nu_{1,2}^* (\vec{u}_2 - \vec{u}_1) + 2\rho_1\vec{u}_1 \times \vec{\Omega}_E$$

Diffusion processes are generally so slow that the inertial, viscosity and Coriolis forces can all be neglected to a very good approximation. In this case, the diffusion flux (the particle flux of the secondary gas relative to the primary gas) is given by

$$\vec{\phi}_{1,2} = n_1 (\vec{u}_1 - \vec{u}_2) = - \frac{1}{m_1\nu_{1,2}^*} (\nabla p_1 - \rho_1\vec{g}) \qquad (A.43)$$

or for the vertical component of interest here

$$(\phi_{1,2})_z = - \frac{1}{m_1\nu_{1,2}^*} \left(\frac{dp_1}{dz} + \rho_1 g \right) \qquad (A.44)$$

This relation is denoted the *diffusion equation* and should not be confused with the time dependent equation (2.73), Fick's second law. It follows from Eq. (A.44) that every density distribution incompatible with the aerostatic equation produces a particle flux that strives to eliminate the existing deviations. This flux ceases only when the individual fluxes induced by pressure gradient and Earth's gravity are of equal magnitude (diffusive equilibrium), in which case we are dealing with a static density distribution in a macroscopic sense.

Differentiating and expanding the result into individual terms, the above diffusion equation may be written as follows

$$(\phi_{1,2})_z = - D_{1,2} \left(\frac{dn_1}{dz} + \frac{n_1}{T_1}\frac{dT_1}{dz} + \frac{n_1}{H_1} \right) \qquad (A.45)$$

where we have made use of the definitions for the diffusion coefficient, $D_{1,2} = kT_1/m_1\nu^*_{1,2}$, and the pressure scale height of the secondary gas. Whereas the first term on the right side describes the diffusion in its strict sense and the third term is the downward flux due to the Earth's gravitational acceleration, the second term corresponds to a diffusion flux produced by a temperature difference

$$\phi^T_z = -\ D_{1,2}\ \frac{n_1}{T_1}\ \frac{dT_1}{dz} \tag{A.46}$$

In order to understand the gas kinetic meaning of this particle flux, we refer once more to the scenario in Fig. 2.23, except that the density gradient is now replaced by a temperature gradient. The particle exchange between volumes 1 and 2 now proceeds with different velocities because $\bar{c}$ is a function of temperature

$$d\phi_1 = \frac{1}{6}\ \frac{n}{l}\ \bar{c}_1\ dz, \qquad d\phi_2 = \frac{1}{6}\ \frac{n}{l}\ \bar{c}_2\ dz$$

Expanding $\bar{c}_2$ in a Taylor series, subtracting the above particle fluxes to obtain the total net flux, and summing over the various net contributions, one obtains

$$\phi^T_z = -\ \frac{1}{6}\ n\ l\ \frac{d\bar{c}}{dz} \tag{A.47}$$

This expression, upon applying the relation $\bar{c} = \sqrt{8\ k\ T/\pi\ m}$, may be transformed into a form similar to Eq. (A.46).

Note that the exact form of the diffusion equation depends on the particular momentum balance equation used. Should this be rearranged with additional force terms or a more precise expression for the frictional force, it is clear that this will be reflected in a more complicated form of the diffusion equation.

A.6 Derivation of the Momentum Balance Equation from the Boltzmann Equation

We first consider the inertial force. Its connection to the distribution function becomes clear from the following manipulations

$$\begin{aligned}(\vec{F}^*)_{inertial} &= \frac{\partial(\rho\vec{u})}{\partial t} = \frac{\partial(\rho\langle\vec{v}\rangle)}{\partial t} = \frac{\partial}{\partial t}\left\{\rho\int_{VS}\vec{v}\ \frac{dN}{\delta N}\right\}\\ &= \frac{\partial}{\partial t}\int_{VS}(m\vec{v})\ f\,d^3v = \int_{VS}(m\vec{v})\ \frac{\partial f}{\partial t}\,d^3v\end{aligned} \tag{A.48}$$

The mean velocity (indicated by angular brackets) is calculated here by weighting each value of the particle velocity $\vec{v}$ with its relative occurrence probability $dN/\delta N$ and then integrating over the entire velocity space VS, see also Eq. (2.93). The last two expressions are based on the Eqs. (2.85) and

(2.87), as well as the fact that $\vec{v}$ and t are independent variables in phase space. The relation (A.48) shows how the momentum balance equation can be derived from the Boltzmann equation (2.88), namely multiply it with $m\vec{v}$ and then integrate over velocity space. This yields

$$\frac{\partial(\rho\vec{u})}{\partial t} = \int_{VS} (m\vec{v}) \frac{\partial f}{\partial t} \, \mathrm{d}^3v = -\int_{VS} (m\vec{v}) \, (\vec{v} \, \nabla f) \, \mathrm{d}^3v$$

$$-\int_{VS} (m\vec{v}) \, (\vec{a} \, \nabla_v f) \, \mathrm{d}^3v + \int_{VS} (m\vec{v}) \left(\frac{\delta f}{\delta t}\right)_{collisions} \mathrm{d}^3v \quad (A.49)$$

This equation can be rearranged in a somewhat more convenient form, the details of which we will skip and refer instead to the literature on gas kinetics and plasma physics given at the end of Chapters 2 and 5. It suffices to note that the first term on the right side of Eq. (A.49) is broken down into three components

$$\int_{VS} (m\vec{v}) \, (\vec{v} \, \nabla f) \, \mathrm{d}^3v = \nabla[\phi^{I(v)}] = \nabla(\rho[\vec{u}\vec{u}]) + \nabla[P]$$

$$= \nabla(\rho[\vec{u}\vec{u}]) + \nabla p + \nabla[\tau] \quad (A.50)$$

Here, $[\phi^{I(v)}]$ is the *momentum flux tensor*

$$[\phi^{I(v)}] = \begin{bmatrix} \phi_{x,x}^{I(v)} & \phi_{x,y}^{I(v)} & \phi_{x,z}^{I(v)} \\ \phi_{y,x}^{I(v)} & \phi_{y,y}^{I(v)} & \phi_{y,z}^{I(v)} \\ \phi_{z,x}^{I(v)} & \phi_{z,y}^{I(v)} & \phi_{z,z}^{I(v)} \end{bmatrix} \quad (A.51)$$

where $\phi_{s,t}^{I(v)}$ stands for the various components of the momentum flux, each of which is dependent on v. The flow gradient force is denoted by $\nabla(\rho[\vec{u}\vec{u}])$, where $[\vec{u}\vec{u}]$ is the dyadic product of the flow velocity

$$[\vec{u}\vec{u}] = \begin{bmatrix} u_x u_x & u_x u_y & u_x u_z \\ u_y u_x & u_y u_y & u_y u_z \\ u_z u_x & u_z u_y & u_z u_z \end{bmatrix} \quad (A.52)$$

If an airflow only exits in the x-direction ($u_y, u_z = 0$), then this term can be simplified using the relation (A.40) to the form given in Eq. (3.69). $[P]$ denotes the *pressure tensor* with the various thermal momentum flux components $\phi_{s,t}^{I(c)}$

$$[P] = \begin{bmatrix} \phi_{x,x}^{I(c)} & \phi_{x,y}^{I(c)} & \phi_{x,z}^{I(c)} \\ \phi_{y,x}^{I(c)} & \phi_{y,y}^{I(c)} & \phi_{y,z}^{I(c)} \\ \phi_{z,x}^{I(c)} & \phi_{z,y}^{I(c)} & \phi_{z,z}^{I(c)} \end{bmatrix} = \begin{bmatrix} \rho\langle c_x c_x \rangle & \rho\langle c_x c_y \rangle & \rho\langle c_x c_z \rangle \\ \rho\langle c_y c_x \rangle & \rho\langle c_y c_y \rangle & \rho\langle c_y c_z \rangle \\ \rho\langle c_z c_x \rangle & \rho\langle c_z c_y \rangle & \rho\langle c_z c_z \rangle \end{bmatrix} \quad (A.53)$$

This may be decomposed into a pressure tensor and a viscosity tensor

$$
[P] = \begin{bmatrix} p & 0 & 0 \\ 0 & p & 0 \\ 0 & 0 & p \end{bmatrix} + \begin{bmatrix} \phi_{x,x}^{I(c)} - p & \phi_{x,y}^{I(c)} & \phi_{x,z}^{I(c)} \\ \phi_{y,x}^{I(c)} & \phi_{y,y}^{I(c)} - p & \phi_{y,z}^{I(c)} \\ \phi_{z,x}^{I(c)} & \phi_{z,y}^{I(c)} & \phi_{z,z}^{I(c)} - p \end{bmatrix} = p[\mathrm{I}] + [\tau] \qquad (A.54)
$$

where p is the thermodynamic pressure , $[\mathrm{I}]$ the unit tensor and $[\tau]$ the *viscosity* or *stress tensor*. Considering further that the second integral on the right side of Eq. (A.49) can be reduced to the simple form $\rho\vec{a}$, this equation may now be written in the following way

$$
\frac{\partial(\rho\vec{u})}{\partial t} = -\nabla(\rho\,[\vec{u}\vec{u}]) - \nabla p - \nabla[\tau] + \rho\vec{a} + \int_{VS} (m\vec{v}) \left(\frac{\delta f}{\delta t}\right)_{collisions} \mathrm{d}^3 v \quad (A.55)
$$

where the divergence of a tensor is calculated with the help of Eq. (A.40). Evidently, the fourth term on the right side of Eq. (A.55) describes the external forces acting on the given gas volume and $\vec{a}$ corresponds to the force per unit mass, i.e. the acceleration. Setting $\vec{a} = \vec{g}$, for example, yields the gravitational force acting on the gas. Finally, the still undeveloped fifth term summarizes the frictional forces acting on the gas volume.

Note that the previous manipulations contain no approximations and the momentum balance equation (A.55) thus commands the same degree of accuracy as the Boltzmann equation itself. Loss of accuracy will first occur when one explicitly calculates the viscosity and friction terms, because the distribution function f is needed for this task. The latter, however, as the solution to the Boltzmann equation, is only approximately known. Let us first assume that the distribution function corresponds to a locally valid Maxwell distribution

$$
f \simeq f_M = n(\vec{r}) \left(\frac{m}{2\pi kT(\vec{r})}\right)^{3/2} \mathrm{e}^{-m(\vec{v}-\vec{u}(\vec{r}))^2/2kT(\vec{r})} \qquad (A.56)
$$

Since this function is symmetrical with respect to the velocity distribution, all mean values of the velocity products $\langle c_s c_t \rangle$ for $s \neq t$ vanish and all diagonal elements of $[P]$ are identical and equal to the pressure p. The pressure is therefore isotropic ($p = \phi_{x,x}^{I(c)} = \phi_{y,y}^{I(c)} = \phi_{z,z}^{I(c)}$) and the viscosity tensor vanishes. Accordingly, a distribution function displaying deviations from a Maxwell distribution is necessary in order to describe viscosity effects at all. We use the following approach

$$
f \simeq f_M(1 + \varepsilon(\vec{r}, \vec{v}, t)) \qquad (A.57)
$$

where the absolute value of the perturbation function ε is small compared with unity. Inserting this modified distribution function into the Boltzmann equation enables an approximate determination of ε and thus f. To a first approximation, the viscosity tensor in this case assumes the following form

$$
[\tau] \simeq -\eta\left([\nabla\vec{u}] + [\nabla\vec{u}]^T - (2/3)(\nabla\vec{u})[\mathrm{I}]\right) \qquad (A.58)
$$

where $[\nabla \vec{u}]^T$ is the transpose of the dyad $[\nabla \vec{u}]$ (rows and columns are interchanged) and [I] is again the unit tensor. Using the value $\eta = f(T)$ given in Eq. (3.72), one obtains the following expression for the viscosity force

$$(\vec{F}^*)_{viscosity} = \nabla[\tau]$$

$$= \eta \, \Delta \vec{u} + (\eta/3) \, \nabla(\nabla \vec{u}) + \frac{\eta}{2T} \nabla T \left\{ [\nabla \vec{u}] + [\nabla \vec{u}]^T - \frac{2}{3}(\nabla \vec{u})[\text{I}] \right\} \quad (A.59)$$

where Δ is again the Laplacian, see Eq. (A.25). One frequently neglects the second and third terms on the right side of this equation, yielding the approximation

$$(\vec{F}^*)_{viscosity} \simeq \eta \, \Delta \, \vec{u} \quad (A.60)$$

An alternative type of simplification is obtained for $u_z \ll u_x$, u_y and $H_z \ll H_x$, H_y, where H is the spatial scale length describing the distances over which the flow velocities in the given directions may vary significantly. Under these conditions the viscosity force (A.59) is reduced to the form

$$(\vec{F}^*)_{viscosity} \simeq \eta \, \frac{\partial^2 \vec{u}_h}{\partial z^2} + \frac{\eta}{2T} \frac{\partial T}{\partial z} \frac{\partial \vec{u}_h}{\partial z} \quad (A.61)$$

where $\vec{u}_h$ denotes the horizontal velocity. The ratio of these two terms is of the order of

$$\frac{\eta(\partial^2 u_h/\partial z^2)}{(\eta/2T)(\partial T/\partial z)(\partial u_h/\partial z)} \simeq \frac{2(H_T)_z}{(H_{u_h})_z} \quad (A.62)$$

The scale height of the vertical temperature variation $(H_T)_z$ is usually larger than that of the horizontal wind variation in the vertical direction $(H_{u_h})_z$, so the second term on the right side of Eq. (A.61) can be frequently neglected and the viscosity force assumes the form given in Eq. (3.80).

The frictional force term only assumes the simple form given in Eq. (3.80) under certain conditions. This form is valid, for example, when the repulsive potential of the interacting particles varies with the fourth power of the distance, as is the case for collisions between neutral gas particles and ions. For all other interaction potentials the frictional force depends on the given distribution function. For example, assuming that the distribution function of the interacting gases can be sufficiently well described by a locally valid Maxwell distribution, then the collision term also reduces to the form given in Eq. (3.80), but only if the relative velocities of the participating gases are not too large. This is usually (but not always) the case for the flow velocities observed in the upper atmosphere.

A.7 Energy Balance Equation of an Adiabatic Gas Flow

Similar to the derivation of the momentum balance equation in Section 3.4.2, we consider the flow of various gases into and out of a stationary volume element. This time we are interested in the temporal changes in the kinetic energy of a gas flowing through this volume. Contributing to the kinetic energy are both the flow motion and the thermal motion of the gas particles. Because the energy balance equation can assume very complicated forms (it is already quite complicated for a *stationary* gas, see Eq. (3.52)), and because we should work with the simplest possible relations for an analytic treatment of dynamic processes, some rather drastic simplifications are made right at the beginning:

1. Production and loss rates of the given gas species are assumed to be negligibly small. Energy in the volume is thus neither gained nor lost by the creation or destruction of particles. This also implies that heat gain or loss by latent energy (e.g. in the form of recombination energy) remains unconsidered. This assumption is usually justified for the neutral gases of the upper atmosphere.
2. Radiation-induced gain or loss of heat also remains unconsidered, i.e. the terms q^W and l^W in the heat balance equation are set equal to zero. This severe restriction, of course, is only acceptable if we are interested in the behavior of an existing flow, not its origin.
3. Only those flows are considered for which, at least to a first approximation, the molecular heat conduction can be neglected.
4. Similarly, energy transport via flow shears (i.e. nonlinear gradients in the flow velocity perpendicular to the flow direction, compare here the derivation of the viscosity force in Section 3.4.2) are not considered. These viscous heating effects are thus neglected.
5. No interaction should occur between the gas under consideration and other gas species. This implies that all gas species have the same flow velocity so that no frictional heat is created, and all gases have the same temperature so that no heat exchange occurs.
6. Gravity is the only external force considered. Coriolis forces play no role, because their acceleration is always perpendicular to the flow direction and hence no work is performed.

Under these conditions the energy balance equation can be written as follows

$$\frac{\partial E_{kin}^*}{\partial t} = \frac{\partial (n(mu^2/2) + fp/2)}{\partial t} = -\nabla \sum_i \vec{\phi}_i^E - \nabla(p\vec{u}) + n\,m\,\vec{g}\,\vec{u}$$

(A.63)

where E_{kin}^* denotes the total kinetic energy of the particles per unit volume, with $n(mu^2/2)$ representing the flow energy component and $n(f(kT/2)) = fp/2$ the thermal energy component. The transport term on the right side of the equation, $-\nabla \sum_i \vec{\phi}_i^E$, summarizes the contributions of the various fluxes to the energy balance of the given gas volume, whereby both flow energy and thermal energy transport are considered. Note the formal analogy with the transport term in the continuity and heat balance equations. With the

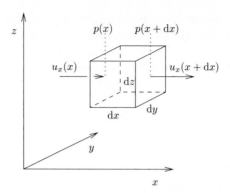

Fig. A.1. Deriving the energy gain or loss of a gas volume from work

particle flux $n\vec{u}$ and the flow energy of a gas particle $mu^2/2$, the flux of the flow energy may be written as

$$\vec{\phi}_{convection}^{flow\ energy} = n\vec{u}\left(\frac{mu^2}{2}\right) \tag{A.64}$$

Molecular heat conduction is neglected, so only the convective part of the heat transport must be considered. Since each gas particle carries an amount of thermal energy $f(kT/2)$, the convective heat flux becomes

$$\vec{\phi}_{convection}^{heat} = n\vec{u}\left(f\,\frac{kT}{2}\right) = f\,\frac{p\vec{u}}{2} \tag{A.65}$$

The second term on the right side of Eq. (A.63), $-\nabla(p\vec{u})$, quantifies the gain or loss of energy in the volume resulting from work. The form of this term can be derived with the help of Fig. A.1. Consider a one-dimensional flow in the x-direction. The external environment evidently performs work when the gas is forced into the volume at the position x against the internal pressure $p(x)$. The work introduced into the volume in the form of internal energy per time interval dt is given by

$$\left(\frac{\mathrm{d}W}{\mathrm{d}t}\right)^{+} = (F_p\mathrm{d}x)/\mathrm{d}t = p(x)\,\mathrm{d}y\,\mathrm{d}z\,u_x(x)$$

At the same time the gas in the volume performs work (this time at the expense of the internal energy), when it forces gas particles out of the volume at the position $x + \mathrm{d}x$ against the external pressure

$$\left(\frac{\mathrm{d}W}{\mathrm{d}t}\right)^{-} = p(x + \mathrm{d}x)\,\mathrm{d}y\,\mathrm{d}z\,u_x(x + \mathrm{d}x)$$

$$\simeq p(x)\,\mathrm{d}y\,\mathrm{d}z\,u_x(x) + \left(\frac{\mathrm{d}(pu_x)}{\mathrm{d}x}\,\mathrm{d}x\right)\mathrm{d}y\,\mathrm{d}z$$

The net gain of energy of the volume element thus becomes

$$\left(\frac{dW}{dt}\right) \simeq \left(\frac{dW}{dt}\right)^{+} - \left(\frac{dW}{dt}\right)^{-} = -\frac{d(pu_x)}{dx}\,dV$$

or, when extending to three dimensions and normalizing to the size of the volume

$$\left(\frac{dW}{dtdV}\right) = -\operatorname{div}(p\vec{u}) \tag{A.66}$$

The form of the last term on the right side of Eq. (A.63) is again immediately understandable. Every gas particle subject to the Earth's gravitational acceleration $\vec{g}$ and moving along the elemental path length $d\vec{s}$ gains an amount of energy given by $(m\vec{g})d\vec{s}$. Normalizing this to the given time interval and number of particles per unit volume, one obtains the final term in Eq. (A.63).

Using the expressions derived above for the energy fluxes, the energy balance equation takes the following form

$$\frac{\partial(\rho u^2/2 + fp/2)}{\partial t} + \nabla\left[\frac{\rho u^2 \vec{u}}{2} + \left(\frac{f}{2}+1\right)p\vec{u}\right] = \rho\vec{g}\vec{u} \tag{A.67}$$

An alternative form of this equation is obtained by differentiating the first and third terms on the left side and appropriately combining the results

$$\frac{\partial(\rho u^2/2)}{\partial t} + \nabla\left(\frac{\rho u^2 \vec{u}}{2}\right) = \frac{mu^2}{2}\left(\frac{\partial n}{\partial t} + \nabla(n\vec{u})\right) + \rho\vec{u}\left(\frac{\partial\vec{u}}{\partial t} + (\vec{u}\nabla)\vec{u}\right)$$

For the case considered here with no particle production or loss (assumption 1), the expression in large parentheses in the first term on the right side is zero (continuity equation!). Furthermore, the expression in large parentheses in the second term on the right side corresponds exactly to the convective derivative of $\vec{u}$. Using the momentum balance equation (3.80) for a nonrotating coordinate system ($2\rho\vec{u} \times \vec{\Omega}_E = 0$) and neglecting viscosity and frictional forces (assumptions 2 and 5), this term may be written as follows

$$\rho\vec{u}\left(\frac{\partial\vec{u}}{\partial t} + (\vec{u}\nabla)\vec{u}\right) = \rho\vec{u}\frac{D\vec{u}}{Dt} = -\vec{u}\nabla p + \rho\vec{u}\vec{g}$$

With these manipulations the energy balance equation assumes the following simple form

$$\frac{f}{2}\frac{\partial p}{\partial t} + \left(\frac{f}{2}+1\right)\nabla(p\vec{u}) - \vec{u}\,\nabla p = 0 \tag{A.68}$$

or

$$\frac{f}{2}\frac{\partial p}{\partial t} + \left(\frac{f}{2}+1\right)p\nabla\vec{u} + \frac{f}{2}\vec{u}\,\nabla p = 0 \tag{A.69}$$

Dividing by $f/2$ and introducing the adiabatic exponent $\gamma = 1+2/f$, collecting the first and third terms together to form the convective derivative Dp/Dt

and applying the relation $Dn/Dt + n\nabla \vec{u} = 0$ (which follows immediately from the continuity equation), one obtains

$$\frac{Dp}{Dt} - \left(\frac{\gamma p}{n}\right) \frac{Dn}{Dt} = 0 \tag{A.70}$$

Finally, multiplication with Dt, separation of variables and subsequent integration yields

$$\frac{p}{p_0} = \left(\frac{n}{n_0}\right)^{\gamma}$$

This, however, corresponds to the adiabatic relation $p\rho^{-\gamma} = $ constant derived in a different way in Section 2.1.2, see Eq. (2.36), which is thus nothing but an extremely simplified form of the energy balance equation.

An alternative derivation of the energy balance equation – as well as all other balance equations – begins with the Boltzmann equation. Multiplying the latter with $mv^2/2$ and integrating over velocity space, the first term of this equation assumes the following form

$$\int_{VS} \frac{mv^2}{2} \frac{\partial f}{\partial t} \, d^3 v = \frac{m}{2} \frac{\partial}{\partial t} \int_{VS} v^2 f \, d^3 v = \frac{m}{2} \frac{\partial}{\partial t} (\overline{nv^2})$$

$$= \frac{\partial}{\partial t} \left(\frac{\rho u^2}{2} + \frac{\overline{\rho c^2}}{2}\right) = \frac{\partial}{\partial t} \left(\frac{\rho u^2}{2} + \frac{fp}{2}\right) = \frac{\partial}{\partial t} E^*_{kin} \tag{A.71}$$

Similar to the derivation of the inertial force term of the momentum balance equation in Appendix A.6, we first interchange the order of integration and differentiation because t and $\vec{v}$ are independent variables in the phase space representation. Integration of the square of the velocity multiplied by the distribution function over velocity space corresponds again to the calculation of a mean value as already described in connection with Eq. (2.93) and Eq. (A.48). In the third step above we have broken the velocity down into its bulk and random components and applied the relation $\langle uc \rangle = u \langle c \rangle = 0$. The final expression is based on the definition of the pressure according to Eq. (2.23) and assuming three degrees of freedom ($f = 3$). Only the translational energy, not the energy associated with the rotational motion of the gas particles or that stored in other degrees of freedom, is included in Eq. (A.71). The remaining terms of the Boltzmann equation, multiplied with $mv^2/2$ and integrated over velocity space, represent the other contributions to the energy balance equation, as shown in the greatly abbreviated form on the right side of Eq. (A.63).

From a casual comparison, one would probably not appreciate that the relations (3.52) and (A.63) are special forms of one and the same energy balance equation. It is thus worthwhile to point out the essential differences of both approaches. One-dimensional, quasi-static conditions are considered to hold for the heat balance equation (3.52). Horizontal dynamics of the thermosphere are ignored, thereby excluding the possibility of converting heat into the kinetic energy of a horizontal flow and thus transporting this heat or the flow kinetic energy itself from one location on Earth to another. The fact that Eq. (3.52) correctly describes the observations, at least to within the correct order of magnitude, indicates that horizontal dynamics is important, but not dominant. The importance of horizontal transport lies in its decisive influence on the amplitude, and particularly the phase, of temperature and density oscillations. The issue of the origin of the flow considered in the energy balance equation (A.63), however, remains completely open. A certain amount of flow energy and heat is available from the beginning and the total budget remains unaltered by energy gains or losses. This approach evidently provides a sufficiently good description of the behavior of a given flow in many situations. In order to simulate the complete behavior of a neutral gas (or a plasma) in the presence of heat sources and sinks, however, one must return to a more general form of the energy balance equation that accounts for all the terms contained in Eq. (3.52) and Eq. (A.63) and also any other previously neglected effects. Such an equation, of course, can only be solved numerically and with a great amount of effort.

A.8 Bernoulli Equation

Consider a one-dimensional stationary gas flow in the x-direction. Assuming that viscosity, friction, Coriolis and external forces may all be neglected, the momentum balance equation is reduced to the simple form (see Eq. (3.79))

$$\rho u \frac{\mathrm{d}u}{\mathrm{d}x} + \frac{\mathrm{d}p}{\mathrm{d}x} = 0$$

from which one also obtains

$$\frac{1}{2} \frac{\mathrm{d}u^2}{\mathrm{d}x} + \frac{1}{\rho} \frac{\mathrm{d}p}{\mathrm{d}x} = 0 \tag{A.72}$$

For simplicity, we have suppressed the directional indexing of the flow velocity, i.e. $u = u_x$. If the gas flow is also considered to be incompressible to a first approximation ($\rho \simeq$ constant), we obtain

$$\frac{\mathrm{d}}{\mathrm{d}x} \left(\frac{u^2}{2} + \frac{p}{\rho} \right) = 0$$

or

$$\frac{u^2}{2} + \frac{p}{\rho} = \text{constant} \tag{A.73}$$

This is the *Bernoulli equation* for *incompressible* gas flows.

It is usually more realistic to assume an adiabatic behavior for the gas flow. With $\rho = \text{constant} \cdot p^{1/\gamma}$, the second term of Eq. (A.72) may be transformed into the following differential quotient

$$\frac{1}{\rho}\frac{\mathrm{d}p}{\mathrm{d}x} = \frac{\gamma}{\gamma - 1}\frac{\mathrm{d}(p/\rho)}{\mathrm{d}x} \qquad (A.74)$$

This enables one to write the entire equation of motion as a differential quotient

$$\frac{\mathrm{d}}{\mathrm{d}x}\left(\frac{u^2}{2} + \frac{\gamma}{\gamma - 1}\frac{p}{\rho}\right) = 0 \qquad (A.75)$$

from which one obtains the Bernoulli equation for *adiabatic* gas flows

$$\frac{u^2}{2} + \frac{\gamma}{\gamma - 1}\frac{p}{\rho} = \text{constant} \qquad (A.76)$$

This equation also holds for three-dimensional adiabatic flows, as long as they are laminar, i.e. free of turbulence, $\nabla \times \vec{u} = 0$. This follows from the vector identity $(\vec{u}\nabla)\vec{u} = \nabla u^2/2 - \vec{u} \times (\nabla \times \vec{u})$, see Eq. (A.34). An alternative form of the Bernoulli equation is obtained upon dividing out the particle number density in the ratio p/ρ. One then obtains the following relation between the velocity and the temperature in an adiabatic flow

$$\frac{m}{2}u^2 + \frac{k\gamma}{\gamma - 1}T = \text{constant} \qquad (A.77)$$

A.9 Rankine-Hugoniot Equations

As shown in the measurements of Fig. 6.31, all state parameters of a gas flow are changed upon passing through a shock wave. A quantitative estimate of these changes may be derived from the conservation laws for mass, momentum and energy. It can be assumed that local thermodynamic equilibrium holds at a distance sufficiently far from the shock (larger than a few mean free paths), and that the previously used equations of gas dynamics are thus valid.

Consider a one-dimensional stationary neutral gas flow in the x-direction, which passes through a shock front oriented perpendicular to the x-direction; see Fig. A.2. If the gas has no sources or sinks, the density balance equation assumes the following simple form

$$\frac{\mathrm{d}(nu)}{\mathrm{d}x} = 0 \qquad (A.78)$$

where indices for the velocity directions have been suppressed ($u = u_x$). We further assume that the viscosity, frictional, Coriolis and external forces can all be neglected. In this case the momentum balance equation (3.79) is reduced to the simple form

$$m\,n\,u\,\frac{\mathrm{d}u}{\mathrm{d}x} + \frac{\mathrm{d}p}{\mathrm{d}x} = 0 \qquad (A.79)$$

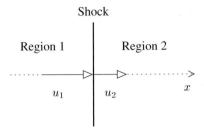

Fig. A.2. Deriving the jump conditions across a perpendicular shock front

Finally, under the conditions given in Appendix A.7, the stationary energy balance equation may be written as follows, see Eq. (A.68)

$$\frac{\gamma}{\gamma-1}\frac{\mathrm{d}(pu)}{\mathrm{d}x} - u\frac{\mathrm{d}p}{\mathrm{d}x} = 0 \tag{A.80}$$

In order to determine the changes in the state parameters caused by the shock, we integrate the three equations (A.78) to (A.80) in the x-direction from a point in the upstream (pre-shock) region 1, through the shock front, and on to a point in the downstream (post-shock) region 2. This is clearly possible – the variations of the state parameters may well be discontinuous, but they will remain finite everywhere. It follows immediately from the density balance equation that

$$\int_{(nu)_1}^{(nu)_2}\frac{\mathrm{d}(nu)}{\mathrm{d}x}\,\mathrm{d}x = n_2 u_2 - n_1 u_1 = 0$$

or

$$n_2 u_2 = n_1 u_1 \tag{A.81}$$

Appending the term $m\,u\,\mathrm{d}(nu)/\mathrm{d}x = 0$ to the momentum balance equation, it may be written in the following form

$$m\frac{\mathrm{d}(nu^2)}{\mathrm{d}x} + \frac{\mathrm{d}p}{\mathrm{d}x} = 0$$

This can be easily integrated and one obtains

$$m\,n_2\,u_2^2 + p_2 = m\,n_1\,u_1^2 + p_1 \tag{A.82}$$

The energy balance equation may also be transformed into the form of a differential quotient

$$\frac{\gamma}{\gamma-1}\frac{\mathrm{d}(u\,p)}{\mathrm{d}x} - u\frac{\mathrm{d}p}{\mathrm{d}x} = \frac{\gamma}{\gamma-1}\frac{\mathrm{d}(u\,p)}{\mathrm{d}x} + m\,n\,u^2\frac{\mathrm{d}u}{\mathrm{d}x}$$

$$= \frac{\gamma}{\gamma-1}\frac{\mathrm{d}(u\,p)}{\mathrm{d}x} + \frac{1}{2}m\frac{\mathrm{d}(nu^3)}{\mathrm{d}x} = 0$$

where we have made use of Eq. (A.79) in the first line and Eq. (A.78) in the second line. Integrating, we obtain the relation

$$\frac{\gamma}{\gamma - 1} \, u_2 \, p_2 + \frac{1}{2} \, m \, n_2 \, u_2^3 \;=\; \frac{\gamma}{\gamma - 1} \, u_1 \, p_1 + \frac{1}{2} \, m \, n_1 \, u_1^3 \tag{A.83}$$

In the following, we want to describe the changes in the state parameters as a function of the Mach numbers $M = u/v_S$ ($v_S = \sqrt{\gamma p/nm}$) in the upstream and downstream regions. Inserting these in Eqs. (A.82) and (A.83), we obtain

$$p_2 \, (1 + \gamma \, M_2^2) \;=\; p_1 \, (1 + \gamma \, M_1^2) \tag{A.84}$$

and

$$u_2 p_2 \left(1 + \frac{(\gamma - 1)}{2} \, M_2^2 \right) \;=\; u_1 p_1 \left(1 + \frac{(\gamma - 1)}{2} \, M_1^2 \right) \tag{A.85}$$

The pressure change across the shock follows directly from Eq. (A.84)

$$\frac{p_2}{p_1} \;=\; \frac{1 + \gamma M_1^2}{1 + \gamma M_2^2} \tag{A.86}$$

Using Eq. (A.81) to replace the velocity u_2 in Eq. (A.85) with the velocity u_1, we obtain the corresponding temperature jump across the shock

$$\frac{u_2 p_2}{u_1 p_1} \;=\; \frac{u_1 n_1 p_2}{u_1 n_2 p_1} \;=\; \frac{T_2}{T_1} \;=\; \frac{2 + (\gamma - 1) \, M_1^2}{2 + (\gamma - 1) \, M_2^2} \tag{A.87}$$

From the definition of the Mach number it follows that the velocity is proportional to the Mach number and to the square root of the temperature, $u \sim M\sqrt{T}$. The jump conditions for the velocity and the density are therefore given by

$$\frac{u_2}{u_1} \;=\; \left(\frac{n_2}{n_1} \right)^{-1} \;=\; \frac{M_2}{M_1} \sqrt{\frac{T_2}{T_1}} \;=\; \frac{M_2}{M_1} \sqrt{\frac{2 + (\gamma - 1) \, M_1^2}{2 + (\gamma - 1) \, M_2^2}} \tag{A.88}$$

In order to derive explicit expressions for the shock-induced changes in the state parameters, we need to know the relation between the two Mach numbers M_1 and M_2. Using Eqs. (A.86) and (A.87), we first determine an alternate expression for the ratio of the velocities

$$\frac{u_2}{u_1} \;=\; \frac{p_1}{p_2} \frac{T_2}{T_1} \;=\; \frac{1 + \gamma M_2^2}{1 + \gamma M_1^2} \frac{2 + (\gamma - 1) \, M_1^2}{2 + (\gamma - 1) \, M_2^2} \tag{A.89}$$

Equating the two relations (A.88) and (A.89), we obtain

$$\frac{1 + \gamma M_2^2}{1 + \gamma M_1^2} \;=\; \frac{M_2}{M_1} \sqrt{\frac{2 + (\gamma - 1) \, M_2^2}{2 + (\gamma - 1) \, M_1^2}}$$

As may be shown by substitution and a series of manipulations, a nontrivial solution of this equation is

$$M_2 = \sqrt{\frac{2 + (\gamma - 1) \, M_1^2}{2\gamma M_1^2 - (\gamma - 1)}} \tag{A.90}$$

In combination with Eqs. (A.86), (A.87) and (A.88), this expression leads to the Rankine-Hugoniot relations for perpendicular shocks introduced in Section 6.4.2.

A.10 Maxwell Equations

Electromagnetic fields play an extremely important role in space physics. They are calculated with the help of a familiar set of equations formulated in the years 1861–64 by J.C. Maxwell. The differential form of these *Maxwell equations* in a vacuum are written as follows

$$\begin{aligned}
&(1) && \nabla \vec{\mathcal{E}} = \rho_L / \varepsilon_0 && \text{(A.91)} \\
&(2) && \nabla \times \vec{\mathcal{E}} = -\frac{\partial \vec{B}}{\partial t} && \text{(A.92)} \\
&(3) && \nabla \vec{B} = 0 && \text{(A.93)} \\
&(4) && \nabla \times \vec{B} = \mu_0 \vec{j} + \mu_0 \varepsilon_0 \frac{\partial \vec{\mathcal{E}}}{\partial t} && \text{(A.94)}
\end{aligned}$$

with

- $\vec{\mathcal{E}}$ the electric field strength
- $\vec{B}$ the magnetic flux density (field strength in our nomenclature)
- $\vec{j}$ the current density
- ρ_L the charge density
- ε_0 the electrical permittivity
- μ_0 the magnetic permeability

The first Maxwell equation says that electrical charges are sources of electric fields, the field lines of which start at positive charges and end on negative charges. The second Maxwell equation (corresponding to the differential form of the Faraday induction law) stipulates that a time-varying magnetic field produces a rotational electric field, the field lines of which encircle the magnetic field. The strength of this induced electric field is proportional to the intensity and rate of change of the magnetic field. The third Maxwell equation prohibits the existence of magnetic charges (monopoles), thereby requiring that magnetic field lines form closed curves without starting or ending points. Finally, the fourth Maxwell equation states that both stationary current densities (i.e. moving charges), as well as time-varying electric fields, produce rotational magnetic fields, the field lines of which encircle the current density and the electric field, respectively. The strength of this induced magnetic field is proportional to the strength of the current density and to the intensity and rate of change of the electric field. Also note that, without the vacuum displacement current density $\varepsilon_0 \, \partial \vec{\mathcal{E}}/\partial t$, the fourth Maxwell equation corresponds to the differential form of Ampère's law.

The Maxwell equations in integral form are

$$(1) \qquad \oint_A \vec{\mathcal{E}} \, \mathrm{d}\vec{A} = \mathcal{Q}_L / \varepsilon_0 \tag{A.95}$$

$$(2) \qquad \oint_{L(A)} \vec{\mathcal{E}} \, \mathrm{d}\vec{l} = -\frac{\mathrm{d}}{\mathrm{d}t} \int_A \vec{\mathcal{B}} \, \mathrm{d}\vec{A} \tag{A.96}$$

$$(3) \qquad \oint_A \vec{\mathcal{B}} \, \mathrm{d}\vec{A} = 0 \tag{A.97}$$

$$(4) \qquad \oint_{L(A)} \vec{\mathcal{B}} \, \mathrm{d}\vec{l} = \mu_0 \mathcal{I} + \mu_0 \varepsilon_0 \frac{\mathrm{d}}{\mathrm{d}t} \int_A \vec{\mathcal{E}} \, \mathrm{d}\vec{A} \tag{A.98}$$

with

$\mathcal{Q}_L$ the electric charge enclosed by the surface A
$\mathcal{I}$ the current strength
$L(A)$ the closed boundary curve of the surface A
$\mathrm{d}\vec{A} = \hat{n} \mathrm{d}A$ the differential element of the surface A
$\mathrm{d}\vec{l}$ the differential element of the boundary curve $L(A)$

Note that Eq. (A.95) corresponds to Gauss's law and Eq. (A.96) to Faraday's induction law. This latter law follows from the definition of electric voltage

$$\mathcal{U}_{L(A)} = \oint_{L(A)} \vec{\mathcal{E}} \, \mathrm{d}\vec{l} \tag{A.99}$$

and the definition of magnetic flux

$$\Phi_A = \int_A \vec{\mathcal{B}} \, \mathrm{d}\vec{A} \tag{A.100}$$

so that we obtain

$$\mathcal{U}_{L(A)} = -\frac{\mathrm{d}\Phi_A}{\mathrm{d}t} \tag{A.101}$$

A.11 Curvature of a Dipole Field Line

For the case of a dipole field, it is sufficient to determine the curvature of a planar field line. We use a Cartesian coordinate system and place the origin

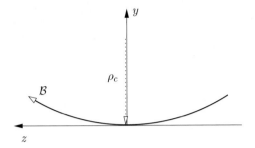

Fig. A.3. Coordinate system used for calculating the curvature of a planar field line

at that point where we want to determine the curvature (see Fig. A.3). The y-axis lies in the plane of the field line and the z-axis points in the direction of the magnetic field at the origin. Designating the field strength at the origin by $\mathcal{B}_{z0}$, the field strength in the neighborhood of the origin may be written as follows

$$\vec{\mathcal{B}}(y, z) = \hat{y}\,\mathcal{B}_y(y, z) + \hat{z}\,(\mathcal{B}_{z0} + \mathcal{B}_z(y, z))$$

Here, $\mathcal{B}_y$ and $\mathcal{B}_z$ are small perturbation fields caused by the curvature that are superposed onto the main field $\mathcal{B}_{z0}$. Since field line elements, by definition, are always directed along the local magnetic field, it is required that

$$\mathrm{d}\vec{s} \times \vec{\mathcal{B}} = 0$$

where in our case the field line element is given by

$$\mathrm{d}\vec{s} = \hat{y}\,\mathrm{d}y + \hat{z}\,\mathrm{d}z$$

Accordingly, the cross product above may be written as

$$\hat{x}\,(\mathrm{d}y\,(\mathcal{B}_{z0} + \mathcal{B}_z) - \mathrm{d}z\mathcal{B}_y) = 0$$

Separation of the variables and subsequent integration leads to the following expression for y

$$y = \int \frac{\mathcal{B}_y\,\mathrm{d}z}{\mathcal{B}_{z0} + \mathcal{B}_z} + \mathrm{const.}$$

In order to evaluate the integral, we expand $\mathcal{B}_y$ in a Taylor series about the origin and truncate after the second term

$$\mathcal{B}_y = \mathcal{B}_y(z = 0) + \left.\frac{\mathrm{d}\mathcal{B}_y}{\mathrm{d}z}\right|_{z=0} \Delta z + \ldots \simeq \left.\frac{\mathrm{d}\mathcal{B}_y}{\mathrm{d}z}\right|_{z=0} z$$

We also neglect the small quantity $\mathcal{B}_z$ with respect to $\mathcal{B}_{z0}$ in the denominator of the integrand and obtain

$$y \simeq \frac{(\mathrm{d}\mathcal{B}_y/\mathrm{d}z)_{z=0}}{\mathcal{B}_{z0}} \int z\,\mathrm{d}z + \mathrm{const.} = \frac{(\mathrm{d}\mathcal{B}_y/\mathrm{d}z)_{z=0}}{2\mathcal{B}_{z0}} z^2 + \mathrm{const.}$$

This approximation thus describes the field line shape as a parabola at the position where the curvature was to be determined. Noting that the radius of curvature of any two dimensional curve $y = f(x)$ is given by the familiar expression

$$\rho_c = \frac{\left(1 + (\mathrm{d}y/\mathrm{d}x)^2\right)^{3/2}}{\mathrm{d}^2 y/\mathrm{d}x^2}$$

we obtain for the case at hand

$$\rho_c(z = 0) = \frac{\mathcal{B}_{z0}}{(\mathrm{d}\mathcal{B}_y/\mathrm{d}z)_{z=0}} \tag{A.102}$$

Furthermore, as shown in Eq. (A.94), $\nabla \times \vec{\mathcal{B}} = 0$ in a static or slowly varying magnetic field with negligibly small local current density. For the case of interest here, this requires that

$$\hat{x}\left(\frac{\mathrm{d}\mathcal{B}_z}{\mathrm{d}y} - \frac{\mathrm{d}\mathcal{B}_y}{\mathrm{d}z}\right) = 0$$

Correspondingly, the radius of curvature may be written as

$$\rho_c = \frac{\mathcal{B}_{z0}}{(\mathrm{d}\mathcal{B}_z/\mathrm{d}y)_{z=0}}$$

or generalized

$$\rho_c = \frac{\mathcal{B}}{|\nabla_\perp \mathcal{B}|} \tag{A.103}$$

where $\nabla_\perp$ denotes the field gradient perpendicular to the magnetic field. Although the derivation will not be shown here, the (vector) expression for the radius of curvature in the more general case of a nonplanar field line is given by

$$\frac{\hat{\rho}_c}{\rho_c} = -(\hat{\mathcal{B}}\nabla)\hat{\mathcal{B}} \tag{A.104}$$

A.12 Gradient Drift Velocity

Consider the drift scenario sketched in Fig. 5.17, but now with a constant gradient rather than a discontinuous jump in the field strength. To simplify the calculation, we assume that the ratio of the gyroradius to the typical spatial scale of magnetic field strength variation is so small that the charge carriers essentially gyrate around the local magnetic field lines and the drift motion represents only a small perturbation. Since the particles only drift in the $\pm x$-direction, the magnetic force in the y-direction averaged over the gyroperiod must vanish

$$\int_{T_B} (F_B)_y \, dt = q \int_{T_B} u_x \, \mathcal{B}(y) \, dt = 0 \qquad (A.105)$$

We approximate the field strength in the vicinity of the gyration orbit by the following expression, a truncated Taylor series

$$\mathcal{B}(y) \simeq \mathcal{B} + \frac{d\mathcal{B}}{dy} y$$

where $\mathcal{B}$ is the magnetic field strength at the position of the guiding center at $x = 0, y = 0$. Substituting this in Eq. (A.105) yields

$$q \int_{T_B} u_x (\mathcal{B} + \frac{d\mathcal{B}}{dy} y) \, dt \simeq 0$$

from which one obtains

$$\int_{T_B} u_x dt = \int_{x(T_B)} dx = \Delta x = -\frac{1}{\mathcal{B}} \frac{d\mathcal{B}}{dy} \int_{x(T_B)} y \, dx$$

Here, $x(T_B)$ is the distance in the x-direction and Δx is the small displacement of the guiding center over one gyroperiod. Since Δx is small and the drift thus induces only a slight distortion of the gyration orbit, the integral on the right side of the equation may be calculated to a first approximation along the undisturbed circular orbit. In this case we have

$$y = f(x) \simeq \pm \sqrt{r_B^2 - x^2}$$

see Eq. (5.28). Accordingly, the integral over y takes the following form for electrons

$$\int_{x(T_B)} y \, dx \simeq \int_{r_B}^{-r_B} \sqrt{r_B^2 - x^2} \, dx + \int_{-r_B}^{r_B} -\sqrt{r_B^2 - x^2} \, dx$$

Using

$$\int \sqrt{a^2 - x^2} \, dx = \frac{1}{2} \left(x\sqrt{a^2 - x^2} + a^2 \arcsin(x/a) \right)$$

the value of the integral becomes $-r_B^2 \pi$ and the displacement of the guiding center is given by

$$\Delta x = \frac{1}{\mathcal{B}} \frac{d\mathcal{B}}{dy} r_B^2 \pi = \frac{E_\perp T_B}{|q| \mathcal{B}^2} \frac{d\mathcal{B}}{dy}$$

Dividing by the gyration period, we arrive at the desired expression for the drift velocity

$$u_D^{gr} = \frac{\Delta x}{T_B} = \frac{E_\perp}{|q| \mathcal{B}^2} \frac{d\mathcal{B}}{dy}$$

the general form of which may be written as

$$\vec{u}_D^{gr} = \frac{E_\perp}{q \, \mathcal{B}^3} \vec{\mathcal{B}} \times \nabla_\perp \mathcal{B} \qquad (A.106)$$

A.13 System of Equations for Ideal Magnetoplasmadynamics

In the following we derive a system of equations that self-consistently describes a magnetically biased, ionized gas mixture. For this purpose we utilize the balance equations for a magnetized plasma, the Maxwell equations, and a generalized form of Ohm's law. The approximations made in this derivation are verified using the solar wind as an example.

A.13.1 Balance Equations of a Magnetoplasma

Corresponding to the properties of the solar wind, we consider a fully ionized, two-component plasma consisting of ions and their associated electrons. The densities of these charge carriers are assumed to be equal everywhere, but not their flow velocities

- $n_i \simeq n_e = n$
- $u_i \neq u_e$

The indices i and e again denote the ion and electron gas components. Similar to a neutral gas mixture, the balance equations for a mixture of charge carrier gases are obtained by adding the balance equations for the single components.

Density Balance Equation. Neglecting production and loss processes, which is possible to a good approximation for the solar wind as well as for other regions of space (such as the magnetosphere), we may write the separate density balance equations of the ion and electron gases as

$$\frac{\partial n_i}{\partial t} = -\nabla(n_i \, \vec{u}_i) \tag{A.107}$$

$$\frac{\partial n_e}{\partial t} = -\nabla(n_e \, \vec{u}_e) \tag{A.108}$$

Multiplying the first equation by m_i, the second equation by m_e, and adding the results, one obtains

$$\frac{\partial(\rho_i + \rho_e)}{\partial t} = -\nabla(\rho_i \, \vec{u}_i + \rho_e \, \vec{u}_e) \tag{A.109}$$

An expression corresponding to the density balance or continuity equation for neutral gas mixtures may be derived if the mass density and flow velocity of the plasma are defined as follows

$$\rho = \rho_i + \rho_e \tag{A.110}$$

and

$$\vec{u} = (\rho_i \, \vec{u}_i + \rho_e \, \vec{u}_e) \, / \, \rho \tag{A.111}$$

The plasma velocity is thus defined as the weighted sum of the flow velocities of the two plasma components – the weighting factor being equal to the relative mass fraction of each component. As such, the plasma velocity corresponds to the mean mass velocity of the plasma and thus usually the velocity of the ion gas. At the same time, this definition guarantees that the kinetic momentum of the plasma is given by the sum of the kinetic momenta of the plasma components, in our case the sum of the momenta of the ion and electron gases. Inserting these definitions into the relation (A.109), we obtain the following density balance equation for a fully ionized plasma in the absence of production or losses

$$\frac{\partial \rho}{\partial t} = -\nabla(\rho \, \vec{u}) \tag{A.112}$$

Momentum Balance Equation. An analogous approach may be used to derive the momentum balance equation of a plasma, i.e. we add the momentum balance equations for the ion and electron gases. Referring to Section 3.4.3, these may be written as

$$\rho_i \frac{D\vec{u}_i}{Dt} = -\nabla p_i + \eta_i \, \underline{\Delta}\vec{u}_i + \rho_i \, \vec{g} + e \, n_i \, \vec{\mathcal{E}} + e \, n_i \, \vec{u}_i \times \vec{B}$$
$$+ \rho_i \, \nu^*_{i,e} (\vec{u}_e - \vec{u}_i) + 2\rho_i \, \vec{u}_i \times \vec{\Omega} \tag{A.113}$$

$$\rho_e \frac{D\vec{u}_e}{Dt} = -\nabla p_e + \eta_e \, \underline{\Delta}\vec{u}_e + \rho_e \, \vec{g} - e \, n_e \, \vec{\mathcal{E}} - e \, n_e \, \vec{u}_e \times \vec{B}$$
$$+ \rho_e \, \nu^*_{e,i} (\vec{u}_i - \vec{u}_e) + 2\rho_e \, \vec{u}_e \times \vec{\Omega} \tag{A.114}$$

No assumptions regarding the type of gas were made in order to derive Eq. (3.80), which is thus valid for both a neutral and an ionized gas. The difference is that for an ionized gas we must account for electric and magnetic forces in addition to the force of gravity. Moreover, due to the absence of a preferred direction, the viscosity force must be extended to three dimensions (see Appendix A.6). Prior to adding them together, the above equations will be simplified to meet our immediate needs.

We first consider a stationary coordinate system in which the Coriolis forces vanish. Viscosity is also neglected and the reasons why this is possible are explained in Section A.13.3. With these simplifications the momentum balance equations of the ion and electron gases assume the following form

$$\rho_i \frac{\partial \vec{u}_i}{\partial t} + \rho_i \, (\vec{u}_i \, \nabla)\vec{u}_i = -\nabla p_i + \rho_i \vec{g} + e \, n_i \, \vec{\mathcal{E}} + e \, n_i \, \vec{u}_i \times \vec{B} + \rho_i \, \nu^*_{i,e} (\vec{u}_e - \vec{u}_i) \tag{A.115}$$

$$\rho_e \frac{\partial \vec{u}_e}{\partial t} + \rho_e \, (\vec{u}_e \nabla)\vec{u}_e = -\nabla p_e + \rho_e \vec{g} - e \, n_e \, \vec{\mathcal{E}} - e \, n_e \vec{u}_e \times \vec{B} + \rho_e \, \nu^*_{e,i} (\vec{u}_i - \vec{u}_e) \tag{A.116}$$

where we have now written out the convective derivative. Adding just the first terms of each equation yields

$$\rho_i \frac{\partial \vec{u}_i}{\partial t} + \rho_e \frac{\partial \vec{u}_e}{\partial t} = (m_i + m_e) n \frac{\partial}{\partial t} \left(\frac{m_i n \vec{u}_i + m_e n \vec{u}_e}{n(m_i + m_e)} \right) = \rho \frac{\partial \vec{u}}{\partial t} \quad (A.117)$$

We run into some difficulty upon adding the second terms of each equation because $\rho_i(\vec{u}_i \nabla)\vec{u}_i + \rho_e(\vec{u}_e \nabla)\vec{u}_e \neq \rho(\vec{u}\nabla)\vec{u}$. We can avoid this problem if we no longer refer the random velocity of the ions and electrons to their own flow velocities $\vec{u}_i$ and $\vec{u}_e$, but rather to the plasma velocity $\vec{u}$ defined in Eq. (A.111) $\vec{c}_i^* = \vec{v}_i - \vec{u}$ and $\vec{c}_e^* = \vec{v}_e - \vec{u}$ (see the references on plasma physics given at the end of Chapter 6). At least for the solar wind, however, we know that the flow velocities of the ions and electrons are nearly equal and vary over the same scale. In this case we can apply the very good approximation $\rho_e(\vec{u}_e \nabla)\vec{u}_e \ll \rho_i(\vec{u}_i \nabla)\vec{u}_i$ and therefore

$$\rho_i(\vec{u}_i \nabla)\vec{u}_i + \rho_e(\vec{u}_e \nabla)\vec{u}_e \simeq \rho(\vec{u}\nabla)\vec{u}$$

Upon adding the terms on the right side of Eqs. (A.115) and (A.116), we introduce the plasma pressure and the current density

$$p = p_i + p_e \quad (A.118)$$

and

$$\vec{j} = e\,(n_i\,\vec{u}_i - n_e\,\vec{u}_e) \simeq e\,n(\vec{u}_i - \vec{u}_e) \quad (A.119)$$

As we know, this latter quantity is defined as the net transport of charge through a surface per unit time and unit area, see also Eq. (5.4). Furthermore, according to Eqs. (2.10), (2.56) and (5.54), we can make use of the following relation between the momentum transfer collision frequencies of the charge carriers

$$\nu_{e,i}^* = \frac{m_i}{m_e} \nu_{i,e}^* \quad (A.120)$$

Accordingly, the frictional forces, as well as the electrical forces, cancel out upon addition (the total momentum must be conserved for elastic collisions), and the momentum balance equation of our fully ionized plasma in the absence of production or losses takes the following form

$$\rho \frac{D\vec{u}}{Dt} = -\nabla p + \rho\vec{g} + \vec{j} \times \vec{B} \quad (A.121)$$

Energy Balance Equation or Equation of State. Greatly simplified, the energy balance equation assumes the form of an adiabatic relation (see Appendix A.7). Applying this approximation, as summarized in Eq. (2.36), both the ion and electron gases must obey the relations

$$p_i = \text{const.}\, \rho_i^\gamma \tag{A.122}$$

$$p_e = \text{const.}\, \rho_e^\gamma$$

For the case of the solar wind, the assumption of adiabatic expansion leads to atypically low temperatures at the Earth's orbit (see Section 6.1.3). On the other hand, the assumption of isothermal conditions according to the ideal gas law $p = nkT = \text{const.}\, \rho$, i.e. direct proportionality between pressure and density, is also unrealistic. The exponent γ^* on the density, denoted the polytropic index, must therefore be less than $\gamma = 5/3 \simeq 1.67$ but greater than 1 in the real solar wind. Indeed, when accounting for heat conduction with

$$p = nkT = nkT_0\, r_0^{2/7}\, r^{-2/7} \simeq nkT_0\, n_0^{-1/7} n^{1/7} = \text{const.}\, \rho^{8/7}$$

one obtains a polytropic index of $\gamma^* = 8/7 \simeq 1.14$, see also Eq. (6.17) and (6.22). We may thus use the more general relations

$$p_i = \alpha_i\, \rho_i^{\gamma^*} \tag{A.123}$$

$$p_e = \alpha_e\, \rho_e^{\gamma^*} \tag{A.124}$$

where α_i and α_e denote constants and it is assumed that both the ions and electrons are described by the same polytropic index. Adding these two equations yields

$$p_i + p_e = \frac{\alpha_i m_i^{\gamma^*} + \alpha_e m_e^{\gamma^*}}{(m_i + m_e)^{\gamma^*}} (m_i + m_e)^{\gamma^*} n^{\gamma^*} \tag{A.125}$$

and one obtains

$$p = \alpha\, \rho^{\gamma^*} \tag{A.126}$$

where α is a constant composite factor. Note that another form of this equation often found in the literature, namely

$$\frac{D(p\rho^{-\gamma^*})}{Dt} = \frac{\partial}{\partial t}(p\rho^{-\gamma^*}) + (\vec{u}\nabla)(p\rho^{-\gamma^*}) = 0$$

corresponds to our Eq. (A.70).

Comparing the number of unknowns appearing in the plasma balance equations (A.112), (A.121) and (A.126) with the number of defining equations, we are confronted with a dilemma. Indeed, only five scalar equations are available for a determination of the eleven scalar unknowns (ρ, p, u_x, u_y, u_z, j_x, j_y, j_z, $\mathcal{B}_x$, $\mathcal{B}_y$, $\mathcal{B}_z$). We are evidently still lacking six additional relations for a determination of all eleven unknowns. Since electromagnetic quantities appear in the momentum balance equation, it is obvious that we should include the Maxwell equations. This certainly helps, but is not enough. In fact, we need a generalized form of Ohm's law to close the system of equations.

A.13.2 Maxwell Equations and the Generalized Ohm's Law

We are interested here in the second and fourth Maxwell equations given in Appendix A.10, namely

$$\nabla \times \vec{\mathcal{E}} = -\frac{\partial \vec{B}}{\partial t} \tag{A.127}$$

and

$$\nabla \times \vec{B} = \mu_0 \vec{j} + \mu_0 \varepsilon_0 \frac{\partial \vec{\mathcal{E}}}{\partial t} \tag{A.128}$$

Since, in the case of the solar wind, we are only interested in quasi-stationary processes, the second term on the right side of Eq. (A.128) can be neglected with respect to the first term. This equation then reduces to Ampère's law

$$\nabla \times \vec{B} = \mu_0 \vec{j} \tag{A.129}$$

Together with the Faraday induction law (A.127), this yields two additional vector equations (i.e. six additional scalar equations) for the determination of the current density and magnetic field, albeit at the cost of another unknown, the electric field. We clearly need an additional relation that couples the electric field to the other variables. We recall that the electric field was part of the separate momentum balance equations for electrons and ions, but then canceled out upon adding these equations. If we instead subtract these two equations, this quantity is retained and we obtain the necessary additional relation. Actually, it is not surprising that we lost information when combining the *two* independent momentum balance equations into *one* plasma balance equation. We recover this information by subtracting the momentum balance equations of the single gases. It is clear that subtraction and addition of the two original equations does yield two independent relations. This follows not only from mathematics, but also from the fact that the new relation contains electric and frictional force terms that are absent in the plasma momentum balance equation.

We again consider the simplified momentum balance equations (A.115) and (A.116). Using the convective derivative operator, multiplying the ion equation with m_e and the electron equation with m_i, these may be written in the following way

$$m_e \rho_i \frac{D\vec{u}_i}{Dt} = -m_e \nabla p_i + m_e \rho_i \vec{g} + e m_e n_i \vec{\mathcal{E}} + e m_e n_i \vec{u}_i \times \vec{B}$$
$$+ m_e \rho_i \nu_{i,e}^* (\vec{u}_e - \vec{u}_i) \tag{A.130}$$

$$m_i \rho_e \frac{D\vec{u}_e}{Dt} = -m_i \nabla p_e + m_i \rho_e \vec{g} - e m_i n_e \vec{\mathcal{E}} - e m_i n_e \vec{u}_e \times \vec{B}$$
$$+ m_i \rho_e \nu_{e,i}^* (\vec{u}_i - \vec{u}_e) \tag{A.131}$$

Subtracting these equations, the left side becomes

$$\frac{m_i\, m_e\, n}{e}\, \frac{D}{Dt}\left(\frac{n\, e\, \vec{u}_i - n\, e\, \vec{u}_e}{n}\right) = \frac{m_i\, m_e\, n}{e}\, \frac{D(\vec{j}/n)}{Dt}$$

The difference of the first three terms on the right side is given by

$$-(m_e\, \nabla\, p_i - m_i\, \nabla\, p_e) + \rho\, e\, \vec{\mathcal{E}}$$

Subtraction of the fourth terms yields

$$e\, n\, (m_e\, \vec{u}_i + m_i\, \vec{u}_e)\, \times\, \vec{B}$$
$$= e\, n\, [m_e\, (\vec{u}_i - \vec{u}_e) + m_e\, \vec{u}_e + m_i\, (\vec{u}_e - \vec{u}_i) + m_i\, \vec{u}_i]\, \times\, \vec{B}$$
$$= e\, n\, (\frac{\rho}{n}\vec{u} - \frac{m_i - m_e}{e\, n}\, \vec{j})\, \times\, \vec{B}$$

In this latter expression we have made use of the definitions for the plasma density, the plasma velocity and the current density. Finally, the difference of the fifth (frictional force) terms is

$$m_e\, n\, m_i(\frac{m_e}{m_i}\, \nu_{e,i}^*)\, (\vec{u}_e - \vec{u}_i) - m_i\, n\, m_e\, \nu_{e,i}^*\, (\vec{u}_i - \vec{u}_e)$$
$$= -(m_e + m_i)\, n\, m_e\, \nu_{e,i}^*(\vec{u}_i - \vec{u}_e)$$
$$= -\frac{\rho\, m_e\, \nu_{e,i}^*}{e\, n}\, \vec{j} = -\frac{\rho\, e}{\sigma_B}\, \vec{j}$$

where in the last step we have introduced the abbreviation

$$\sigma_B = \frac{e^2\, n}{m_e\, \nu_{e,i}^*} \tag{A.132}$$

Collecting all terms, the difference of the momentum balance equations assumes the following form

$$\frac{m_i\, m_e\, n}{e}\, \frac{D(\vec{j}/n)}{Dt} = -\,(m_e\, \nabla\, p_i - m_i\, \nabla p_e) + \rho\, e\, \vec{\mathcal{E}}$$
$$+ e\, n\, (\frac{\rho}{n}\vec{u} - \frac{m_i - m_e}{e\, n}\, \vec{j})\, \times\, \vec{B} - \frac{\rho\, e}{\sigma_B}\, \vec{j}$$

Since the mass of the electrons is small compared to that of the ions, it can be neglected in the third term on the right side of the above equation. Furthermore, knowing from solar wind observations that the temperature of the ions and its gradient are of the same order of magnitude as their electron counterparts, we have $m_e\, \nabla p_i \ll m_i\, \nabla p_e$. Applying these approximations, the above difference equation may be rearranged to yield an equation for the electric field

$$\vec{\mathcal{E}} \simeq -\vec{u}\, \times\, \vec{B} + \frac{\vec{j}}{\sigma_B} - \frac{1}{e\, n}\nabla p_e + \frac{1}{e\, n}\vec{j}\, \times\, \vec{B} + \frac{m_e}{e^2}\frac{D(\vec{j}/n)}{Dt} \tag{A.133}$$

Knowing the plasma state parameters and the magnetic field, this equation provides a connection between the electric field and the current density. As such, it can be considered as a *generalized form of Ohm's law*. Of immediate interest is the physical meaning of the individual terms contributing to the electric field. The first term on the right side clearly describes a dynamo field produced by the motion of the conducting plasma across the magnetic field. Since σ_B corresponds to the conductivity of a fully ionized plasma parallel to the magnetic field (see Eq. (7.11) of Section 7.3.3), the second term on the right side represents the electric field, parallel to the magnetic field, that maintains the current density flowing through the specific resistance $1/\sigma_B$. The third term corresponds to an electric polarization field produced by the difference between the pressure gradient forces acting on the ion and electron gases. Setting p_e equal to a barometric height distribution, for example, this expression leads to the familiar Pannekoek-Rosseland polarization field, see Eq. (4.43). The fourth term describes the electric field associated with Hall currents (i.e. currents flowing perpendicular to the electric and magnetic fields). Finally, the fifth and last term is an electric field arising from inertial effects of the charge carriers (here the electrons). In the following we examine the consequences of a rather drastic simplification, namely: all terms except the first are neglected. That this is possible for the case considered here, will be shown in the following section. Under this approximation the generalized Ohm's law assumes the following simple form

$$\vec{\mathcal{E}} \simeq -\vec{u} \times \vec{B} \tag{A.134}$$

The validity of this relation is one of the fundamental assumptions of 'ideal' magnetoplasmadynamics and has important consequences for the interactions between electric fields, plasma motions and magnetic fields (see Section 6.2.7). Using it to eliminate the electric field from Eq. (A.127), we obtain

$$\frac{\partial \vec{B}}{\partial t} = \nabla \times (\vec{u} \times \vec{B}) \tag{A.135}$$

This relation, together with the Eqs. (A.112), (A.121), (A.126) and (A.129), represents the system of equations of ideal magnetoplasmadynamics introduced in Section 6.2.7. It still remains to be shown that the approximations inherent to this system are, in fact, justified. Their validity will be checked using actual solar wind data.

A.13.3 Validity Test of the Approximations

We begin by checking the consequence of neglecting the viscosity in the momentum balance equation. As explained earlier, viscous forces in a magnetoplasma are effective parallel to the magnetic field. The thermal motion of the charge carriers, and thus any thermally associated momentum transport, is suppressed by magnetic forces in the direction perpendicular to the

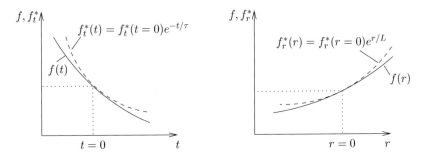

Fig. A.4. Estimating time and space derivatives of physical quantities

magnetic field. In the case of the solar wind, this means that only velocity differences along the interplanetary magnetic field need be considered. Since the solar wind quickly reaches a constant asymptotic value, these differences are generally small. Neglect of viscosity is thus usually possible to a very good approximation.

Justification of the radical simplification of the generalized Ohm's law (A.133) proves to be a more involved task. In this case we really need to assess the relative order of magnitude of all the terms appearing on the right side of this equation. This process starts by first estimating the magnitude of the time and space derivatives contained in these terms. For this purpose we approximate the function to be differentiated by a product of exponential functions as shown in Fig. A.4. The free parameters τ and L are determined from the requirement that the approximating function $f^* = f_t^* \cdot f_r^*$ not only has the same ordinate value, but also the same slope as the original function f at the point where the derivative is evaluated. An advantage of the exponential functions used for this approximation is that they enable a direct estimate of the variation scale lengths. The value of the function f^* changes by a factor e or $1/e$ in the time $t = \tau$, or over a distance $r = L$, and this variation is deemed significant for the case at hand. Accordingly, τ and L can be considered as typical time and distance scale lengths, over which f^* – and thereby also f – significantly varies. A further advantage of exponential functions, of course, is the simple form of their derivative. For differentiation with respect to time, for example, we have

$$\left| \frac{\partial f}{\partial t} \right| \simeq \left| \frac{\partial f^*}{\partial t} \right| = \left| -\frac{f^*}{\tau} \right| \simeq \frac{f}{\tau}$$

This approach was used earlier to estimate ionospheric time constants (see Section 4.5.1). Correspondingly, one obtains the following relations for spatial derivatives in association with the nabla operator

$$|\nabla f| \simeq |\nabla f^*| \rightarrow \left| \frac{\partial f^*}{\partial r} \right| = \frac{f^*}{L} \simeq \frac{f}{L}$$

The general rule is thus

$$\frac{\partial}{\partial t} \rightarrow \frac{1}{\tau} \;, \qquad \nabla \rightarrow \frac{1}{L} \tag{A.136}$$

We use these approximations in the following to estimate the relative magnitude $R_i, i = 2$ to 5 of all the terms neglected on the right side of Eq. (A.133). The absolute magnitude of the one term not neglected in this approximation, $|-\vec{u} \times \vec{B}| \simeq u\,B$, serves as a reference value. For the second term we obtain

$$R_2 = \frac{j/\sigma_B}{u\,B} \simeq \frac{(B/\mu_0\,L)/\sigma_B}{u\,B} \simeq \frac{10^8 \ln \Lambda}{u\,L\,T_{\mathrm e}^{3/2}} \tag{A.137}$$

where we have made use of Eq. (A.129) and Eq. (7.12). Applying this process to the third term, we obtain

$$R_3 = \frac{\nabla p_{\mathrm e}/(e\,n)}{u\,B} \simeq \frac{n\,k\,T_{\mathrm e}/(e\,n\,L)}{u\,B} \simeq \frac{10^{-4}\,T_{\mathrm e}}{u\,B\,L} \tag{A.138}$$

Correspondingly, the fourth term may be written as

$$R_4 = \frac{j\,B/(e\,n)}{u\,B} \simeq \frac{B/(\mu_0\,L)}{e\,n\,u} \simeq \frac{5 \cdot 10^{24}B}{n\,u\,L} \tag{A.139}$$

and finally the same procedure for the fifth term yields

$$R_5 = \frac{(m_{\mathrm e}/e^2)\,\mathrm{D}(j/n)/\mathrm{Dt}}{u\,B} = \frac{(m_{\mathrm e}/e^2)\,[\partial(j/n)/\partial t + (u\,\nabla)(j/n)]}{u\,B}$$
$$\simeq \frac{(m_{\mathrm e}/e^2)\,[j/(n\,\tau) + j\,u/(nL)]}{u\,B} \simeq \frac{2\,m_{\mathrm e}}{e^2\,\mu_0\,n\,L^2} \simeq \frac{6 \cdot 10^{13}}{n\,L^2} \tag{A.140}$$

where we have utilized the relation $L = u\,\tau$ and Eq. (A.129). For an explicit estimate of these terms, we consider typical solar wind conditions at the orbit of the Earth (see Tables 6.1 and 6.2) and take a rather conservative value for the scale length, $L \simeq 10^3$ km. This yields the following relative magnitudes for the various terms: $R_2 < 10^{-9}$, $R_3 < 10^{-1}$, $R_4 < 10^{-2}$ and $R_5 \simeq 10^{-5}$. These values are certainly small enough to justify the original approximations.

A.14 Two Theorems of Magnetoplasmadynamics

In order to understand the configuration of the interplanetary magnetic field, we have invoked the following two theorems of ideal magnetoplasmadynamics

- The magnetic flux passing through a closed curve that moves with the plasma flow velocity remains constant   (A.141)

- Plasma elements connected at any point in time by a common magnetic field line remain connected by a common field line   (A.142)

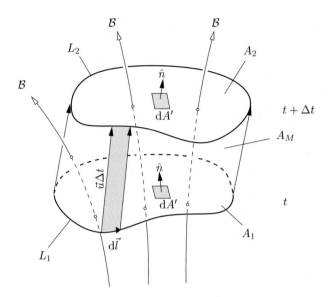

Fig. A.5. A magnetized plasma ring embedded in a plasma flow at the times t (index 1) and $t + \Delta t$ (index 2)

see Eqs. (6.29) and (6.30). To prove the first theorem, we consider a closed curve (a ring) of arbitrary shape that connects contiguous plasma elements and moves with the flow. The magnetic flux through the ring is given by

$$\Phi = \int_A \vec{\mathcal{B}} \, \hat{n} \, dA' \tag{A.143}$$

where A is the area of the surface bounded by the ring L, and $\hat{n}$ is the unit normal to an element dA' of this surface. The first theorem may thus be expressed as follows:

$$\frac{D'\Phi}{D't} = 0 \tag{A.144}$$

Here, the operator $D'/D't$ is a special type of convective derivative that accounts for the temporal change of the magnetic flux through the ring as it moves with the plasma (Lagrange approach). The total change contains a temporal as well as a spatial component, similar to the convective derivative D/Dt introduced in Section 3.4.3. Whereas that operator was used in the context of a locally defined quantity, we are dealing here with an extended ring of plasma. In order to formulate the convective derivative applicable in this case, we consider the ring of plasma at two nearby instants of time, t and $t + \Delta t$. For this scenario (illustrated in Fig. A.5), where A_1 and A_2 are the surfaces bounded by the plasma ring at times t and $t + \Delta t$, we obtain

$$\frac{D'\Phi}{D't} = \lim_{\Delta t \to 0} \frac{\Phi(A_2, t + \Delta t) - \Phi(A_1, t)}{\Delta t} \tag{A.145}$$

In order to manipulate this expression, we exploit the fact that the magnetic flux through a *closed* surface is equal to zero. Lacking sources in the form of magnetic monopoles, all field lines that enter the volume enclosed by the surface must exit the same. In other words, the sum of all contributions to the magnetic flux must cancel out upon integration over the surface of the volume. This result may be obtained formally with help of Gauss's law (A.41) and the third Maxwell equation (A.93)

$$\oint_{A(V)} \vec{B}\hat{n}\,\mathrm{d}A' = \int_V \nabla\vec{B}\,\mathrm{d}V' = 0 \qquad (\text{A.146})$$

We apply this relation to the volume sketched in Fig. A.5, which is bounded by the surfaces A_1, A_2, and A_M. The cylindrical surface A_M is defined by the motion of the plasma ring. According to Eqs. (A.143) and (A.146), it must hold that

$$-\Phi(A_1, t+\Delta t) + \Phi(A_2, t+\Delta t) + \Phi(A_M, t+\Delta t) = 0$$

The minus sign on the first term accounts for the fact that the surface normal must always be directed out of the volume, as prescribed for the validity of the relation (A.41) or (A.146). Substituting this in Eq. (A.145), we obtain

$$\frac{\mathrm{D}'\Phi}{\mathrm{D}'t} = \lim_{\Delta t\to 0} \frac{\Phi(A_1, t+\Delta t) - \Phi(A_1, t)}{\Delta t} - \lim_{\Delta t\to 0} \frac{\Phi(A_M, t+\Delta t)}{\Delta t}$$

The first term is evidently just the partial derivative of Φ with respect to time at the position A_1, for which

$$\lim_{\Delta t\to 0} \frac{\Phi(A_1, t+\Delta t) - \Phi(A_1, t)}{\Delta t} = \left.\frac{\partial\Phi}{\partial t}\right|_{A_1} = \int_{A_1} \frac{\partial\vec{B}}{\partial t}\hat{n}\,\mathrm{d}A'$$

The contribution of the second term may be estimated as follows

$$\lim_{\Delta t\to 0} \frac{\Phi(A_M, t+\Delta t)}{\Delta t} \simeq \lim_{\Delta t\to 0} \oint_{L_1} \frac{\vec{B}(\mathrm{d}\vec{l} \times \vec{u}\Delta t)}{\Delta t}$$

$$= \oint_{L_1} (\vec{u} \times \vec{B})\,\mathrm{d}\vec{l} = \int_{A_1} \nabla \times (\vec{u} \times \vec{B})\hat{n}\,\mathrm{d}A'$$

where, in the first step, we have replaced the directed surface element of the cylindrical surface by the cross product $\mathrm{d}\vec{l} \times \vec{u}\Delta t$ and the circumference of the cylindrical surface by the length of the ring L_1. Furthermore, we have approximated the value of the magnetic field and flow velocity on the cylindrical surface at the time $t+\Delta t$ by their values at the location of the ring L_1 at time t. These are second order approximations and vanish for $\Delta t \to 0$. In the second and third steps, we have made use of the vector identity (A.18), and Stokes's theorem (A.42), respectively. We thus obtain

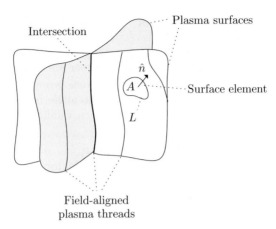

Fig. A.6. Field-aligned plasma thread formed by the intersection of two field line parallel plasma surfaces

$$\frac{\mathrm{D}'\varPhi}{\mathrm{D}'t} \simeq \int_{A_1} \left(\frac{\partial \vec{B}}{\partial t} - \nabla \times (\vec{u} \times \vec{B}) \right) \hat{n}\,\mathrm{d}A' \qquad (\text{A.147})$$

The parenthetical expression in the integrand, however, vanishes in the case of ideal magnetohydrodynamics, see Eq. (A.135), and we obtain $\mathrm{D}'\varPhi/\mathrm{D}'t = 0$, or $\varPhi = \text{const.}$, which is what we originally set out to prove.

To prove the second theorem, we consider two intersecting surfaces at an arbitrary point in time that are formed by contiguous, field aligned plasma threads; see Fig. A.6. Since only one field line may pass through a given point (neutral points are excluded), the line defined by the intersection of the surfaces must also correspond to a field-aligned plasma thread. As time goes on, the plasma surfaces may become warped because of differing flow velocities, but their intersection must remain one and the same plasma thread. In order that this plasma thread continues to follow a local magnetic field line, the surfaces, no matter how much they are distorted, must stay aligned with the magnetic field. This implies that the magnetic flux passing through an arbitrary plasma ring embedded in one of these surfaces must always be zero. Only then is it guaranteed that all magnetic field lines run along the surface and thus do not penetrate either the surface, or the embedded plasma ring, at an oblique angle. We just proved, however, that the magnetic flux through a plasma ring moving with the flow must remain constant (zero in our case). As a result, the second theorem given above must also be valid. Beyond the basic content of this theorem, one may visualize the magnetic fields to stick to the plasma threads such that both move in unison. Because magnetic field lines are only fictitious entities, there is nothing wrong with this often quite useful notion.

A.15 Magnetoplasma Waves

In order to derive wave solutions for a magnetoplasma, the system of equations for ideal magnetoplasmadynamics is further simplified and linearized using a perturbation approach. Solutions are then presented for two particularly simple constellations with respect to the angle between propagation direction and magnetic field. This is followed by a validity test of the approximations used in this derivation.

A.15.1 Simplification of the System of Equations

Waves, if they are to exist at all, must obey the equations describing the propagation medium. For a neutral gas these are the density, momentum and energy balance equations, the latter often in the form of an adiabatic relation. Indeed, all three equations are needed to derive the properties of acoustic waves (see Section 3.5.2). Here we are interested in waves in a magnetoplasma, a propagation medium described in a simplified form by the system of equations of ideal magnetoplasmadynamics (see Section 6.2.7). In order to derive wave solutions analytically, however, this system of equations must be simplified even further.

We first neglect the inertial and gravitational acceleration terms in the momentum balance equation (6.44). The validity of these approximations, as well as some other simplifications already introduced into our system of equations, will be checked in Section A.15.4. Moreover, only adiabatic changes in the state of the plasma are considered, i.e. the polytropic index is replaced by the adiabatic exponent. Finally, we eliminate four of the eleven unknowns in our system of equations by substituting the relations (6.45) and (6.46) into Eq. (6.44). This yields

$$\frac{\partial \rho}{\partial t} + \nabla(\rho \vec{u}) = 0 \tag{A.148}$$

$$\rho \frac{\partial \vec{u}}{\partial t} + \nabla(\alpha \rho^\gamma) - \frac{1}{\mu_0}\left(\nabla \times \vec{B}\right) \times \vec{B} = 0 \tag{A.149}$$

$$\frac{\partial \vec{B}}{\partial t} - \nabla \times (\vec{u} \times \vec{B}) = 0 \tag{A.150}$$

This system of equations can be solved analytically if only waves of small amplitude are considered. As applied successfully for the derivation of acoustic waves, we thus perform a perturbation calculation, starting with the familiar approach

$$
\begin{array}{llll}
\rho = \rho_0 + \rho_1 & \text{with} & \rho_1 \ll \rho_0, & \rho_0 \neq f(\vec{r}, t) \\
p = p_0 + p_1 & \text{with} & p_1 \ll p_0, & p_0 \neq f(\vec{r}, t) \\
\vec{B} = \vec{B}_0 + \vec{B}_1 & \text{with} & B_1 \ll B_0, & B_0 \neq f(\vec{r}, t) \\
\vec{u} = \vec{u}_1 & \text{and} & \vec{u}_0 = 0 & \\
\vec{j} = \vec{j}_1 & \text{and} & \vec{j}_0 = 0 & \\
\vec{\mathcal{E}} = \vec{\mathcal{E}}_1 & \text{and} & \vec{\mathcal{E}}_0 = 0 &
\end{array} \tag{A.151}
$$

where the quantities with index '0' describe the time invariant, homogeneous background plasma at rest and those with index '1' are the small amplitude perturbations of the background plasma due to the wave. In order that the background plasma be at rest, it is necessary in the case of the solar wind to select a coordinate system moving with the flow. The interplanetary medium is then assumed to be homogeneous and independent of time. Inserting the appropriate equations of the above set into the density balance equation (A.148) yields

$$\frac{\partial(\rho_0 + \rho_1)}{\partial t} + (\rho_0 + \rho_1)\nabla\vec{u}_1 + \vec{u}_1\nabla(\rho_0 + \rho_1) = 0$$

where we have made use of the relation (A.31). With $\partial\rho_0/\partial t = 0$, $\rho_0\nabla\vec{u}_1 \gg \rho_1\nabla\vec{u}_1$ and $\nabla\rho_0 = 0$, we may write this equation as follows

$$\frac{\partial\rho_1}{\partial t} + \rho_0\nabla\vec{u}_1 + \vec{u}_1\nabla\rho_1 = 0 \tag{A.152}$$

A simple estimate shows that the term $\vec{u}_1\nabla\rho_1$, which is nonlinear in perturbation quantities, may also be neglected (see Section A.15.4). The density balance equation therefore assumes the following form

$$\frac{\partial\rho_1}{\partial t} + \rho_0\nabla\vec{u}_1 = 0 \tag{A.153}$$

Using $p = \alpha\rho^\gamma, \nabla(\alpha\rho^\gamma) = (\gamma p/\rho)\nabla\rho$ and the perturbation approach (A.151), the momentum balance equation (A.149) may be written as

$$(\rho_0 + \rho_1)\frac{\partial\vec{u}_1}{\partial t} + \frac{\gamma(p_0 + p_1)}{\rho_0 + \rho_1}\nabla(\rho_0 + \rho_1) + \frac{1}{\mu_0}(\vec{B}_0 + \vec{B}_1)\times(\nabla\times(\vec{B}_0 + \vec{B}_1)) = 0$$

We neglect ρ_1 with respect to ρ_0 in the first term and the same for $\vec{B}_1$ with respect to $\vec{B}_0$ in the third term. We further note that $\nabla\rho_0$ and $\nabla\times\vec{B}_0$ are zero. Finally, we write the factor in the second term in the following way

$$\frac{\gamma(p_0 + p_1)}{\rho_0 + \rho_1} \simeq \frac{\gamma p_0}{\rho_0} = v_{PS}^2 \tag{A.154}$$

where we have introduced the quantity $v_{PS} = \sqrt{\gamma p_0/\rho_0}$. Under these conditions the momentum balance equation assumes the form

$$\rho_0\frac{\partial\vec{u}_1}{\partial t} + v_{PS}^2\nabla\rho_1 + \frac{1}{\mu_0}\vec{B}_0\times(\nabla\times\vec{B}_1) = 0 \tag{A.155}$$

Finally, the modified induction law (A.150) is also simplified upon invoking the linearization procedure (A.151). We obtain

$$\frac{\partial(\vec{B}_0 + \vec{B}_1)}{\partial t} - \nabla\times(\vec{u}_1\times(\vec{B}_0 + \vec{B}_1)) = 0$$

or

$$\frac{\partial \vec{B}_1}{\partial t} - \nabla \times (\vec{u}_1 \times \vec{B}_0) - \nabla \times (\vec{u}_1 \times \vec{B}_1) = 0$$

where the magnitude of the third term is usually negligible with respect to that of the second term

$$|\nabla \times (\vec{u}_1 \times \vec{B}_1)| \ll |\nabla \times (\vec{u}_1 \times \vec{B}_0)|$$

Care should be exercised, however, because the right side of this inequality vanishes for $\vec{u}_1 \parallel \vec{B}_0$. If the inequality is fulfilled, we may write the induction law in the form

$$\frac{\partial \vec{B}_1}{\partial t} - \nabla \times (\vec{u}_1 \times \vec{B}_0) = 0 \tag{A.156}$$

A.15.2 Wave Propagation Parallel to a Magnetic Field

The greatly simplified linearized system of equations resulting from the perturbation approach is satisfied by more than one wave solution. We first examine the possibility of wave propagation parallel to an externally imposed magnetic field $\vec{B}_0$. To simplify the calculation (and anticipating the result), we assume that the magnetic perturbation field of the wave, $\vec{B}_1$, is perpendicular to the external magnetic field. We use a coordinate system with a z-axis directed along the external magnetic field (and thus in the direction of wave propagation) and a y-axis pointing in the direction of the magnetic perturbation field; see Fig. A.7. Considering only plane, harmonic waves of small amplitude, we have

$$a_1(t, z) = a_{10} \cos(\omega t - k_z z) \tag{A.157}$$

where a_1 stands for any one of the five remaining unknown perturbation quantities of the wave, a_{10} its amplitude, ω its angular frequency and k_z

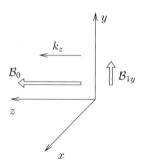

Fig. A.7. Coordinate system used for the description of waves propagating parallel to the magnetic field

its wavenumber. If necessary, this test solution can be augmented by an arbitrary, constant shift in phase. Furthermore,

$$\partial\,a_1\,/\,\partial x \,=\, 0, \quad \partial\,a_1\,/\,\partial y \,=\, 0 \tag{A.158}$$

and

$$\vec{B}_0 \,=\, \hat{z}\,B_0, \quad \vec{B}_1 \,=\, \hat{y}\,B_{1y} \tag{A.159}$$

where $\hat{z}$ and $\hat{y}$ denote unit vectors. Inserting this test solution into the density balance equation (A.153), we obtain

$$-\,\omega\,\rho_{10}\,\sin(\omega\,t \,-\, k_z\,z) \,-\, \rho_0\,(-k_z)\,(u_{10})_z\,\sin(\omega\,t \,-\, k_z\,z) \,=\, 0$$

so that

$$\frac{\rho_{10}}{\rho_0} \,=\, \frac{k_z}{\omega}\,(u_{10})_z \tag{A.160}$$

Turning to the modified induction law (A.156), we first determine a simpler form for the curl term. Applying Eqs. (A.16) and (A.23), we obtain

$$\nabla \times (\vec{u}_1 \times \vec{B}_0) \,=\, \nabla \times (\hat{x}\,u_{1y}\,B_0 \,-\, \hat{y}\,u_{1x}B_0)$$
$$=\, \hat{x}\,B_0\,\frac{\partial u_{1x}}{\partial z} \,+\, \hat{y}\,B_0\,\frac{\partial u_{1y}}{\partial z} \tag{A.161}$$

which, when substituted into Eq. (A.156), yields the relation

$$\hat{y}\,\frac{\partial B_{1y}}{\partial t} \,-\, \hat{x}\,B_0\,\frac{\partial u_{1x}}{\partial z} \,-\, \hat{y}\,B_0\,\frac{\partial u_{1y}}{\partial z} \,=\, 0 \tag{A.162}$$

from which it directly follows that

$$u_{1x} \,=\, 0 \tag{A.163}$$

The possibility that $u_{1x} = $ constant must be excluded because of the requirement that $\vec{u}_0 = 0$. Inserting the wave solution (A.157) into the two remaining terms, we obtain

$$-\,\omega(B_{10})_y\,\sin(\omega\,t \,-\, k_z\,z) \,-\, B_0\,k_z\,(u_{10})_y\,\sin(\omega\,t \,-\, k_z\,z) \,=\, 0$$

from which it must hold that

$$\frac{(B_{10})_y}{B_0} \,=\, -\frac{k_z}{\omega}\,(u_{10})_y \tag{A.164}$$

Finally, using

$$\vec{u}_1 \,=\, \hat{y}\,u_{1y} \,+\, \hat{z}\,u_{1z}$$

and

$$\vec{B}_0 \times (\nabla \times \vec{B}_1) \,=\, \hat{z}\,B_0 \times (-\hat{x}\,\frac{\partial B_{1y}}{\partial z}) \,=\, -\hat{y}B_0\,\frac{\partial B_{1y}}{\partial z}$$

the momentum balance equation (A.155) may be written as

$$\hat{y}\,\rho_0\,\frac{\partial u_{1y}}{\partial t} + \hat{z}\,\rho_0\,\frac{\partial u_{1z}}{\partial t} + \hat{z}\,v_{PS}^2\,\frac{\partial \rho_1}{\partial z} - \hat{y}\,\frac{B_0}{\mu_0}\,\frac{\partial B_{1y}}{\partial z} = 0 \qquad (A.165)$$

from which we obtain for each of the two components

$$\rho_0\,\frac{\partial u_{1z}}{\partial t} + v_{PS}^2\,\frac{\partial \rho_1}{\partial z} = 0 \qquad (A.166)$$

and

$$\rho_0\,\frac{\partial u_{1y}}{\partial t} - \frac{B_0}{\mu_0}\,\frac{\partial B_{1y}}{\partial z} = 0 \qquad (A.167)$$

In the first case, the wave test solution (A.157) yields the relation

$$-\rho_0\,\omega\,(u_{10})_z\,\sin(\omega\,t - k_z\,z) + v_{PS}^2\,k_z\,\rho_{10}\,\sin(\omega\,t - k_z\,z) = 0$$

thereby requiring that

$$\frac{\rho_{10}}{\rho_0} = \frac{\omega}{k_z}\,\frac{1}{v_{PS}^2}\,(u_{10})_z \qquad (A.168)$$

Comparing this expression with Eq. (A.160), one arrives at the dispersion relation

$$\omega - v_{PS}\,k_z = 0 \qquad (A.169)$$

In the second case, the wave solution yields the relation

$$-\rho_0\,\omega\,(u_{10})_y\,\sin(\omega\,t - k_z) - (B_0/\mu_0)\,k_z\,(B_{10})_y\,\sin(\omega\,t - k_z\,z) = 0$$

which leads to the requirement that

$$\frac{(B_{10})_y}{B_0} = -\frac{\mu_0\,\rho_0}{B_0^2}\,\frac{\omega}{k_z}\,(u_{10})_y = -\frac{1}{v_A^2}\,\frac{\omega}{k_z}\,(u_{10})_y \qquad (A.170)$$

where we have introduced the quantity $v_A = B_0\,/\,\sqrt{\mu_0\,\rho_0}$. Again comparing this with the expression (A.164), one arrives at a different dispersion relation

$$\omega - v_A\,k_z = 0 \qquad (A.171)$$

For the case $v_{PS} \neq v_A$, only one of the two dispersion relations (A.169) or (A.171) can be valid, not both simultaneously. Referring back to Eq. (A.165), it is clear that the dispersion relation (A.169) is coupled with the condition

$$u_{1y} = 0 \quad \text{and} \quad B_{1y} = 0 \qquad (A.172)$$

and the dispersion relation (A.171) is subject to the condition

$$u_{1z} = 0 \quad \text{and} \quad \rho_1 = 0 \qquad (A.173)$$

These two possibilities correspond to two completely different types of waves. The first type, defined by the relations (A.169) and (A.172), are designated as *plasma acoustic waves*. The second type, subject to (A.171) and (A.173), are known as *Alfvén waves*. In the following the properties of these waves are discussed with respect to their effects on the plasma velocity, density, plasma pressure, electromagnetic field and current density. In this context it is helpful to relate all perturbations to that of the velocity.

Plasma Acoustic Waves. In this case the velocity takes the following form

$$u_{1x} = 0, \quad u_{1y} = 0 \quad \text{and} \quad u_{1z} = (u_{10})_z \cos(\omega t - k_z z) \tag{A.174}$$

Using Eq. (A.160), the density perturbation is found to be

$$\rho_1 = \rho_0 \frac{k_z}{\omega} u_{1z} \tag{A.175}$$

The pressure perturbation is determined with the help of the adiabatic relation. The derivative of this relation with respect to z is

$$\frac{\partial p}{\partial z} = \frac{\partial}{\partial z}(\alpha \rho^\gamma) = (\gamma p/\rho)\frac{\partial \rho}{\partial z} \tag{A.176}$$

It follows from the perturbation approach (A.151) and the relation (A.154) that

$$\frac{\partial p_1}{\partial z} \simeq v_{PS}^2 \frac{\partial \rho_1}{\partial z}$$

which yields

$$p_1 = v_{PS}^2 \rho_1 = v_{PS}^2 \rho_0 (k_z/\omega) u_{1z} \tag{A.177}$$

The magnetic perturbation field may be derived from Eq. (A.164) as

$$\mathcal{B}_{1y} = -\mathcal{B}_0 (k_z/\omega) u_{1y} = 0 \tag{A.178}$$

The current density due to the wave, as follows from Eq. (A.129), is

$$\vec{j}_1 = \frac{1}{\mu_0} \nabla \times \hat{y} \mathcal{B}_{1y} = -\hat{x} \frac{1}{\mu_0} \frac{\partial \mathcal{B}_{1y}}{\partial z} = \hat{x} \frac{\mathcal{B}_0}{\mu_0} \frac{k_z}{\omega} \frac{\partial u_{1y}}{\partial z} = 0 \tag{A.179}$$

We obtain the electric field from Eq. (A.134) as

$$\vec{\mathcal{E}}_1 = -\hat{z} u_{1z} \times (\hat{z}\mathcal{B}_0 + \hat{y}\mathcal{B}_{1y}) = \hat{x}\mathcal{B}_{1y}u_{1z} = 0 \tag{A.180}$$

Finally, the dispersion relation may be used to yield the phase velocity of plasma acoustic waves

$$(v_{ph})_{plasma\ acoustic} = v_{PS} = \omega/k_z = \sqrt{\gamma p_0/\rho_0} \tag{A.181}$$

Alfvén Waves. The velocity for this case is given by

$$u_{1x} = 0, \quad u_{1y} = (u_{10})_y \cos(\omega t - k_z z) \quad \text{and} \quad u_{1z} = 0 \tag{A.182}$$

Using Eq. (A.160) we obtain the density perturbation

$$\rho_1 = \rho_0 \frac{k_z}{\omega} u_{1z} = 0 \tag{A.183}$$

According to Eq. (A.177), the pressure perturbation is found to be

$$p_1 = v_{PS}^2 \rho_1 = 0 \tag{A.184}$$

The magnetic perturbation field from Eq. (A.164) is

$$\mathcal{B}_{1y} = -\mathcal{B}_0 \left(k_z/\omega \right) u_{1y} = -\sqrt{\mu_0 \rho_0}\, u_{1y} \tag{A.185}$$

Thus, for the case of wave propagation in the direction of the background magnetic field, magnetic field and velocity perturbations are anticorrelated. According to Eq. (A.179), the wave-induced current density is

$$\vec{j}_1 = \hat{x} \frac{\mathcal{B}_0}{\mu_0} \frac{k_z}{\omega} \frac{\partial u_{1y}}{\partial z} = \hat{x} \frac{\mathcal{B}_0 k_z^2}{\mu_0 \omega} (u_{10})_y \cos(\omega t - k_z z - \pi/2) = \hat{x} j_{1x} \tag{A.186}$$

The current density is thus phase shifted by a quarter period with respect to the velocity. The electric field, taken from Eq. (A.134), is given by

$$\vec{\mathcal{E}}_1 = -\hat{y} u_{1y} \times (\hat{z} \mathcal{B}_0 + \hat{y} \mathcal{B}_{1y}) = -\hat{x} \mathcal{B}_0 u_{1y} = -\hat{x} \mathcal{E}_{1x} \tag{A.187}$$

Finally, the phase velocity of the Alfvén wave is

$$(v_{ph})_{Alfvén} = v_A = \omega/k_z = \sqrt{\mathcal{B}_0^2/\mu_0 \rho_0} \tag{A.188}$$

A.15.3 Wave Propagation Perpendicular to a Magnetic Field

Magnetosonic waves are a third type of long period waves in a magneto-plasma. In order to derive their properties, we examine the possibility of wave propagation perpendicular to a given external magnetic field. To simplify matters (and anticipating the result), we allow only magnetic field perturbations parallel to the given external field. The trial solution, based on the coordinate system sketched in Fig. A.8, is thus given by

$$a_1 = a_{10} \cos(\omega t - k_y y)$$

$$\partial a_1 / \partial x = 0 \quad , \quad \partial a_1 / \partial z = 0 \tag{A.189}$$

and

$$\vec{\mathcal{B}}_0 = \hat{z} \mathcal{B}_0 \quad , \quad \vec{\mathcal{B}}_1 = \hat{z} \mathcal{B}_{1z}$$

where a_1 again stands for any perturbation quantity and k_y is the wavenumber in the y-direction. If necessary, phase shifts can be accounted for by an appropriate phase constant. Inserting this test solution into the density

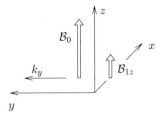

Fig. A.8. Coordinate system used for the description of magnetosonic waves

balance equation (A.153), the momentum balance equation (A.155) and the induction law (A.156), it can be shown that waves of this type do indeed exist. Without going into the details of the calculation, we content ourselves here with a brief description of the derived wave properties. The velocity for this type of wave is given by

$$u_{1x} = 0, \quad u_{1y} = (u_{10})_y \cos(\omega t - k_y y) \quad \text{and} \quad u_{1z} = 0 \qquad (A.190)$$

Similar to plasma acoustic waves, the density and pressure perturbations are found to be

$$\rho_1 = \rho_0 \frac{k_y}{\omega} u_{1y} \qquad (A.191)$$

$$p_1 = v_{PS}^2 \rho_1 = v_{PS}^2 \rho_0 (k_y/\omega) u_{1y} \qquad (A.192)$$

and similar to Alfvén waves, the perturbations in magnetic field, current density and electric field are given by

$$B_{1z} = B_0 (k_y/\omega) u_{1y} \qquad (A.193)$$

$$\vec{j}_1 = \hat{x} \frac{B_0}{\mu_0} \frac{k_y}{\omega} \frac{\partial u_{1y}}{\partial y} = \hat{x} \frac{B_0 k_y^2}{\mu_0 \omega} (u_{10})_y \cos(\omega t - k_y y - \pi/2) = \hat{x} j_{1x} \qquad (A.194)$$

$$\vec{\mathcal{E}}_1 = -\hat{x} B_0 u_{1y} = -\hat{x} \mathcal{E}_{1x} \qquad (A.195)$$

Finally, the dispersion relation yields the following expression for the phase velocity of a magnetosonic wave

$$(v_{ph})_{magnetosonic} = v_{MS} = \omega/k_y = \sqrt{v_{PS}^2 + v_A^2} \qquad (A.196)$$

A.15.4 Validity Test of the Approximations

A number of terms were neglected in our derivation of the defining equations for magnetoplasma waves. In this section we justify these simplifications by appropriate order-of-magnitude comparisons. This requires suitable estimates for the size of the numerous temporal and spatial derivatives. Consider the variation of an arbitrary wave parameter a_1

$$a_1 = a_{10} \cos(\omega t - k_i x_i)$$

where, as before, a_{10} denotes the amplitude of the wave perturbation, $\omega = 2\pi/\tau$ is the angular frequency, and $k_i = 2\pi/\lambda$ is the wavenumber in the propagation direction x_i. The order of magnitude of the temporal and spatial derivative of the perturbation parameters is thus given by

$$(\partial a_1 / \partial t) = (-\omega\, a_{10} \sin(\omega t - k_i x_i)) = (2\pi/\tau)(a_1)$$

$$(\partial a_1 / \partial x_i) = (k_i\, a_{10} \sin(\omega t - k_i x_i)) = (2\pi/\lambda)(a_1)$$

where the orders of magnitude of the sine and cosine functions are considered to be equal. Time and space derivatives, the latter in form of nabla operators, can thus be approximated as follows

$$\partial/\partial t \rightarrow 2\pi/\tau, \quad \nabla \rightarrow 2\pi/\lambda \qquad (A.197)$$

We first examine the simplification of the momentum balance equation (6.44) or (A.121), in which both the inertial and the gravitational acceleration terms were neglected. We justify this by comparing their magnitude relative to those terms that were retained. For example, the ratio of the spatial acceleration to the temporal acceleration is given by

$$\frac{(u_1 \nabla)u_1}{\partial u_1/\partial t} \simeq \frac{u_1 \tau}{\lambda} = \frac{u_1}{v_{ph}} \ll 1 \qquad (A.198)$$

where u_1 is the velocity oscillation produced by the wave and v_{ph} is the phase velocity. Recalling the perturbation conditions and Eqs. (A.175) and (A.185), it is clear that the ratio of these two quantities is much smaller than unity. Justifying our original neglect of the gravitational acceleration also presents no problem. Comparing this with the acceleration due to the magnetic field, for example, one obtains near the Earth ($h \geq 10\, R_E$)

$$\frac{\rho\, g}{n\, e\, u_1\, B} \leq 10^{-4} \qquad (A.199)$$

where we have set $u_1 = 0.1 v_A$, i.e. one tenth of the local Alfvén velocity.

We next justify our neglect of the displacement current term in the fourth Maxwell equation. When describing waves, of course, we are no longer subject to quasi-stationary conditions. Comparing the second term on the right side of Eq. (A.128) with the left side of the equation, one obtains

$$\left(\frac{\mu_0 \varepsilon_0 \partial \vec{\mathcal{E}}/\partial t}{\nabla \times \vec{B}}\right) \simeq \frac{v_{ph}}{c_0^2} \frac{\mathcal{E}}{B} < \frac{v_{ph}}{c_0} \ll 1 \qquad (A.200)$$

where we have made use of the definition of the speed of light, $c_0 = 1/\sqrt{\varepsilon_0 \mu_0}$, in the second step, and the ratio of electric to magnetic field strength for electromagnetic waves in vacuum, $\mathcal{E}/B = c_0$, in the third step, see Eq. (4.92).

Note that the displacement current term is considered to be dominant in the derivation of electromagnetic waves and the ratio $\mathcal{E}/\mathcal{B}$ assumes a correspondingly large value in this case. Since the phase velocity of magnetoplasma waves is much smaller than the speed of light, it is clear that neglecting the displacement current term is an excellent approximation.

In a third step, the radical simplifications of the generalized Ohm's law should be checked. For this purpose, similar to the procedure used in Section A.13.3, we calculate the relative magnitudes of all terms on the right side of Eq. (A.133). The one term not neglected, $-\vec{u} \times \vec{\mathcal{B}}$, again serves as our reference. The following ratios are obtained for each term

$$R_2 = \frac{j / \sigma_B}{u_1 \, \mathcal{B}} \simeq \frac{(2 \pi \mathcal{B} / \mu_0 \, \lambda) / \sigma_B}{u_1 \, \mathcal{B}} \simeq \frac{10^{10}}{u_1 \, \lambda \, T_e^{3/2}} = \frac{10^{10}}{u_1 \, v_{ph} \, \tau \, T_e^{3/2}} \quad (A.201)$$

$$R_3 = \frac{\nabla p_e / (e \, n)}{u_1 \, \mathcal{B}} \simeq \frac{2 \pi \, n \, k \, T_e / (e \, n \, \lambda)}{u_1 \, \mathcal{B}} = \frac{5 \cdot 10^{-4} \, T_e}{u_1 \, v_{ph} \, \tau \, \mathcal{B}} \quad (A.202)$$

$$R_4 = \frac{j \, \mathcal{B} / (e \, n)}{u_1 \, \mathcal{B}} \simeq \frac{2 \pi \, \mathcal{B} / (\mu_0 \, \lambda)}{e \, n \, u_1} \simeq \frac{3 \cdot 10^{25} \mathcal{B}}{u_1 \, v_{ph} \, \tau \, n} \quad (A.203)$$

and

$$R_5 = \frac{(m_e / e^2) \, \partial(j / n) / \partial t}{u_1 \, \mathcal{B}} \simeq \frac{(m_e / e^2) \, 2 \pi \, j / (n \, \tau)}{u_1 \, \mathcal{B}} \simeq \frac{10^{15}}{u_1 \, v_{ph} \, \tau^2 \, n} \quad (A.204)$$

For an explicit estimate of these terms, we assume typical interplanetary values near the Earth's orbit: $T_e = T_i = T \simeq 10^5$ K, $n = 6 \cdot 10^6$ m^{-3}, $\mathcal{B} \simeq$ 3.5 nT. We assume the phase velocity is approximately equal to the Alfvén velocity and assign a typical value for our conditions of $v_{ph} \simeq v_A \simeq 30$ km/s. We further let the velocity fluctuation produced by the wave reach a value of one tenth the Alfvén velocity: $u_1 \simeq v_{ph} / 10$. Finally, we select a typical wave period of $\tau = 30$ minutes. Using these values, the relative magnitudes of the individual terms are found to be $R_2 < 10^{-8}$, $R_3 \simeq R_4 \simeq 0.1$ and $R_5 < 10^{-7}$. All of these are thus small in comparison to the leading term and can be neglected to a first approximation.

In a final test, we turn to the simplification of the density balance equation (A.152). In this case we have

$$\frac{u_1 \, \nabla \rho_1}{\rho_0 \, \nabla u_1} \simeq \frac{\rho_1}{\rho_0} \ll 1 \quad (A.205)$$

and

$$\frac{u_1 \, \nabla \rho_1}{\partial \rho_1 / \partial t} \simeq \frac{u_1}{v_{ph}} \ll 1 \quad (A.206)$$

so that neglect of the term nonlinear in perturbation quantities, $u_1 \nabla \rho_1$, is justified. Expressed more precisely – and this pertains to all the estimates of this section – the results obtained from our derivation are consistent with the *a priori* assumptions.

A.16 Plasma Instabilities

Instabilities are disturbances that quickly grow in amplitude. As the name implies, they arise in systems or media that are in an instable state. Any deviation from this state releases energy that reinforces the deviation and induces a runaway increase in the amplitude of the disturbance. Many such instabilities are observed in plasmas. The classification distinguishes between micro-, or velocity space, instabilities and macro-, or configuration space, instabilities. The former draw their energy from velocity distributions that strongly deviate from an equilibrium distribution (i.e. from a Maxwell distribution). The instability is nature's way of enabling a return to a velocity distribution of lower energy and thus higher entropy. Such microinstabilities are the prime suspects deemed responsible for an entire suite of otherwise poorly understood phenomena such as reconnection events, spontaneous interruption of thin current sheets, anomalous resistivity in good conducting plasmas along magnetic field lines, temperature assimilation in collisionless plasmas, and many more. Since, by necessity, a treatment of this type of instability is rather involved, it will not be pursued further here. The interested reader is referred to the textbooks cited at the end of Chapter 6. We content ourselves here with a qualitative description of why plasma instabilities occur. This works somewhat better for the case of macroinstabilities, two of which are described in the following.

Rayleigh-Taylor Instabilities in the Equatorial Ionosphere. As mentioned in Section 4.6, the low latitude ionosphere is known for its susceptibility to plasma instabilities. Of particular importance is the Rayleigh-Taylor instability. In order to understand its occurrence, we consider the bottom side of the equatorial ionosphere, i.e. a region with a positive density gradient. Because of the Earth's attraction, all charge carriers in the ionosphere are subject to a gravitational drift. If we are above the dynamo region and collisions can be neglected, Eq. (5.40) is valid to a good approximation and the drift velocity amounts to

$$(\vec{u}_D^g)_s \simeq \frac{m_s \vec{g} \times \vec{B}}{q_s B^2}$$

where s again stands for either ions or electrons, but essentially only the ions do the drifting because $m_i \gg m_e$. We now assume that some external source excites a sinusoidal density oscillation in the east-west direction; see Fig. A.9. The ion drift on the west slope of a density crest then produces a decrease in density, and on the east slope an increase in ion density. This follows directly from the continuity equation for this species

$$\frac{\partial n_i}{\partial t} = -\frac{\partial \left(n_i(-u_D^g)_i \right)}{\partial x} = (u_D^g)_i \frac{\partial n_i}{\partial x}$$

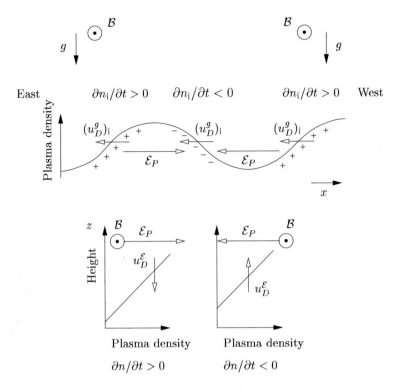

Fig. A.9. Origin of the Rayleigh-Taylor instability on the bottom side of the equatorial ionosphere. An east-west section of the ionosphere is viewed here from the North. Annotated are the gravitational drift of the ions $(u_D^g)_i$, the $\vec{\mathcal{E}} \times \vec{B}$ drift $u_D^{\mathcal{E}}$, the ion density n_i, the plasma density $n (= n_i)$, the geomagnetic field B, the acceleration of gravity g, and the electrical polarization field strength $\mathcal{E}_P$.

The deficit of positive charges on the west slope and the positive surplus on the east slope result in an electric polarization field that is directed from east to west near the density crests and from west to east near the density troughs. Combined with the geomagnetic field, this induces an $\vec{\mathcal{E}} \times \vec{B}$ drift that points downward in the density crests and upward in the density troughs. As shown in the lower part of Fig. A.9, this drift leads to a density increase in the crests and a density decrease in the troughs

$$\frac{\partial n}{\partial t} = - \left(\pm u_D^{\mathcal{E}} \right) \frac{\partial n}{\partial z}$$

Strengthening the pre-existing density oscillations, this positive feedback can result in spectacular fluctuations in the density (*plasma bubbles* or *plasma holes*). Direct evidence for this type of instability is provided by the multiple reflections that lead to smearing of radio sounding echoes and have become known in the technical nomenclature as *spread F*.

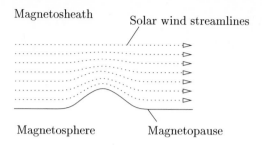

Fig. A.10. Origin of the Kelvin-Helmholtz instability on the flanks of the magnetosphere

Kelvin-Helmholtz Instabilities on the Magnetopause. Kelvin-Helmholtz instabilities were mentioned in Section 5.7 as possible sources of ULF waves. In order to understand the occurrence of this instability, we consider the flank regions of the magnetopause. A pressure balance exists here between the magnetosphere and the solar wind of the magnetosheath, flowing swiftly along the magnetopause. Should some natural fluctuation create an outward bulge in the magnetopause, the solar wind is forced to flow around this obstacle; see Fig. A.10. Because of the bend in the streamline, the solar wind feels an outward-directed centrifugal force that leads to a weakening of the external pressure on the magnetopause. This pressure decrease may be formally derived from the Bernoulli equation (A.76). The cross sectional area of the magnetosheath is diminished by the bulge, as suggested by the higher density of the streamlines in Fig. A.10. Given a constant solar wind flux, this requires a corresponding increase in the flow velocity. Whatever approach one uses to visualize the effect, a decrease in the external pressure must be accompanied by an increase in the bulge. The original perturbation is strengthened and can reach considerable amplitudes. Familiar examples of this type of instability are water waves, which can be excited by winds blowing at the surface. The energy of these instabilities is provided by the flow energy of the (solar) wind.

Appendix B
Figure and Table References

Akasofu, S.-I., The aurora, in *The Physics of Everyday Phenomena* (Scientific American), 15, Freeman and Co., San Francisco, 1979

Akasofu, S.-I., and S. Chapman, *Solar-Terrestrial Physics*, Clarendon Press, Oxford, 1972

Alfvén, H., *Cosmic Plasma*, Reidel Publ. Co., Dordrecht, 1981

Alfvén, H., and C.-G. Fälthammar, *Cosmical Electrodynamics*, Clarendon Press, Oxford, 1963

Banks, P.M., and G. Kockarts, *Aeronomy B*, Academic Press, New York, 1973

Banks, P.M., C.R. Chappell and A.F. Nagy, A new model for the interaction of auroral electrons with the atmosphere: Spectral degradation, backscatter, optical emission, and ionization, *J. Geophys. Res.*, *79*, 1459, 1974

Belcher, J.W., L. Davis, Jr. and E.J. Smith, Large-amplitude Alfvén waves in the interplanetary medium: Mariner 5, *J. Geophys. Res.*, *74*, 2302, 1969

Broadfoot, A.L., and K.R. Kendall, The airglow spectrum, 3100 – 10,000 Å, *J. Geophys. Res.*, *73*, 426, 1968

Bronshtén, V.A., The structure of the far outer corona of 19 June 1936, *Soviet Astronomy*, *3*, 821, 1960

Carruthers, G.R., T. Page and R.R. Meier, Apollo 16 Lyman-alpha imagery of the hydrogen geocorona, *J. Geophys. Res.*, *81*, 1664, 1976

Chapman, S., and J. Bartels, *Geomagnetism I*, Clarendon Press, Oxford, 1962

Cowley, S.W.H., Magnetic reconnection, in *Solar System Magnetic Fields* (E.R. Priest, ed.), 121, Reidel Publ. Co., Dordrecht, 1985

Craven, J.D., and L.A. Frank, The temporal evolution of a small auroral substorm as viewed from high altitudes with Dynamics Explorer 1, *Geophys. Res. Lett.*, *12*, 465, 1985

Eastman, T.E., Transition regions in solar system and astrophysical plasmas, *IEEE Trans. Plasma Sci.*, *18*, 18, 1990

Eccles, D., and J.W. King, A review of topside sounder studies of the equatorial ionosphere, *Proc. IEEE*, *57*, 1012, 1969

Fahr, H.J., and B. Shizgal, Modern exospheric theories and their observational relevance, *Rev. Geophys. Space Phys.*, *21*, 75, 1983

Fairfield, D.H., Structure of the geomagnetic tail, in *Magnetotail Physics* (A.T.Y. Lui, ed.), Johns Hopkins University Press, Baltimore, 1987

Fichtner, H., and H.J. Fahr, Plasma expansion from diverging magnetic field configurations: The plasma-magnetic field interaction, *Planet. Space Sci.*, *37*, 987, 1989

Frank, L.A., and J.D. Craven, Imaging results from Dynamics Explorer 1, *Rev. Geophys.*, *26.* 249, 1988

Frisch, P.C., The galactic environment of the sun, *J. Geophys. Res.*, *105*, 10279, 2000

Glaßmeier, K.-H., ULF Pulsations, in *Handbook of Atmospheric Electrodynamics II* (H. Volland, ed.), 463, CRC Press, Boca Raton, 1995

Hedin, A.E., MSIS-86 thermospheric model, *J. Geophys. Res.*, *92*, 4649, 1987

Hedin, A.E., E.L. Fleming, A.H. Manson, F.J. Schmidlin, S.K. Avery, R.R. Clark, S.J. Franke, G.J. Fraser, T. Tsuda, F. Vial, and R.A. Vincent, Empirical wind model for the upper, middle and lower atmosphere, *J. Atmos. Terr. Phys.*, *58*, 1421, 1996

Heroux, L., and H.E. Hinteregger, Aeronomical reference spectrum for solar UV below 2000 Å, *J. Geophys. Res.*, *83*, 5305, 1978

Hess, W.N., *The Radiation Belt and Magnetosphere*, Blaisdell Publ. Co., Waltham, MA, 1968

Hones, E.W., Substorm processes in the magnetotail: Comments on 'On hot tenuous plasmas, fireballs, and boundary layers in the earth's magnetotail' by L.A. Frank, K.L. Ackerson and R.P Lepping, *J. Geophys. Res.*, *82*, 5633, 1977

Hundhausen, A.J., The origin and propagation of coronal mass ejections, in *Proceedings of the Sixth International Solar Wind Conference I* (V.J. Pizzo, T.E. Holzer and D.G. Sime, eds.), 181, HAO/NCAR, Boulder, CO, 1988

Jacchia, L.G., Thermospheric temperature, density, and composition: New models, *Smithsonian Astrophys. Obs. Spec. Report*, *375*, Cambridge, MA, 1977

Johnson, C.Y., Ionospheric composition and density from 90 to 1200 kilometers at solar minimum, *J. Geophys. Res.*, *71*, 330, 1966

Kane, S.R., Impulsive (flash) phase of solar flares: Hard, X-ray, microwave, EUV, and optical observations, in *Coronal Disturbances* (G. Newkirk, Jr., ed.), 105, Reidel Publ. Co., Dordrecht, 1974

Kertz, W., *Einführung in die Geophysik I*, B.I.-Wissenschaftsverlag, Mannheim, 1985

Kockarts, G., Effects of solar variations on the upper atmosphere, *Solar Physics*, *74*, 295, 1981

Köhnlein, W., A model of the electron and ion temperatures in the ionosphere, *Planet. Space Sci.*, *34*, 609, 1986

Lang, K.R., *Astrophysical Data: Planets and Stars*, Springer-Verlag, Berlin, 1992

Larson, D.J., and R.L. Kaufmann, Structure of the magnetotail current sheet, *J. Geophys. Res.*, *101*, 21447, 1996

Lean, J., Solar EUV irradiances and indices, *Adv. Space Res.*, *8*, *No.5*, 263, 1988

Lepping, R.P., J.A. Jones and L.F. Burlaga, Magnetic field structure of interplanetary magnetic clouds at 1 AU, *J. Geophys. Res.*, *95*, 11957, 1990

Linde, T.J., T.I. Gombosi, Ph.L. Roe, K.G. Powell and D.L. DeZeeuw, Heliosphere in the magnetized local interstellar medium: Results of a three-dimensional MHD simulation, *J. Geophys. Res.*, *103*, 1889, 1998

Malitson, H.H., The solar energy spectrum, *Sky and Telescope*, *29*, 162, 1965

Mariani, F., and F.M. Neubauer, The interplanetary magnetic field, in *Physics of the Inner Heliosphere I* (R. Schwenn and E. Marsch, eds.), 183, Springer-Verlag, Berlin, 1990

Marsch, E., K.-H. Mühlhäuser, R. Schwenn, H. Rosenbauer, W.G. Pilipp and F.M. Neubauer, Solar wind protons: Three-dimensional velocity distributions and derived plasma parameters measured between 0.3 and 1 AU, *J. Geophys. Res.*, *87*, 52, 1982

Matuura, N., Reaction rates in the F-region, *Rept. Ionosph. Space Res. Japan*, *21*, 289, 1966

McComas, D.J., B.L. Barraclough, H.O. Funsten, J.T. Gosling, E. Santiago-Muñoz, R.M. Skoug, B.E. Goldstein, M. Neugebauer, P. Riley and A. Balogh, Solar wind observations over Ulysses' first polar orbit, *J. Geophys. Res.*, *105*, 10419, 2000

McPherron, R.L., C.T. Russell and M.P. Aubry, Satellite studies of magnetospheric substorms on August 15, 1968, 9. Phenomenological model for substorms, *J. Geophys. Res.*, *78*, 3131, 1973

Meyer, P., R. Ramaty and W.R. Webber, Cosmic rays - astronomy with energetic particles, *Physics Today*, *27*, *No.10*, 23, 1974

Olsen, P.W., The geomagnetic field and its extension into space, *Adv. Space Res.*, *2*, *No.1*, 13, 1982

Parker, E.N., *Interplanetary Dynamical Processes*, Interscience Publishers, John Wiley and Sons, New York, 1963

Prölss, G.W., Ionospheric F-region storms, in *Handbook of Atmospheric Electrodynamics, Vol. 2* (H. Volland, ed.), 195, CRC Press, Boca Raton, 1995

Prölss, G.W., Magnetic storm associated perturbations of the upper atmosphere, in *Magnetic Storms* (B.T. Tsurutani, W.D. Gonzalez, Y. Kamide and J.K. Arballo, eds.), AGU Monograph No. 98, 227, Washington, 1997

Ratcliffe, J.A., *An Introduction to the Ionosphere and Magnetosphere*, Cambridge University Press, Cambridge, 1972

Roble, R.G., E.C. Ridley and R.E. Dickinson, On the global mean structure of the thermosphere, *J. Geophys. Res.*, *92*, 8745, 1987

Schield, M.A., J.W. Freeman and A.J. Dessler, A source for field-aligned currents at auroral latitudes, *J. Geophys. Res.*, *74*, 247, 1969

Schunk, R.W., The terrestrial ionosphere, in *Solar-Terrestrial Physics* (R.L. Carovillano and J.M. Forbes, eds.), 609, Reidel Publ. Co., Dordrecht, 1983

Schwenn, R., Large-scale structure of the interplanetary medium, in *Physics of the Inner Heliosphere I* (R. Schwenn and E. Marsch, eds.), 99, Springer-Verlag, Berlin, 1990

Sckopke, N., G. Paschmann, A.L. Brinca, C.W. Carlson and H. Lühr, Ion thermalization in quasi-perpendicular shocks involving reflected ions, *J. Geophys. Res.*, *95*, 6337, 1990

Shiokawa, K., W. Baumjohann, G. Haerendel, G. Paschmann, J.F. Fennel, E. Friis-Christensen, H. Lühr, G.D. Reeves, C.T. Russell, P.R. Sutcliffe and K. Takahashi, High speed ion flow, substorm current wedge, and multiple Pi 2 pulsations, *J. Geophys. Res.*, *103*, 4491, 1998

Solar-Geophysical Data, No. 367, Part II, 40, NOAA, Boulder, CO, 1975

Spreiter, J.R., A.L. Summers and A.Y. Alksne, Hydromagnetic flow around the magnetosphere, *Planet. Space Sci.*, *14*, 223, 1966

Stern, D.P., The origins of Birkeland currents, *Rev. Geophys. Space Phys.*, *21*, 125, 1983

Torr, M.R., and D.G. Torr, Ionization frequencies for solar cycle 21: Revised, *J. Geophys. Res.*, *90*, 6675, 1985

Torr, M.R., D.G. Torr, R.A. Ong and H.E. Hinteregger, Ionization frequencies for major thermospheric constituents as a function of solar cycle 21, *Geophys. Res. Lett.*, *6*, 771, 1979

Tsyganenko, N.A., Quantitative models of the magnetospheric magnetic field: Methods and results, *Space Sci. Rev.*, *54*, 75, 1990

van de Hulst, H.C., The chromosphere and the corona, in *The Sun* (G.P. Kuiper, ed.), 207, University of Chicago Press, Chicago, 1953

Vsekhsvjatsky, S.K., The structure of the solar corona and the corpuscular streams, in *The Solar Corona* (J.W. Evans, ed.), 271, Academic Press, New York, 1963

Walker, A.B.C., Jr., Multispectral observations complementary to the study of high-energy solar phenomena, *Solar Physics, 118*, 209, 1988

Weimer, D.R., Substorm time constants, *J. Geophys. Res.*, *99*, 11005, 1994

Willis, D.M., Structure of the magnetopause, *Rev. Geophys. Space Phys.*, *9*, 953, 1971

Wright, J.W., Dependence of the ionospheric F-region on the solar cycle, *Nature, 194*, 461, 1962

Index

RETURN TO: PHYSICS LIBRARY

351 LeConte Hall   510-642-3122

LOAN PERIOD 1	2	3
1-MONTH		
4	5	6

ALL BOOKS MAY BE RECALLED AFTER 7 DAYS.

Renewable by telephone.

DUE AS STAMPED BELOW.

This book will be held
in PHYSICS LIBRARY
until   MAR 0 1 2005

SEP 1 8 2008

FORM NO. DD 22
500   4-03

UNIVERSITY OF CALIFORNIA, BERKELEY
Berkeley, California 94720–6000